NEW ENGLAND INSTITUTE
OF TECHNOLOGY
LIBRARY

The Electric Power Engineering Handbook

POWER SYSTEM STABILITY AND CONTROL

THIRD EDITION

The Electric Power Engineering Handbook
Third Edition

Edited by
Leonard L. Grigsby

Electric Power Generation, Transmission, and Distribution, Third Edition
Edited by Leonard L. Grigsby

Electric Power Transformer Engineering, Third Edition
Edited by James H. Harlow

Electric Power Substations Engineering, Third Edition
Edited by John D. McDonald

Power Systems, Third Edition
Edited by Leonard L. Grigsby

Power System Stability and Control, Third Edition
Edited by Leonard L. Grigsby

The Electric Power Engineering Handbook

POWER SYSTEM STABILITY AND CONTROL

THIRD EDITION

EDITED BY
LEONARD L. GRIGSBY

NEW ENGLAND INSTITUTE
OF TECHNOLOGY
LIBRARY

CRC Press is an imprint of the
Taylor & Francis Group, an **informa** business

CRC Press
Taylor & Francis Group
6000 Broken Sound Parkway NW, Suite 300
Boca Raton, FL 33487-2742

© 2012 by Taylor & Francis Group, LLC
CRC Press is an imprint of Taylor & Francis Group, an Informa business

No claim to original U.S. Government works

Printed in the United States of America on acid-free paper
Version Date: 20120202

International Standard Book Number: 978-1-4398-8320-4 (Hardback)

This book contains information obtained from authentic and highly regarded sources. Reasonable efforts have been made to publish reliable data and information, but the author and publisher cannot assume responsibility for the validity of all materials or the consequences of their use. The authors and publishers have attempted to trace the copyright holders of all material reproduced in this publication and apologize to copyright holders if permission to publish in this form has not been obtained. If any copyright material has not been acknowledged please write and let us know so we may rectify in any future reprint.

Except as permitted under U.S. Copyright Law, no part of this book may be reprinted, reproduced, transmitted, or utilized in any form by any electronic, mechanical, or other means, now known or hereafter invented, including photocopying, microfilming, and recording, or in any information storage or retrieval system, without written permission from the publishers.

For permission to photocopy or use material electronically from this work, please access www.copyright.com (http://www.copyright.com/) or contact the Copyright Clearance Center, Inc. (CCC), 222 Rosewood Drive, Danvers, MA 01923, 978-750-8400. CCC is a not-for-profit organization that provides licenses and registration for a variety of users. For organizations that have been granted a photocopy license by the CCC, a separate system of payment has been arranged.

Trademark Notice: Product or corporate names may be trademarks or registered trademarks, and are used only for identification and explanation without intent to infringe.

Library of Congress Cataloging-in-Publication Data

Power system stability and control / editor, Leonard L. Grigsby. -- 3rd ed.
     p. cm.
  Includes bibliographical references and index.
  ISBN 978-1-4398-8320-4 (alk. paper)
  1. Electric power system stability. 2. Electric power systems--Control. I. Grigsby, Leonard L.

TK1010.P68 2012
621.31--dc23   2011044127

Visit the Taylor & Francis Web site at
http://www.taylorandfrancis.com

and the CRC Press Web site at
http://www.crcpress.com

Contents

Preface ix
Editor xi
Contributors xiii

PART I   Power System Protection

Miroslav M. Begovic I-1

1 Transformer Protection 1-1
   *Alexander Apostolov, John Appleyard, Ahmed Elneweihi, Robert Haas,
   and Glenn W. Swift*

2 The Protection of Synchronous Generators 2-1
   Gabriel Benmouyal

3 Transmission Line Protection 3-1
   Stanley H. Horowitz

4 System Protection 4-1
   Miroslav M. Begovic

5 Digital Relaying 5-1
   James S. Thorp

6 Use of Oscillograph Records to Analyze System Performance 6-1
   John R. Boyle

7 Systems Aspects of Large Blackouts 7-1
   Ian Dobson

PART II   Power System Dynamics and Stability

Prabha S. Kundur II-1

8 Power System Stability 8-1
   Prabha S. Kundur

v

9 Transient Stability 9-1
Kip Morison

10 Small-Signal Stability and Power System Oscillations 10-1
John Paserba, Juan Sanchez-Gasca, Lei Wang, Prabha S. Kundur, Einar Larsen, and Charles Concordia

11 Voltage Stability 11-1
Yakout Mansour and Claudio Cañizares

12 Direct Stability Methods 12-1
Vijay Vittal

13 Power System Stability Controls 13-1
Carson W. Taylor

14 Power System Dynamic Modeling 14-1
William W. Price and Juan Sanchez-Gasca

15 Wide-Area Monitoring and Situational Awareness 15-1
Manu Parashar, Jay C. Giri, Reynaldo Nuqui, Dmitry Kosterev, R. Matthew Gardner, Mark Adamiak, Dan Trudnowski, Aranya Chakrabortty, Rui Menezes de Moraes, Vahid Madani, Jeff Dagle, Walter Sattinger, Damir Novosel, Mevludin Glavic, Yi Hu, Ian Dobson, Arun Phadke, and James S. Thorp

16 Assessment of Power System Stability and Dynamic Security Performance 16-1
Lei Wang and Pouyan Pourbeik

17 Power System Dynamic Interaction with Turbine Generators 17-1
Bajarang L. Agrawal, Donald G. Ramey, and Richard G. Farmer

18 Wind Power Integration in Power Systems 18-1
Reza Iravani

19 Flexible AC Transmission Systems (FACTS) 19-1
Rajiv K. Varma and John Paserba

PART III Power System Operation and Control

Bruce F. Wollenberg III-1

20 Energy Management 20-1
Neil K. Stanton, Jay C. Giri, and Anjan Bose

21 Generation Control: Economic Dispatch and Unit Commitment 21-1
Charles W. Richter Jr.

22 State Estimation 22-1
Jason G. Lindquist and Danny Julian

23 Optimal Power Flow 23-1
Mohamed E. El-Hawary

24 Security Analysis 24-1
Nouredine Hadjsaid

Index **Index**-1

Preface

The generation, delivery, and utilization of electric power and energy remain one of the most challenging and exciting fields of electrical engineering. The astounding technological developments of our age are highly dependent upon a safe, reliable, and economic supply of electric power. The objective of the Electric Power Engineering Handbook is to provide a contemporary overview of this far-reaching field as well as a useful guide and educational resource for its study. It is intended to define electric power engineering by bringing together the core of knowledge from all of the many topics encompassed by the field. The chapters are written primarily for the electric power engineering professional who seeks factual information, and secondarily for the professional from other engineering disciplines who wants an overview of the entire field or specific information on one aspect of it.

The first and second editions of this handbook were well received by readers worldwide. Based upon this reception and the many recent advances in electric power engineering technology and applications, it was decided that the time was right to produce a third edition. Because of the efforts of many individuals, the result is a major revision. There are completely new chapters covering such topics as FACTS, smart grid, energy harvesting, distribution system protection, electricity pricing, linear machines. In addition, the majority of the existing chapters have been revised and updated. Many of these are major revisions.

The handbook consists of a set of five books. Each is organized into topical parts and chapters in an attempt to provide comprehensive coverage of the generation, transformation, transmission, distribution, and utilization of electric power and energy as well as the modeling, analysis, planning, design, monitoring, and control of electric power systems. The individual chapters are different from most technical publications. They are not journal-type articles nor are they textbooks in nature. They are intended to be tutorials or overviews providing ready access to needed information while at the same time providing sufficient references for more in-depth coverage of the topic.

This book is devoted to the subjects of power system protection, power system dynamics and stability, and power system operation and control. If your particular topic of interest is not included in this list, please refer to the list of companion books referred to at the beginning.

In reading the individual chapters of this handbook, I have been most favorably impressed by how well the authors have accomplished the goals that were set. Their contributions are, of course, key to the success of the book. I gratefully acknowledge their outstanding efforts. Likewise, the expertise and dedication of the editorial board and section editors have been critical in making this handbook possible. To all of them I express my profound thanks.

They are as follows:

- Nonconventional Power Generation   Saifur Rahman
- Conventional Power Generation   Rama Ramakumar
- Transmission Systems   George G. Karady
- Distribution Systems   William H. Kersting

- Electric Power Utilization — Andrew P. Hanson
- Power Quality — S. Mark Halpin
- *Transformer Engineering* (a complete book) — James H. Harlow
- *Substations Engineering* (a complete book) — John D. McDonald
- Power System Analysis and Simulation — Andrew P. Hanson
- Power System Transients — Pritindra Chowdhuri
- Power System Planning (Reliability) — Gerry Sheblé
- Power Electronics — R. Mark Nelms
- Power System Protection — Miroslav M. Begovic[*]
- Power System Dynamics and Stability — Prabha S. Kundur[†]
- Power System Operation and Control — Bruce Wollenberg

I wish to say a special thank-you to Nora Konopka, engineering publisher for CRC Press/Taylor & Francis, whose dedication and diligence literally gave this edition life. I also express my gratitude to the other personnel at Taylor & Francis who have been involved in the production of this book, with a special word of thanks to Jessica Vakili. Their patience and perseverance have made this task most pleasant.

Finally, I thank my longtime friend and colleague—Mel Olken, editor, the *Power and Energy Magazine*—for graciously providing the picture for the cover of this book.

---

[*] Arun Phadke for the first and second editions.
[†] Richard Farmer for the first and second editions.

Editor

Leonard L. ("Leo") Grigsby received his BS and MS in electrical engineering from Texas Tech University, Lubbock, Texas and his PhD from Oklahoma State University, Stillwater, Oklahoma. He has taught electrical engineering at Texas Tech University, Oklahoma State University, and Virginia Polytechnic Institute and University. He has been at Auburn University since 1984, first as the Georgia power distinguished professor, later as the Alabama power distinguished professor, and currently as professor emeritus of electrical engineering. He also spent nine months during 1990 at the University of Tokyo as the Tokyo Electric Power Company endowed chair of electrical engineering. His teaching interests are in network analysis, control systems, and power engineering.

During his teaching career, Professor Grigsby received 13 awards for teaching excellence. These include his selection for the university-wide William E. Wine Award for Teaching Excellence at Virginia Polytechnic Institute and University in 1980, the ASEE AT&T Award for Teaching Excellence in 1986, the 1988 Edison Electric Institute Power Engineering Educator Award, the 1990–1991 Distinguished Graduate Lectureship at Auburn University, the 1995 IEEE Region 3 Joseph M. Beidenbach Outstanding Engineering Educator Award, the 1996 Birdsong Superior Teaching Award at Auburn University, and the IEEE Power Engineering Society Outstanding Power Engineering Educator Award in 2003.

Professor Grigsby is a fellow of the Institute of Electrical and Electronics Engineers (IEEE). During 1998–1999, he was a member of the board of directors of IEEE as the director of Division VII for power and energy. He has served the institute in 30 different offices at the chapter, section, regional, and international levels. For this service, he has received seven distinguished service awards, such as the IEEE Centennial Medal in 1984, the Power Engineering Society Meritorious Service Award in 1994, and the IEEE Millennium Medal in 2000.

During his academic career, Professor Grigsby has conducted research in a variety of projects related to the application of network and control theory to modeling, simulation, optimization, and control of electric power systems. He has been the major advisor for 35 MS and 21 PhD graduates. With his students and colleagues, he has published over 120 technical papers and a textbook on introductory network theory. He is currently the series editor for the Electrical Engineering Handbook Series published by CRC Press. In 1993, he was inducted into the Electrical Engineering Academy at Texas Tech University for distinguished contributions to electrical engineering.

Contributors

Mark Adamiak
General Electric
Wayne, Pennsylvania

Bajarang L. Agrawal
Arizona Public Service Company
Phoenix, Arizona

Alexander Apostolov
OMICRON Electronics
Los Angeles, California

John Appleyard
S&C Electric Company
and
Quanta Technology
Cary, North Carolina

Miroslav M. Begovic
Georgia Institute of Technology
Atlanta, Georgia

Gabriel Benmouyal
Schweitzer Engineering Laboratories, Ltd.
Pullman, Washington

Anjan Bose
Washington State University
Pullman, Washington

John R. Boyle
Power System Analysis
Signal Mountain, Tennessee

Claudio Cañizares
Department of Electrical and Computer
    Engineering
University of Waterloo
Waterloo, Ontario, Canada

Aranya Chakrabortty
Department of Electrical and Computer
    Engineering
North Carolina State University
Raleigh, North Carolina

Charles Concordia
Consultant

Jeff Dagle
Pacific Northwest National Laboratory
Richland, Washington

Ian Dobson
Department of Electrical and Computer
    Engineering
Iowa State University
Ames, Iowa

Mohamed E. El-Hawary
Department of Electrical and Computer
    Engineering
Dalhousie University
Halifax, Nova Scotia, Canada

Ahmed Elneweihi
British Columbia Hydro
    and Power Authority
Vancouver, British Columbia, Canada

Richard G. Farmer
School of Electrical, Computer and Energy
    Engineering
Arizona State University
Tempe, Arizona

R. Matthew Gardner
Dominion Virginia Power
Richmond, Virginia

Jay C. Giri
ALSTOM Grid, Inc.
Redmond, Washington

Mevludin Glavic
Quanta Technology
Raleigh, North Carolina

Robert Haas
Haas Engineering
and
KY RESC
Villa Hills, Kentucky

Nouredine Hadjsaid
Institut National Polytechnique
 de Grenoble
Grenoble, France

Stanley H. Horowitz
Consultant
Columbus, Ohio

Yi Hu
Quanta Technology
Raleigh, North Carolina

Reza Iravani
Department of Electrical and Computer
 Engineering
University of Toronto
Toronto, Ontario, Canada

Danny Julian
ABB Inc.
Raleigh, North Carolina

Dmitry Kosterev
Bonneville Power Administration
Portland, Oregon

Prabha S. Kundur
Kundur Power Systems Solutions, Inc.
Toronto, Ontario, Canada

Einar Larsen
General Electric Energy
Schenectady, New York

Jason G. Lindquist
Siemens Energy Automation
Minneapolis, Minnesota

Vahid Madani
Pacific Gas & Electric
San Francisco, California

Yakout Mansour
California Independent System Operator
Folsom, California

Rui Menezes de Moraes
Universidade Federal Fluminense
Rio de Janeiro, Brazil

Kip Morison
British Columbia Hydro
 and Power Authority
Vancouver, British Columbia, Canada

Damir Novosel
Quanta Technology
Raleigh, North Carolina

Reynaldo Nuqui
Asea Brown Boveri
Cary, North Carolina

Manu Parashar
ALSTOM Grid, Inc.
Redmond, Washington

John Paserba
Mitsubishi Electric Power Products, Inc.
Warrendale, Pennsylvania

Arun Phadke
Virginia Tech
Blacksburg, Virginia

Pouyan Pourbeik
Electric Power Research Institute
Palo Alto, California

William W. Price
Consultant
Livingston, Texas

Donald G. Ramey (retired)
Siemens Corporation
Apex, North Carolina

Contributors

Charles W. Richter Jr.
Charles Richter Associates, LLC
Kenmore, Washington

Juan Sanchez-Gasca
General Electric Energy
Schenectady, New York

Walter Sattinger
Department of System Management Support
Swiss Grid
Laufenburg, Switzerland

Neil K. Stanton
Stanton Associates
Medina, Washington

Glenn W. Swift
APT Power Technologies
Winnipeg, Manitoba, Canada

Carson W. Taylor (retired)
Bonneville Power Administration
Portland, Oregon

James S. Thorp
Virginia Tech
Blacksburg, Virginia

Dan Trudnowski
Department of Electrical Engineering
Montana Tech
Butte, Montana

Rajiv K. Varma
Department of Electrical and Computer Engineering
University of Western Ontario
London, Ontario, Canada

Vijay Vittal
School of Electrical, Computer and Energy Engineering
Arizona State University
Tempe, Arizona

Lei Wang
Powertech Labs Inc.
Surrey, British Columbia, Canada

Bruce F. Wollenberg
Department of Electrical and Computer Engineering
University of Minnesota
Minneapolis, Minnesota

I

Power System Protection

Miroslav M. Begovic

1 **Transformer Protection** *Alexander Apostolov, John Appleyard, Ahmed Elneweihi, Robert Haas, and Glenn W. Swift* 1-1
Types of Transformer Faults • Types of Transformer Protection • Special Considerations • Special Applications • Restoration • References

2 **The Protection of Synchronous Generators** *Gabriel Benmouyal* 2-1
Review of Functions • Differential Protection for Stator Faults (87G) • Protection against Stator Winding Ground Fault • Field Ground Protection • Loss-of-Excitation Protection (40) • Current Imbalance (46) • Anti-Motoring Protection (32) • Overexcitation Protection (24) • Overvoltage (59) • Voltage Imbalance Protection (60) • System Backup Protection (51V and 21) • Out-of-Step Protection • Abnormal Frequency Operation of Turbine-Generator • Protection against Accidental Energization • Generator Breaker Failure • Generator Tripping Principles • Impact of Generator Digital Multifunction Relays • References

3 **Transmission Line Protection** *Stanley H. Horowitz* 3-1
Nature of Relaying • Current Actuated Relays • Distance Relays • Pilot Protection • Relay Designs • Reference

4 **System Protection** *Miroslav M. Begovic* 4-1
Introduction • Disturbances: Causes and Remedial Measures • Transient Stability and Out-of-Step Protection • Overload and Underfrequency Load Shedding • Voltage Stability and Undervoltage Load Shedding • Special Protection Schemes • Modern Perspective: Technology Infrastructure • Future Improvements in Control and Protection • Acknowledgments • References

5 **Digital Relaying** *James S. Thorp* 5-1
Sampling • Antialiasing Filters • Sigma-Delta A/D Converters • Phasors from Samples • Symmetrical Components • Algorithms • References

6 **Use of Oscillograph Records to Analyze System Performance** *John R. Boyle*............. 6-1

7 **Systems Aspects of Large Blackouts** *Ian Dobson* 7-1
References

Miroslav M. Begovic is a professor in the School of Electrical and Computer Engineering at Georgia Tech, Atlanta, Georgia. He received his PhD in electrical engineering from Virginia Tech, Blacksburg, Virginia and MS and Dipl.-Ing. in electrical engineering from Belgrade University, Belgrade, Serbia. Dr. Begovic's research interests are in monitoring, analysis, and control of power systems, as well as development and applications of renewable and sustainable energy systems. He has been a member of the IEEE PES Power System Relaying Committee for almost 20 years and chaired its first working group on wide area protection and emergency control, as well as a number of other working groups within the System Protection Subcommittee. The Working Group on Protective Aids for Voltage Stability, of which he was a member, received the IEEE PES Working Group Award for Best Report. Dr. Begovic is the chair of the Electric Energy Group in the School of ECE at Georgia Institute of Technology, former chair of the Emerging Technologies Coordinating Committee of IEEE PES, distinguished lecturer of IEEE PES, and elected treasurer of the IEEE Power and Energy Society in 2010–2011. He is also a fellow of the IEEE and member of Sigma Xi, Eta Kappa Nu, Phi Kappa Phi, and Tau Beta Pi.

1

Transformer Protection

Alexander Apostolov
OMICRON Electronics

John Appleyard
S&C Electric Company

Ahmed Elneweihi
British Columbia Hydro and Power Authority

Robert Haas
Haas Engineering

Glenn W. Swift
APT Power Technologies

1.1 Types of Transformer Faults 1-1
1.2 Types of Transformer Protection 1-2
  Electrical • Mechanical • Thermal
1.3 Special Considerations 1-6
  Current Transformers • Magnetizing Inrush (Initial, Recovery, and Sympathetic) • Primary–Secondary Phase Shift • Turn-to-Turn Faults • Through Faults • Backup Protection
1.4 Special Applications 1-8
  Shunt Reactors • Zigzag Transformers • Phase Angle Regulators and Voltage Regulators • Unit Systems • Single-Phase Transformers • Sustained Voltage Unbalance
1.5 Restoration 1-10
  History • Oscillographs, Event Recorders, and Gas Monitors • Date of Manufacture • Magnetizing Inrush • Relay Operations
References 1-11

1.1 Types of Transformer Faults

Any number of conditions have been the reason for an electrical transformer failure. Statistics show that winding failures most frequently cause transformer faults (ANSI/IEEE, 1985). Insulation deterioration, often the result of moisture, overheating, vibration, voltage surges, and mechanical stress created during transformer through faults, is the major reason for winding failure.

Voltage regulating load tap changers, when supplied, rank as the second most likely cause of a transformer fault. Tap-changer failures can be caused by a malfunction of the mechanical switching mechanism, high resistance load contacts, insulation tracking, overheating, or contamination of the insulating oil.

Transformer bushings are the third most likely cause of failure. General aging, contamination, cracking, internal moisture, and loss of oil can all cause a bushing to fail. Two other possible reasons are vandalism and animals that externally flash over the bushing.

Transformer core problems have been attributed to core insulation failure, an open ground strap, or shorted laminations.

Other miscellaneous failures have been caused by current transformers (CTs), oil leakage due to inadequate tank welds, oil contamination from metal particles, overloads, and overvoltage.

1.2 Types of Transformer Protection

1.2.1 Electrical

1.2.1.1 Fuse

Power fuses have been used for many years to provide transformer fault protection. Generally it is recommended that transformers sized larger than 10 MVA be protected with more sensitive devices such as the differential relay discussed later in this section. Fuses provide a low-maintenance, economical solution for protection. Protection and control devices, circuit breakers, and station batteries are not required.

There are some drawbacks. Fuses provide limited protection for some internal transformer faults. A fuse is also a single-phase device. Certain system faults may only operate one fuse. This will result in single-phase service to connected three-phase customers.

Fuse selection criteria include adequate interrupting capability, calculating load currents during peak and emergency conditions, performing coordination studies that include source and low-side protection equipment, and expected transformer size and winding configuration (ANSI/IEEE, 1985).

1.2.1.2 Overcurrent Protection

Overcurrent relays generally provide the same level of protection as power fuses. Higher sensitivity and fault clearing times can be achieved in some instances by using an overcurrent relay connected to measure residual current. This application allows pickup settings to be lower than expected maximum load current. It is also possible to apply an instantaneous overcurrent relay set to respond only to faults within the first 75% of the transformer. This solution, for which careful fault current calculations are needed, does not require coordination with low-side protective devices.

Overcurrent relays do not have the same maintenance and cost advantages found with power fuses. Protection and control devices, circuit breakers, and station batteries are required. The overcurrent relays are a small part of the total cost and when this alternative is chosen, differential relays are generally added to enhance transformer protection. In this instance, the overcurrent relays will provide backup protection for the differentials.

1.2.1.3 Differential

The most widely accepted device for transformer protection is called a restrained differential relay. This relay compares current values flowing into and out of the transformer windings. To assure protection under varying conditions, the main protection element has a multislope restrained characteristic. The initial slope ensures sensitivity for internal faults while allowing for up to 15% mismatch when the power transformer is at the limit of its tap range (if supplied with a load tap changer). At currents above rated transformer capacity, extra errors may be gradually introduced as a result of CT saturation.

However, misoperation of the differential element is possible during transformer energization. High inrush currents may occur, depending on the point on wave of switching as well as the magnetic state of the transformer core. Since the inrush current flows only in the energized winding, differential current results. The use of traditional second-harmonic restraint to block the relay during inrush conditions may result in a significant slowing of the relay during heavy internal faults due to the possible presence of second harmonics as a result of saturation of the line CTs. To overcome this, some relays use a waveform recognition technique to detect the inrush condition. The differential current waveform associated with magnetizing inrush is characterized by a period of each cycle where its magnitude is very small, as shown in Figure 1.1. By measuring the time of this period of low current, an inrush condition can be identified. The detection of inrush current in the differential current is used to inhibit that phase of the low set restrained differential algorithm. Another high-speed method commonly used to detect high-magnitude faults in the unrestrained instantaneous unit is described later in this section.

When a load is suddenly disconnected from a power transformer, the voltage at the input terminals of the transformer may rise by 10%–20% of the rated value, causing an appreciable increase in transformer

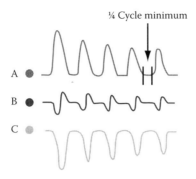

FIGURE 1.1 Transformer inrush current waveforms.

steady-state excitation current. The resulting excitation current flows in one winding only and hence appears as differential current that may rise to a value high enough to operate the differential protection. A waveform of this type is characterized by the presence of fifth harmonic. A Fourier technique is used to measure the level of fifth harmonic in the differential current. The ratio of fifth harmonic to fundamental is used to detect excitation and inhibits the restrained differential protection function. Detection of overflux conditions in any phase blocks that particular phase of the low set differential function.

Transformer faults of a different nature may result in fault currents within a very wide range of magnitudes. Internal faults with very high fault currents require fast fault clearing to reduce the effect of CT saturation and the damage to the protected transformer. An unrestrained instantaneous high set differential element ensures rapid clearance of such faults. Such an element essentially measures the peak value of the input current to ensure fast operation for internal faults with saturated CTs. Restrained units generally calculate an rms current value using more waveform samples. The high set differential function is not blocked under magnetizing inrush or overexcitation conditions; hence, the setting must be set such that it will not operate for the largest inrush currents expected.

At the other end of the fault spectrum are low current winding faults. Such faults are not cleared by the conventional differential function. Restricted ground fault protection gives greater sensitivity for ground faults and hence protects more of the winding. A separate element based on the high impedance circulating current principle is provided for each winding.

Transformers have many possible winding configurations that may create a voltage and current phase shift between the different windings. To compensate for any phase shift between two windings of a transformer, it is necessary to provide phase correction for the differential relay (see Section 1.3).

In addition to compensating for the phase shift of the protected transformer, it is also necessary to consider the distribution of primary zero sequence current in the protection scheme. The necessary filtering of zero sequence current has also been traditionally provided by appropriate connection of auxiliary CTs or by delta connection of primary CT secondary windings. In microprocessor transformer protection relays, zero sequence current filtering is implemented in software when a delta CT connection would otherwise be required. In situations where a transformer winding can produce zero sequence current caused by an external ground fault, it is essential that some form of zero sequence current filtering is employed. This ensures that ground faults out of the zone of protection will not cause the differential relay to operate in error. As an example, an external ground fault on the wye side of a delta/wye connected power transformer will result in zero sequence current flowing in the CTs associated with the wye winding but, due to the effect of the delta winding, there will be no corresponding zero sequence current in the CTs associated with the delta winding, that is, differential current flow will cause the relay to operate. When the virtual zero sequence current filter is applied within the relay, this undesired trip will not occur.

Some of the most typical substation configurations, especially at the transmission level, are breaker-and-a-half or ring-bus. Not that common, but still used are two-breaker schemes. When a power transformer is connected to a substation using one of these breaker configurations, the transformer protection is connected to three or more sets of CTs. If it is a three-winding transformer or an autotransformer with a tertiary connected to a lower-voltage sub-transmission system, four or more sets of CTs may be available.

It is highly recommended that separate relay input connections be used for each set used to protect the transformer. Failure to follow this practice may result in incorrect differential relay response. Appropriate testing of a protective relay for such configuration is another challenging task for the relay engineer.

1.2.1.4 Overexcitation

Overexcitation can also be caused by an increase in system voltage or a reduction in frequency. It follows, therefore, that transformers can withstand an increase in voltage with a corresponding increase in frequency but not an increase in voltage with a decrease in frequency. Operation cannot be sustained when the ratio of voltage to frequency exceeds more than a small amount.

Protection against overflux conditions does not require high-speed tripping. In fact, instantaneous tripping is undesirable, as it would cause tripping for transient system disturbances, which are not damaging to the transformer.

An alarm is triggered at a lower level than the trip setting and is used to initiate corrective action. The alarm has a definite time delay, while the trip characteristic generally has a choice of definite time delay or inverse time characteristic.

1.2.2 Mechanical

There are two generally accepted methods used to detect transformer faults using mechanical methods. These detection methods provide sensitive fault detection and compliment protection provided by differential or overcurrent relays.

1.2.2.1 Accumulated Gases

The first method accumulates gases created as a by-product of insulating oil decomposition created from excessive heating within the transformer. The source of heat comes from either the electrical arcing or a hot area in the core steel. This relay is designed for conservator tank transformers and will capture gas as it rises in the oil. The relay, sometimes referred to as a Buchholz relay, is sensitive enough to detect very small faults.

1.2.2.2 Pressure Relays

The second method relies on the transformer internal pressure rise that results from a fault. One design is applicable to gas-cushioned transformers and is located in the gas space above the oil. The other design is mounted well below minimum liquid level and responds to changes in oil pressure. Both designs employ an equalizing system that compensates for pressure changes due to temperature (ANSI/IEEE, 1985).

1.2.3 Thermal

1.2.3.1 Hot-Spot Temperature

In any transformer design, there is a location in the winding that the designer believes to be the *hottest* spot within that transformer (ANSI/IEEE, 1995). The significance of the "hot-spot temperature" measured at this location is an assumed relationship between the temperature level and the rate of degradation of the cellulose insulation. An instantaneous alarm or trip setting is often used, set at a judicious level above the full load rated hot-spot temperature (110°C for 65°C rise transformers). (Note that "65°C rise" refers to the full load rated *average* winding temperature rise.) Also, a relay or monitoring system can mathematically integrate the rate of degradation, that is, rate of loss of life of the insulation for overload assessment purposes.

1.2.3.2 Heating due to Overexcitation

Transformer core flux density (B), induced voltage (V), and frequency (f) are related by the following formula:

$$B = k_1 \cdot \frac{V}{f} \tag{1.1}$$

where k_1 is a constant for a particular transformer design. As B rises above about 110% of normal, that is, when saturation starts, significant heating occurs due to stray flux eddy currents in the nonlaminated structural metal parts, including the tank. Since it is the voltage/hertz quotient in Equation 1.1 that defines the level of B, a relay sensing this quotient is sometimes called a "volts-per-hertz" relay. The expressions "overexcitation" and "overfluxing" refer to this same condition. Since temperature rise is proportional to the integral of power with respect to time (neglecting cooling processes), it follows that an inverse-time characteristic is useful, that is, *volts-per-hertz* versus *time*. Another approach is to use definite-time-delayed alarm or trip at specific per unit flux levels.

1.2.3.3 Heating due to Current Harmonic Content (ANSI/IEEE, 1993)

One effect of nonsinusoidal currents is to cause current rms magnitude (I_{RMS}) to be incorrect if the method of measurement is not "true-rms":

$$I_{RMS}^2 = \sum_{n=1}^{N} I_n^2 \tag{1.2}$$

where
  n is the harmonic order
  N is the highest harmonic of significant magnitude
  I_n is the harmonic current rms magnitude

If an overload relay determines the I^2R heating effect using the fundamental component of the current only [I_1], then it will underestimate the heating effect. Bear in mind that "true-rms" is only as good as the passband of the antialiasing filters and sampling rate, for numerical relays.

A second effect is heating due to high-frequency eddy-current loss in the copper or aluminum of the windings. The winding eddy-current loss due to each harmonic is proportional to the square of the harmonic amplitude and the square of its frequency as well. Mathematically,

$$P_{EC} = P_{EC\text{-}RATED} \cdot \sum_{n=1}^{N} I_n^2 n^2 \tag{1.3}$$

where
  P_{EC} is the winding eddy-current loss
  $P_{EC\text{-}RATED}$ is the rated winding eddy-current loss (pure 60 Hz)
  I_n is the *n*th harmonic current in per-unit based on the fundamental

Notice the fundamental difference between the effect of harmonics in Equation 1.2 and their effect in Equation 1.3. In the latter, higher harmonics have a proportionately greater effect because of the n^2 factor. IEEE Standard C57.110-2008 (R1992), *Recommended Practice for Establishing Transformer Capability When Supplying Nonsinusoidal Load Currents* gives two empirically based methods for calculating the derating factor for a transformer under these conditions.

1.2.3.4 Heating due to Solar-Induced Currents

Solar magnetic disturbances cause geomagnetically induced currents (GIC) in the earth's surface (EPRI, 1993). These DC currents can be of the order of tens of amperes for tens of minutes, and flow into the neutrals of grounded transformers, biasing the core magnetization. The effect is worst in single-phase units and negligible in three-phase core-type units. The core saturation causes second-harmonic content in the current, resulting in increased *security* in second-harmonic-restrained transformer differential relays, but decreased *sensitivity*. Sudden gas pressure relays could provide the necessary alternative internal fault tripping. Another effect is increased stray heating in the transformer, protection for which can be accomplished using gas accumulation relays for transformers with conservator oil systems. Hot-spot tripping is not sufficient because the commonly used hot-spot simulation model does not account for GIC.

1.2.3.5 Load Tap-Changer Overheating

Damaged current carrying contacts within an underload tap-changer enclosure can create excessive heating. Using this heating symptom, a way of detecting excessive wear is to install magnetically mounted temperature sensors on the tap-changer enclosure and on the main tank. Even though the method does not accurately measure the internal temperature at each location, the *difference* is relatively accurate, since the error is the same for each. Thus, excessive wear is indicated if a relay/monitor detects that the temperature difference has changed significantly over time.

1.3 Special Considerations

1.3.1 Current Transformers

CT ratio selection and performance require special attention when applying transformer protection. Unique factors associated with transformers, including its winding ratios, magnetizing inrush current, and the presence of winding taps or load tap changers, are sources of difficulties in engineering a dependable and secure protection scheme for the transformer. Errors resulting from CT saturation and load tap changers are particularly critical for differential protection schemes where the currents from more than one set of CTs are compared. To compensate for the saturation/mismatch errors, overcurrent relays must be set to operate above these errors.

1.3.1.1 CT Current Mismatch

Under normal, non-fault conditions, a transformer differential relay should ideally have identical currents in the secondaries of all CTs connected to the relay so that no current would flow in its operating coil. It is difficult, however, to match CT ratios exactly to the transformer winding ratios. This task becomes impossible with the presence of transformer off-load and on-load taps or load tap changers that change the voltage ratios of the transformer windings depending on system voltage and transformer loading.

The highest secondary current mismatch between all CTs connected in the differential scheme must be calculated when selecting the relay operating setting. If time-delayed overcurrent protection is used, the time-delay setting must also be based on the same consideration. The mismatch calculation should be performed for maximum load and through-fault conditions.

1.3.1.2 CT Saturation

CT saturation could have a negative impact on the ability of the transformer protection to operate for internal faults (dependability) and not to operate for external faults (security).

For internal faults, dependability of the harmonic-restraint-type relays could be negatively affected if current harmonics generated in the CT secondary circuit due to CT saturation are high enough to restrain the relay. With a saturated CT, second and third harmonics predominate initially, but the even harmonics gradually disappear with the decay of the DC component of the fault current. The relay may then operate eventually when the restraining harmonic component is reduced. These relays usually include an instantaneous overcurrent element that is not restrained by harmonics, but is set very high (typically 20 times transformer rating). This element may operate on severe internal faults.

For external faults, security of the differentially connected transformer protection may be jeopardized if the CTs' unequal saturation is severe enough to produce error current above the relay setting. Relays equipped with restraint windings in each CT circuit would be more secure. The security problem is particularly critical when the CTs are connected to bus breakers rather than the transformer itself. External faults in this case could be of very high magnitude as they are not limited by the transformer impedance.

1.3.2 Magnetizing Inrush (Initial, Recovery, and Sympathetic)

1.3.2.1 Initial

When a transformer is energized after being de-energized, a transient magnetizing or exciting current that may reach instantaneous peaks of up to 30 times full load current may flow. This can cause operation of overcurrent or differential relays protecting the transformer. The magnetizing current flows in only one winding; thus, it will appear to a differentially connected relay as an internal fault.

Techniques used to prevent differential relays from operating on inrush include detection of current harmonics and zero current periods, both being characteristics of the magnetizing inrush current. The former takes advantage of the presence of harmonics, especially the second harmonic, in the magnetizing inrush current to restrain the relay from operation. The latter differentiates between the fault and inrush currents by measuring the zero current periods, which will be much longer for the inrush than for the fault current.

1.3.2.2 Recovery Inrush

A magnetizing inrush current can also flow if a voltage dip is followed by recovery to normal voltage. Typically, this occurs upon removal of an external fault. The magnetizing inrush is usually less severe in this case than in initial energization as the transformer was not totally de-energized prior to voltage recovery.

1.3.2.3 Sympathetic Inrush

A magnetizing inrush current can flow in an energized transformer when a nearby transformer is energized. The offset inrush current of the bank being energized will find a parallel path in the energized bank. Again, the magnitude is usually less than the case of initial inrush.

Both the recovery and sympathetic inrush phenomena suggest that restraining the transformer protection on magnetizing inrush current is required at all times, not only when switching the transformer in service after a period of de-energization.

1.3.3 Primary–Secondary Phase Shift

For transformers with standard delta–wye connections, the currents on the delta and wye sides will have a 30° phase shift relative to each other. CTs used for traditional differential relays must be connected in wye–delta (opposite of the transformer winding connections) to compensate for the transformer phase shift.

Phase correction is often internally provided in microprocessor transformer protection relays via software virtual interposing CTs for each transformer winding and, as with the ratio correction, will depend upon the selected configuration for the restrained inputs. This allows the primary CTs to all be connected in wye.

1.3.4 Turn-to-Turn Faults

Fault currents resulting from a turn-to-turn fault have low magnitudes and are hard to detect. Typically, the fault will have to evolve and affect a good portion of the winding or arc over to other parts of the transformer before being detected by overcurrent or differential protection relays.

For early detection, reliance is usually made on devices that can measure the resulting accumulation of gas or changes in pressure inside the transformer tank.

1.3.5 Through Faults

Through faults could have an impact on both the transformer and its protection scheme. Depending on their severity, frequency, and duration, through-fault currents can cause mechanical transformer damage, even though the fault is somewhat limited by the transformer impedance.

For transformer differential protection, CT mismatch and saturation could produce operating currents on through faults. This must be taken into consideration when selecting the scheme, CT ratio, relay sensitivity, and operating time. Differential protection schemes equipped with restraining windings offer better security for these through faults.

1.3.6 Backup Protection

Backup protection, typically overcurrent or impedance relays applied to one or both sides of the transformer, perform two functions. One function is to back up the primary protection, most likely a differential relay, and operate in the event of its failure to trip.

The second function is protection for thermal or mechanical damage to the transformer. Protection that can detect these external faults and operate in time to prevent transformer damage should be considered. The protection must be set to operate before the through-fault withstand capability of the transformer is reached. If, because of its large size or importance, only differential protection is applied to a transformer, clearing of external faults before transformer damage can occur by other protective devices must be ensured.

1.4 Special Applications

1.4.1 Shunt Reactors

Shunt reactor protection will vary depending on the type of reactor, size, and system application. Protective relay application will be similar to that used for transformers.

Differential relays are perhaps the most common protection method (Blackburn, 1987). Relays with separate phase inputs will provide protection for three single-phase reactors connected together or for a single three-phase unit. CTs must be available on the phase and neutral end of each winding in the three-phase unit.

Phase and ground overcurrent relays can be used to back up the differential relays. In some instances, where the reactor is small and cost is a factor, it may be appropriate to use overcurrent relays as the only protection. The ground overcurrent relay would not be applied on systems where zero sequence current is negligible.

As with transformers, turn-to-turn faults are most difficult to detect since there is little change in current at the reactor terminals. If the reactor is oil filled, a sudden pressure relay will provide good protection. If the reactor is an ungrounded dry type, an overvoltage relay (device 59) applied between the reactor neutral and a set of broken delta connected voltage transformers can be used (ABB, 1994).

Negative sequence and impedance relays have also been used for reactor protection but their application should be carefully researched (ABB, 1994).

1.4.2 Zigzag Transformers

The most common protection for zigzag (or grounding) transformers is three overcurrent relays that are connected to CTs located on the primary phase bushings. These CTs must be connected in delta to filter out unwanted zero sequence currents (ANSI/IEEE, 1985).

It is also possible to apply a conventional differential relay for fault protection. CTs in the primary phase bushings are paralleled and connected to one input. A neutral CT is used for the other input (Blackburn, 1987).

An overcurrent relay located in the neutral will provide backup ground protection for either of these schemes. It must be coordinated with other ground relays on the system.

Sudden pressure relays provide good protection for turn-to-turn faults.

1.4.3 Phase Angle Regulators and Voltage Regulators

Protection of phase angle and voltage regulators varies with the construction of the unit. Protection should be worked out with the manufacturer at the time of order to insure that CTs are installed inside the unit in the appropriate locations to support planned protection schemes. Differential, overcurrent, and sudden pressure relays can be used in conjunction to provide adequate protection for faults (Blackburn, 1987; ABB, 1994).

1.4.4 Unit Systems

A unit system consists of a generator and associated step-up transformer. The generator winding is connected in wye with the neutral connected to ground through a high impedance grounding system. The step-up transformer low-side winding on the generator side is connected delta to isolate the generator from system contributions to faults involving ground. The transformer high side winding is connected in wye and solidly grounded. Generally there is no breaker installed between the generator and transformer.

It is common practice to protect the transformer and generator with an overall transformer differential that includes both pieces of equipment. It may be appropriate to install an additional differential to protect only the transformer. In this case, the overall differential acts as secondary or backup protection for the transformer differential. There will most likely be another differential relay applied specifically to protect the generator.

A volts-per-hertz relay, whose pickup is a function of the ratio of voltage to frequency, is often recommended for overexcitation protection. The unit transformer may be subjected to overexcitation during generator start-up and shutdown when it is operating at reduced frequencies or when there is major loss of load that may cause both overvoltage and overspeed (ANSI/IEEE, 1985).

As with other applications, sudden pressure relays provide sensitive protection for turn-to-turn faults that are typically not initially detected by differential relays.

Backup protection for phase faults can be provided by applying either impedance or voltage controlled overcurrent relays to the generator side of the unit transformer. The impedance relays must be connected to respond to faults located in the transformer (Blackburn, 1987).

1.4.5 Single-Phase Transformers

Single-phase transformers are sometimes used to make up three-phase banks. Standard protection methods described earlier in this section are appropriate for single-phase transformer banks as well. If one or both sides of the bank are connected in delta and CTs located on the transformer bushings are to be used for protection, the standard differential connection cannot be used. To provide proper ground fault protection, CTs from each of the bushings must be utilized (Blackburn, 1987).

1.4.6 Sustained Voltage Unbalance

During sustained unbalanced voltage conditions, wye-connected core-type transformers without a delta-connected tertiary winding may produce damaging heat. In this situation, the transformer case may produce damaging heat from sustained circulating current. It is possible to detect this situation by using either a thermal relay designed to monitor tank temperature or applying an overcurrent relay connected to sense "effective" tertiary current (ANSI/IEEE, 1985).

1.5 Restoration

Power transformers have varying degrees of importance to an electrical system depending on their size, cost, and application, which could range from generator step-up to a position in the transmission/distribution system, or perhaps as an auxiliary unit.

When protective relays trip and isolate a transformer from the electric system, there is often an immediate urgency to restore it to service. There should be a procedure in place to gather system data at the time of trip as well as historical information on the individual transformer, so an informed decision can be made concerning the transformer's status. No one should reenergize a transformer when there is evidence of electrical failure.

It is always possible that a transformer could be incorrectly tripped by a defective protective relay or protection scheme, system backup relays, or by an abnormal system condition that had not been considered. Often system operators may try to restore a transformer without gathering sufficient evidence to determine the exact cause of the trip. An operation should always be considered as legitimate until proven otherwise.

The more vital a transformer is to the system, the more sophisticated the protection and monitoring equipment should be. This will facilitate the accumulation of evidence concerning the outage.

1.5.1 History

Daily operation records of individual transformer maintenance, service problems, and relayed outages should be kept to establish a comprehensive history. Information on relayed operations should include information on system conditions prior to the trip out. When no explanation for a trip is found, it is important to note all areas that were investigated. When there is no damage determined, there should still be a conclusion as to whether the operation was correct or incorrect. Periodic gas analysis provides a record of the normal combustible gas value.

1.5.2 Oscillographs, Event Recorders, and Gas Monitors

System monitoring equipment that initiates and produces records at the time of the transformer trip usually provides information necessary to determine if there was an electrical short circuit involving the transformer or if it was a "through-fault" condition.

1.5.3 Date of Manufacture

Transformers manufactured before 1980 were likely not designed or constructed to meet the severe through-fault conditions outlined in ANSI/IEEE C57.109, (1993). Maximum through-fault values should be calculated and compared to short-circuit values determined for the trip out. Manufacturers should be contacted to obtain documentation for individual transformers in conformance with ANSI/IEEE C57.109.

1.5.4 Magnetizing Inrush

Differential relays with harmonic restraint units are typically used to prevent trip operations upon transformer energizing. However, there are nonharmonic restraint differential relays in service that use time delay and/or percentage restraint to prevent trip on magnetizing inrush. Transformers so protected may have a history of falsely tripping on energizing inrush, which may lead system operators to attempt restoration without analysis, inspection, or testing. There is always the possibility that an electrical fault can occur upon energizing, which is masked by historical data.

Relay harmonic restraint circuits are either factory set at a threshold percentage of harmonic inrush or the manufacturer provides predetermined settings that should prevent an unwanted operation upon transformer energization. Some transformers have been manufactured in recent years using a grain-oriented steel and a design that results in very low percentages of the restraint harmonics in the inrush current. These values are, in some cases, less than the minimum manufacture recommended threshold settings.

1.5.5 Relay Operations

Transformer protective devices not only trip but prevent reclosing of all sources energizing the transformer. This is generally accomplished using an auxiliary "lockout" relay. The lockout relay requires manual resetting before the transformer can be energized. This circuit encourages manual inspection and testing of the transformer before reenergization decisions are made.

Incorrect trip operations can occur due to relay failure, incorrect settings, or coordination failure. New installations that are in the process of testing and wire checking are most vulnerable. Backup relays, by design, can cause tripping for upstream or downstream system faults that do not otherwise clear properly.

References

ANSI/IEEE C57.109-1993, *IEEE Guide for Transformer through Fault Current Duration*, 1993.
ANSI/IEEE Std. 62-1995, *IEEE Guide for Diagnostic Field Testing of Electric Power Apparatus—Part 1: Oil Filled Power Transformers, Regulators, and Reactors*, 1995.
ANSI/IEEE C57.91-1995/2002, *IEEE Guide for Loading Mineral Oil-Immersed Transformers*, 1995/2002.
ANSI/IEEE C57.104-2008, *Guide for the Interpretation of Gases Generated in Oil-Immersed Transformers*, 2008.
ANSI/IEEE C37.91-2008, *IEEE Transformer Protection Guide*, 2008.
ANSI/IEEE C57.12.10-2010, *IEEE Standard Requirements for Liquid-Immersed Power, Transformers*, 2010.
Blackburn, J.L., *Protective Relaying: Principles and Applications*, Marcel Decker, Inc., New York, 1987.
Elmore, W.A., ed., *Protective Relaying, Theory and Application*, ABB, Marcel Dekker, Inc., New York, 1994.
GEC Measurements, *Protective Relays Application Guide*, Stafford, England, 1975.
IEEE Std. C57.110-2008, *IEEE Recommended Practice for Establishing Liquid-Filled and Dry Type Power and Distribution Transformer Capability When Supplying Nonsinusoidal Load Currents*, 2008.
Mason, C.R., *The Art and Science of Protective Relaying*, John Wiley & Sons, New York, 1996.
Rockefeller, G. et al., Differential relay transient testing using EMTP simulations, in *Paper Presented to the 46th Annual Protective Relay Conference (Georgia Tech.)*, Atlanta, GA, April 29–May 1, 1992.
Solar magnetic disturbances/geomagnetically-induced current and protective relaying, Electric Power Research Institute Report TR-102621, Project 321-04, EPRI, Palo Alto, CA, August 1993.
Warrington, A.R. van C., *Protective Relays, Their Theory and Practice*, Vol. 1, Wiley, New York, 1963, Vol. 2, Chapman and Hall Ltd., London, U.K., 1969.

2
The Protection of Synchronous Generators

2.1	Review of Functions	2-2
2.2	Differential Protection for Stator Faults (87G)	2-2
2.3	Protection against Stator Winding Ground Fault	2-4
2.4	Field Ground Protection	2-6
2.5	Loss-of-Excitation Protection (40)	2-6
2.6	Current Imbalance (46)	2-7
2.7	Anti-Motoring Protection (32)	2-9
2.8	Overexcitation Protection (24)	2-9
2.9	Overvoltage (59)	2-10
2.10	Voltage Imbalance Protection (60)	2-11
2.11	System Backup Protection (51V and 21)	2-12
2.12	Out-of-Step Protection	2-14
2.13	Abnormal Frequency Operation of Turbine-Generator	2-15
2.14	Protection against Accidental Energization	2-16
2.15	Generator Breaker Failure	2-17
2.16	Generator Tripping Principles	2-18
2.17	Impact of Generator Digital Multifunction Relays	2-18
	Improvements in Signal Processing • Improvements in Protective Functions	
	References	2-20

Gabriel Benmouyal
Schweitzer Engineering Laboratories, Ltd.

In an apparatus protection perspective, generators constitute a special class of power network equipment because faults are very rare but can be highly destructive and therefore very costly when they occur. If for most utilities, generation integrity must be preserved by avoiding erroneous tripping, removing a generator in case of a serious fault is also a primary if not an absolute requirement. Furthermore, protection has to be provided for out-of-range operation normally not found in other types of equipment such as overvoltage, overexcitation, limited frequency or speed range, etc.

It should be borne in mind that, similar to all protective schemes, there is to a certain extent a "philosophical approach" to generator protection and all utilities and all protective engineers do not have the same approach. For instance, some functions like overexcitation, backup impedance elements, loss-of-synchronism, and even protection against inadvertent energization may not be applied by some organizations and engineers. It should be said, however, that with the digital multifunction generator protective packages presently available, a complete and extensive range of functions exists within the same "relay": and economic reasons for not installing an additional protective element is a tendency which must disappear.

The nature of the prime mover will have some definite impact on the protective functions implemented into the system. For instance, little or no concern at all will emerge when dealing with the abnormal frequency operation of hydraulic generators. On the contrary, protection against underfrequency operation of steam turbines is a primary concern.

The sensitivity of the motoring protection (the capacity to measure very low levels of negative real power) becomes an issue when dealing with both hydro and steam turbines. Finally, the nature of the prime mover will have an impact on the generator tripping scheme. When delayed tripping has no detrimental effect on the generator, it is common practice to implement sequential tripping with steam turbines as described later.

The purpose of this chapter is to provide an overview of the basic principles and schemes involved in generator protection. For further information, the reader is invited to refer to additional resources dealing with generator protection. The ANSI/IEEE guides (ANSI/IEEE C37.106-1987, 1987; ANSI/IEEE C37.102; ANSI/IEEE C37.101) are particularly recommended. The *IEEE Tutorial on the Protection of Synchronous Generators* (IEEE, 1995) is a detailed presentation of North American practices for generator protection. All these references have been a source of inspiration in this writing.

2.1 Review of Functions

Table 2.1 provides a list of protective relays and their functions most commonly found in generator protection schemes. These relays are implemented as shown on the single-line diagram of Figure 2.1.

As shown in the Relay Type column, most protective relays found in generator protection schemes are not specific to this type of equipment but are more generic types.

2.2 Differential Protection for Stator Faults (87G)

Protection against stator phase faults are normally covered by a high-speed differential relay covering the three phases separately. All types of phase faults (phase-phase) will be covered normally by this type of protection, but the phase-ground fault in a high-impedance grounded generator will not be covered. In this case, the phase current will be very low and therefore below the relay pickup.

TABLE 2.1 Most Commonly Found Relays for Generator Protection

Identification Number	Function Description	Relay Type
87G	Generator phase windings protection	Differential protection
87T	Step-up transformer differential protection	Differential protection
87U	Combined differential transformer and generator protection	Differential protection
40	Protection against the loss of field voltage or current supply	Offset Offset-mho relay
46	Protection against current imbalance. Measurement of phase negative sequence current	Time-overcurrent relay
32	Anti-motoring protection	Reverse-power relay
24	Overexcitation protection	Volt/Hertz relay
59	Phase overvoltage protection	Overvoltage relay
60	Detection of blown voltage transformer fuses	Voltage balance relay
81	Under- and over-frequency protection	Frequency relays
51V	Backup protection against system faults	Voltage controlled or voltage-restrained time overcurrent relay
21	Backup protection against system faults	Distance relay
78	Protection against loss of synchronization	Combination of offset offset-mho and blinders

The Protection of Synchronous Generators

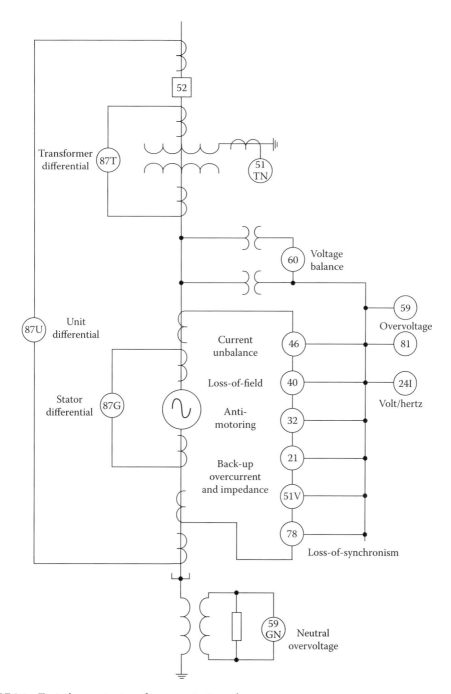

FIGURE 2.1 Typical generator-transformer protection scheme.

Contrary to transformer differential applications, no inrush exists on stator currents and no provision is implemented to take care of overexcitation. Therefore, stator differential relays do not include harmonic restraint (second and fifth harmonic). Current transformer (CT) saturation is still an issue, however, particularly in generating stations because of the high X/R ratio found near generators.

The most common type of stator differential is the percentage differential, the main characteristics of which are represented in Figure 2.2.

FIGURE 2.2 Single, dual, and variable-slope percentage differential characteristics.

FIGURE 2.3 Stator winding current configuration.

For a stator winding, as shown in Figure 2.3, the restraint quantity will very often be the absolute sum of the two incoming and outgoing currents as in

$$Irestraint = \frac{|IA\_in| + |IA\_out|}{2}, \qquad (2.1)$$

whereas the operate quantity will be the absolute value of the difference:

$$Ioperate = |IA\_in - IA\_out| \qquad (2.2)$$

The relay will output a fault condition when the following inequality is verified:

$$Irestraint \geq K \cdot Ioperate \qquad (2.3)$$

where K is the differential percentage. The dual and variable slope characteristics will intrinsically allow CT saturation for an external fault without the relay picking up.

An alternative to the percentage differential relay is the high-impedance differential relay, which will also naturally surmount any CT saturation. For an internal fault, both currents will be forced into a high-impedance voltage relay. The differential relay will pickup when the tension across the voltage element gets above a high-set threshold. For an external fault with CT saturation, the saturated CT will constitute a low-impedance path in which the current from the other CT will flow, bypassing the high-impedance voltage element which will not pick up.

Backup protection for the stator windings will be provided most of the time by a transformer differential relay with harmonic restraint, the zone of which (as shown in Figure 2.1) will cover both the generator and the step-up transformer.

An impedance element partially or totally covering the generator zone will also provide backup protection for the stator differential.

2.3 Protection against Stator Winding Ground Fault

Protection against stator-to-ground fault will depend to a great extent upon the type of generator grounding. Generator grounding is necessary through some impedance in order to reduce the current level of a phase-to-ground fault. With solid generator grounding, this current will reach destructive levels. In order to avoid this, at least low impedance grounding through a resistance or a reactance is required. High-impedance through a distribution transformer with a resistor connected across the secondary winding will limit the current level of a phase-to-ground fault to a few primary amperes.

The Protection of Synchronous Generators

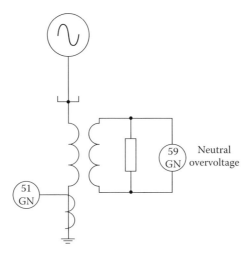

FIGURE 2.4 Stator-to-ground neutral overvoltage scheme.

The most common and minimum protection against a stator-to-ground fault with a high-impedance grounding scheme is an overvoltage element connected across the grounding transformer secondary, as shown in Figure 2.4.

For faults very close to the generator neutral, the overvoltage element will not pick up because the voltage level will be below the voltage element pick-up level. In order to cover 100% of the stator windings, two techniques are readily available:

1. Use of the third harmonic generated at the neutral and generator terminals
2. Voltage injection technique

Looking at Figure 2.5, a small amount of third harmonic voltage will be produced by most generators at their neutral and terminals. The level of these third harmonic voltages depends upon the generator operating point as shown in Figure 2.5a. Normally they would be higher at full load. If a fault develops near the neutral, the third harmonic neutral voltage will approach 0 and the terminal voltage will increase. However, if a fault develops near the terminals, the terminal third harmonic voltage will reach 0 and the neutral voltage will increase. Based on this, three possible schemes have been devised. The relays available to cover the three possible choices are

1. Use of a third harmonic undervoltage at the neutral. It will pick up for a fault at the neutral.
2. Use of a third harmonic overvoltage at the terminals. It will pick up for a fault near the neutral.
3. The most sensitive schemes are based on third harmonic differential relays that monitor the ratio of third harmonic at the neutral and the terminals (Yin et al., 1990).

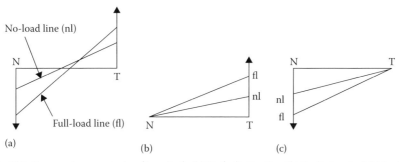

FIGURE 2.5 Third harmonic on neutral and terminals. (a) No fault situation. (b) Fault at neutral. (c) Fault at terminal.

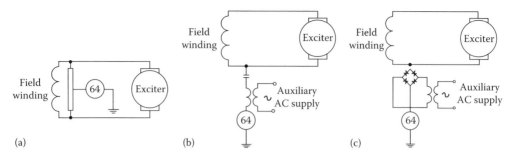

FIGURE 2.6 Various techniques for field-ground protection. (a) Voltage divider method. (b) AC injection method. (c) DC injection method.

2.4 Field Ground Protection

A generator field circuit (field winding, exciter, and field breaker) is a DC circuit that does not need to be grounded. If a first earth fault occurs, no current will flow and the generator operation will not be affected. If a second ground fault at a different location occurs, a current will flow that is high enough to cause damage to the rotor and the exciter. Furthermore, if a large section of the field winding is short-circuited, a strong imbalance due to the abnormal air-gap fluxes could result on the forces acting on the rotor with a possibility of serious mechanical failure. In order to prevent this situation, a number of protecting devices exist. Three principles are depicted in Figure 2.6.

The first technique (Figure 2.6a) involves connecting a resistor in parallel with the field winding. The resistor centerpoint is connected the ground through a current sensitive relay. If a field circuit point gets grounded, the relay will pick up by virtue of the current flowing through it. The main shortcoming of this technique is that no fault will be detected if the field winding centerpoint gets grounded.

The second technique (Figure 2.6b) involves applying an AC voltage across one point of the field winding. If the field winding gets grounded at some location, an AC current will flow into the relay and causes it to pick up.

The third technique (Figure 2.6c) involves injecting a DC voltage rather than an AC voltage. The consequence remains the same if the field circuit gets grounded at some point.

The best protection against field-ground faults is to move the generator out of service as soon as the first ground fault is detected.

2.5 Loss-of-Excitation Protection (40)

A loss-of-excitation on a generator occurs when the field current is no longer supplied. This situation can be triggered by a variety of circumstances and the following situation will then develop:

1. When the field supply is removed, the generator real power will remain almost constant during the next seconds. Because of the drop in the excitation voltage, the generator output voltage drops gradually. To compensate for the drop in voltage, the current increases at about the same rate.
2. The generator then becomes underexcited and it will absorb increasingly negative reactive power.
3. Because the ratio of the generator voltage over the current becomes smaller and smaller with the phase current leading the phase voltage, the generator positive sequence impedance as measured at its terminals will enter the impedance plane in the second quadrant. Experience has shown that the positive sequence impedance will settle to a value between Xd and Xq.

The most popular protection against a loss-of-excitation situation uses an offset-mho relay as shown in Figure 2.7. The relay is supplied with generator terminals voltages and currents and is normally associated with a definite time delay. Many modern digital relays will use the positive sequence voltage and current to evaluate the positive sequence impedance as seen at the generator terminal.

The Protection of Synchronous Generators

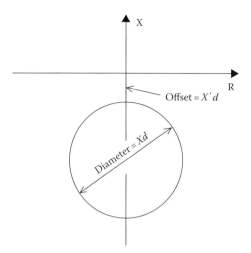

FIGURE 2.7 Loss-of-excitation offset-mho characteristic.

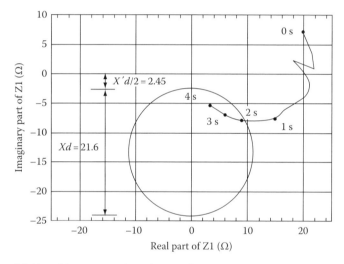

FIGURE 2.8 Loss-of-field positive sequence impedance trajectory.

Figure 2.8 shows the digitally emulated positive sequence impedance trajectory of a 200 MVA generator connected to an infinite bus through an 8% impedance transformer when the field voltage was removed at 0 s time.

2.6 Current Imbalance (46)

Current imbalance in the stator with its subsequent production of negative sequence current will be the cause of double-frequency currents on the surface of the rotor. This, in turn, may cause excessive overheating of the rotor and trigger substantial thermal and mechanical damages (due to temperature effects).

The reasons for temporary or permanent current imbalance are numerous:

- System asymmetries
- Unbalanced loads
- Unbalanced system faults or open circuits
- Single-pole tripping with subsequent reclosing

The energy supplied to the rotor follows a purely thermal law and is proportional to the square of the negative sequence current. Consequently, a thermal limit K is reached when the following integral equation is solved:

$$K = \int_0^t I_2^2 \, dt \qquad (2.4)$$

where
 K is the constant depending upon the generator design and size
 I_2 is the RMS value of negative sequence current
 t is the time

The integral equation can be expressed as an inverse time-current characteristic where the maximum time is given as the negative sequence current variable:

$$t = \frac{K}{I_2^2} \qquad (2.5)$$

In this expression the negative sequence current magnitude will be entered most of the time as a percentage of the nominal phase current and integration will take place when the measured negative sequence current becomes greater than a percentage threshold.

Thermal capability constant, K, is determined by experiment by the generator manufacturer. Negative sequence currents are supplied to the machine on which strategically located thermocouples have been installed. The temperature rises are recorded and the thermal capability is inferred.

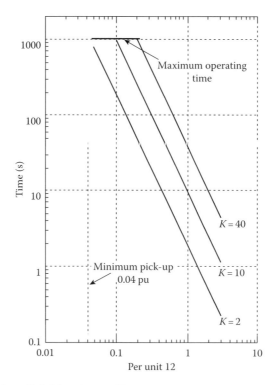

FIGURE 2.9 Typical static or digital time-inverse 46 curve.

Forty-six (46) relays can be supplied in all three technologies (electromechanical, static, or digital). Ideally the negative sequence current should be measured in rms magnitude. Various measurement principles can be found. Digital relays could measure the fundamental component of the negative sequence current because this could be the basic principle for phasor measurement. Figure 2.9 represents a typical relay characteristic.

2.7 Anti-Motoring Protection (32)

A number of situations exist where a generator could be driven as a motor. Anti-motoring protection will more specifically apply in situations where the prime-mover supply is removed for a generator supplying a network at synchronous speed with the field normally excited. The power system will then drive the generator as a motor.

A motoring condition may develop if a generator is connected improperly to the power system. This will happen if the generator circuit breaker is closed inadvertently at some speed less than synchronous speed. Typical situations are when the generator is on turning gear, slowing down to a standstill, or has reached standstill. This motoring condition occurs during what is called "generator inadvertent energization." The protection schemes that respond to this situation are different and will be addressed later in this article.

Motoring will cause adverse effects, particularly in the case of steam turbines. The basic phenomenon is that the rotation of the turbine rotor and the blades in a steam environment will cause windage losses. Windage losses are a function of rotor diameter, blade length, and are directly proportional to the density of the enclosed steam. Therefore, in any situation where the steam density is high, harmful windage losses could occur. From the preceding discussion, one may conclude that the anti-motoring protection is more of a prime-mover protection than a generator protection.

The most obvious means of detecting motoring is to monitor the flow of real power into the generator. If that flow becomes negative below a preset level, then a motoring condition is detected. Sensitivity and setting of the power relay depends upon the energy drawn by the prime mover considered now as a motor.

With a gas turbine, the large compressor represents a substantial load that could reach as high as 50% of the unit nameplate rating. Sensitivity of the power relay is not an issue and is definitely not critical. With a diesel type engine (with no firing in the cylinders), load could reach as high as 25% of the unit rating and sensitivity, once again, is not critical. With hydroturbines, if the blades are below the tail-race level, the motoring energy is high. If above, the reverse power gets as low as 0.2%–2% of the rated power and a sensitive reverse power relay is then needed. With steam turbines operating at full vacuum and zero steam input, motoring will draw 0.5%–3% of unit rating. A sensitive power relay is then required.

2.8 Overexcitation Protection (24)

When generator or step-up transformer magnetic core iron becomes saturated beyond rating, stray fluxes will be induced into nonlaminated components. These components are not designed to carry flux and therefore thermal or dielectric damage can occur rapidly.

In dynamic magnetic circuits, voltages are generated by the Lenz Law:

$$V = K \frac{d\phi}{dt} \tag{2.6}$$

Measured voltage can be integrated in order to get an estimate of the flux. Assuming a sinusoidal voltage of magnitude V_p and frequency f, and integrating over a positive or negative half-cycle interval:

$$\phi = \frac{1}{K} \int_0^{T/2} V_p \sin(\omega t + \theta)\, dt = \frac{V_p}{2\pi fK}(-\cos\omega t)\Big|_0^{T/2} \tag{2.7}$$

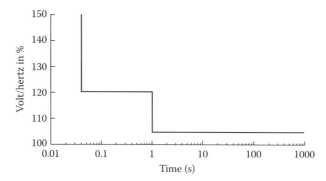

FIGURE 2.10 Dual definite-time characteristic.

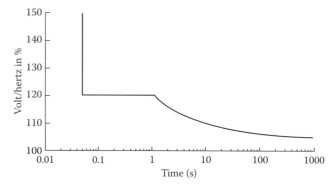

FIGURE 2.11 Combined definite and inverse-time characteristics.

one derives an estimate of the flux that is proportional to the value of peak voltage over the frequency. This type of protection is then called volts per hertz.

$$\phi \approx \frac{V_p}{f} \quad (2.8)$$

The estimated value of the flux can then be compared to a maximum value threshold. With static technology, volt/hertz relays would practically integrate the monitored voltage over a positive or negative (or both) half-cycle period of time and develop a value that would be proportional to the flux. With digital relays, since measurement of the frequency together with the magnitudes of phase voltages are continuously available, a direct ratio computation as shown in Equation 2.8 would be performed.

ANSI/IEEE standard limits are 1.05 pu for generators and 1.05 for transformers (on transformer secondary base, at rated load, 0.8 power factor or greater; 1.1 pu at no-load). It has been traditional to supply either definite time or inverse-time characteristics as recommended by the ANSI/IEEE guides and standards. Figure 2.10 represents a typical dual definite-time characteristic whereas Figure 2.11 represents a combined definite and inverse-time characteristic.

One of the primary requirements of a volt/hertz relay is that it should measure both voltage magnitude and frequency over a broad range of frequency.

2.9 Overvoltage (59)

An overvoltage condition could be encountered without exceeding the volt/hertz limits. For that reason, an overvoltage relay is recommended. Particularly for hydro-units, C37-102 recommends both an instantaneous and an inverse element. The instantaneous should be set to 130%–150% of rated voltage and the inverse element should have a pick-up voltage of 110% of the rated voltage. Coordination with the voltage regulator should be verified.

2.10 Voltage Imbalance Protection (60)

The loss of a voltage phase signal can be due to a number of causes. The primary cause for this nuisance is a blown-out fuse in the voltage transformer circuit. Other causes can be a wiring error, a voltage transformer (VT) failure, a contact opening, a misoperation during maintenance, etc.

Since the purpose of these VTs is to provide voltage signals to the protective relays and the voltage regulator, the immediate effect of a loss of VT signal will be the possible misoperation of some protective relays and the cause for generator overexcitation by the voltage regulator. Among the protective relays to be impacted by the loss of VT signal are

- Function 21: Distance relay. Backup for system and generator zone phase faults.
- Function 32: Reverse power relay. Anti-motoring function, sequential tripping and inadvertent energization functions.
- Function 40: Loss-of-field protection.
- Function 51V: Voltage-restrained time overcurrent relay.

Normally these functions should be blocked if a condition of fuse failure is detected.

It is common practice for large generators to use two sets of voltage transformers for protection, voltage regulation, and measurement. Therefore, the most common practice for loss of VT signals detection is to use a voltage balance relay as shown in Figure 2.12 on each pair of secondary phase voltage. When a fuse blows, the voltage relationship becomes imbalanced and the relay operates. Typically, the voltage imbalance will be set at around 15%.

The advent of digital relays has allowed the use of sophisticated algorithms based on symmetrical components to detect the loss of VT signal. When a situation of loss of one or more of the VT signals occurs, the following conditions develop:

- There will be a drop in the positive sequence voltage accompanied by an increase in the negative sequence voltage magnitude. The magnitude of this drop will depend upon the number of phases impacted by a fuse failure.
- In case of a loss of VT signal and contrary to a fault condition, there should not be any change in the current's magnitudes and phases. Therefore, the negative and zero sequence currents should remain below a small tolerance value. A fault condition can be distinguished from a loss of VT signal by monitoring the changes in the positive and negative current levels. In case of a loss of VT signals, these changes should remain below a small tolerance level.

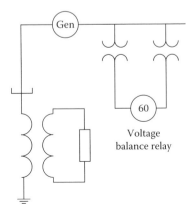

FIGURE 2.12 Example of voltage balance relay.

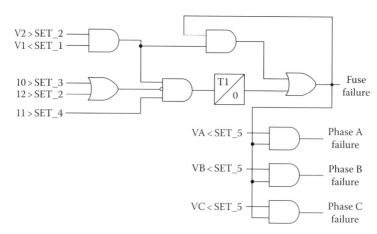

FIGURE 2.13 Symmetrical component implementation of fuse failure detection.

All the above conditions can be incorporated into a complex logic scheme to determine if indeed a there has been a condition of loss of VT signal or a fault. Figure 2.13 represents the logic implementation of a voltage transformer single and double fuse failure based on symmetrical components.

If the following conditions are met in the same time (and condition) during a time delay longer than T1:

- The positive sequence voltage is below a voltage set-value SET_1.
- The negative sequence voltage is above a voltage set-value SET_2.
- There exists a small value of current such that the positive sequence current $I1$ is above a small set-value SET_4 and the negative and zero sequence currents $I2$ and $I2$ do not exceed a small set-value SET_3.

then a fuse failure condition will pick up to one and remain in that state thanks to the latch effect. Fuse failure of a specific phase can be detected by monitoring the level voltage of each phase and comparing it to a set-value SET_5. As soon as the positive sequence voltage returns to a value greater than the set-value SET_1 and the negative sequence voltage disappears, the fuse failure condition returns to a zero state.

2.11 System Backup Protection (51V and 21)

Generator backup protection is not applied to generator faults but rather to system faults that have not been cleared in time by the system primary protection, but which require generator removal in order for the fault to be eliminated. By definition, these are time-delayed protective functions that must coordinate with the primary protective system.

System backup protection (Figure 2.14) must provide protection for both phase faults and ground faults. For the purpose of protecting against phase faults, two solutions are most commonly applied: the use of overcurrent relays with either voltage restraint or voltage control, or impedance-type relays.

The basic principle behind the concept of supervising the overcurrent relay by voltage is that a fault external to the generator and on the system will have the effect of reducing the voltage at the generator terminal. This effect is being used in both types of overcurrent applications: the voltage controlled overcurrent relay will block the overcurrent element unless the voltage gets below a pre-set value, and the voltage restraint overcurrent element will have its pick-up current reduced by an amount proportional to the voltage reduction (see Figure 2.15).

The impedance type backup protection could be applied to the low or high side of the step-up transformer. Normally, three 21 elements will cover all types of phase faults on the system as in a line relay.

As shown in Figure 2.16, a reverse offset is allowed in the mho element in order for the backup to partially or totally cover the generator windings.

The Protection of Synchronous Generators

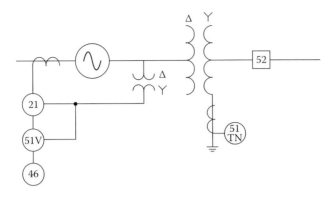

FIGURE 2.14 Backup protection basic scheme.

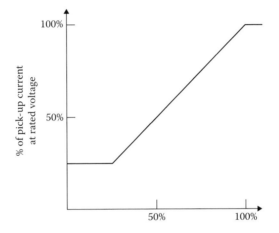

FIGURE 2.15 Voltage restraint overcurrent relay principle.

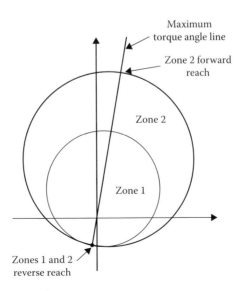

FIGURE 2.16 Typical 21 elements application.

2.12 Out-of-Step Protection

When there is an equilibrium between generation and load on an electrical network, the network frequency will be stable and the internal angle of the generators will remain constant with respect to each other. If an imbalance (loss of generation, sudden addition of load, network fault, etc.) occurs, however, the internal angle of a generator will undergo some changes and two situations might develop: a new stable state will be reached after the disturbance has faded away, or the generator internal angle will not stabilize and the generator will run synchronously with respect to the rest of the network (moving internal angle and different frequency). In the latter case, an out-of-step protection is implemented to detect the situation.

That principle can be visualized by considering the two-source network of Figure 2.17.

If the angle between the two sources is θ and the ratio between the voltage magnitudes is $n = E_G/E_S$, then the positive sequence impedance seen from location will be

$$Z_R = \frac{n(Z_G + Z_T + Z_S)(n - \cos\theta - j\sin\theta)}{(n - \cos\theta)^2 + \sin^2\theta} - Z_G \qquad (2.9)$$

If n is equal to 1, Equation 2.9 simplifies to

$$Z_R = \frac{n(Z_G + Z_T + Z_S)(1 - j\cot g\theta/2)}{2} - Z_G \qquad (2.10)$$

The impedance locus represented by this equation is a straight line, perpendicular to and crossing the vector $Z_S + Z_T + Z_G$ at its middle point. If n is different from 1, the loci become circles as shown in Figure 2.18. The angle θ between the two sources is the angle between the two segments joining Z_R to the base

FIGURE 2.17 Elementary two-source network.

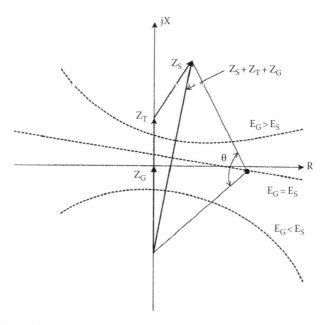

FIGURE 2.18 Impedance locus for different source angles.

The Protection of Synchronous Generators

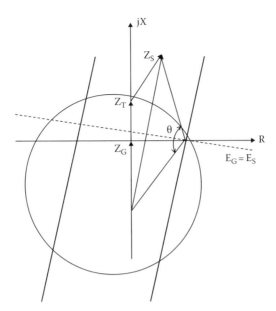

FIGURE 2.19 Out-of-step mho detector with blinders.

of Z_G and the summit of Z_S. Normally, that angle will take a small value. In an out-of-step condition, it will assume a bigger value and when it reaches 180°, it crosses $Z_S + Z_T + Z_G$ at its middle point.

Normally, because of the machine's inertia, the impedance Z_R moves slowly. The phenomenon can be taken advantage of and an out-of-step condition will very often be detected by the combination a mho relay and two blinders as shown in Figure 2.19. In this application, an out-of-step condition will be assumed to be detected when the impedance locus enters the mho circle and remains between the two blinders for an interval of time longer than a preset definite time delay. Implicit in this scheme is the fact that the angle between the two sources is assumed to take a large value when Z_R crosses the blinders. Implementation of an out-of-step protection will normally require some careful studies and eventually will require some stability simulations in order to determine the nature and the locus of the stable and the unstable swings. One of the paramount requirement of an out-of-step protection is not to trip the generator in case of a stable wing.

2.13 Abnormal Frequency Operation of Turbine-Generator

Although it is not a concern for hydraulic generators, the protection against abnormal frequency operation becomes an issue with steam turbine-graters. If the turbine is rotated at a frequency other than synchronous, the blades in the low pressure turbine element could resonate at their natural frequency. Blading mechanical fatigue could result with subsequent damage and failure.

Figure 2.20 (ANSI/IEEE C37.106-1987, 1987) represents a typical steam turbine operating limitation curve. Continuous operation is allowed around 60 Hz. Time-limited zones exist above and below the continuous operation regions. Prohibited operation regions lie beyond.

With the advent of modern generator microprocessor-based relays (IEEE, 1989), there does not seem to be a consensus emerging among the relay and turbine manufacturers, regarding the digital implementation of underfrequency turbine protection. The following points should, however, be taken into account:

- Measurement of frequency is normally available on a continuous basis and over a broad frequency range. Precision better than 0.01 Hz in the frequency measurement has been achieved.
- In practically all products, a number of independent over- or under-frequency definite time functions can be combined to form a composite curve.

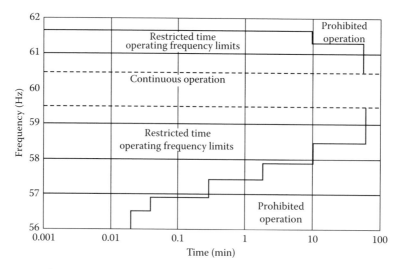

FIGURE 2.20 Typical steam turbine operating characteristic. (Modified from Guide for abnormal frequency protection for power generating plant, ANSI/IEEE C37.106-1987, 1987, Figure 6.)

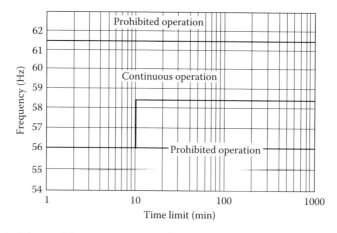

FIGURE 2.21 Typical abnormal frequency protection characteristic.

Therefore, with digital technology, a typical over/underfrequency scheme, as shown in Figure 2.21, comprising one definite-time over-frequency and two definite-time under-frequency elements is readily implementable.

2.14 Protection against Accidental Energization

A number of catastrophic failures have occurred in the past when synchronous generators have been accidentally energized while at standstill. Among the causes for such incidents were human errors, breaker flashover, or control circuitry malfunction.

A number of protection schemes have been devised to protect the generator against inadvertent energization. The basic principle is to monitor the out-of-service condition and to detect an accidental energizing immediately following that state. As an example, Figure 2.22 shows an application using an over-frequency relay supervising three single phase instantaneous overcurrent elements. When the generator is put out of service or the over-frequency element drops out, the timer will pick up. If inadvertent

The Protection of Synchronous Generators

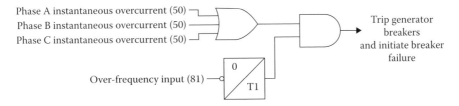

FIGURE 2.22 Frequency supervised overcurrent inadvertent energizing protection.

energizing occurs, the over-frequency element will pick up, but because of the timer drop-out delay, the instantaneous overcurrent elements will have the time to initiate the generator breakers opening. The supervision could also be implemented using a voltage relay.

Accidental energizing caused by a single or three-phase breaker flashover occurring during the generator synchronizing process will not be detected by the logic of Figure 2.22. In such an instance, by the time the generator has been closed to the synchronous speed, the overcurrent element outputs would have been blocked.

2.15 Generator Breaker Failure

Generator breaker failure follows the general pattern of the same function found in other applications: once a fault has been detected by a protective device, a timer will monitor the removal of the fault. If, after a time delay, the fault is still detected, conclusion is reached that the breaker(s) have not opened and a signal to open the backup breakers will be sent.

Figure 2.23 shows a conventional breaker failure diagram where provision has been added to detect a flashover occurring before the synchronizing of the generator: in addition to the protective relays detecting a fault, a flashover condition is detected by using an instantaneous overcurrent relay installed on the neutral of the step-up transformer. If this relay picks up and the breaker position contact (52b) is closed (breaker open), then a flashover condition is asserted and breaker failure is initiated.

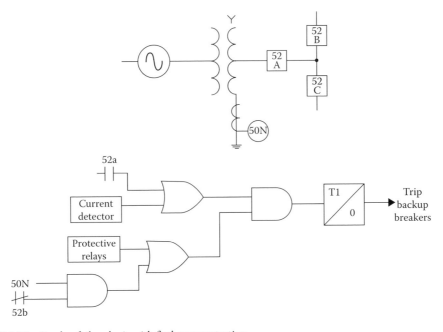

FIGURE 2.23 Breaker failure logic with flashover protection.

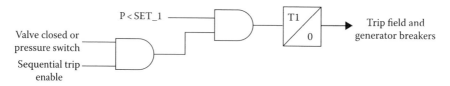

FIGURE 2.24 Implementation of a sequential tripping function.

2.16 Generator Tripping Principles

A number of methods for isolating a generator once a fault has been detected are commonly being implemented. They fall into four groups:

- Simultaneous tripping involves simultaneously shutting the prime mover down by closing its valves and opening the field and generator breakers. This technique is highly recommended for severe internal generator faults.
- Generator tripping involves simultaneously opening both the field and generator breakers.
- Unit separation involves opening the generator breaker only.
- Sequential tripping is applicable to steam turbines and involves first tripping the turbine valves in order to prevent any overspeeding of the unit. Then, the field and generator breakers are opened. Figure 2.24 represents a possible logical scheme for the implementation of a sequential tripping function. If the following three conditions are met, (1) the real power is below a negative pre-set threshold SET_1, (2) the steam valve or a differential pressure switch is closed (either condition indicating the removal of the prime-mover), (3) the sequential tripping function is enabled, then a trip signal will be sent to the generator and field breakers.

2.17 Impact of Generator Digital Multifunction Relays*

The latest technological leap in generator protection has been the release of digital multifunction relays by various manufacturers (Benmouyal, 1988, Benmouyal et al., 1994; Yalla, 1992; Yip, 1994). With more sophisticated characteristics being available through software algorithms, generator protective function characteristics can be improved. Therefore, multifunction relays have many advantages, most of which stem from the technology on which they are based.

2.17.1 Improvements in Signal Processing

Most multifunction relays use a full-cycle discrete Fourier transform (DFT) algorithm for acquisition of the fundamental component of the current and voltage phasors. Consequently, they will benefit from the inherent filtering properties provided by the algorithms, such as

- Immunity from DC component and good suppression of exponentially decaying offset due to the large value of X/R time constants in generators
- Immunity to harmonics
- Nominal response time of one cycle for the protective functions requiring fast response

Since sequence quantities are computed mathematically from the voltage and current phasors, they will also benefit from the above advantages.

However, it should be kept in mind that fundamental phasors of waveforms are not the only parameters used in digital multifunction relays. Other parameters like peak or rms values of waveforms can be equally acquired through simple algorithms, depending upon the characteristics of a particular algorithm.

---

* This section was published previously in a modified form in Working Group J-11 of PSRC (1999).

A number of techniques have been used to make the measurement of phasor magnitudes independent of frequency, and therefore achieve stable sensitivities over large frequency excursions. One technique is known as frequency tracking and consists of having a number of samples in one cycle that is constant, regardless of the value of the frequency or the generator's speed. A software digital phase-locked loop allows implementation of such a scheme and will inherently provide a direct measurement of the frequency or the speed of the generator (Benmouyal, 1989). A second technique keeps the sampling period fixed, but varies the time length of the data window to follow the period of the generator frequency. This results in a variable number of samples in the cycles (Hart et al., 1997). A third technique consists of measuring the root-mean square value of a current or voltage waveform. The variation of this quantity with frequency is very limited, and therefore, this technique allows measurement of the magnitude of a waveform over a broad frequency range.

A further improvement consists of measuring the generator frequency digitally. Precision, in most cases, will be one hundredth of a hertz or better, and good immunity to harmonics and noise is achievable with modern algorithms.

2.17.2 Improvements in Protective Functions

The following functions will benefit from some inherent advantages of the digital processing capability:

- A number of improvements can be attributed to stator differential protection. The first is the detection of CT saturation in case of external faults that would cause the protection relay to trip. When CT ratios do not match perfectly, the difference can be either automatically or manually introduced into the algorithm in order to suppress the difference.
- It is no longer necessary to provide a Δ-Y conversion for the backup 21 elements in order to cover the phase fault on the high side of the voltage transformer. That conversion can be accomplished mathematically inside the relay.
- In the area of detection of voltage transformer blown fuses, the use of symmetrical components allows identification of the faulted phase. Therefore, complex logic schemes can be implemented where only the protection function impacted by the phase will be blocked. As an example, if a 51V is implemented on all three phases independently, it will be sufficient to block the function only on the phase on which a fuse has been detected as blown. Furthermore, contrary to the conventional voltage balance relay scheme, a single VT will suffice when using this modern algorithm.
- Because of the different functions recording their characteristics over a large frequency interval, it is no longer necessary to monitor the frequency in order to implement start-up or shut-down protection.
- The 100% stator-ground protection can be improved by using third-harmonic voltage measurements both at the phase and neutral.
- The characteristic of an offset mho impedance relay in the R–X plane can be made to be independent of frequency by using one of the following two techniques: the frequency-tracking algorithm previously mentioned, or the use of the positive sequence voltage and current because their ratio is frequency-independent.
- Functions which are inherently three-phase phenomena can be implemented by using the positive sequence voltage and current quantities. The loss-of-field or loss-of-synchronism are examples.
- In the reverse power protection, improved accuracy and sensitivity can be obtained with digital technology.
- Digital technology allows the possibility of tailoring inverse volt/hertz curves to the user's needs. Full programmability of these same curves is readily achievable. From that perspective, volt/hertz protection is improved by a closer match between the implemented curve and the generator or step-up transformer damage curve.

Multifunction generator protection packages have other functions that make use of the inherent capabilities of microprocessor devices. These include: oscillography and event recording, time synchronization, multiple settings, metering, communications, self-monitoring, and diagnostics.

References

Benmouyal, G., Design of a universal protection relay for synchronous generators, CIGRE Session, No. 34–09, 1988.

Benmouyal, G., An adaptive sampling interval generator for digital relaying, *IEEE Transactions on Power Delivery*, 4(3), 1602–1609, July 1989.

Benmouyal, G., Adamiak, M.G., Das, D.P., and Patel, S.C., Working to develop a new multifunction digital package for generator protection, *Electricity Today*, 6(3), March 1994.

Berdy, J., Loss-of-excitation for synchronous generators, *IEEE Transactions on Power Apparatus and Systems*, PAS-94(5), September/October 1975.

Guide for abnormal frequency protection for power generating plant, ANSI/IEEE C37.106-1987, 1987.

Guide for AC generator protection, ANSI/IEEE C37.102.

Guide for generator ground protection, ANSI/IEEE C37.101.

Hart, D., Novosel, D., Hu, Y., Smith, R., and Egolf, M., A new frequency tracking and phasor estimation algorithm for generator protection, *IEEE Transactions on Power Delivery*, 12(3), 1064–1073, July 1997.

Ilar, M. and Wittwer, M., Numerical generator protection offers new benefits of gas turbines, *International Gas Turbine and Aeroengine Congress and Exposition*, Cologne, Germany, June 1992.

Institute of Electrical and Electronics Engineers (IEEE), IEEE recommended practice for protection and coordination of industrial and commercial power systems, ANSI/IEEE 242-1986, 1986.

Institute of Electrical and Electronics Engineers (IEEE), Power Engineering Education Committee, IEEE Power Engineering Society, and Power Systems Relaying Committee, *IEEE Tutorial on the Protection of Synchronous Generators*, Piscataway, NJ: IEEE Service Center, IEEE Catalog No. 95TP102, 1995.

Mozina, C.J., Arehart, R.F., Berdy, J., Bonk, J.J., Conrad, S.P., Darlington, A.N., Elmore, W.A. et al., Inadvertent energizing protection of synchronous generators, *IEEE Transactions on Power Delivery*, 4(2), 965–977, April 1989.

Wimmer, W., Fromm, W., Muller, P., and IIar, F., Fundamental considerations on user-configurable multifunctional numerical protection, 34–202, CIGRE Session, 1996.

Working Group J-11 of Power System Relaying Committee (PSRC), Application of multifunction generator protection systems, *IEEE Transactions on Power Delivery*, 14(4), 1285–1294, October 1999.

Yalla, M.V.V.S., A digital multifunction protection relay, *IEEE Transactions on Power Delivery*, 7(1), 193–201, January 1992.

Yin, X.G., Malik, O.P., Hope, G.S., and Chen, D.S., Adaptive ground fault protection schemes for turbogenerator based on third harmonic voltages, *IEEE Transactions on Power Delivery*, 5(2), 595–601, July 1990.

Yip, H.T., *An Integrated Approach to Generator Protection*, Canadian Electrical Association, Toronto, ON, Canada, March 1994.

3
Transmission Line Protection

	3.1	Nature of Relaying..3-2
		Reliability • Zones of Protection • Relay Speed • Primary and Backup Protection • Reclosing • System Configuration
	3.2	Current Actuated Relays..3-5
		Fuses • Inverse Time-Delay Overcurrent Relays • Instantaneous Overcurrent Relays • Directional Overcurrent Relays
	3.3	Distance Relays 3-8
		Impedance Relay • Admittance Relay • Reactance Relay
	3.4	Pilot Protection 3-10
		Directional Comparison • Transfer Tripping • Phase Comparison • Pilot Wire • Current Differential
Stanley H. Horowitz	3.5	Relay Designs..3-11
Consultant		Electromechanical Relays • Solid-State Relays • Computer Relays
		Reference 3-13

The study of transmission line protection presents many fundamental relaying considerations that apply, in 1° or another, to the protection of other types of power system protection. Each electrical element, of course, will have problems unique to itself, but the concepts of reliability, selectivity, local and remote backup, and zones of protection, coordination, and speed, which may be present in the protection of one or more other electrical apparatus, are all present in the considerations surrounding transmission line protection.

Since transmission lines are also the links to adjacent lines or connected equipment, transmission line protection must be compatible with the protection of all of these other elements. This requires coordination of settings, operating times, and characteristics.

The purpose of power system protection is to detect faults or abnormal operating conditions and to initiate corrective action. Relays must be able to evaluate a wide variety of parameters to establish that corrective action is required. Obviously, a relay cannot prevent the fault. Its primary purpose is to detect the fault and take the necessary action to minimize the damage to the equipment or to the system. The most common parameters that reflect the presence of a fault are the voltages and currents at the terminals of the protected apparatus or at the appropriate zone boundaries. The fundamental problem in power system protection is to define the quantities that can differentiate between normal and abnormal conditions. This problem is compounded by the fact that "normal" in the present sense means outside the zone of protection. This aspect, which is of the greatest significance in designing a secure relaying system, dominates the design of all protection systems.

3.1 Nature of Relaying

3.1.1 Reliability

Reliability, in system protection parlance, has special definitions that differ from the usual planning or operating usage. A relay can misoperate in two ways: it can fail to operate when it is required to do so, or it can operate when it is not required or desirable for it to do so. To cover both situations, there are two components in defining reliability:

Dependability, which refers to the certainty that a relay will respond correctly for all faults for which it is designed and applied to operate

Security, which is the measure that a relay will not operate incorrectly for any fault

Most relays and relay schemes are designed to be dependable since the system itself is robust enough to withstand an incorrect tripout (loss of security), whereas a failure to trip (loss of dependability) may be catastrophic in terms of system performance.

3.1.2 Zones of Protection

The property of security is defined in terms of regions of a power system—called zones of protection—for which a given relay or protective system is responsible. The relay will be considered secure if it responds only to faults within its zone of protection. Figure 3.1 shows typical zones of protection with transmission lines, buses, and transformers, each residing in its own zone. Also shown are "closed zones," in which all power apparatus entering the zone is monitored, and "open" zones, the limit of which varies with the fault current. Closed zones are also known as "differential," "unit," or "absolutely selective," and open zones are "nonunit," "unrestricted," or "relatively selective."

The zone of protection is bounded by the current transformers (CTs), which provide the input to the relays. While a CT provides the ability to detect a fault within its zone, the circuit breaker (CB) provides the ability to isolate the fault by disconnecting all of the power equipment inside its zone. When a CT is part of the CB, it becomes a natural zone boundary. When the CT is not an integral part of the CB, special attention must be paid to the fault detection and fault interruption logic. The CTs still define the zone of protection, but a communication channel must be used to implement the tripping function.

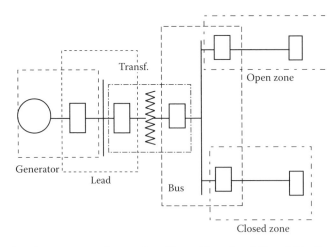

FIGURE 3.1 Closed and open zones of protection. (From Horowitz, S.H. and Phadke, A.G., *Power System Relaying*, 3rd edn., John Wiley & Sons, Ltd., Chichester, U.K., 2009. With permission.)

3.1.3 Relay Speed

It is, of course, desirable to remove a fault from the power system as quickly as possible. However, the relay must make its decision based upon voltage and current waveforms, which are severely distorted due to transient phenomena that follow the occurrence of a fault. The relay must separate the meaningful and significant information contained in these waveforms upon which a secure relaying decision must be based. These considerations demand that the relay take a certain amount of time to arrive at a decision with the necessary degree of certainty. The relationship between the relay response time and its degree of certainty is an inverse one and is one of the most basic properties of all protection systems.

Although the operating time of relays often varies between wide limits, relays are generally classified by their speed of operation as follows:

1. Instantaneous: These relays operate as soon as a secure decision is made. No intentional time delay is introduced to slow down the relay response.
2. Time delay: An intentional time delay is inserted between the relay decision time and the initiation of the trip action.
3. High speed: A relay that operates in less than a specified time. The specified time in present practice is 50 ms (three cycles on a 60 Hz system).
4. Ultrahigh speed: This term is not included in the relay standards but is commonly considered to be operation in 4 ms or less.

3.1.4 Primary and Backup Protection

The main protection system for a given zone of protection is called the primary protection system. It operates in the fastest time possible and removes the least amount of equipment from service. On extra high voltage (EHV) systems, that is, 345 kV and above, it is common to use duplicate primary protection systems in case a component in one primary protection chain fails to operate. This duplication is, therefore, intended to cover the failure of the relays themselves. One may use relays from a different manufacturer, or relays based on a different principle of operation to avoid common-mode failures. The operating time and the tripping logic of both the primary and its duplicate system are the same.

It is not always practical to duplicate every element of the protection chain. On high voltage (HV) and EHV systems, the costs of transducers and CBs are very expensive and the cost of duplicate equipment may not be justified. On lower voltage systems, even the relays themselves may not be duplicated. In such situations, a backup set of relays will be used. Backup relays are slower than the primary relays and may remove more of the system elements than is necessary to clear the fault.

Remote backup: These relays are located in a separate location and are completely independent of the relays, transducers, batteries, and CBs that they are backing up. There are no common failures that can affect both sets of relays. However, complex system configurations may significantly affect the ability of a remote relay to "see" all faults for which backup is desired. In addition, remote backup may remove more sources of the system than can be allowed.

Local backup: These relays do not suffer from the same difficulties as remote backup, but they are installed in the same substation and use some of the same elements as the primary protection. They may then fail to operate for the same reasons as the primary protection.

3.1.5 Reclosing

Automatic reclosing infers no manual intervention but probably requires specific interlocking such as a full or check synchronizing, voltage or switching device checks, or other safety or operating constraints. Automatic reclosing can be high speed or delayed. High speed reclosing (HSR) allows only enough time for the arc products of a fault to dissipate, generally 15–40 cycles on a 60 Hz base, whereas time-delayed reclosings have a specific coordinating time, usually 1 s or more. HSR has the possibility of generator shaft torque damage and should be closely examined before applying it.

It is common practice in the United States to trip all three phases for all faults and then reclose the three phases simultaneously. In Europe, however, for single line-to-ground faults, it is not uncommon to trip only the faulted phase and then reclose that phase. This practice has some applications in the United States, but only in rare situations. When one phase of a three-phase system is opened in response to a single phase-to-ground fault, the voltage and current in the two healthy phases tend to maintain the fault arc after the faulted phase is de-energized. Depending on the length of the line, load current, and operating voltage, compensating reactors may be required to extinguish this "secondary arc."

3.1.6 System Configuration

Although the fundamentals of transmission line protection apply in almost all system configurations, there are different applications that are more or less dependent upon specific situations.

Operating voltages: Transmission lines will be those lines operating at 138 kV and above, subtransmission lines are 34.5–138 kV, and distribution lines are below 34.5 kV. These are not rigid definitions and are only used to generically identify a transmission system and connote the type of protection usually provided. The higher voltage systems would normally be expected to have more complex, hence more expensive, relay systems. This is so because higher voltages have more expensive equipment associated with them and one would expect that this voltage class is more important to the security of the power system. The higher relay costs, therefore, are more easily justified.

Line length: The length of a line has a direct effect on the type of protection, the relays applied, and the settings. It is helpful to categorize the line length as "short," "medium," or "long" as this helps establish the general relaying applications although the definition of "short," "medium," and "long" is not precise. A short line is one in which the ratio of the source to the line impedance (SIR) is large (>4, for example), the SIR of a long line is 0.5 or less and a medium line's SIR is between 4 and 0.5. It must be noted, however, that the per-unit impedance of a line varies more with the nominal voltage of the line than with its physical length or impedance. So a "short" line at one voltage level may be a "medium" or "long" line at another.

Multiterminal lines: Occasionally, transmission lines may be tapped to provide intermediate connections to additional sources without the expense of a CB or other switching device. Such a configuration is known as a multiterminal line and, although it is an inexpensive measure for strengthening the power system, it presents special problems for the protection engineer. The difficulty arises from the fact that a relay receives its input from the local transducers, that is, the current and voltage at the relay location. Referring to Figure 3.2, the current contribution to a fault from the intermediate source is not monitored. The total fault current is the sum of the local current plus the contribution from the intermediate source, and the voltage at the relay location is the sum of the two voltage drops, one of which is the product of the unmonitored current and the associated line impedance.

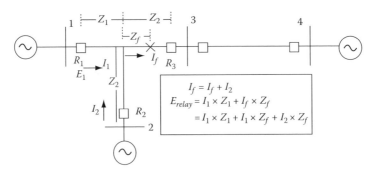

FIGURE 3.2 Effect of infeed on local relays. (From Horowitz, S.H. and Phadke, A.G., *Power System Relaying*, 3rd edn., John Wiley & Sons, Ltd., Chichester, U.K., 2009. With permission.)

3.2 Current Actuated Relays

3.2.1 Fuses

The most commonly used protective device in a distribution circuit is the fuse. Fuse characteristics vary considerably from one manufacturer to another and the specifics must be obtained from their appropriate literature. Figure 3.3 shows the time-current characteristics, which consist of the minimum melt and total clearing curves.

Minimum melt is the time between initiation of a current large enough to cause the current responsive element to melt and the instant when arcing occurs. Total clearing time (TCT) is the total time elapsing from the beginning of an overcurrent to the final circuit interruption; that is, TCT is minimum melt plus arcing time.

In addition to the different melting curves, fuses have different load-carrying capabilities. Manufacturer's application tables show three load-current values: continuous, hot-load pickup, and cold-load pickup. Continuous load is the maximum current that is expected for 3 h or more for which the fuse will not be damaged. Hot load is the amount that can be carried continuously, interrupted, and immediately reenergized without melting. Cold load follows a 30 min outage and is the high current that is the result in the loss of diversity when service is restored. Since the fuse will also cool down during this period, the cold-load pickup and the hot-load pickup may approach similar values.

3.2.2 Inverse Time-Delay Overcurrent Relays

The principal application of time-delay overcurrent (TDOC) relays is on a radial system where they provide both phase and ground protection. A basic complement of relays would be two phase and one ground relay. This arrangement will protect the line for all combinations of phase and ground faults using the minimum number of relays. Adding a third phase relay, however, provides complete backup protection, that is, two relays for every type of fault, and is the preferred practice. TDOC relays are usually used in industrial systems and on subtransmission lines that cannot justify more expensive protection such as distance or pilot relays.

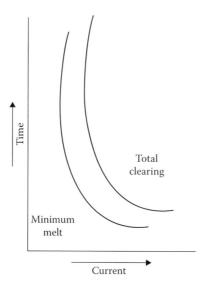

FIGURE 3.3 Fuse time-current characteristic. (From Horowitz, S.H. and Phadke, A.G., *Power System Relaying*, 3rd edn., John Wiley & Sons, Ltd., Chichester, U.K., 2009. With permission.)

There are two settings that must be applied to all TDOC relays: the pickup and the time delay. The pickup setting is selected so that the relay will operate for all short circuits in the line section for which it is to provide protection. This will require margins above the maximum load current, usually twice the expected value, and below the minimum fault current, usually 1/3 the calculated phase-to-phase or phase-to-ground fault current. If possible, this setting should also provide backup for an adjacent line section or adjoining equipment. The time-delay function is an independent parameter that is obtained in a variety of ways, either the setting of an induction disk lever or an external timer. The purpose of the time delay is to enable relays to coordinate with each other. Figure 3.4 shows the family of curves of a single TDOC model. The ordinate is time in milliseconds or seconds depending on the relay type; the abscissa is in multiples of pickup to normalize the curve for all fault current values. Figure 3.5 shows how TDOC relays on a radial line coordinate with each other.

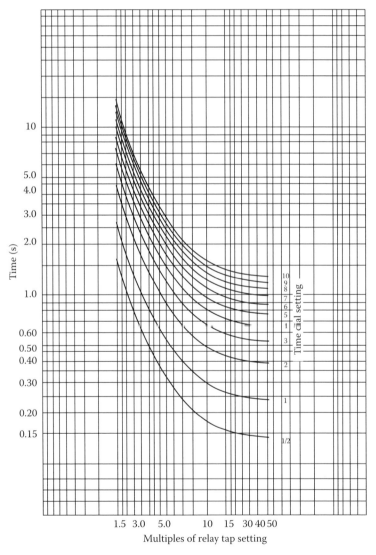

FIGURE 3.4 Family of TDOC time-current characteristics. (From Horowitz, S.H. and Phadke, A.G., *Power System Relaying*, 3rd edn., John Wiley & Sons, Ltd., Chichester, U.K., 2009. With permission.)

Transmission Line Protection

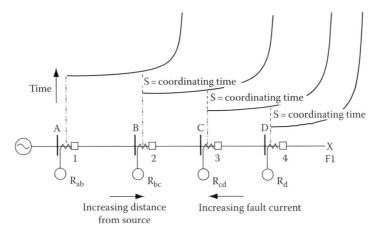

FIGURE 3.5 Coordination of TDOC relays. (From Horowitz, S.H. and Phadke, A.G., *Power System Relaying*, 3rd edn., John Wiley & Sons, Ltd., Chichester, U.K., 2009. With permission.)

3.2.3 Instantaneous Overcurrent Relays

Figure 3.5 also shows why the TDOC relay cannot be used without additional help. The closer the fault is to the source, the greater the fault current magnitude, yet the longer the tripping time. The addition of an instantaneous overcurrent relay makes this system of protection viable. If an instantaneous relay can be set to "see" almost up to, but not including, the next bus, all of the fault clearing times can be lowered as shown in Figure 3.6. In order to properly apply the instantaneous overcurrent relay, there must be a substantial reduction in short-circuit current as the fault moves from the relay toward the far end of the line. However, there still must be enough of a difference in the fault current between the near and far end faults to allow a setting for the near end faults. This will prevent the relay from operating for faults beyond the end of the line and still provide high-speed protection for an appreciable portion of the line.

Since the instantaneous relay must not see beyond its own line section, the values for which it must be set are very much higher than even emergency loads. It is common to set an instantaneous relay about 125%–130% above the maximum value that the relay will see under normal operating situations and about 90% of the minimum value for which the relay should operate.

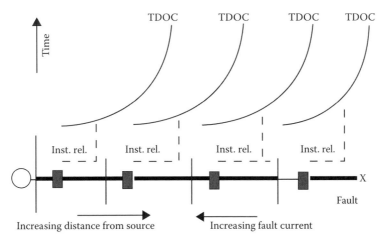

FIGURE 3.6 Effect of instantaneous relays. (From Horowitz, S.H. and Phadke, A.G., *Power System Relaying*, 3rd edn., John Wiley & Sons, Ltd., Chichester, U.K., 2009. With permission.)

3.2.4 Directional Overcurrent Relays

Directional overcurrent relaying is necessary for multiple source circuits when it is essential to limit tripping for faults in only one direction. If the same magnitude of fault current could flow in either direction at the relay location, coordination cannot be achieved with the relays in front of, and, for the same fault, the relays behind the nondirectional relay, except in very unusual system configurations.

Polarizing quantities: To achieve directionality, relays require two inputs—the operating current and a reference, or polarizing, quantity that does not change with fault location. For phase relays, the polarizing quantity is almost always the system voltage at the relay location. For ground directional indication, the zero-sequence voltage ($3E_0$) can be used. The magnitude of $3E_0$ varies with the fault location and may not be adequate in some instances. An alternative and generally preferred method of obtaining a directional reference is to use the current in the neutral of a wye-grounded/delta power transformer. When there are several transformer banks at a station, it is common practice to parallel all of the neutral CTs.

3.3 Distance Relays

Distance relays respond to the voltage and current, that is, the impedance, at the relay location. The impedance per mile is fairly constant so these relays respond to the distance between the relay location and the fault location. As the power systems become more complex and the fault current varies with changes in generation and system configuration, directional overcurrent relays become difficult to apply and to set for all contingencies, whereas the distance relay setting is constant for a wide variety of changes external to the protected line.

There are three general distance relay types as shown in Figure 3.7. Each is distinguished by its application and its operating characteristic.

3.3.1 Impedance Relay

The impedance relay has a circular characteristic centered at the origin of the R–X diagram. It is nondirectional and is used primarily as a fault detector.

3.3.2 Admittance Relay

The admittance relay is the most commonly used distance relay. It is the tripping relay in pilot schemes and as the backup relay in step distance schemes. Its characteristic passes through the origin of the R–X

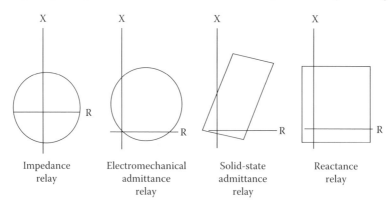

FIGURE 3.7 Distance relay characteristics. (From Horowitz, S.H. and Phadke, A.G., *Power System Relaying*, 3rd edn., John Wiley & Sons, Ltd., Chichester, U.K., 2009. With permission.)

diagram and is therefore directional. In the electromechanical design it is circular, and in the solid-state design, it can be shaped to correspond to the transmission line impedance.

3.3.3 Reactance Relay

The reactance relay is a straight-line characteristic that responds only to the reactance of the protected line. It is nondirectional and is used to supplement the admittance relay as a tripping relay to make the overall protection independent of resistance. It is particularly useful on short lines where the fault arc resistance is the same order of magnitude as the line length.

Figure 3.8 shows a three-zone step distance relaying scheme that provides instantaneous protection over 80%–90% of the protected line section (Zone 1) and time-delayed protection over the remainder of the line (Zone 2) plus backup protection over the adjacent line section. Zone 3 also provides backup protection for adjacent line sections.

In a three-phase power system, 10 types of faults are possible: three single-phase-to-ground faults, three phase-to-phase faults, three double-phase-to-ground faults, and one three-phase fault. It is essential that the relays provided have the same setting regardless of the type of fault. This is possible if the relays are connected to respond to delta voltages and currents. The delta quantities are defined as the difference between any two phase quantities, for example, $E_a - E_b$ is the delta quantity between phases a and b. In general, for a multiphase fault between phases x and y,

$$\frac{Ex - Ey}{Ix - Iy} = Z_1 \tag{3.1}$$

where
  x and y can be a, b, or c
  Z_1 is the positive sequence impedance between the relay location and the fault

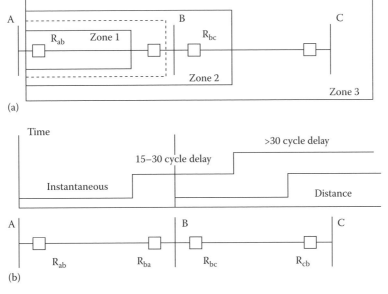

FIGURE 3.8 Three-zone step distance relaying to protect 100% of a line and backup the neighboring line. (a) Distance measurements. (b) Operating times. (From Horowitz, S.H. and Phadke, A.G., *Power System Relaying*, 3rd edn., John Wiley & Sons, Ltd., Chichester, U.K., 2009. With permission.)

For ground distance relays, the faulted phase voltage and a compensated faulted phase current must be used:

$$\frac{Ex}{Ix + mI_0} = Z_1 \qquad (3.2)$$

where

m is a constant depending on the line impedances
I_0 is the zero sequence current in the transmission line

A full complement of relays consists of three phase distance relays and three ground distance relays. This is the preferred protective scheme for HV and EHV systems.

3.4 Pilot Protection

As can be seen from Figure 3.8, step distance protection does not offer instantaneous clearing of faults over 100% of the line segment. In most cases this is unacceptable due to system stability considerations. To cover the 10%–20% of the line not covered by Zone 1, the information regarding the location of the fault is transmitted from each terminal to the other terminal(s). A communication channel is used for this transmission. These pilot channels can be over power line carrier, microwave, fiber optics, or wire pilot. Although the underlying principles are the same regardless of the pilot channel, there are specific design details that are imposed by this choice.

Power line carrier uses the protected line itself as the channel, superimposing a high frequency signal on top of the 60 Hz power frequency. Since the line being protected is also the medium used to actuate the protective devices, a blocking signal is used. This means that a trip will occur at both ends of the line unless a signal is received from the remote end.

Microwave or fiber-optic channels are independent of the transmission line being protected so a tripping signal can be used.

Wire pilot channels are limited by the impedance of the copper wire and are used at lower voltages where the distance between the terminals is not great, usually less than 10 miles.

3.4.1 Directional Comparison

The most common pilot relaying scheme in the United States is the directional comparison blocking scheme, using power line carrier. The fundamental principle upon which this scheme is based utilizes the fact that, at a given terminal, the direction of a fault either forward or backward is easily determined by a directional relay. By transmitting this information to the remote end, and by applying appropriate logic, both ends can determine whether a fault is within the protected line or external to it. Since the power line itself is used as the communication medium, a blocking signal is used because if the line itself is damaged a tripping signal could not be transmitted.

3.4.2 Transfer Tripping

If the communication channel is independent of the power line, a tripping scheme is a viable protection scheme. Using the same directional relay logic to determine the location of a fault, a tripping signal is sent to the remote end. To increase security, there are several variations possible. A direct tripping signal can be sent, or additional underreaching or overreaching directional relays can be used to supervise the tripping function and increase security. An underreaching relay sees less than 100% of the protected line, that is, Zone 1. An overreaching relay sees beyond the protected line such as Zone 2 or 3.

3.4.3 Phase Comparison

Phase comparison is a differential scheme that compares the phase angle between the currents at the ends of the line. If the currents are essentially in phase, there is no fault in the protected section. If these currents are essentially 180° out of phase, there is a fault within the line section. Any communication link can be used.

3.4.4 Pilot Wire

Pilot wire relaying is a form of differential line protection similar to phase comparison, except that the phase currents are compared over a pair of metallic wires. The pilot channel is often a rented circuit from the local telephone company. However, as the telephone companies are replacing their wired facilities with microwave or fiber optics, this protection must be closely monitored. It is becoming more common for the power company to install its own pilot cable.

3.4.5 Current Differential

More recently, the improvements in communication technology has made it possible to implement a long-distance current differential scheme similar to the schemes used locally for transformers and generators. In a current differential scheme a true differential measurement is made. Ideally, the difference should be zero or equal to any tapped load on the line. In practice, this may not be practical due to CT errors, ratio mismatch or any line charging currents. Information concerning both the phase and magnitude of the current at each terminal must be made available at all terminals at in order to prevent operation on external faults. Thus, a communication medium must be provided that is suitable for the transmission of these data.

3.5 Relay Designs

3.5.1 Electromechanical Relays

Early relay designs utilized actuating forces that were produced by electromagnetic interaction between currents and fluxes, much as in a motor. These forces were created by a combination of input signals, stored energy in springs, and dash pots. The plunger-type relays are usually driven by a single actuating quantity while an induction-type relay may be activated by a single or multiple inputs (see Figures 3.9 and 3.10). Traditionally, the electromechanical relay has been the major protective device applied to new construction.

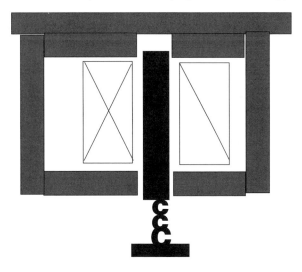

FIGURE 3.9 Plunger-type relay.

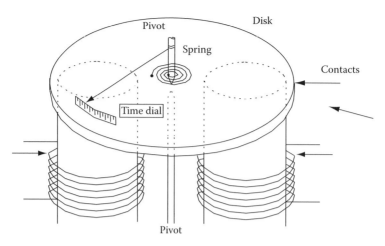

FIGURE 3.10 Principle of construction of an induction disk relay. Shaded poles and damping magnets are omitted for clarity.

The present improvements, however, in the application, ability and versatility and price in computer relays has made it the protection of choice in virtually all new installations.

3.5.2 Solid-State Relays

The expansion and growing complexity of modern power systems have brought a need for protective relays with a higher level of performance and more sophisticated characteristics. This has been made possible by the development of semiconductors and other associated components, which can be utilized in many designs, generally referred to as solid-state or static relays. All of the functions and characteristics available with electromechanical relays are available with solid-state relays. They use low-power components but have limited capability to tolerate extremes of temperature, humidity, overvoltage, or overcurrent. Their settings are more repeatable and hold to closer tolerances and their characteristics can be shaped by adjusting the logic elements as opposed to the fixed characteristics of electromechanical relays. This can be a distinct advantage in difficult relaying situations. Solid-state relays are designed, assembled, and tested as a system that puts the overall responsibility for proper operation of the relays on the manufacturer. Figure 3.11 shows a solid-state instantaneous overcurrent relay. However, with the increased use of computer relays, discussed in Section 3.5.3, solid-state relays are not the relays of choice in most new installations.

3.5.3 Computer Relays

It has been noted that a relay is basically an analog computer. It accepts inputs, processes them electromechanically or electronically to develop a torque or a logic output, and makes a decision resulting in a contact closure or output signal. With the advent of rugged, high-performance microprocessors, it is

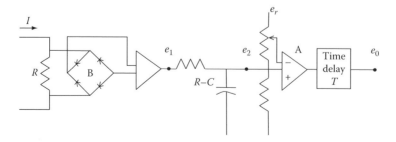

FIGURE 3.11 Possible circuit configuration for a solid-state instantaneous overcurrent delay.

Transmission Line Protection

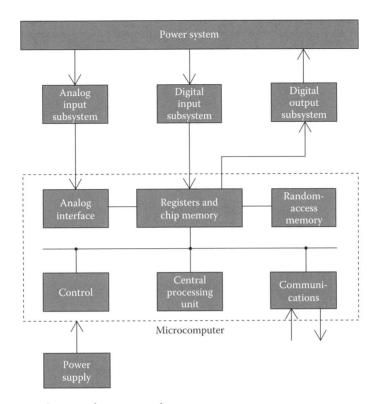

FIGURE 3.12 Major subsystem of a computer relay.

obvious that a digital computer can perform the same function. Since the usual relay inputs consist of power system voltages and currents, it is necessary to obtain a digital representation of these parameters. This is done by sampling the analog signals, and using an appropriate algorithm to create suitable digital representations of the signals. The functional blocks in Figure 3.12 represent a possible configuration for a digital relay.

In the early stages of their development, computer relays were designed to replace existing protection functions, such as transmission line and transformer or bus protection. Some relays used microprocessors to make the relay decision from digitized analog signals; others continue to use analog functions to make the relaying decisions and digital techniques for the necessary logic and auxiliary functions. In all cases, however, a major advantage of the digital relay was its ability to diagnose itself, a capability that could only be obtained, if at all, with great effort, cost, and complexity. In addition, the digital relay provides a communication capability to warn system operators when it is not functioning properly, permitting remote diagnostics and possible correction.

As digital relay investigations continued another dimension was added. The ability to adapt itself, in real time, to changing system conditions is an inherent, and important, feature in the software-dominated relay. This adaptive feature is rapidly becoming a vital aspect of future system reliability.

As computer relays become the primary protection device, two industry standards are of particular interest for protection systems. These are COMTRADE (IEEE Standard and a parallel IEC standard) and SYNCHROPHASOR (IEEE standard).

Reference

Horowitz, S.H. and Phadke, A.G., *Power System Relaying*, 3rd edn., 2009, John Wiley & Sons, Ltd., Chichester, U.K.

4
System Protection

4.1	Introduction	4-1
4.2	Disturbances: Causes and Remedial Measures	4-1
4.3	Transient Stability and Out-of-Step Protection	4-2
4.4	Overload and Underfrequency Load Shedding	4-3
4.5	Voltage Stability and Undervoltage Load Shedding	4-4
4.6	Special Protection Schemes	4-6
4.7	Modern Perspective: Technology Infrastructure	4-7
	Phasor Measurement Technology • Communication Technology	
4.8	Future Improvements in Control and Protection	4-9
	Acknowledgments	4-10
	References	4-10

Miroslav M. Begovic
*Georgia Institute
of Technology*

4.1 Introduction

While most of the protective system designs are made around individual components, system-wide disturbances in power systems are becoming a frequent and challenging problem for the electric utilities. The occurrence of major disturbances in power systems requires coordinated protection and control actions to stop the system degradation, restore the normal state, and minimize the impact of the disturbance. Local protection systems are often not capable of protecting the overall system, which may be affected by the disturbance. Among the phenomena, which create the power system, disturbances are various types of system instability, overloads, and power system cascading [1–5].

The power system planning has to account for tight operating margins, with less redundancy, because of new constraints placed by restructuring of the entire industry. The advanced measurement and communication technology in wide area monitoring and control are expected to provide new, faster, and better ways to detect and control an emergency [6].

4.2 Disturbances: Causes and Remedial Measures [7]

Phenomena that create power system disturbances are divided, among others, into the following categories: transient instabilities, voltage instabilities, overloads, power system cascading, etc. They are mitigated using a variety of protective relaying and emergency control measures.

Out-of-step protection has the objective to eliminate the possibility of damage to generators as a result of an out-of-step condition. In case the power system separation is imminent, it should separate the system along the boundaries, which will form islands with balanced load and generation. Distance relays are often used to provide an out-of-step protection function, whereby they are called upon to provide blocking or tripping signals upon detecting an out-of-step condition.

The most common predictive scheme to combat loss of synchronism is the equal-area criterion and its variations. This method assumes that the power system behaves like an equivalent two-machine model where one area oscillates against the rest of the system. Whenever the underlying assumption holds true, the method has potential for fast detection.

Voltage instabilities in power systems arise from heavy loading, inadequate reactive support resources, unforeseen contingencies and/or mis-coordinated action of the tap-changing transformers. Such incidents can lead to system-wide blackouts (which have occurred in the past and have been documented in many power systems world-wide).

The risk of voltage instability increases as the transmission system becomes more heavily loaded. The typical scenario of these instabilities starts with a high system loading, followed by a relay action due to a fault, a line overload, or operation beyond an excitation limit.

Overload of one, or a few power system elements may lead to a cascading overload of many more elements, mostly transmission lines, and ultimately, it may lead to a complete power system blackout.

A quick, simple, and reliable way to reestablish active power balance is to shed load by underfrequency relays. There are a large variety of practices in designing load shedding schemes based on the characteristics of a particular system and the utility practices.

While the system frequency is a consequence of the power deficiency, the rate of change of frequency is an instantaneous indicator of power deficiency and can enable incipient recognition of the power imbalance. However, change of the machine speed is oscillatory by nature, due to the interaction among the generators. These oscillations depend on location of the sensors in the island and the response of the generators. A system having smaller inertia causes a larger peak-to-peak value for oscillations, requiring enough time for the relay to calculate the actual rate of change of frequency reliably. Measurements at load buses close to the electrical center of the system are less susceptible to oscillations (smaller peak-to-peak values) and can be used in practical applications. A system having smaller inertia causes a higher frequency of oscillations, which enables faster calculation of the actual rate of change of frequency. However, it causes faster rate of change of frequency and consequently, a larger frequency drop. Adaptive settings of frequency and frequency derivative relays may enable implementation of a frequency derivative function more effectively and reliably.

4.3 Transient Stability and Out-of-Step Protection

Every time when a fault or a topological change affects the power balance in the system, the instantaneous power imbalance creates oscillations between the machines. Stable oscillations lead to transition from one (prefault) to another (postfault) equilibrium point, whereas unstable ones allow machines to oscillate beyond the acceptable range. If the oscillations are large, then the stations' auxiliary supplies may undergo severe voltage fluctuations, and eventually trip [1]. Should that happen, the subsequent resynchronization of the machines might take a long time. It is, therefore, desirable to trip the machine exposed to transient unstable oscillations while preserving the plant auxiliaries energized.

The frequency of the transient oscillations is usually between 0.5 and 2 Hz. Since the fault imposes almost instantaneous changes on the system, the slow speed of the transient disturbances can be used to distinguish between the two. For the sake of illustration, let us assume that a power system consists of two machines, A and B, connected by a transmission line. Figure 4.1 represents the trajectories of the stable and unstable swings between the machines, as well as a characteristic of the mho relay covering the line between them, shown in the impedance plane. The stable swing moves from the distant stable operating point toward the trip zone of the relay, and may even encroach it, then leave again. The unstable trajectory may pass through the entire trip zone of the relay. The relaying tasks are to detect, and then trip (or block) the relay, depending on the situation. Detection is accomplished by out-of-step

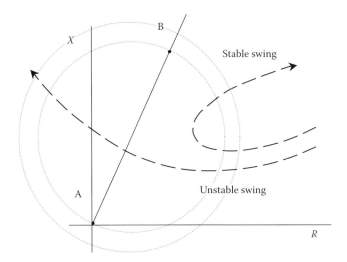

FIGURE 4.1 Trajectories of stable and unstable swings in the impedance plane.

relays, which have multiple characteristics. When the trajectory of the impedance seen by the relays enters the outer zone (a circle with a larger radius), the timer is activated, and depending on the speed at which the impedance trajectory moves into the inner zone (a circle with a smaller radius), or leaves the outer zone, a tripping (or blocking) decision can be made. The relay characteristic may be chosen to be straight lines, known as "blinders," which prevent the heavy load to be misrepresented as a fault, or instability. Another information that can be used in detection of transient swings is that they are symmetrical, and do not create any zero, or negative sequence currents.

In the case when power system separation is imminent, out-of-step protection should take place along boundaries, which will form islands with matching load and generation. Distance relays are often used to provide an out-of-step protection function, whereby they are called upon to provide blocking or tripping signals upon detecting an out-of-step condition. The most common predictive scheme to combat loss of synchronism is the equal-area criterion and its variations. This method assumes that the power system behaves like a two-machine model where one area oscillates against the rest of the system. Whenever the underlying assumption holds true, the method has potential for fast detection.

4.4 Overload and Underfrequency Load Shedding

Outage of one or more power system components due to the overload may result in overload of other elements in the system. If the overload is not alleviated in time, the process of power system cascading may start, leading to power system separation. When a power system separates, islands with an imbalance between generation and load are formed. One consequence of the imbalance is deviation of frequency from the nominal value. If the generators cannot handle the imbalance, load or generation shedding is necessary. A special protection system or out-of-step relaying can also start the separation.

A quick, simple, and reliable way to reestablish active power balance is to shed load by underfrequency relays. The load shedding is often designed as a multistep action, and the frequency settings and blocks of load to be shed are carefully selected to maximize the reliability and dependability of the action. There are a large variety of practices in designing load shedding schemes based on the characteristics of a particular system and the utility practices. While the system frequency is a final result of the power deficiency, the rate of change of frequency is an instantaneous indicator of power deficiency and can enable incipient recognition of the power imbalance. However, change of the machine speed is oscillatory by nature, due to the interaction among generators. These oscillations depend on location of the sensors in the island and the response of the generators. The problems regarding the rate of change of frequency function are

- Systems having small inertia may cause larger oscillations. Thus, enough time must be allowed for the relay to calculate the actual rate of change of frequency reliably. Measurements at load buses close to the electrical center of the system are less susceptible to oscillations (smaller peak-to-peak values) and can be used in practical applications. Smaller system inertia causes a higher frequency of oscillations, which enables faster calculation of the actual rate of change of frequency. However, it causes a faster rate of change of frequency and consequently, a larger frequency drop.
- Even if rate of change of frequency relays measure the average value throughout the network, it is difficult to set them properly, unless typical system boundaries and imbalance can be predicted. If this is the case (e.g., industrial and urban systems), the rate of change of frequency relays may improve a load shedding scheme (scheme can be more selective and/or faster).

4.5 Voltage Stability and Undervoltage Load Shedding

Voltage stability is defined by the "System Dynamic Performance Subcommittee of the IEEE Power System Engineering Committee" as being the ability of a system to maintain voltage such that when load admittance is increased, load power will increase, so that both power and voltage are controllable. Also, voltage collapse is defined as being the process by which voltage instability leads to a very low voltage profile in a significant part of the system. It is accepted that this instability is caused by the load characteristics, as opposed to the angular instability, which is caused by the rotor dynamics of generators.

Voltage stability problems are manifested by several distinguishing features: low system voltage profiles, heavy reactive line flows, inadequate reactive support, and heavily loaded power systems. The voltage collapse typically occurs abruptly, after a symptomatic period that may last in the time frames of a few seconds to several minutes, sometimes hours. The onset of voltage collapse is often precipitated by low-probability single or multiple contingencies. The consequences of collapse often require long system restoration, while large groups of customers are left without supply for extended periods of time. Schemes which mitigate against collapse need to use the symptoms to diagnose the approach of the collapse in time to initiate corrective action.

Analysis of voltage collapse models can be divided into two main categories, static or dynamic:

- *Fast*: Disturbances of the system structure, which may involve equipment outages, or faults followed by equipment outages. These disturbances may be similar to those which are consistent with transient stability symptoms, and sometimes the distinction is hard to make, but the mitigation tools for both types are essentially similar, making it less important to distinguish between them.
- *Slow*: Load disturbances, such as fluctuations of the system load. Slow load fluctuations may be treated as inherently static. They cause the stable equilibrium of the system to move slowly, which makes it possible to approximate voltage profile changes by a discrete sequence of steady states rather than a dynamic model.

Figure 4.2 shows a symbolic depiction of the process of coalescing of the stable and unstable power system equilibria (saddle node bifurcation) through slow load variations, which leads to a voltage collapse (a precipitous departure of the system state along the center manifold at the moment of coalescing). VPQ curve (see Figure 4.2) represents the trajectory of the load voltage V of a two-bus system model when active (P) and reactive (Q) powers of the load can change arbitrarily.

Figure 4.2 represents a trajectory of the load voltage V when active (P) and reactive (Q) powers change independently. It also shows the active and reactive power margins as projections of the distances. The voltage stability boundary is represented by a projection onto the PQ plane (a bold curve). It can be observed that: (a) there may be many possible trajectories to (and points of) voltage collapse; (b) active and reactive power margins depend on the initial operating point and the trajectory to collapse.

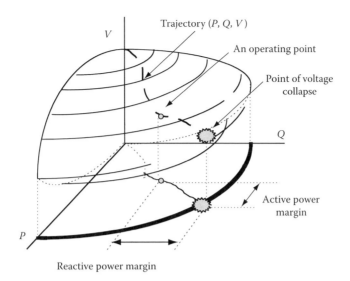

FIGURE 4.2 Relationship between voltages, active and reactive powers of the load and voltage collapse.

There have been numerous attempts to use the observations and find accurate voltage collapse proximity indicators. They are usually based on measurement of the state of a given system under stress and derivation of certain parameters which indicate the stability or proximity to instability of that system.

Parameters based on measurement of system condition are useful for planning and operating purposes to avoid the situation where a collapse might occur. However, it is difficult to calculate the system condition and derive the parameters in real time. Rapid derivation and analysis of these parameters are important to initiate automatic corrective actions fast enough to avoid collapse under emergency conditions, which arise due to topological changes or very fast load changes.

It is preferable if a few critical parameters that can be directly measured could be used in real time to quickly indicate proximity to collapse. An example of such indicator is the sensitivity of the generated reactive powers with respect to the load parameters (active and reactive powers of the loads). When the system is close to a collapse, small increases in load result in relatively large increases in reactive power absorption in the system. These increases in reactive power absorption must be supplied by dynamic sources of reactive power in the region. At the point of collapse, the rate of change of generated reactive power at key sources with respect to load increases at key busses tends to infinity.

The sensitivity matrix of the generated reactive powers with respect to loading parameters is relatively easy to calculate in off-line studies, but could be a problem in real-time applications, because of the need for system-wide measurement information. Large sensitivity factors reveal both critical generators (those required to supply most of the newly needed reactive power) and critical loads (those whose location in the system topology imposes the largest increase in reactive transmission losses, even for the modest changes of their own load parameters). The norm of such a sensitivity matrix represents a useful proximity indicator, but one that is still relatively difficult to interpret. It is not the generated reactive power, but its derivatives with respect to loading parameters which become infinite at the point of imminent collapse.

Voltage instability can be alleviated by a combination of the following remedial measures: adding reactive compensation near load centers, strengthening the transmission lines, varying the operating conditions such as voltage profile and generation dispatch, coordinating relays and controls, and load shedding. Most utilities rely on planning and operation studies to guard against voltage instability. Many utilities utilize localized voltage measurements in order to achieve load shedding as a measure against incipient voltage instability. The efficiency of the load shedding depends on the selected voltage thresholds, locations of pilot points in which the voltages are monitored, locations and sizes of the blocks of load to be shed, as well as the operating conditions, which may activate the shedding. The wide variety of conditions that may lead to voltage instability suggests that the most accurate decisions should imply the adaptive relay settings, but such applications are still in the stage of early development.

4.6 Special Protection Schemes

Increasingly popular over the past several years are the so-called special protection systems, sometimes also referred to as remedial action schemes [8,9]. Depending on the power system in question, it is sometimes possible to identify the contingencies, or combinations of operating conditions, which may lead to transients with extremely disastrous consequences [10]. Such problems include, but are not limited to, transmission line faults, the outages of lines and possible cascading that such an initial contingency may cause, outages of the generators, rapid changes of the load level, problems with high voltage DC (HVDC) transmission or flexible AC transmission systems (FACTS) equipment, or any combination of those events.

Among the many varieties of special protection schemes (SPS), several names have been used to describe the general category [2]: special stability controls, dynamic security controls, contingency arming schemes, remedial action schemes, adaptive protection schemes, corrective action schemes, security enhancement schemes, etc. In the strict sense of protective relaying, we do not consider any control schemes to be SPS, but only those protective relaying systems, which possess the following properties [9]:

- SPS can be operational (armed), or out of service (disarmed), in conjunction with the system conditions.
- SPS are responding to very low-probability events; hence they are active rarely more than once a year.
- SPS operate on simple, predetermined control laws, often calculated based on extensive off-line studies.
- Often times, SPS involve communication of remotely acquired measurement data (supervisory control and data acquisition [SCADA]) from more than one location in order to make a decision and invoke a control law.

The SPS design procedure is based on the following [2]:

- *Identification of critical conditions*: On the grounds of extensive off-line steady state studies on the system under consideration, a variety of operating conditions and contingencies are identified as potentially dangerous, and those among them which are deemed the most harmful are recognized as the critical conditions. The issue of their continuous monitoring, detection, and mitigation is resolved through off-line studies.
- *Recognition triggers*: Those are the measurable signals that can be used for detection of critical conditions. Often times, such detection is accomplished through a complicated heuristic logical reasoning, using the logic circuits to accomplish the task: "If event A and event B occur together, or event C occurs, then…" inputs for the decision making logic are called recognition triggers, and can be status of various relays in the system, sometimes combined with a number of (SCADA) measurements.
- *Operator control*: In spite of extensive simulations and studies done in the process of SPS design, it is often necessary to include the human intervention, i.e., to include human interaction in the feedback loop. This is necessary because SPS are not needed all the time, and the decision to arm or disarm them remains in the hands of an operator.

Among the SPS reported in the literature [8,9], the following schemes are represented:

- Generator rejection
- Load rejection
- Underfrequency load shedding
- System separation

System Protection

- Turbine valve control
- Stabilizers
- HVDC controls
- Out-of-step relaying
- Dynamic braking
- Generator runback
- VAR compensation
- Combination of schemes

Some of them have already been described in the above text. A general trend continues toward more complex schemes, capable of outperforming the present solutions, and taking advantage of the most recent technological developments, and advances in systems analysis. Some of the trends are described in the following text [6].

4.7 Modern Perspective: Technology Infrastructure

4.7.1 Phasor Measurement Technology [7]

The technology of synchronized phasor measurements is well established, and is rapidly gaining acceptance as a platform for monitoring systems. It provides an ideal measurement system with which to monitor and control a power system, in particular during stressed conditions. The essential feature of the technique is that it measures positive sequence voltages and currents of a power system in real time with precise time synchronization. This allows accurate comparison of measurements over widely separated locations as well as potential real-time measurement based control actions. Very fast recursive discrete Fourier transform (DFT) calculations are normally used in phasor calculations.

The synchronization is achieved through a global positioning satellite (GPS) system. GPS is a U.S. Government sponsored program that provides world-wide position and time broadcasts free of charge. It can provide continuous precise timing at better than the 1 μs level. It is possible to use other synchronization signals, if these become available in the future, provided that a sufficient accuracy of synchronization could be maintained. Local, proprietary systems can be used such as a sync signal broadcast over microwave or fiber optics. Two other precise positioning systems, global navigation satellite system (GLONASS), a Russian system, and Galileo, a proposed European system, are also capable of providing precise time. The GPS transmission is obtained by the receiver, which delivers a phase-locked sampling clock pulse to the analog-to-digital converter system. The sampled data are converted to a complex number which represents the time-tagged phasor of the sampled waveform. Phasors of all three phases are combined to produce the positive sequence measurement.

Any computer-based relay which uses sampled data is capable of developing the positive sequence measurement. By using an externally derived synchronizing pulse, such as from a GPS receiver, the measurement could be placed on a common time reference. Thus, potentially all computer-based relays could furnish the synchronized phasor measurement. When currents are measured in this fashion, it is important to have a high enough resolution in the analog-to-digital converter to achieve sufficient accuracy of representation at light loads. A 16 bit A/D converter generally provides adequate resolution to read light load currents, as well as fault currents.

For the most effective use of phasor measurements, some kind of a data concentrator is required. The simplest is a system that will retrieve files recorded at the measurement site and then correlate files from different sites by the recording time stamps. This allows doing system and event analysis utilizing the precision of phasor measurement. For real-time applications, continuous data acquisition is required. Phasor concentrator inputs phasor measurement data broadcast from a large number of PMUs, and performs data checks, records disturbances, and rebroadcasts the combined data stream to other monitor and control applications. This type of unit fulfills the need for both hard and soft real-time applications as well as saving data for system analysis. Tests performed using this phasor monitoring unit–phasor data concentrator (PMU–PDC) technology have shown the time intervals from measurement to data availability at a central controller can be as fast as 60 ms for a direct link and 200 ms for secondary links. These times meet the requirements for many types of wide area controls.

A broader effort is the wide area measurement system (WAMS) concept. It includes all types of measurements that can be useful for system analysis over the wide area of an interconnected system. Real-time performance is not required for this type of application, but is no disadvantage. The main elements are time tags with enough precision to unambiguously correlate data from multiple sources and the ability to all data to a common format. Accuracy and timely access to data are important as well. Certainly with its system-wide scope and precise time tags, phasor measurements are a prime candidate for WAMS.

4.7.2 Communication Technology [7]

Communications systems are a vital component of a wide area relay system. These systems distribute and manage the information needed for operation of the wide area relay and control system. However, because of potential loss of communication, the relay system must be designed to detect and tolerate failures in the communication system. It is important also that the relay and communication systems be independent and subject as little as possible to the same failure modes. This has been a serious source of problems in the past.

To meet these difficult requirements, the communications network needs to be designed for fast, robust, and reliable operation. Among the most important factors to consider in achieving these objectives are type and topology of the communications network, communications protocols, and media used. These factors will in turn effect communication system bandwidth, usually expressed in bits per second (BPS), latency in data transmission, reliability, and communication error handling.

Presently, electrical utilities use a combination of analog and digital communications systems for their operations consisting of power line carrier, radio, microwave, leased phone lines, satellite systems, and fiber optics. Each of these systems has applications, where it is the best solution. The advantages and disadvantages of each are briefly summarized in the following paragraph.

Power line carrier is generally rather inexpensive, but has limited distance of coverage and low bandwidth. It is best suited to station-to-station protection and communications to small stations that are hard to access otherwise. Company owned microwave is cost effective and reliable but requires substantial maintenance. It is good for general communications for all types of applications. Radio tends to be narrower band but is good for mobile applications or locations hard to access otherwise. Satellite systems likewise are effective for reaching hard to access locations, but not good where the long delay is a problem. They also tend to be expensive. Leased phone lines are very effective where a one solid link is needed at a site served by a standard carrier. They tend to be expensive in the long-term, so are usually not the best solution where many channels area required. Fiber optic systems are the newest option. They are expensive to install and provision, but are expected to be very cost effective. They have the advantage of using existing right-of-way and delivering communications directly between points of use. In addition they have the very high bandwidth needed for modern data communications.

Several types of communication protocols are used with optical systems. Two of the most common are synchronous optical networks (SONET/SDH) and asynchronous transfer mode (ATM). Wideband Ethernet is also gaining popularity, but is not often used for backbone systems. SONET systems are channel oriented, where each channel has a time slot whether it is needed or not. If there is no data for a particular channel at a particular time, the system just stuffs in a null packet. ATM by contrast puts data on the system as it arrives in private packets. Channels are reconstructed from packets as they come through. It is more efficient as there are no null packets sent, but has the overhead of prioritizing and sorting the packets. Each system has different system management options for coping with problems.

SONET are well established in electrical utilities throughout the world and are available under two similar standards: (a) SONET is the American System under ANSI T1.105 and Bellcore GR standards; (b) synchronous digital hierarchy (SDH) under the international telecommunications union (ITU) standards.

SONET and SDH networks are based on a ring topology. This topology is a bidirectional ring with each node capable of sending data in either direction; data can travel in either direction around the ring to connect any two nodes. If the ring is broken at any point, the nodes detect where the break is relative to the other nodes and automatically reverse transmission direction if necessary. A typical network, however, may consist of a mix of tree, ring, and mesh topologies rather than strictly rings with only the main backbone being rings.

Self-healing (or survivability) capability is a distinctive feature of SONET/SDH networks made possible because it is a ring topology. This means that if communication between two nodes is lost, the traffic among them switches over to the protected path of the ring. This switching to the protected path is made as fast as 4 ms, perfectly acceptable to any wide area protection and control.

Communication protocols are an intrinsic part of modern digital communications. Most popular protocols found in the electrical utility environment and suitable for wide area relaying and control are the distributed network protocol (DNP), Modbus, IEC870-5, and utility communication architecture/manufacturing message specifications (UCA/MMS). Transmission control protocol/Internet protocol (TCP/IP), probably the most extensively used protocol and will undoubtedly find applications in wide area relaying.

UCA/MMS protocol is the result of an effort between utilities and vendors (coordinated by Electric Power Research Institute). It addresses all communication needs of an electric utility. Of particular interest is its "peer to peer" communications capability that allows any node to exchange real-time control signals with any other node in a wide area network. DNP and Modbus are also real-time type protocols suitable for relay applications. TCP on Ethernet lacks a real-time type requirement, but over a system with low traffic performs as well as the other protocols. Other slower speed protocols like Inter Control Center Protocol (ICCP) (America) or TASEII (Europe) handle higher level but slower applications like SCADA. Many other protocols are available but are not commonly used in the utility industry.

4.8 Future Improvements in Control and Protection

Existing protection/control systems may be improved and new protection/control systems may be developed to better adapt to prevailing system conditions during system-wide disturbance. While improvements in the existing systems are mostly achieved through advancement in local measurements and development of better algorithms, improvements in new systems are based on remote communications. However, even if communication links exist, conventional systems that utilize only local information may still need improvement since they are supposed to serve as fall back positions. The increased functions and communication ability in today's SCADA systems provide the opportunity for an intelligent and adaptive control, and protection system for system-wide disturbance. This in turn can make possible full utilization of the network, which will be less vulnerable to a major disturbance.

Out-of-step relays have to be fast and reliable. The present technology of out-of-step tripping or blocking distance relays is not capable of fully dealing with the control and protection requirements of power systems. Central to the development effort of an out-of-step protection system is the investigation of the multi-area out-of-step situation. The new generation of out-of-step relays has to utilize more measurements, both local and remote, and has to produce more outputs. The structure of the overall relaying system has to be distributed and coordinated through a central control. In order for the relaying system to manage complexity, most of the decisions have to be taken locally. The relay system is preferred to be adaptive, in order to cope with system changes. To deal with out-of-step prediction, it is necessary to start with a system-wide approach, find out what sets of information are crucial, how to process information with acceptable speed and accuracy.

The protection against voltage instability should also be addressed as a part of hierarchical structure. The sound approach for designing the new generation of voltage instability protection is to first design a voltage instability relay with only local signals. The limitations of local signals should be identified in order to be in a position to select appropriate communicated signals. However, a minimum set of communicated signals should always be known in order to design a reliable protection, and it requires the following: (a) determining the algorithm for gradual reduction of the number of necessary measurement sites with minimum loss of information necessary for voltage stability monitoring, analysis, and control; (b) development of methods (i.e., sensitivity analysis), which should operate *concurrent* with any existing local protection techniques, and possessing superior performance, both in terms of security and dependability.

Acknowledgments

Portions of the material presented in this chapter were obtained from the IEEE Special Publication [7], which the author chaired. The author would like to acknowledge the Working Group members for their contribution to the report [7]: Alex Apostolov, Ernest Baumgartner, Bob Beckwith, Miroslav Begovic (Chairman), Stuart Borlase, Hans Candia, Peter Crossley, Jaime De La Ree Lopez, Tom Domin, Olivier Faucon, Adly Girgis, Fred Griffin, Charlie Henville, Stan Horowitz, Mohamed Ibrahim, Daniel Karlsson, Mladen Kezunovic, Ken Martin, Gary Michel, Jay Murphy, Damir Novosel, Tony Seegers, Peter Solanics, James Thorp, Demetrios Tziouvaras.

References

1. Horowitz, S.H. and Phadke, A.G., *Power System Relaying*, John Wiley & Sons, Inc., New York, 1992.
2. Elmore, W.A., ed., *Protective Relaying Theory and Applications*, ABB and Marcel Dekker, New York, 1994.
3. Blackburn, L., *Protective Relaying*, Marcel Dekker, New York, 1987.
4. Phadke, A.G. and Thorp, J.S., *Computer Relaying for Power Systems*, John Wiley & Sons, New York, 1988.
5. Anderson, P.M., *Power System Protection*, McGraw Hill and IEEE Press, New York, 1999.
6. Begovic, M., Novosel, D., and Milisavljevic, M., Trends in power system protection and control, *Proceedings 1999 HICSS Conference*, Maui, Hawaii, January 4–7, 1999.
7. Begovic, M. and Working Group C-6 of the IEEE Power System Relaying Committee, Wide area protection and emergency control, Special Publication of IEEE Power Engineering Society at a IEEE Power System Relaying Committee, May 2002. Published electronically at http://www.pespsrc.org/
8. Anderson, P.M. and LeReverend, B.K., Industry experience with special protection schemes, IEEE/CIGRE Committee Report, *IEEE Transactions on Power Systems, PWRS*, 11, 1166–1179, August 1996.
9. McCalley, J. and Fu, W., Reliability of special protection schemes, IEEE Power Engineering Society paper PE-123-PWRS-0-10-1998.
10. Tamronglak, S., Horowitz, S., Phadke, A., and Thorp, J., Anatomy of power system blackouts: Preventive relaying strategies, *IEEE Transactions on Power Delivery, PWRD*, 11, 708–715, April 1996.

5
Digital Relaying

5.1	Sampling...	5-2
5.2	Antialiasing Filters	5-2
5.3	Sigma-Delta A/D Converters...	5-3
5.4	Phasors from Samples	5-4
5.5	Symmetrical Components..	5-6
5.6	Algorithms...	5-7
	Parameter Estimation • Least Squares Fitting • DFT • Differential Equations • Kalman Filters • Wavelet Transforms • Neural Networks	
	References...	5-19

James S. Thorp
Virginia Tech

Digital relaying had its origins in the late 1960s and early 1970s with pioneering papers by Rockefeller (1969), Mann and Morrison (1971), and Poncelet (1972) and an early field experiment (Gilcrest et al., 1972; Rockefeller and Udren, 1972). Because of the cost of the computers in those times, a single high-cost minicomputer was proposed by Rockefeller (1969) to perform multiple relaying calculations in the substation. In addition to having high cost and high power requirements, early minicomputer systems were slow in comparison with modern systems and could only perform simple calculations. The well-founded belief that computers would get smaller, faster, and cheaper combined with expectations of benefits of computer relaying kept the field moving. The third IEEE tutorial on microprocessor protection (Sachdev, 1997) lists more then 1100 publications in the area since 1970. Nearly two thirds of the papers are devoted to developing and comparing algorithms. It is not clear this trend should continue. Issues beyond algorithms should receive more attention in the future.

The expected benefits of microprocessor protection have largely been realized. The ability of a digital relay to perform self-monitoring and checking is a clear advantage over the previous technology. Many relays are called upon to function only a few cycles in a year. A large percentage of major disturbances can be traced to "hidden failures" in relays that were undetected until the relay was exposed to certain system conditions (Tamronglak et al., 1996). The ability of a digital relay to detect a failure within itself and remove itself from service before an incorrect operation occurs is one of the most important advantages of digital protection.

The microprocessor revolution has created a situation in which digital relays are the relays of choice because of economic reasons. The cost of conventional (analog) relays has increased while the hardware cost of the most sophisticated digital relays has decreased dramatically. Even including substantial software costs, digital relays are the economic choice and have the additional advantage of having lower wiring costs. Prior to the introduction of microprocessor-based systems, several panels of space and considerable wiring was required to provide all the functions needed for each zone of transmission line protection. For example, an installation requiring phase distance protection for phase-to-phase and three-phase faults, ground distance, ground-overcurrent, a pilot scheme, breaker failure, and reclosing logic demanded redundant wiring, several hundred watts of power, and a lot of panel space. A single microprocessor system is a single box, with a 10-W power requirement and with only direct wiring, has replaced the old system.

Modern digital relays can provide SCADA, metering, and oscillographic records. Line relays can also provide fault location information. All of this data can be available by modem or on a WAN. A LAN in the substation connecting the protection modules to a local host is also a possibility. Complex multifunction relays can have an almost bewildering number of settings. Techniques for dealing with setting management are being developed. With improved communication technology, the possibility of involving microprocessor protection in wide-area protection and control is being considered.

5.1 Sampling

The sampling process is essential for microprocessor protection to produce the numbers required by the processing unit to perform calculations and reach relaying decisions. Both 12 and 16 bit A/D converters are in use. The large difference between load and fault current is a driving force behind the need for more precision in the A/D conversion. It is difficult to measure load current accurately while not saturating for fault current with only 12 bits. It should be noted that most protection functions do not require such precise load current measurement. Although there are applications, such as hydro generator protection, where the sampling rate is derived from the actual power system frequency, most relay applications involve sampling at a fixed rate that is a multiple of the *nominal* power system frequency.

5.2 Antialiasing Filters

ANSI/IEEE Standard C37.90, provides the standard for the Surge Withstand Capability (SWC) to be built into protective relay equipment. The standard consists of both an oscillatory and transient test. Typically the surge filter is followed by an antialiasing filter before the A/D converter. Ideally the signal x(t) presented to the A/D converter x(t) is band-limited to some frequency ω_c, i.e., the Fourier Transform of x(t) is confined to a low-pass band less that ω_c such as shown in Figure 5.1. Sampling the low-pass signal at a frequency of ω_s produces a signal with a transform made up of shifted replicas of the low-pass transform as shown in Figure 5.2. If $\omega_s - \omega_c > \omega_c$, i.e., $\omega_s > 2\omega_c$ as shown, then an ideal low pass filter applied to z(t) can recover the original signal x(t). The frequency of twice the highest frequency present in the signal to be sampled is the Nyquist sampling rate. If $\omega_s < 2\omega_c$ the sampled signal is said to be "aliased" and the output of the low-pass filter is not the original signal. In some applications the

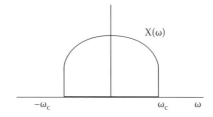

FIGURE 5.1 The Fourier Transform of a band-limited function.

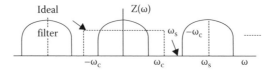

FIGURE 5.2 The Fourier Transform of a sampled version of the signal x(t).

frequency content of the signal is known and the sampling frequency is chosen to avoid aliasing (music CDs), while in digital relaying applications the sampling frequency is specified and the frequency content of the signal is controlled by filtering the signal before sampling to insure its highest frequency is less than half the sampling frequency. The filter used is referred to as an antialiasing filter.

Aliasing also occurs when discrete sequences are sampled or decimated. For example, if a high sampling rate such as 7200 Hz is used to provide data for oscillography, then taking every tenth sample provides data at 720 Hz to be used for relaying. The process of taking every tenth sample (decimation) will produce aliasing unless a digital antialiasing filter with a cut-off frequency of 360 Hz is provided.

5.3 Sigma-Delta A/D Converters

There is an advantage in sampling at rates many times the Nyquist rate. It is possible to exchange speed of sampling for bits of resolution. So called Sigma-Delta A/D converters are based on 1 bit sampling at very high rates. Consider a signal x(t) sampled at a high rate T = 1/fs, i.e., x[n] = x(nT) with the difference between the current sample and α times the last sample given by

$$d[n] = x[n] - \alpha x[n-1] \tag{5.1}$$

If d[n] is quantized through a 1-bit quantizer with a step size of Δ, then

$$x_q[n] = \alpha x_q[n-1] + d_q[n] \tag{5.2}$$

The quantization is called delta modulation and is represented in Figure 5.3. The z^{-1} boxes are unit delays while the 1 bit quantizer is shown as the box with d[n] as input and $d_q[n]$ as output. The output $x_q[n]$ is a staircase approximation to the signal x(t) with stairs that are spaced at T seconds and have height Δ. The delta modulator output has two types of errors: one when the maximum slope Δ/T is too small for rapid changes in the input (shown on Figure 5.3) and the second, a sort of chattering when the signal x(t) is slowly varying. The feedback loop below the quantizer is a discrete approximation to an integrator with α = 1. Values of α less than one correspond to an imperfect integrator. A continuous form of the delta modulator is also shown in Figure 5.4. The low pass filter (LPF) is needed because of the high frequency content of the staircase. Shifting the integrator from in front of the LPF to before the delta modulator improves both types of error. In addition, the two integrators can be combined.

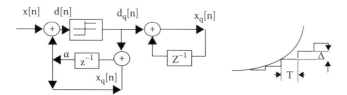

FIGURE 5.3 Delta modulator and error.

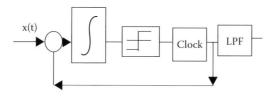

FIGURE 5.4 Sigma-Delta modulator.

The modulator can be thought of as a form of voltage follower circuit. Resolution is increased by oversampling to spread the quantization noise over a large bandwidth. It is possible to shape the quantization noise so it is larger at high frequencies and lower near DC. Combining the shaped noise with a very steep cut-off in the digital low pass filter, it is possible to produce a 16-bit result from the 1 bit comparator. For example, a 16-bit answer at 20 kHz can be obtained with an original sampling frequency of 400 kHz.

5.4 Phasors from Samples

A phasor is a complex number used to represent sinusoidal functions of time such as AC voltages and currents. For convenience in calculating the power in AC circuits from phasors, the phasor magnitude is set equal to the rms value of the sinusoidal waveform. A sinusoidal quantity and its phasor representation are shown in Figure 5.5, and are defined as follows:

$$\text{Sinusoidal quantity} \qquad \text{Phasor}$$
$$y(t) = Y_m \cos(\omega t + \phi) \qquad Y = \frac{Y_m}{\sqrt{2}} e^{j\phi} \tag{5.3}$$

A phasor represents a single frequency sinusoid and is not directly applicable under transient conditions. However, the idea of a phasor can be used in transient conditions by considering that the phasor represents an estimate of the fundamental frequency component of a waveform observed over a finite window. In case of N samples y_k, obtained from the signal $y(t)$ over a period of the waveform:

$$Y = \frac{1}{\sqrt{2}} \frac{2}{N} \sum_{k=1}^{N} y_k e^{-jk 2\pi/N} \tag{5.4}$$

or,

$$Y = \frac{1}{\sqrt{2}} \frac{2}{N} \left\{ \sum_{k=1}^{N} y_k \cos\left(\frac{k 2\pi}{N}\right) - j \sum_{k=1}^{N} y_k \sin\left(\frac{k 2\pi}{N}\right) \right\} \tag{5.5}$$

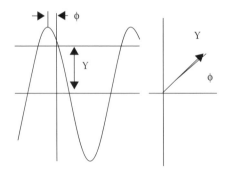

FIGURE 5.5 Phasor representation.

Using θ for the sampling angle $2\pi/N$, it follows that

$$Y = \frac{1}{\sqrt{2}} \frac{2}{N}(Y_c - jY_s) \tag{5.6}$$

where

$$Y_C = \sum_{k=1}^{N} y_k \cos(k\theta)$$
$$Y_S = \sum_{k=1}^{N} y_k \sin(k\theta) \tag{5.7}$$

Note that the input signal y(t) must be band-limited to $N\omega/2$ to avoid aliasing errors. In the presence of white noise, the fundamental frequency component of the Discrete Fourier Transform (DFT) given by Equations 5.4 through 5.7 can be shown to be a least-squares estimate of the phasor. If the data window is not a multiple of a half cycle, the least-squares estimate is some other combination of Y_c and Y_s, and is no longer given by Equation 5.6. Short window (less than one period) phasor computations are of interest in some digital relaying applications. For the present, we will concentrate on data windows that are multiples of a half cycle of the nominal power system frequency.

The data window begins at the instant when sample number 1 is obtained as shown in Figure 5.5. The sample set y_k is given by

$$y_k = Y_m \cos(k\theta + \phi) \tag{5.8}$$

Substituting for y_k from Equation 5.8 in Equation 5.4,

$$Y = \frac{1}{\sqrt{2}} \frac{2}{N} \sum_{k=1}^{N} Y_m \cos(k\theta + \phi) e^{-jk\theta} \tag{5.9}$$

or

$$Y = \frac{1}{\sqrt{2}} Y_m e^{j\phi} \tag{5.10}$$

which is the familiar expression Equation 5.3, for the phasor representation of the sinusoid in Equation 5.3. The instant at which the first data sample is obtained defines the orientation of the phasor in the complex plane. The reference axis for the phasor, i.e., the horizontal axis in Figure 5.5, is specified by the first sample in the data window.

Equations 5.6 and 5.7 define an algorithm for computing a phasor from an input signal. A recursive form of the algorithm is more useful for real-time measurements. Consider the phasors computed from two adjacent sample sets: y_k {k = 1, 2, ..., N} and, y'_k {k = 2, 3, ..., N + 1}, and their corresponding phasors Y^1 and $Y^{2'}$ respectively:

$$Y^1 = \frac{1}{\sqrt{2}} \frac{2}{N} \sum_{k=1}^{N} y_k e^{-jk\theta} \tag{5.11}$$

$$Y^{2\prime} = \frac{1}{\sqrt{2}} \frac{2}{N} \sum_{k=1}^{N} y_{k+1} e^{-jk\theta} \qquad (5.12)$$

We may modify Equation 5.12 to develop a recursive phasor calculation as follows:

$$Y^2 = Y^{2\prime} e^{-j\theta} = Y^1 + \frac{1}{\sqrt{2}} \frac{2}{N} (y_{N+1} - y_1) e^{-j\theta} \qquad (5.13)$$

Since the angle of the phasor $Y^{2\prime}$ is greater than the angle of the phasor Y^1 by the sampling angle θ, the phasor Y^2 has the same angle as the phasor Y^1. When the input signal is a constant sinusoid, the phasor calculated from Equation 5.13 is a constant complex number. In general, the phasor Y, corresponding to the data y_k {k = r, r + 1, r + 2, ..., N + r − 1} is recursively modified into Y^{r+1} according to the formula

$$Y^{r+1} = Y^r e^{-j\theta} = Y^r + \frac{1}{\sqrt{2}} \frac{2}{N} (y_{N+r} - y_r) e^{-j\theta} \qquad (5.14)$$

The recursive phasor calculation as given by Equation 5.13 is very efficient. It regenerates the new phasor from the old one and utilizes most of the computations performed for the phasor with the old data window.

5.5 Symmetrical Components

Symmetrical components are linear transformations on voltages and currents of a three phase network. The symmetrical component transformation matrix S transforms the phase quantities, taken here to be voltages E_ϕ, (although they could equally well be currents), into symmetrical components E_S:

$$E_s = \begin{bmatrix} E_0 \\ E_1 \\ E_2 \end{bmatrix} = SE_\phi = \frac{1}{3} \begin{bmatrix} 1 & 1 & 1 \\ 1 & \alpha & \alpha^2 \\ 1 & \alpha^2 & \alpha \end{bmatrix} \begin{bmatrix} E_a \\ E_b \\ E_c \end{bmatrix} \qquad (5.15)$$

where $(1, \alpha, \alpha^2)$ are the three cube-roots of unity. The symmetrical component transformation matrix S is a similarity transformation on the impedance matrices of balanced three phase circuits, which diagonalizes these matrices. The symmetrical components, designated by the subscripts (0,1,2) are known as the zero, positive, and negative sequence components of the voltages (or currents). The negative and zero sequence components are of importance in analyzing unbalanced three phase networks. For our present discussion, we will concentrate on the positive sequence component E_1 (or I_1) only. This component measures the balanced, or normal voltages and currents that exist in a power system. Dealing with positive sequence components only allows the use of single-phase circuits to model the three-phase network, and provides a very good approximation for the state of a network in quasi-steady state. All power generators generate positive sequence voltages, and all machines work best when energized by positive sequence currents and voltages. The power system is specifically designed to produce and utilize almost pure positive sequence voltages and currents in the absence of faults or other abnormal imbalances. It follows from Equation 5.15 that the positive sequence component of the phase quantities is given by

$$Y_1 = \frac{1}{3}(Y_a + \alpha Y_b + \alpha^2 Y_c) \qquad (5.16)$$

Digital Relaying

Or, using the recursive form of the phasors given by Equation 5.14,

$$Y_1^{r+1} = Y_1^r + \frac{1}{\sqrt{2}}\frac{2}{N}\left[(x_{a,N+r} - x_{a,r})e^{-jr\theta} + \alpha(x_{b,N+r} - x_{b,r})e^{-jr\theta} + \alpha^2(x_{c,N+r} - x_{c,r})e^{-jr\theta}\right] \quad (5.17)$$

Recognizing that for a sampling rate of 12 times per cycle, α and α^2 correspond to $\exp(j4\theta)$ and $\exp(j8\theta)$, respectively, it can be seen from Equation 5.17 that

$$Y_1^{r+1} = Y_1^r + \frac{1}{\sqrt{2}}\frac{2}{N}\left[(x_{a,N+r} - x_{a,r})e^{-jr\theta} + (x_{b,N+r} - x_{b,r})e^{j(4-r)\theta} + (x_{c,N+r} - x_{c,r})e^{j(8-r)\theta}\right] \quad (5.18)$$

With a carefully chosen sampling rate—such as a multiple of three times the nominal power system frequency—very efficient symmetrical component calculations can be performed in real time. Equations similar to (5.18) hold for negative and zero sequence components also. The sequence quantities can be used to compute a distance to the fault that is independent of fault type. Given the 10 possible faults in a three-phase system (three line-ground, three phase-phase, three phase-phase-ground, and three phase), early microprocessor systems were taxed to determine the fault type before computing the distance to the fault. Incorrect fault type identification resulted in a delay in relay operation. The symmetrical component relay solved that problem. With advances in microprocessor speed it is now possible to simultaneously compute the distance to all six phase-ground and phase-phase faults in order to solve the fault classification problem.

The positive sequence calculation is still of interest because of the use of synchronized phasor measurements. Phasors, representing voltages and currents at various buses in a power system, define the state of the power system. If several phasors are to be measured, it is essential that they be measured with a common reference. The reference, as mentioned in the previous section, is determined by the instant at which the samples are taken. In order to achieve a common reference for the phasors, it is essential to achieve synchronization of the sampling pulses. The precision with which the time synchronization must be achieved depends upon the uses one wishes to make of the phasor measurements. For example, one use of the phasor measurements is to estimate, or validate, the state of the power systems so that crucial performance features of the network, such as the power flows in transmission lines could be determined with a degree of confidence. Many other important measures of power system performance, such as contingency evaluation, stability margins, etc., can be expressed in terms of the state of the power system, i.e., the phasors. Accuracy of time synchronization directly translates into the accuracy with which phase angle differences between various phasors can be measured. Phase angles between the ends of transmission lines in a power network may vary between a few degrees, and may approach 180° during particularly violent stability oscillations. Under these circumstances, assuming that one may wish to measure angular differences as little as 1°, one would want the accuracy of measurement to be better than 0.1°. Fortunately, synchronization accuracies of the order of 1 µs are now achievable from the global positioning system (GPS) satellites. One microsecond corresponds to 0.022° for a 60 Hz power system, which more than meets our needs. Real-time phasor measurements have been applied in static state estimation, frequency measurement, and wide area control.

5.6 Algorithms

5.6.1 Parameter Estimation

Most relaying algorithms extract information about the waveform from current and voltage waveforms in order to make relaying decisions. Examples include: current and voltage phasors that can be used to compute impedance, the rms value, the current that can be used in an overcurrent relay, and the harmonic content of a current that can be used to form a restraint in transformer protection. An approach that unifies a number of algorithms is that of parameter estimation. The samples are assumed to be of a current or voltage that has a known form with some unknown parameters. The simplest such signal can be written as

$$y(t) = Y_c \cos \omega_0 t + Y_s \sin \omega_0 t + e(t) \tag{5.19}$$

where
- ω_0 is the nominal power system frequency
- Y_c and Y_s are unknown quantities
- $e(t)$ is an error signal (all the things that are not the fundamental frequency signal in this simple model)

It should be noted that in this formulation, we assume that the power system frequency is known. If the numbers, Y_c and Y_s were known, we could compute the fundamental frequency phasor. With samples taken at an interval of T seconds,

$$y_n = y(nT) = Y_c \cos n\theta + Y_s \sin n\theta + e(nT) \tag{5.20}$$

where $\theta = \omega_0 T$ is the sampling angle. If signals other than the fundamental frequency signal were present, it would be useful to include them in a formulation similar to Equation 5.19 so that they would be included in $e(t)$. If, for example, the second harmonic were included, Equation 5.19 could be modified to

$$y_n = Y_{1c} \cos n\theta + Y_{1s} \sin n\theta + Y_{2c} \cos 2n\theta + Y_{2s} \sin 2n\theta + e(nT) \tag{5.21}$$

It is clear that more samples are needed to estimate the parameters as more terms are included. Equation 5.21 can be generalized to include any number of harmonics (the number is limited by the sampling rate), the exponential offset in a current, or any known signal that is suspected to be included in the post-fault waveform. No matter how detailed the formulation, $e(t)$ will include unpredictable contributions from

- The transducers (CTs and PTs)
- Fault arc
- Traveling wave effects
- A/D converters
- The exponential offset in the current
- The transient response of the antialiasing filters
- The power system itself

The current offset is not an error signal for some algorithms and is removed separately for some others. The power system generated signals are transients depending on fault location, the fault incidence angle, and the structure of the power system. The power system transients are low enough in frequency to be present after the antialiasing filter.

We can write a general expression as

$$y_n = \sum_{k=1}^{K} s_k(nT) Y_k + e_n \tag{5.22}$$

If y represents a vector of N samples, and Y a vector of K unknown coefficients, then there are N equations in K unknowns in the form

$$y = SY + e \qquad (5.23)$$

The matrix S is made up of samples of the signals s_K.

$$S = \begin{bmatrix} s_1(T) & s_2(T) & \cdots & s_K(T) \\ s_1(2T) & s_2(2T) & \cdots & s_K(2T) \\ \vdots & \vdots & & \vdots \\ s_1(NT) & s_2(NT) & \cdots & s_K(NT) \end{bmatrix} \qquad (5.24)$$

The presence of the error e and the fact that the number of equations is larger than the number of unknowns (N > K) makes it necessary to estimate Y.

5.6.2 Least Squares Fitting

One criterion for choosing the estimate $\hat{Y}$ is to minimize the scalar formed as the sum of the squares of the error term in Equation 5.23, viz.

$$e^T e = (y - SY)^T (y - SY) = \sum_{n=1}^{N} e_n^2 \qquad (5.25)$$

It can be shown that the minimum least squared error (the minimum value of Equation 5.25) occurs when

$$\hat{Y} = (S^T S)^{-1} S^T y = By \qquad (5.26)$$

where $B = (S^T S)^{-1} S^T$. The calculations involving the matrix S can be performed off-line to create an "algorithm," i.e., an estimate of each of the K parameters is obtained by multiplying the N samples by a set of stored numbers. The rows of Equation 5.26 can represent a number of different algorithms depending on the choice of the signals $s_K(nT)$ and the interval over which the samples are taken.

5.6.3 DFT

The simplest form of Equation 5.26 is when the matrix $S^T S$ is diagonal. Using a signal alphabet of cosines and sines of the first N harmonics of the fundamental frequency over a window of one cycle of the fundamental frequency, the familiar Discrete Fourier Transform (DFT) is produced. With

$$\begin{aligned} s_1(t) &= \cos(\omega_0 t) \\ s_2(t) &= \sin(\omega_0 t) \\ s_3(t) &= \cos(2\omega_0 t) \\ s_4(t) &= \sin(2\omega_0 t) \\ &\vdots \\ s_{N-1}(t) &= \cos(N\omega_0 t/2) \\ s_N(t) &= \sin(N\omega_0 t/2) \end{aligned} \qquad (5.27)$$

The estimates are given by:
$$\hat{Y}_{Cp} = \frac{2}{N}\sum_{n=0}^{N-1} y_n \cos(pn\theta)$$
$$\hat{Y}_{Sp} = \frac{2}{N}\sum_{n=0}^{N-1} y_n \sin(pn\theta) \qquad (5.28)$$

Note that the harmonics are also estimated by Equation 5.28. Harmonics have little role in line relaying but are important in transformer protection. It can be seen that the fundamental frequency phasor can be obtained as

$$Y = \frac{2}{N\sqrt{2}}(Y_{C1} - jY_{S1}) \qquad (5.29)$$

The normalizing factor in Equation 5.29 is omitted if the ratio of phasors for voltage and current are used to form impedance.

5.6.4 Differential Equations

Another kind of algorithm is based on estimating the values of parameters of a physical model of the system. In line protection, the physical model is a series R-L circuit that represents the faulted line. A similar approach in transformer protection uses the magnetic flux circuit with associated inductance and resistance as the model. A differential equation is written for the system in both cases.

5.6.4.1 Line Protection Algorithms

The series R-L circuit of Figure 5.6 is the model of a faulted line. The offset in the current is produced by the circuit model and hence will not be an error signal.

$$v(t) = Ri(t) + L\frac{di(t)}{dt} \qquad (5.30)$$

Looking at the samples at k, k + 1, k + 2

$$\int_{t_0}^{t_1} v(t)dt = R\int_{t_0}^{t_1} i(t)dt + L(i(t_1) - i(t_0)) \qquad (5.31)$$

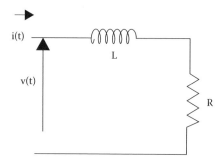

FIGURE 5.6 Model of a faulted line.

$$\int_{t_1}^{t_2} v(t)\,dt = R\int_{t_1}^{t_2} i(t)\,dt + L\bigl(i(t_2) - i(t_1)\bigr) \tag{5.32}$$

Using trapezoidal integration to evaluate the integrals (assuming t is small)

$$\int_{t_1}^{t_2} v(t)\,dt = R\int_{t_1}^{t_2} i(t)\,dt + L\bigl(i(t_2) - i(t_1)\bigr) \tag{5.33}$$

$$\int_{t_1}^{t_2} v(t)\,dt = R\int_{t_1}^{t_2} i(t)\,dt + L\bigl(i(t_2) - i(t_1)\bigr) \tag{5.34}$$

R and L are given by

$$R = \left[\frac{(v_{k+1} + v_k)(i_{k+2} - i_{k+1}) - (v_{k+2} + v_{k+1})(i_{k+1} - i_k)}{2\bigl(i_k i_{k+2} - i_{k+1}^2\bigr)}\right] \tag{5.35}$$

$$L = \frac{T}{2}\left[\frac{(v_{k+2} + v_{k+1})(i_{k+1} + i_k) - (v_{k+1} + v_k)(i_{k+2} + i_{k+1})}{2\bigl(i_k i_{k+2} - i_{k+1}^2\bigr)}\right] \tag{5.36}$$

It should be noted that the sample values occur in both numerator and denominator of Equations 5.35 and 5.36. The denominator is not constant but varies in time with local minima at points where both the current and the derivative of the current are small. For a pure sinusoidal current, the current and its derivative are never both small but when an offset is included there is a possibility of both being small once per period.

Error signals for this algorithm include terms that do not satisfy the differential equation such as the currents in the shunt elements in the line model required by long lines. In intervals where the denominator is small, errors in the numerator of Equations 5.35 and 5.36 are amplified. The resulting estimates can be quite poor. It is also difficult to make the window longer than three samples. The complexity of solving such equations for a larger number of samples suggests that the short window results be post processed. Simple averaging of the short-window estimates is inappropriate, however.

A counting scheme was used in which the counter was advanced if the estimated R and L were in the zone and the counter was decreased if the estimates lay outside the zone (Chen and Breingan, 1979). By requiring the counter to reach some threshold before tripping, secure operation can be assured with a cost of some delay. For example, if the threshold were set at 6 with a sampling rate of 16 times a cycle, the fastest trip decision would take a half cycle. Each "bad" estimate would delay the decision by two additional samples. The actual time for a relaying decision is variable and depends on the exact data.

The use of a median filter is an alternate to the counting scheme (Akke and Thorp, 1997). The median operation ranks the input values according to their amplitude and selects the middle value as the output. Median filters have an odd number of inputs. A length five median filter has an input–output relation between input x[n] and output y[n] given by

$$y[n] = \text{median}\{x[n-2], x[n-1], x[n], x[n+1], x[n+2]\} \tag{5.37}$$

Median filters of length five, seven, and nine have been applied to the output of the short window differential equation algorithm (Akke and Thorp, 1997). The median filter preserves the essential features of the input while removing isolated noise spikes. The filter length rather than the counter scheme, fixes the time required for a relaying decision.

5.6.4.2 Transformer Protection Algorithms

Virtually all algorithms for the protection of power transformers use the principle of percentage differential protection. The difference between algorithms lies in how the algorithm restrains the differential trip for conditions of overexcitation and inrush. Algorithms based on harmonic restraint, which parallel existing analog protection, compute the second and fifth harmonics using Equation 5.10 (Thorp and Phadke, 1982). These algorithms use current measurements only and cannot be faster than one cycle because of the need to compute the second harmonic. The harmonic calculation provides for secure operation since the transient event produces harmonic content which delays relay operation for about a cycle.

In an integrated substation with other microprocessor relays, it is possible to consider transformer protection algorithms that use voltage information. Shared voltage samples could be a result of multiple protection modules connected in a LAN in the substation. The magnitude of the voltage itself can be used as a restraint in a digital version of a "tripping suppressor" (Harder and Marter, 1948). A physical model similar to the differential equation model for a faulted line can be constructed using the flux in the transformer. The differential equation describing the terminal voltage, v(t), the winding current, i(t), and the flux linkage $\Lambda(t)$ is

$$v(t) - L\frac{di(t)}{dt} = \frac{d\Lambda(t)}{dt} \tag{5.38}$$

where L is the leakage inductance of the winding.

Using trapezoidal integration for the integral in Equation 5.38

$$\int_{t_1}^{t_2} v(t)dt - L\left[i(t_2) - i(t_1)\right] = \Lambda(t_2) - \Lambda(t_1) \tag{5.39}$$

gives

$$\Lambda(t_2) - \Lambda(t_1) = \frac{T}{2}\left[v(t_2) + v(t_1)\right] - L\left[i(t_2) - i(t_1)\right] \tag{5.40}$$

or

$$\Lambda_{k+1} = \Lambda_k + \frac{T}{2}\left[v_{k+1} + v_k\right] - L\left[i_{k+1} - i_k\right] \tag{5.41}$$

Since the initial flux Λ_0 in Equation 5.41 cannot be known without separate sensing, the slope of the flux current curve is used

$$\left(\frac{d\Lambda}{di}\right)_k = \frac{T}{2}\left[\frac{[v_k + v_{k-1}]}{i_k - i_{k-1}}\right] - L \tag{5.42}$$

Digital Relaying

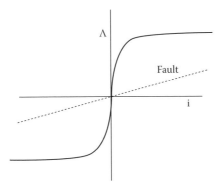

FIGURE 5.7 The flux-current characteristic compared to fault conditions.

The slope of the flux current characteristic shown in Figure 5.7 is different depending on whether there is a fault or not. The algorithm must then be able to differentiate between inrush (the slope alternates between large and small values) and a fault (the slope is always small). The counting scheme used for the differential equation algorithm for line protection can be adapted to this application. The counter increases if the slope is less than a threshold and the differential current indicates trip, and the counter decreases if the slope is greater than the threshold or the differential does not indicate trip.

5.6.5 Kalman Filters

The Kalman filter provides a solution to the estimation problem in the context of an evolution of the parameters to be estimated according to a state equation. It has been used extensively in estimation problems for dynamic systems. Its use in relaying is motivated by the filter's ability to handle measurements that change in time. To model the problem so that a Kalman filter may be used, it is necessary to write a state equation for the parameters to be estimated in the form

$$x_{k+1} = \Phi_k x_k + \Gamma_k w_k \tag{5.43}$$

$$z_k = H_k x_k + v_k \tag{5.44}$$

where Equation 5.43 (the state equation) represents the evolution of the parameters in time and Equation 5.44 represents the measurements. The terms w_k and v_k are discrete time random processes representing state noise, i.e., random inputs in the evolution of the parameters, and measurement errors, respectively. Typically w_k and v_k are assumed to be independent of each other and uncorrelated from sample to sample. If w_k and v_k have zero means, then it is common to assume that

$$\begin{aligned} E\{w_k w_j^T\} &= Q_k : k = j \\ &= 0; \quad k \neq j \end{aligned} \tag{5.45}$$

The matrices Q_k and R_k are the covariance matrices of the random processes and are allowed to change as k changes. The matrix Φ_k in Equation 5.43 is the state transition matrix. If we imagine sampling a pure sinusoid of the form

$$y(t) = Y_c \cos(\omega t) + Y_s \sin(\omega t) \tag{5.46}$$

at equal intervals corresponding to $\omega\Delta\tau = \Psi$, then the state would be

$$x_k = \begin{bmatrix} Y_C \\ Y_S \end{bmatrix} \tag{5.47}$$

and the state transition matrix

$$\Phi_k = \begin{bmatrix} 1 & 0 \\ 0 & 1 \end{bmatrix} \tag{5.48}$$

In this case, H_k, the measurement matrix, would be

$$H_k = \begin{bmatrix} \cos(k\Psi) & \sin(k\Psi) \end{bmatrix} \tag{5.49}$$

Simulations of a 345 kV line connecting a generator and a load (Girgis and Brown, 1981) led to the conclusion that the covariance of the noise in the voltage and current decayed in time. If the time constant of the decay is comparable to the decision time of the relay, then the Kalman filter formulation is

$$x = \begin{bmatrix} Y_C \\ Y_S \\ Y_0 \end{bmatrix} \quad \Phi_k = \begin{bmatrix} 1 & 0 & 0 \\ 0 & 1 & 0 \\ 0 & 0 & e^{-\beta t} \end{bmatrix}$$

appropriate for the estimation problem. The voltage was modeled as in Equations 5.48 and 5.49. The current was modeled with three states to account for the exponential offset.

and

$$H_k = \begin{bmatrix} \cos(k\Psi) & \sin(k\Psi)1 \end{bmatrix} \tag{5.50}$$

The measurement covariance matrix was

$$R_k = Ke^{-k\Delta t/T} \tag{5.51}$$

with T chosen as half the line time constant and different Ks for voltage and current. The Kalman filter estimates phasors for voltage and current as the DFT algorithms. The filter must be started and terminated using some other software. After the calculations begin, the data window continues to grow until the process is halted. This is different from fixed data windows such as a one cycle Fourier calculation. The growing data window has some advantages, but has the limitation that if started incorrectly, it has a hard time recovering if a fault occurs after the calculations have been initiated.

The Kalman filter assumes an initial statistical description of the state x, and recursively updates the estimate of state. The initial assumption about the state is that it is a random vector independent of

Digital Relaying

the processes w_k and v_k and with a known mean and covariance matrix, P_0. The recursive calculation involves computing a gain matrix K_k. The estimate is given by

$$\hat{x}_{k+1} = \Phi_k \hat{x}_k + K_{k+1}[z_{k+1} - H_{k+1} \hat{x}_k] \tag{5.52}$$

The first term in Equation 5.52 is an update of the old estimate by the state transition matrix while the second is the gain matrix K_{k+1} multiplying the observation residual. The bracketed term in Equation 5.52 is the difference between the actual measurement, z_k, and the predicted value of the measurement, i.e., the residual in predicting the measurement. The gain matrix can then be computed recursively. The amount of computation involved depends on the state vector dimension. For the linear problem described here, these calculations can be performed off-line. In the absence of the decaying measurement error, the Kalman filter offers little other than the growing data window. It has been shown that at multiples of a half cycle, the Kalman filter estimate for a constant error covariance is the same as that obtained from the DFT.

5.6.6 Wavelet Transforms

The Wavelet Transform is a signal processing tool that is replacing the Fourier Transform in many applications including data compression, sonar and radar, communications, and biomedical applications. In the signal processing community there is considerable overlap between wavelets and the area of filter banks. In applications in which it is used, the Wavelet Transform is viewed as an improvement over the Fourier Transform because it deals with time-frequency resolution in a different way. The Fourier Transform provides a decomposition of a time function into exponentials, $e^{j\omega t}$, which exist for all time. We should consider the signal that is processed with the DFT calculations in the previous sections as being extended periodically for all time. That is, the data window represents one period of a periodic signal. The sampling rate and the length of the data window determine the frequency resolution of the calculations. While these limitations are well understood and intuitive, they are serious limitations in some applications such as compression. The Wavelet Transform introduces an alternative to these limitations.

The Fourier Transform can be written

$$X(\omega) = \int_{-\infty}^{\infty} x(t) e^{-j\omega t} \, dt \tag{5.53}$$

The effect of a data window can be captured by imagining that the signal $x(t)$ is windowed before the Fourier Transform is computed. The function $h(t)$ represents the windowing function such as a one-cycle rectangle.

$$X(\omega, t) = \int_{-\infty}^{\infty} x(\tau) h(t - \tau) e^{-j\omega \tau} \, d\tau \tag{5.54}$$

The Wavelet Transform is written

$$X(s, t) = \int_{-\infty}^{\infty} x(\tau) \left[\frac{1}{\sqrt{s}} h\left(\frac{\tau - t}{s} \right) \right] d\tau \tag{5.55}$$

where
  s is a scale parameter
  t is a time shift

The scale parameter is an alternative to the frequency parameter of the Fourier Transform. If h(t) has Fourier Transform H(ω), then h(t/s) has Fourier Transform H(sω). Note that for a fixed h(t) that large, s compresses the transform while small s spreads the transform in frequency. There are a few requirements on a signal h(t) to be the "mother wavelet" (essentially that h(t) have finite energy and be a bandpass signal). For example, h(t) could be the output of a bandpass filter. It is also true that it is only necessary to know the Wavelet Transform at discrete values of s and t in order to be able to represent the signal. In particular

$$s = 2^m, \quad t = n2^m \quad m = \ldots, -2, 0, 1, 2, 3, \ldots$$
$$n = \ldots, -2, 0, 1, 2, 3, \ldots$$

where lower values of m correspond to smaller values of s or higher frequencies.

If x(t) is limited to a band B Hz, then it can be represented by samples at $T_s = 1/2B$ seconds.

$$x(n) = x(nT_s)$$

Using a mother wavelet corresponding to an ideal bandpass filter illustrates a number of ideas. Figure 5.8 shows the filters corresponding to m = 0, 1, 2, and 3 and Figure 5.9 shows the corresponding time functions. Since x(t) has no frequencies above B Hz, only positive values of m are necessary. The structure of the process can be seen in Figure 5.10. The boxes labeled LPF_R and HPF_R are low and high pass filters with cutoff frequencies of R Hz. The circle with the down arrow and a 2 represents the process of taking every other sample. For example, on the first line the output of

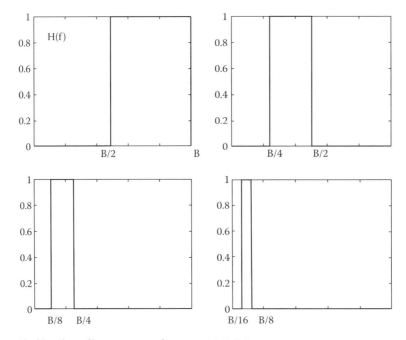

FIGURE 5.8 Ideal bandpass filters corresponding to m = 0, 1, 2, 3.

Digital Relaying

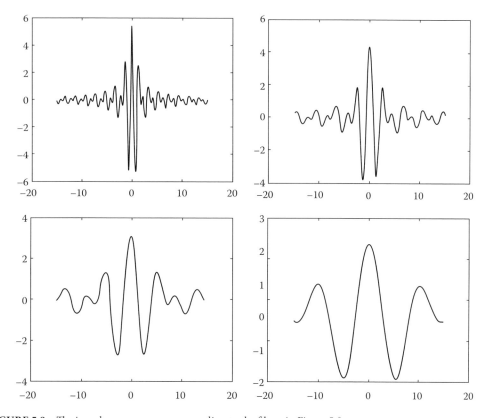

FIGURE 5.9 The impulse responses corresponding to the filters in Figure 5.8.

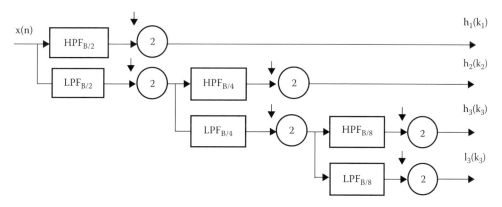

FIGURE 5.10 Cascade filter structure.

the bandpass filter only has a bandwidth of B/2 Hz and the samples at T_S seconds can be decimated to samples at $2T_S$ seconds.

Additional understanding of the compression process is possible if we take a signal made of eight numbers and let the low pass filter be the average of two consecutive samples $(x(n) + x(n+1))/2$ and the high pass filter to be the difference $(x(n) - x(n+1))/2$ (Gail and Nielsen, 1999). For example, with

$$x(n) = [-2 -28 -46 -44 -20\ 12\ 32\ 30]$$

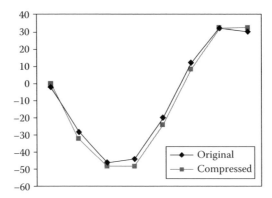

FIGURE 5.11 Original and compressed signals.

we get

$$h_1(k_1) = [13 -1 -161]$$
$$h_2(k_2) = [7 - 8.5]$$
$$h_3(k_3) = [7.75]$$
$$l_3(k_3) = [-0.75]$$

If we truncate to form

$$h_1(k_1) = [16\ 0 -160]$$
$$h_2(k_2) = [8 -8]$$
$$h_3(k_3) = [8]$$
$$l_3(k_3) = [0]$$

and reconstruct the original sequence

$$\tilde{x}(n) = [0 -32 -48 -48 -24\ 8\ 32\ 32]$$

The original and reconstructed compressed waveform is shown in Figure 5.11. Wavelets have been applied to relaying for systems grounded through a Peterson coil where the form of the wavelet was chosen to fit unusual waveforms the Peterson coil produces (Chaari et al., 1996).

5.6.7 Neural Networks

Artificial Neural Networks (ANNs) had their beginning in the "perceptron," which was designed to recognize patterns. The number of papers suggesting relay application have soared. The attraction is the use of ANNs as pattern recognition devices that can be trained with data to recognize faults, inrush, or other protection effects. The basic feed forward neural net is composed of layers of neurons as shown in Figure 5.12.

The function Φ is either a threshold function or a saturating function such as a symmetric sigmoid function. The weights w_i are determined by training the network. The training process is the most

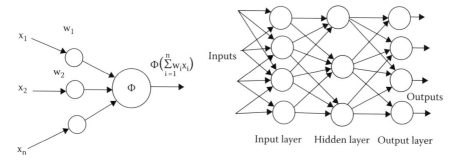

FIGURE 5.12 One neuron and a neural network.

difficult part of the ANN process. Typically, simulation data such as that obtained from EMTP is used to train the ANN. A set of cases to be executed must be identified along with a proposed structure for the net. The structure is described in terms of the number of inputs, neuron in layers, various layers, and outputs. An example might be a net with 12 inputs, and a 4, 3, 1 layer structure. There would be 4×12 plus 4×3 plus 3×1 or 63 weights to be determined. Clearly, a lot more than 60 training cases are needed to learn 63 weights. In addition, some cases not used for training are needed for testing. Software exists for the training process but judgment in determining the training sequences is vital. Once the weights are learned, the designer is frequently asked how the ANN will perform when some combination of inputs are presented to it. The ability to answer such questions is very much a function of the breadth of the training sequence.

The protective relaying application of ANNs include high-impedance fault detection (Eborn et al., 1990), transformer protection (Perez et al., 1994), fault classification (Dalstein and Kulicke, 1995), fault direction determination, adaptive reclosing (Aggarwal et al., 1994), and rotating machinery protection (Chow and Yee, 1991).

References

Aggarwal, R.K., Johns, A.T., Song, Y.H., Dunn, R.W., and Fitton, D.S., Neural-network based adaptive single-pole autoreclosure technique for EHV transmission systems, *IEEE Proceedings—C*, 141, 155, 1994.

Akke, M. and Thorp, J.S., Improved estimates from the differential equation algorithm by median post-filtering, *IEEE Sixth International Conference on Development in Power System Protection*, Nottingham, U.K., March 1997.

Chaari, O., Neunier, M., and Brouaye, F., Wavelets: A new tool for the resonant grounded power distribution system relaying, *IEEE Transactions on Power Delivery*, 11, 1301, July 1996.

Chen, M.M. and Breingan, W.D., Field experience with a digital system with transmission line protection, *IEEE Transactions on Power Apparatus and Systems*, 98, 1796, September/October 1979.

Chow, M. and Yee, S.O., Methodology for on-line incipient fault detection in single-phase squirrel-cage induction motors using artificial neural networks, *IEEE Transactions on Energy Conversion*, 6, 536, September 1991.

Dalstein, T. and Kulicke, B., Neural network approach to fault classification for high speed protective relaying, *IEEE Transactions on Power Delivery*, 10, 1002, April 1995.

Eborn, S., Lubkeman, D.L., and White, M., A neural network approach to the detection of incipient faults on power distribution feeders, *IEEE Transactions on Power Delivery*, 5, 905, April 1990.

Gail, A.W. and Nielsen, O.M., Wavelet analysis for power system transients, *IEEE Computer Applications in Power*, 12, 16, January 1999.

Gilcrest, G.B., Rockefeller, G.D., and Udren, E.A., High-speed distance relaying using a digital computer, Part I: System description, *IEEE Transactions on Power Apparatus and Systems*, 91, 1235, May/June 1972.

Girgis, A.A. and Brown, R.G., Application of Kalman filtering in computer relaying, *IEEE Transactions on Power Apparatus and Systems*, 100, 3387, July 1981.

Harder, E.L. and Marter, W.E., Principles and practices of relaying in the United States, *AIEE Transactions*, 67, Part II, 1005, 1948.

Mann, B.J. and Morrison, I.F., Relaying a three-phase transmission line with a digital computer, *IEEE Transactions on Power Apparatus and Systems*, 90, 742, March/April 1971.

Perez, L.G., Flechsiz, A.J., Meador, J.L., and Obradovic, A., Training an artificial neural network to discriminate between magnetizing inrush and internal faults, *IEEE Transactions on Power Delivery*, 9, 434, January 1994.

Poncelet, R., *The Use of Digital Computers for Network Protection*, International Council on Large Electric Systems (CIGRE), Paris, France, pp. 32–98, August 1972.

Rockefeller, G.D., Fault protection with a digital computer, *IEEE Transactions on Power Apparatus and Systems*, 88, 438, April 1969.

Rockefeller, G.D. and Udren, E.A., High-speed distance relaying using a digital computer, Part II. Test results, *IEEE Transactions*, 91, 1244, May/June 1972.

Sachdev, M.S. (Coordinator), Advancements in microprocessor based protection and communication, IEEE Tutorial Course Text Publication, Publication No. 97TP120-0, 1997.

Tamronglak, S., Horowitz, S.H., Phadke, A.G., and Thorp, J.S., Anatomy of power system blackouts: Preventive relaying strategies, *IEEE Transactions on Power Delivery*, 11, 708, April 1996.

Thorp, J.S. and Phadke, A.G., A microprocessor based voltage-restraint three-phase transformer differential relay, *Proceedings of the South Eastern Symposium on Systems Theory*, 312, April 1982.

6
Use of Oscillograph Records to Analyze System Performance

John R. Boyle
Power System Analysis

Protection of present-day power systems is accomplished by a complex system of extremely sensitive relays that function only during a fault in the power system. Because relays are extremely fast, automatic oscillographs installed at appropriate locations can be used to determine the performance of protective relays during abnormal system conditions. Information from oscillographs can be used to detect the:

1. Presence of a fault
2. Severity and duration of a fault
3. Nature of a fault (A phase to ground, A–B phases to ground, etc.)
4. Location of line faults
5. Adequacy of relay performance
6. Effective performance of circuit breakers in circuit interruption
7. Occurrence of repetitive faults
8. Persistency of faults
9. Dead time required to dissipate ionized gases
10. Malfunctioning of equipment
11. Cause and possible resolution of a problem

Another important aspect of analyzing oscillograms is that of collecting data for statistical analysis. This would require a review of all oscillograms for every fault. The benefits would be to detect incipient problems and correct them before they become serious problems causing multiple interruptions or equipment damage.

An analysis of an oscillograph record shown in Figure 6.1 should consider the nature of the fault. Substation **Y** is comprised of two lines and a transformer. The high side winding is connected to ground. Oscillographic information is available from the bus potential transformers, the line currents from breaker **A** on line 1, and the transformer neutral current. An "**A**" phase-to-ground fault is depicted on line 1. The oscillograph reveals a significant drop in "**A**" phase voltage accompanied with a rise in "**A**" phase line 1 current and a similar rise in the transformer neutral current. The "**A**" phase breaker cleared the fault in three cycles (good). The received carrier on line 1 was "off" during the fault (good) permitting high-speed tripping at both terminals (breakers A and B). There is no evidence of AC or DC current transformer (CT) saturation of either the phase CTs or the transformer neutral CT. The received carrier signal on line 2 was "on" all during the fault to block breaker "D" from tripping at terminal "**X**."

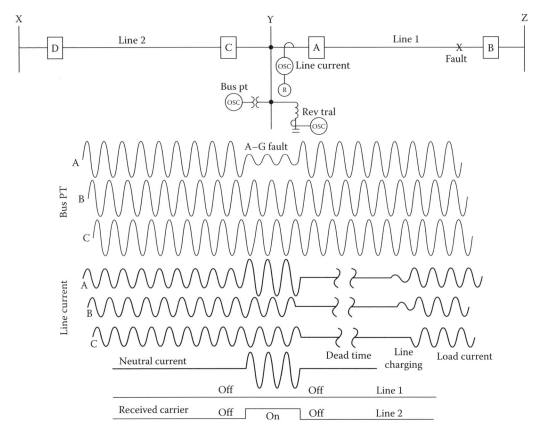

FIGURE 6.1 Analysis of an oscillograph record.

This would indicate that the carrier ground relays on the number 2 line performed properly. This type of analysis may not be made because of budget and personnel constraints. Oscillographs are still used extensively to analyze known cases of trouble (breaker failure, transformer damage, etc.), but oscillograph analysis can also be used as a maintenance tool to prevent equipment failure.

The use of oscillograms as a maintenance tool can be visualized by classifying operations as good (A) or questionable (B) as shown in Figure 6.2. The first fault current waveform (upper left) is classified as A because it is sinusoidal in nature and cleared in three cycles. This could be a four or five cycle fault clearing time and still be classified as A depending upon the breaker characteristics (four or five cycle breaker, etc.) The DC offset wave form can also be classified as A because it indicates a four cycle fault clearing time and a sinusoidal waveform with no saturation.

An example of a questionable waveform (B) is shown on the right side of Figure 6.2. The upper right is one of current magnitude which would have to be determined by use of fault studies. Some breakers have marginal interrupting capabilities and should be inspected whenever close-in faults occur that generate currents that approach or exceed their interrupting capabilities. The waveform in the lower right is an example of a breaker restrike that requires a breaker inspection to prevent a possible breaker failure of subsequent operations.

Carrier performance on critical transmission lines is important because it impacts fast fault clearing, successful high-speed reclosing, high-speed tripping upon reclosure, and delayed breaker failure response for permanent faults upon reclosure, and a "stuck" breaker. In Figure 6.3 two waveforms are shown that depict adequate carrier response for internal and external faults. The first waveform shows a three cycle fault and its corresponding carrier response. A momentary burst of carrier is cut off quickly

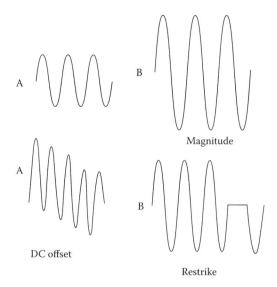

FIGURE 6.2 Use of oscillograms as a maintenance tool.

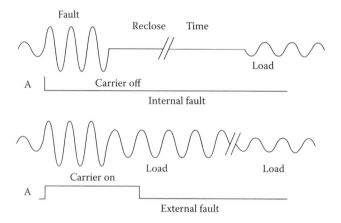

FIGURE 6.3 Two waveforms that depict adequate carrier response for internal and external faults.

allowing the breaker to trip in three cycles. Upon reclosing, load current is restored. The bottom waveform depicts the response of carrier on an adjacent line for the same fault. Note that carrier was "off" initially and cut "on" shortly after fault initiation. It stayed "on" for a few cycles after the fault cleared and stayed "off" all during the reclose "dead" time and after restoration of load current. Both of these waveforms would be classified as "good" and would not need further analysis.

An example of a questionable carrier response for an internal fault is shown in Figure 6.4. Note that the carrier response was good for the initial three cycle fault, but during the reclose dead time, carrier came back "on" and was "on" upon reclosing. This delayed tripping an additional two cycles. Of even greater concern is a delay in the response of breaker-failure clearing time for a stuck breaker. Breaker failure initiation is predicated upon relay initiation which, in the case shown, is delayed two cycles. This type of "bad" carrier response may go undetected if oscillograms are not reviewed. In a similar manner, a delayed carrier response for an internal fault can result in delayed tripping for the initial fault as shown in Figure 6.5. However, a delayed carrier response on an adjacent line can be more serious because it will result in two or more line interruptions. This is shown in Figure 6.6. A fault on line 1 in Figure 6.1 should be accompanied by acceptable carrier blocking signals on all external lines that receive a strong enough

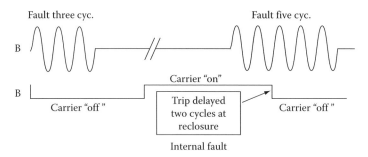

FIGURE 6.4 A questionable carrier response for an internal fault.

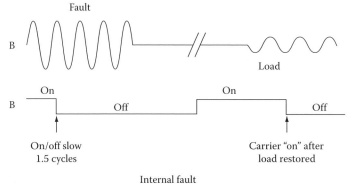

FIGURE 6.5 A delayed carrier response for an internal fault that resulted in delayed tripping for the initial fault.

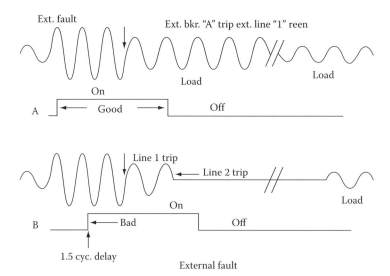

FIGURE 6.6 A delayed carrier response on an adjacent line can be more serious because it will result in two or more line interruptions.

signal to trip if not accompanied by an appropriate carrier blocking signal. Two conditions are shown. A good ("A") block signal and questionable ("B") block signal. The good block signal is shown as one that blocks (comes "on") within a fraction of a cycle after the fault is detected and unblocks (goes "off") a few cycles after the fault is cleared. The questionable block signal shown at the bottom of the waveform in Figure 6.6 is late in going from "off" to "on" (1.5 cycles). The race between the trip element and the block

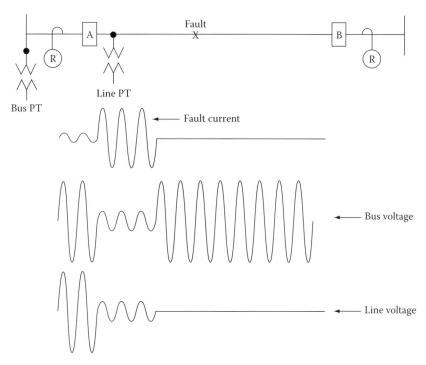

FIGURE 6.7 Bus or line potentials behave differently after a fault has been cleared.

element is such that a trip signal was initiated first and breaker "D" tripped 1.5 cycles after the fault was cleared by breaker **A** in three cycles. This would result in a complete station interruption at station "**Y**."

Impedance relays receive restraint from either bus or line potentials. These two potentials behave differently after a fault has been cleared. This is shown in Figure 6.7. After breakers "A" and "B" open and the line is de-energized, the bus potential restores to its full value thereby applying full restraint to all impedance relays connected to the bus. The line voltage goes to zero after the line is de-energized. Normally this is not a problem because relays are designed to accommodate this condition. However, there are occasions when the line potential restraint voltage can cause a relay to trip when a breaker recloses. This condition usually manifests itself when shunt reactors are connected on the line. Under these conditions an oscillatory voltage will exist on the terminals of the line side potential devices after both breakers "A" and "B" have opened. A waveform example is shown in Figure 6.8. Note that the voltage is not a 60 Hz wave shape. Normally it is less than 60 Hz depending on the degree of compensation. This oscillatory voltage is more pronounced at high voltages because of the higher capacitance charge on the line. On lines that have flat spacing, the two outside voltages transfer energy between each other that results in oscillations that are mirror images of each other. The voltage on the center phase is usually a constant decaying decrement. These oscillations can last up to 400 cycles or more. This abnormal voltage is applied to the relays at the instant of reclosure and has been known to cause a breaker (for example, "**A**") to trip because of the lack of coordination between the voltage restraint circuit and the overcurrent monitoring element. Another more prevalent problem is multiple restrikes across an insulator during the oscillatory voltage on the line. These restrikes prevent the ionized gases from dissipating sufficiently at the time of reclosure. Thus a fault is reestablished when breaker "**A**" and/or "**B**" recloses. This phenomena can readily be seen on oscillograms. Action taken might be to look for defective insulators or lengthen the reclose cycle.

The amount of "dead time" is critical to successful reclosures. For example, at 161 kV a study was made to determine the amount of dead time required to dissipate ionized gases to achieve a 90% reclose success rate. In general, on a good line (clean insulators), at least 13 cycles of dead time are required.

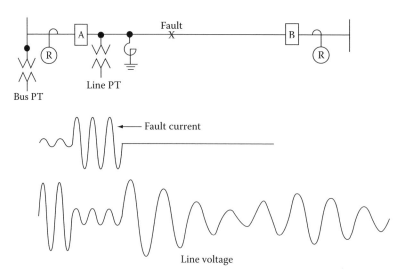

FIGURE 6.8 A waveform example after both "A" and "B" breakers have been opened.

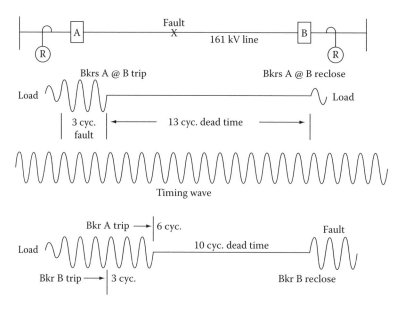

FIGURE 6.9 Depicts the performance of good breaker operations (top waveform).

Contrast this to 10 cycles dead time where the reclose success rate went down to approximately 50%. Oscillograms can help determine the dead time and the cause of unsuccessful reclosures. Note the dead time is a function of the performance of the breakers at both ends of the line. Figure 6.9 depicts the performance of good breaker operations (top waveform). Here, both breakers trip in 3 cycles and reclose successfully in 13 cycles. The top waveform depicts a slow breaker "A" tripping in 6 cycles. This results in an unsuccessful reclosure because the overall dead time is reduced to 10 cycles. Note, the oscillogram readily displays the problem. The analysis would point to possible relay or breaker trouble associated with breaker "A."

Figure 6.10 depicts CT saturation. This phenomenon is prevalent in current circuits and can cause problems in differential and polarizing circuits. The top waveform is an example of a direct current (DC) offset waveform with no evidence of saturation. That is to say that the secondary waveform replicates the primary waveform. Contrast this with a DC offset waveform (lower) that clearly indicates saturation.

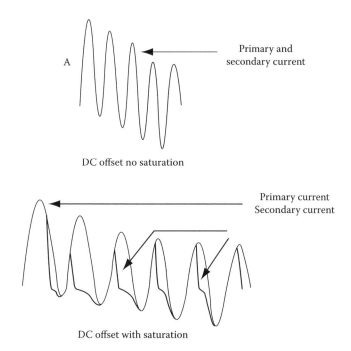

FIGURE 6.10 Depicts current transformer (CT) saturation.

If two sets of CTs are connected differentially around a transformer and the high side CTs do not saturate (upper waveform) and the low side CTs do saturate (lower waveform), the difference current will flow through the operate coil of the relay which may result in de-energizing the transformer when no trouble exists in the transformer. The solution may be the replacement of the offending low side CT with one that has a higher "C" classification, desensitizing the relay or reducing the magnitude of the fault current. Polarizing circuits are also adversely affected by CTs that saturate. This occurs where a residual circuit is compared with a neutral polarizing circuit to obtain directional characteristics and the apparent shift in the polarizing current results in an unwanted trip.

Current reversals can result in an unwanted two-line trip if carrier transmission from one terminal to another does not respond quickly to provide the desired block function of a trip element. This is shown in a step-by-step sequence in Figures 6.11 through 6.14. Consider a line 1 fault at the terminals of breaker "B" (Figure 6.11). For this condition, 2000 A of ground fault current is shown to flow on each line from terminal "**X**" to terminal "**Y**." Since fault current flow is towards the fault at breakers "**A**" and "**B**," *neither* will receive a signal (carrier "off") to initiate tripping. However, it is assumed that both breakers do not open at the same time (breaker "B" opens in three cycles and breaker "A" opens in four cycles). The response of the relays on line 2 is of prime concern. During the initial fault when breakers "**A**" and "**B**" are both closed, a block carrier signal must be sent from breaker "D" to breaker "C" to prevent the tripping of breaker "C." This is shown as a correct "on" carrier signal for three cycles in the bottom oscillogram trace in Figure 6.14. However, when breaker "B" trips in three cycles, the fault current in line 2 increases to 4000 A and, more importantly, it reverses direction to flow from terminal "**Y**" to terminal "**X**." This instantaneous current reversal requires that the directional relays on breaker "C" pickup to initiate a carrier block signal to breaker "D." Failure to accomplish this may result in a trip of breaker "C" if its own carrier signal does not rise rapidly to prevent tripping through its previously made up trip directional elements. This is shown in Figure 6.13 and oscillogram record Figure 6.14. An alternate undesirable operation would be the tripping of breaker "D" if its trip directional elements make up before the carrier block signal from breaker "C" is received at breaker "D." The end result is the same (tripping line 2 for a fault on line 1).

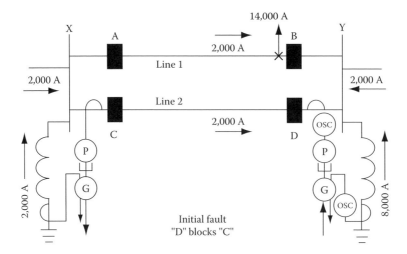

FIGURE 6.11 A line 1 fault at the terminals of breaker "B." Figures 6.11 through 6.14 demonstrate step-by-step sequence.

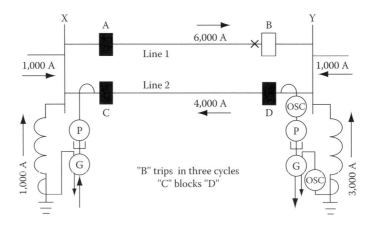

FIGURE 6.12 Second step in sequence.

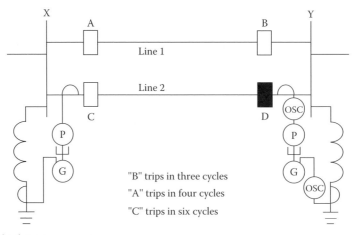

FIGURE 6.13 Third step in sequence.

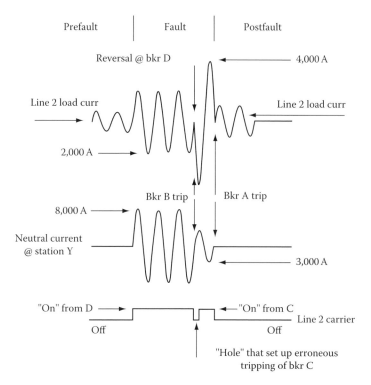

FIGURE 6.14 Final step in sequence.

Restrikes in breakers can result in an explosive failure of the breaker. Oscillograms can be used to prevent breaker failures if the first restrike within the interrupter can be detected before a subsequent restrike around the interrupter results in the destruction of the breaker. This is shown diagrammatically in Figure 6.15. The upper waveform restrike sequence depicts a 1/2 cycle restrike that is successfully extinguished within the interrupter. The lower waveform depicts a restrike that goes around the interrupter.

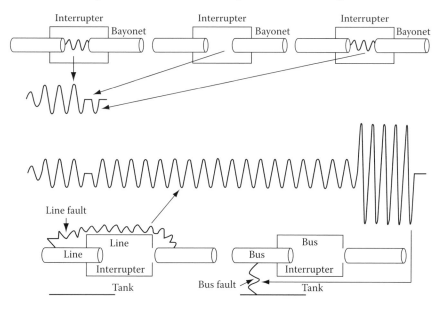

FIGURE 6.15 Diagrams the first restrike within the interrupter.

This restrike cannot be extinguished and will last until the oil becomes badly carbonized and a subsequent fault occurs between the bus breaker terminal and the breaker tank (ground). In Figure 6.15 the interrupter bypass fault lasted 18 cycles. Depending upon the rate of carbonization, the arc time could last longer or less before the flashover to the tank. The result would be the same. A bus fault that could have devastating affects. One example resulted in the loss of eight generators, thirteen 161 kV lines, and three 500 kV lines. The reason for the extensive loss was the result of burning oil that drifted up into adjacent busses steel causing multiple bus and line faults that de-energized all connected equipment in the station. The restrike phenomena is a result of a subsequent lightning strikes across the initial fault (insulator). In the example given above, lightning arresters were installed on the line side of each breaker and no additional restrikes or breaker failures occurred after the initial destructive failures.

Oscillography in microprocessor relays can also be used to analyze system problems. The problem in Figure 6.16 involves a microprocessor differential relay installation that depicts the failure to energize a large motor. The CTs on both sides of the transformer were connected wye–wye but the low

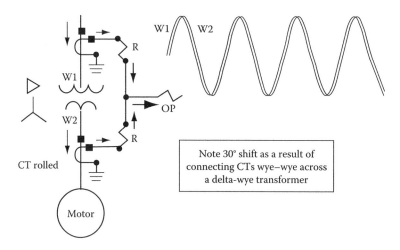

FIGURE 6.16 A microprocessor differential relay installation that depicts the failure to energize a large motor.

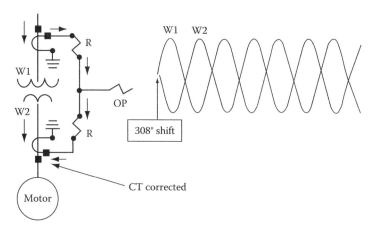

FIGURE 6.17 Corrected connection.

side CTs were rolled. The 30° shift was corrected in the relay and was accurately portrayed by oscillography in the microprocessor relay but the rolled CTs produced current in the operate circuit that resulted in an erroneous trip. Note that with the low side CTs rolled, the high and low side currents W1 and W2 are in phase (incorrect). The oscillography output clearly pin-pointed the problem. The corrected connection is shown in Figure 6.17 together with the correct oscillography (W1 and W2 180° out of phase).

7
Systems Aspects of Large Blackouts

Ian Dobson
Iowa State University

References 7-3

Large blackouts are infrequent but very costly to our society. The power system is sometimes subject to large initial disturbances from extreme weather, such as high winds or ice storms. Moreover, even a small initial power system disturbance can cascade into a complicated chain of dependent failures leading to a widespread blackout. On August 10, 1996, a blackout started in the Northwest United States and spread by cascading to disconnect power to about 7,500,000 customers (Kosterev et al., 1999). On August 14, 2003, a blackout started in Ohio and spread to disconnect power to about 50 million customers in Northeastern United States and Canada (U.S.-Canada Power System Outage Task Force, 2004). Although such extreme events are infrequent, the direct costs are estimated to be in the billions of dollars, disrupting commerce and vital infrastructure. Large blackouts also have a strong effect on shaping the way power systems are regulated and the reputation of the power industry.

While the direct costs of large blackouts to society are substantial, the indirect costs to society and industry are also important. The restoration of power after a large blackout can take a long time, and further interactions complicating and extending the impact can occur as the power remains off. Other infrastructures, such as communications, transport, and water supplies may be affected. For example, cell phone tower backup batteries may only last for several hours, and gas station fuel pumps may be inoperable over a wide area. In general, extended blackouts reveal adverse interactions, including impacts on other essential infrastructures, which do not occur or can be easily mitigated for small blackouts. Also, some blackouts involve social disruptions, such as looting, and these social effects, when they occur, substantially multiply the economic damage. In addition to economic effects, the hardship to people and possible deaths underline the engineer's responsibility to work to avoid blackouts.

Cascading failure is a sequence of dependent events that successively weaken the power system. The successive weakening implies that further events are more likely to occur because of the previous events. Thus, if during cascading events, load has been disconnected to cause a medium-size blackout, then there is an increased chance that the cascade will proceed further and cause a large blackout. This property of cascading causes large blackouts, although still infrequent, to have a significant probability of occurrence. The statistics for large blackouts have correspondingly "heavy tails," indicating that blackouts of all sizes, including extreme blackouts, can occur. One can contrast this heavy tail situation with the statistics for large events that are caused by large numbers of independent small events. These large events have such vanishingly small probabilities (the probability of the large event is the multiplication of the many small probabilities of the independent small events) that the possibility of their occurrence is negligible.

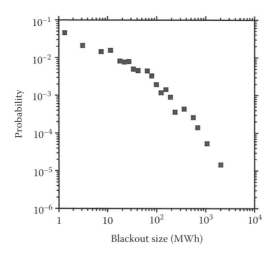

FIGURE 7.1 North American blackout size probability distribution.

For example, Figure 7.1 shows statistics of major North American power transmission blackouts from 1984 to 1998 obtained from the North American Electrical Reliability Council (Carreras et al., 2004; Hines et al., 2009). Figure 7.1 plots on a log–log scale the empirical probability distribution of energy unserved in the blackouts. As the energy unserved increases, blackouts of that size becomes less likely. The decrease of probability as blackout size increases is approximately a power law dependence with an exponent between −1 and −2. (A power law dependence with exponent −1 appears as a straight line of slope −1 on a log–log plot and implies that doubling the blackout size only halves the probability.) The power law region is of course limited in extent in a practical power system because there is a largest possible blackout in which the entire interconnection blacks out. Similar forms of blackout size statistics have been observed in several countries (Dobson et al., 2007).

Blackout risk can be defined as the product of blackout probability and blackout cost. One simple assumption is that blackout cost is roughly proportional to blackout size, although larger blackouts may well have costs (especially indirect costs) that increase faster than linearly (Newman et al., 2011). In the case of a power law dependence, the larger blackouts can become rarer at a similar rate as costs increase, and then the risk of large blackouts is comparable to or even exceeding the risk of small blackouts (Newman et al., 2011). Thus, the power law region in the distribution of blackout size significantly affects the risk of large blackouts. The standard probabilistic techniques that assume independence between events imply exponential tails in blackout size distributions and are not applicable to the risk analysis of systems that exhibit power laws.

Power transmission grids are not static; they are slowly and continually being modified and upgraded in hardware, software, and operating procedures to respond to the opposing requirements to transmit power reliably and economically. In particular, the power system reliability slowly evolves, and it has been suggested that it evolves as a complex system that adjusts itself in response to any changes made: If the power system margins of operation are large, blackouts tend to be more infrequent, the focus is on extracting economic value from the investment in the transmission system, and the power system margins and the reliability will gradually erode. On the other hand, if the power system margins are small, blackouts tend to be more likely, and when the blackouts do happen, substantial upgrades will be made and margins and reliability will be improved. That is, in addition to lack of reliability causing blackouts, blackouts cause reliability. Note that engineers and their responses to blackouts are a vital part of these socioeconomic processes! It has been suggested that these processes tend to shape the power system reliability toward the observed heavy tail blackout statistics (Dobson et al., 2007; Ren et al., 2008).

We now consider the implications for framing the problem of managing blackout risk. The heavy tail in blackout statistics implies that large blackouts will occasionally happen, and, while blackout risk should and must be mitigated, it is not feasible (economically or otherwise) to eliminate large blackouts. In particular, large blackouts may not be dismissed as "perfect storms." Despite the substantial challenges of assessing the risk of large blackouts, the problem of avoiding blackouts should be framed as the problem of jointly manipulating the frequency of blackouts of all sizes rather than simply avoiding blackouts in general. The reason is that well-intentioned efforts to reduce the frequency of small blackouts could, at least in the long run, lead to an increased frequency of large blackouts, and care should be taken that the combined risk from blackouts of all sizes is reduced (Newman et al., 2011). One possible scenario is that a reduction in small blackouts eventually allows the power system to be operated closer to the edge for genuine economic advantage, but at increased risk of large blackouts (Kirschen and Strbac, 2004; Newman et al., 2011).

The implication is that, especially when better methods of assessing cascading failure risk are developed, upgrades to the power grid hardware, software, or procedures should be assessed for their broad impact on reliability, including both small and large blackouts. For example, protection systems play a significant role in most blackouts. (Note that this is not the same as asserting that protection systems are a primary cause of blackouts, since blackouts typically involve multiple systems; it is simply that the list of systems involved in a blackout often includes some form of protection misoperation or unanticipated operation.) The current engineering challenge is to design protection systems toward the (sometimes conflicting) objectives of protecting equipment and avoiding the simpler, small blackouts of the system. In the future, we hope that methods can be developed to provide designs that are robust and resilient to all sizes of blackout.

References

Carreras, B.A., D.E. Newman, I. Dobson, and A.B. Poole. 2004. Evidence for self-organized criticality in a time series of electric power system blackouts. *IEEE Transactions Circuits and Systems*, Part I. 51(9): 1733–1740.

Dobson, I., B.A. Carreras, V.E. Lynch, and D.E. Newman. 2007. Complex systems analysis of series of blackouts: Cascading failure, critical points, and self-organization. *Chaos*, 17: 026103.

Hines, P., J. Apt, and S. Talukdar. 2009. Large blackouts in North America: Historical trends and policy implications. *Energy Policy*, 37(12): 5249–5259.

Kirschen, D.S. and G. Strbac. 2004. Why investments do not prevent blackouts. *The Electricity Journal*, 17(2): 29–36.

Kosterev, D., C. Taylor, and W. Mittelstadt. 1999. Model validation for the August 10, 1996 WSCC system outage. *IEEE Transactions on Power Systems*, 14: 967–979.

Newman, D.E., B.A. Carreras, V.E. Lynch, and I. Dobson. 2011. Exploring complex systems aspects of blackout risk and mitigation. *IEEE Transactions on Reliability*, 60(1): 134–143.

Ren, H., I. Dobson, and B.A. Carreras. 2008. Long-term effect of the n-1 criterion on cascading line outages in an evolving power transmission grid. *IEEE Transactions Power Systems*, 23(3): 1217–1225.

U.S.-Canada Power System Outage Task Force. 2004. Final Report on the August 14, 2003 Blackout in the United States and Canada. US-Canada Power System Outage Task Force, Toronto, Ontario, Canada.

II

Power System Dynamics and Stability

Prabha S. Kundur

8 Power System Stability *Prabha S. Kundur* 8-1
Basic Concepts • Classification of Power System Stability • Historical Review of Stability Problems • Consideration of Stability in Power System Design and Operation • Acknowledgments • References

9 Transient Stability *Kip Morison* 9-1
Introduction • Basic Theory of Transient Stability • Methods of Analysis of Transient Stability • Factors Influencing Transient Stability • Transient Stability Considerations in System Design • Transient Stability Considerations in System Operation • References

10 Small-Signal Stability and Power System Oscillations *John Paserba, Juan Sanchez-Gasca, Lei Wang, Prabha S. Kundur, Einar Larsen, and Charles Concordia* 10-1
Nature of Power System Oscillations • Criteria for Damping • Study Procedure • Mitigation of Power System Oscillations • Higher-Order Terms for Small-Signal Analysis • Modal Identification • Summary • References

11 Voltage Stability *Yakout Mansour and Claudio Cañizares* 11-1
Basic Concepts • Analytical Framework • Mitigation of Voltage Stability Problems • References

12 Direct Stability Methods *Vijay Vittal* 12-1
Review of Literature on Direct Methods • The Power System Model • The Transient Energy Function • Transient Stability Assessment • Determination of the Controlling UEP • The Boundary Controlling UEP Method • Applications of the TEF Method and Modeling Enhancements • References

13 Power System Stability Controls *Carson W. Taylor* 13-1
Review of Power System Synchronous Stability Basics • Concepts of Power System Stability Controls • Types of Power System Stability Controls and Possibilities for Advanced Control • Dynamic Security Assessment • "Intelligent" Controls • Wide-Area Stability Controls • Effect of Industry Restructuring on Stability Controls • Experience from Recent Power Failures • Summary • References

14 **Power System Dynamic Modeling** *William W. Price and Juan Sanchez-Gasca* **14**-1
Modeling Requirements • Generator Modeling • Excitation System Modeling • Prime Mover Modeling • Load Modeling • Transmission Device Models • Dynamic Equivalents • References

15 **Wide-Area Monitoring and Situational Awareness** *Manu Parashar, Jay C. Giri, Reynaldo Nuqui, Dmitry Kosterev, R. Matthew Gardner, Mark Adamiak, Dan Trudnowski, Aranya Chakrabortty, Rui Menezes de Moraes, Vahid Madani, Jeff Dagle, Walter Sattinger, Damir Novosel, Mevludin Glavic, Yi Hu, Ian Dobson, Arun Phadke, and James S. Thorp* **15**-1
Introduction • WAMS Infrastructure • WAMS Monitoring Applications • WAMS in North America • WAMS Worldwide • WAMS Deployment Roadmap • References

16 **Assessment of Power System Stability and Dynamic Security Performance** *Lei Wang and Pouyan Pourbeik* **16**-1
Definitions and Historical Perspective • Phenomena of Interest • Security Criteria • Modeling • Analysis Methods • Control and Enhancements • Off-Line DSA • Online DSA • Status and Summary • References

17 **Power System Dynamic Interaction with Turbine Generators** *Bajarang L. Agrawal, Donald G. Ramey, and Richard G. Farmer* **17**-1
Introduction • Subsynchronous Resonance • Device-Dependent Subsynchronous Oscillations • Supersynchronous Resonance • Device-Dependent Supersynchronous Oscillations • Transient Shaft Torque Oscillations • References

18 **Wind Power Integration in Power Systems** *Reza Iravani* **18**-1
Introduction • Background • Structure of Wind Turbine Generator Units • Wind Power Plant Systems • Models and Control for WPPs • References

19 **Flexible AC Transmission Systems (FACTS)** *Rajiv K. Varma and John Paserba* **19**-1
Introduction • Concepts of FACTS • Reactive Power Compensation in Transmission Lines • Static var Compensator • Thyristor-Controlled Series Compensation • Static Synchronous Compensator • Static Series Synchronous Compensator • Unified Power Flow Controller • FACTS Controllers with Energy Storage • Coordinated Control of FACTS Controllers • FACTS Installations to Improve Power System Dynamic Performance • Conclusions • References

Dr. Prabha S. Kundur received his PhD in electrical engineering from the University of Toronto, Toronto, Ontario, Canada and has over 40 years of experience in the electric power industry. He is currently the president of Kundur Power System Solutions Inc., Toronto, Ontario, Canada. He served as the president and CEO of Powertech Labs Inc., the research and technology subsidiary of BC Hydro, from 1994 to 2006. Prior to joining Powertech, he worked at Ontario Hydro for nearly 25 years and held senior positions involving power system planning and design.

Dr. Kundur has also served as an adjunct professor at the University of Toronto since 1979 and at the University of British Columbia from 1994 to 2006. He is the author of the book *Power System Stability and Control* (McGraw-Hill, New York, 1994), which is a standard modern reference for the subject. He has performed extensive international consulting related to power system planning and design and has delivered technical courses for utilities, manufacturers, and universities around the world.

Dr. Kundur has had a long record of service and leadership in the IEEE. He has chaired numerous committees and working groups of the IEEE Power & Energy Society (PES) and was elected a fellow of the IEEE in 1985. He is the past chairman of the IEEE Power System Dynamic Performance Committee. From 2005 to 2010, he served as a member of the IEEE PES Executive Committee and as the PES vice president for education.

Dr. Kundur is the recipient of several IEEE awards, including the 1997 IEEE Nikola Tesla Award, the 2005 IEEE PES Charles Concordia Power System Engineering Award, and the 2010 IEEE Medal in Power Engineering. He has also been active in CIGRE for many years. He served as the chairman of the CIGRE Study Committee C4 on "System Technical Performance" from 2002 to 2006 and as a member of the CIGRE Administrative Council from 2006 to 2010. He received the CIGRE Technical Committee Award in 1999. He was also awarded the title of Honorary Member by CIGRE in 2006.

Dr. Kundur was elected as a fellow of the Canadian Academy of Engineering in 2003 and as a foreign associate of the U.S. National Academy of Engineering in 2011. He has been awarded the honorary degrees of Doctor Honoris Causa by the University Politechnica of Bucharest, Romania, in 2003, and Doctor of Engineering, Honoris Causa by the University of Waterloo, Ontario, Canada, in 2004.

8
Power System Stability

8.1	Basic Concepts	8-1
8.2	Classification of Power System Stability	8-2
	Need for Classification • Rotor Angle Stability • Voltage	
	Stability • Frequency Stability • Comments on Classification	
8.3	Historical Review of Stability Problems	8-7
8.4	Consideration of Stability in Power System Design and Operation	8-9
	Acknowledgments	8-10
	References	8-10

Prabha S. Kundur
*Kundur Power Systems
Solutions, Inc.*

This introductory chapter provides a general description of the power system stability phenomena including fundamental concepts, classification, and definition of associated terms. A historical review of the emergence of different forms of stability problems as power systems evolved and of the developments of methods for their analysis and mitigation is presented. Requirements for consideration of stability in system design and operation are discussed.

8.1 Basic Concepts

Power system stability denotes the ability of an electric power system, for a given initial operating condition, to regain a state of operating equilibrium after being subjected to a physical disturbance, with most system variables bounded so that system integrity is preserved. Integrity of the system is preserved when practically the entire power system remains intact with no tripping of generators or loads, except for those disconnected by isolation of the faulted elements or intentionally tripped to preserve the continuity of operation of the rest of the system. Stability is a condition of equilibrium between opposing forces; instability results when a disturbance leads to a sustained imbalance between a set of opposing forces.

The power system is a highly nonlinear system that operates in a constantly changing environment; loads, generator outputs, topology, and key operating parameters change continually. When subjected to a transient disturbance, the stability of the system depends on the nature of the disturbance as well as the initial operating condition. The disturbance may be small or large. Small disturbances in the form of load changes occur continually, and the system adjusts to the changing conditions. The system must be able to operate satisfactorily under these conditions and successfully meet the load demand. It must also be able to survive numerous disturbances of a severe nature, such as a short circuit on a transmission line or loss of a large generator.

Following a transient disturbance, if the power system is stable, it will reach a new equilibrium state with practically the entire system intact; the actions of automatic controls and possibly human operators will eventually restore the system to normal state. On the other hand, if the system is unstable, it will result in a runaway or rundown situation; for example, a progressive increase in angular separation of generator rotors, or a progressive decrease in bus voltages. Depending on the network conditions, instability in one part of the power system could lead to cascading outages and a shutdown of a major portion of the power system.

The response of the power system to a disturbance may involve much of the equipment. For instance, a fault on a critical element followed by its isolation by protective relays will cause variations in power flows, network bus voltages, and machine rotor speeds; the voltage variations will actuate both generator and transmission network voltage regulators; the generator speed variations will actuate prime mover governors; and the voltage and frequency variations will affect the system loads to varying degrees depending on their individual characteristics. Further, devices used to protect individual equipment may respond to variations in system variables and thereby affect the power system performance. A typical modern power system is thus a very high-order multivariable process whose dynamic performance is influenced by a wide array of devices with different response rates and characteristics. Hence, instability in a power system may occur in many different ways depending on the system topology, operating mode, and the form of the disturbance.

Traditionally, the stability problem has been one of maintaining synchronous operation. Since power systems rely largely on synchronous machines for generation of electrical power, a necessary condition for satisfactory system operation is that all synchronous machines remain in synchronism or, colloquially, "in step." This aspect of stability is influenced by the dynamics of generator rotor angles and power–angle relationships.

Instability may also be encountered without the loss of synchronism. For example, a system consisting of a generator feeding an induction motor can become unstable due to collapse of load voltage. In this instance, it is the stability and control of voltage that is the issue, rather than the maintenance of synchronism. This type of instability can also occur in the case of loads covering an extensive area in a large system.

In the event of a significant load/generation mismatch within an area, generator and prime mover controls become important, as well as system controls and special protections. If not properly coordinated, it is possible for the system frequency to become unstable, and generating units and/or loads may ultimately be tripped possibly leading to a system blackout. This is another case where units may remain in synchronism (until tripped by such protections as underfrequency), but the system becomes unstable.

Because of the high dimensionality and complexity of stability problems, it is essential to make simplifying assumptions and to analyze specific types of problems using the right degree of detail of system representation.

8.2 Classification of Power System Stability

8.2.1 Need for Classification

Power system stability is a single problem; however, it is impractical to deal with it as such. Instability of the power system can take different forms and is influenced by a wide range of factors. Analysis of stability problems, including identifying essential factors that contribute to instability and devising methods of improving stable operation, is greatly facilitated by classification of stability into appropriate categories. These are based on the following considerations (Kundur, 1994; Kundur and Morison, 1997):

- The physical nature of the resulting instability related to the main system parameter in which instability can be observed.
- The size of the disturbance considered indicates the most appropriate method of calculation and prediction of stability.
- The devices, processes, and the time span that must be taken into consideration in order to determine stability.

Figure 8.1 shows the classification of power system stability into various categories and subcategories. The following are descriptions of the corresponding forms of stability phenomena.

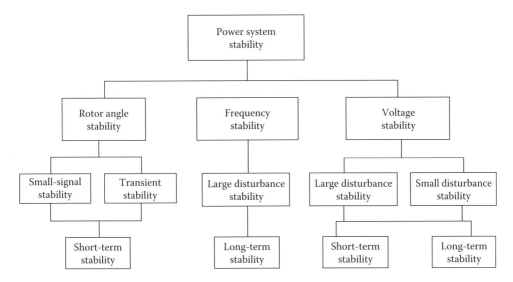

FIGURE 8.1 Classification of power system stability.

8.2.2 Rotor Angle Stability

Rotor angle stability is concerned with the ability of interconnected synchronous machines of a power system to remain in synchronism under normal operating conditions and after being subjected to a disturbance. It depends on the ability to maintain/restore equilibrium between electromagnetic torque and mechanical torque of each synchronous machine in the system. Instability that may result occurs in the form of increasing angular swings of some generators leading to their loss of synchronism with other generators.

The rotor angle stability problem involves the study of the electromechanical oscillations inherent in power systems. A fundamental factor influencing this category of system stability is the manner in which the power outputs of synchronous machines vary as their rotor angles change. The mechanism by which interconnected synchronous machines maintain synchronism with one another is through restoring forces, which act whenever there are forces tending to accelerate or decelerate one or more machines with respect to other machines. Under steady-state conditions, there is equilibrium between the input mechanical torque and the output electrical torque of each machine, and the speed remains constant. If the system is perturbed, this equilibrium is upset, resulting in acceleration or deceleration of the rotors of the machines according to the laws of motion of a rotating body. If one generator temporarily runs faster than another, the angular position of its rotor relative to that of the slower machine will advance. The resulting angular difference transfers part of the load from the slow machine to the fast machine, depending on the power–angle relationship. This tends to reduce the speed difference and hence the angular separation. The power–angle relationship, as discussed earlier, is highly nonlinear. Beyond a certain limit, an increase in angular separation is accompanied by a decrease in power transfer; this increases the angular separation further and leads to instability. For any given situation, the stability of the system depends on whether or not the deviations in angular positions of the rotors result in sufficient restoring torques.

It should be noted that loss of synchronism can occur between one machine and the rest of the system, or between groups of machines, possibly with synchronism maintained within each group after separating from each other.

The change in electrical torque of a synchronous machine following a perturbation can be resolved into two components:

- *Synchronizing torque* component, in phase with a rotor angle perturbation
- *Damping torque* component, in phase with the speed deviation

System stability depends on the existence of both components of torque for each of the synchronous machines. Lack of sufficient synchronizing torque results in *aperiodic* or *non-oscillatory instability*, whereas lack of damping torque results in *oscillatory instability*.

For convenience in analysis and for gaining useful insight into the nature of stability problems, it is useful to characterize rotor angle stability in terms of the following two subcategories:

1. *Small disturbance (or small-signal) rotor angle stability* is concerned with the ability of the power system to maintain synchronism under small disturbances. The disturbances are considered to be sufficiently small that linearization of system equations is permissible for purposes of analysis. Such disturbances are continually encountered in normal system operation, such as small changes in load.
    a. Small-signal stability depends on the initial operating state of the system. Instability that may result can be of two forms: (i) increase in rotor angle through a non-oscillatory or aperiodic mode due to lack of synchronizing torque or (ii) rotor oscillations of increasing amplitude due to lack of sufficient damping torque.
    b. In today's practical power systems, small-signal stability is largely a problem of insufficient damping of oscillations. The time frame of interest in small-signal stability studies is on the order of 10–20 s following a disturbance. The stability of the following types of oscillations is of concern:
        i. *Local modes or machine-system modes*, associated with the swinging of units at a generating station with respect to the rest of the power system. The term "local" is used because the oscillations are localized at one station or a small part of the power system.
        ii. *Interarea modes*, associated with the swinging of many machines in one part of the system against machines in other parts. They are caused by two or more groups of closely coupled machines that are interconnected by weak ties.
        iii. *Control modes*, associated with generating units and other controls. Poorly tuned exciters, speed governors, HVDC (high voltage direct current) converters, and static var compensators are the usual causes of instability of these modes.
        iv. *Torsional modes*, associated with the turbine-generator shaft system rotational components. Instability of torsional modes may be caused by interaction with excitation controls, speed governors, HVDC controls, and series-capacitor-compensated lines.
2. *Large* disturbance *rotor angle stability or transient stability*, as it is commonly referred to, is concerned with the ability of the power system to maintain synchronism when subjected to a severe transient disturbance. The resulting system response involves large excursions of generator rotor angles and is influenced by the nonlinear power–angle relationship.

Transient stability depends on both the initial operating state of the system and the severity of the disturbance. Usually, the disturbance alters the system such that the post-disturbance steady-state operation will be different from that prior to the disturbance. Instability is in the form of aperiodic drift due to insufficient synchronizing torque, and is referred to as *first swing stability*. In large power systems, transient instability may not always occur as first swing instability associated with a single mode; it could be as a result of increased peak deviation caused by superposition of several modes of oscillation causing large excursions of rotor angle beyond the first swing.

The time frame of interest in transient stability studies is usually limited to 3–5 s following the disturbance. It may extend to 10 s for very large systems with dominant inter-area swings.

As identified in Figure 8.1, small-signal stability, as well as transient stability, is categorized as short-term phenomena.

8.2.3 Voltage Stability

Voltage stability is concerned with the ability of a power system to maintain steady voltages at all buses in the system under normal operating conditions, and after being subjected to a disturbance. Instability that may result occurs in the form of a progressive fall or rise of voltage of some buses. The possible outcome of voltage instability is loss of load in the area where voltages reach unacceptably low values, or a loss of integrity of the power system.

Progressive drop in bus voltages can also be associated with rotor angles going out of step. For example, the gradual loss of synchronism of machines as rotor angles between two groups of machines approach or exceed 180° would result in very low voltages at intermediate points in the network close to the electrical center (Kundur, 1994). In contrast, the type of sustained fall of voltage that is related to voltage instability occurs where rotor angle stability is not an issue.

The main factor contributing to voltage instability is usually the higher reactive power losses and the resulting voltage drops that occur when high levels of active power and reactive power flow through inductive reactances associated with the transmission network; this limits the capability of transmission network for power transfer. The power transfer and voltage support are further limited when some of the generators hit their field current limits. The driving force for voltage instability is usually the loads; in response to a disturbance resulting in network-wide voltage reduction, power consumed by the loads tends to be restored by the actions of distribution voltage regulators, tap-changing transformers, and thermostats. Restored loads increase the stress on the high voltage network causing more voltage reduction. A rundown situation causing voltage instability occurs when load dynamics attempts to restore power consumption beyond the capability of the transmission system and the connected generation (Kundur, 1994; Taylor, 1994; Van Cutsem and Vournas, 1998).

While the most common form of voltage instability is the progressive drop in bus voltages, the possibility of overvoltage instability also exists and has been experienced at least on one system (Van Cutsem and Mailhot, 1997). It can occur when EHV transmission lines are loaded significantly below surge impedance loading and underexcitation limiters prevent generators and/or synchronous condensers from absorbing the excess reactive power. Under such conditions, transformer tap changers, in their attempt to control load voltage, may cause voltage instability.

Voltage stability problems may also be experienced at the terminals of HVDC links. They are usually associated with HVDC links connected to weak AC systems (CIGRE Working Group 14.05, 1992). The HVDC link control strategies have a very significant influence on such problems.

As in the case of rotor angle stability, it is useful to classify voltage stability into the following subcategories:

1. *Large disturbance voltage stability* is concerned with a system's ability to control voltages following large disturbances such as system faults, loss of generation, or circuit contingencies. This ability is determined by the system-load characteristics and the interactions of both continuous and discrete controls and protections. Determination of large disturbance stability requires the examination of the nonlinear dynamic performance of a system over a period of time sufficient to capture the interactions of such devices as under-load transformer tap changers and generator field-current limiters. The study period of interest may extend from a few seconds to tens of minutes. Therefore, long-term dynamic simulations are required for analysis (Van Cutsem et al., 1995).
2. *Small disturbance voltage stability* is concerned with a system's ability to control voltages following small perturbations such as incremental changes in system load. This form of stability is determined by the characteristics of loads, continuous controls, and discrete controls at a given instant of time. This concept is useful in determining, at any instant, how the system voltage will respond to small system changes. The basic processes contributing to small disturbance voltage instability are essentially of a steady-state nature. Therefore, static analysis can be effectively used

to determine stability margins, identify factors influencing stability, and examine a wide range of system conditions and a large number of post-contingency scenarios (Gao et al., 1992). A criterion for small disturbance voltage stability is that, at a given operating condition for every bus in the system, the bus voltage magnitude increases as the reactive power injection at the same bus is increased. A system is voltage unstable if, for at least one bus in the system, the bus voltage magnitude (V) decreases as the reactive power injection (Q) at the same bus is increased. In other words, a system is voltage stable if V–Q sensitivity is positive for every bus and unstable if V–Q sensitivity is negative for at least one bus.

The time frame of interest for voltage stability problems may vary from a few seconds to tens of minutes. Therefore, voltage stability may be either a short-term or a long-term phenomenon. It is often more effective to classify and analyze voltage stability based on the duration of the resulting phenomenon as follows:

1. *Short-term voltage stability* involves dynamics of fast-acting devices such as induction motors, electronically controlled loads, and HVDC converters. A typical scenario is a large disturbance such as a fault near a load center, when the power system is operating in a stressed condition during hot weather with a high level of air-conditioning loads. When the fault occurs, air-conditioning compressor motors decelerate drawing high current. Following fault clearing with transmission/distribution line tripping, motors draw high current while attempting to reaccelerate. Motors may stall if power system is weak. Massive loss of load and possibly voltage collapse across the load center may follow. This is a short-term phenomenon involving a study period of a few seconds. Time-domain simulation involving dynamic modeling of loads and voltage control devices is essential for analyzing such problems (Diaz de Leon and Taylor, 2000).
2. *Long-term voltage stability* involves slower acting devices such as tap-changing transformers, thermostatically controlled loads, and generator field current limiters. The study period may extend to several minutes. A typical scenario is loss of a heavily loaded transmission line, when the EHV network is operating in a stressed condition with many of the lines heavily loaded and reactive power reserves at minimum. Immediately following the disturbance there would be considerable reduction of voltages at adjacent EHV buses, which in turn would be reflected into the distribution system. With the restoration of distribution system voltages by substation tap-changing transformers and feeder voltage regulators, power consumed by the loads would be restored, thereby causing further increase in loading of EHV lines and overall reactive power demand in the power system. If the reactive power reserves are at a minimum, nearby generators hit their field or armature current time-overload limits. With fewer generators on voltage control, the power system is much more prone to voltage instability.

For the analysis of long-term voltage stability, the study period may extend to several minutes and long-term simulations are required for the analysis of system dynamic performance. However, stability is usually determined by the resulting outage of equipment rather than the severity of the initial disturbance. In many cases, static analysis techniques can be used to estimate stability margins, identify factors influencing stability, and investigate a large number of scenarios (Morison et al., 1993; Kundur, 1994; Gao et al., 1996).

8.2.4 Frequency Stability

Frequency stability is concerned with the ability of a power system to maintain steady frequency within a nominal range following a severe system upset resulting in a significant imbalance between the overall system generation and load. It depends on the ability to restore balance between system generation and load, with minimum loss of load.

Severe system upsets generally result in large excursions of frequency, power flows, voltage, and other system variables, thereby invoking the actions of processes, controls, and protections that are not modeled in conventional transient stability or voltage stability studies. These processes may be very slow, such as boiler dynamics, or only triggered for extreme system conditions, such as volts/hertz protection tripping generators. In large interconnected power systems, this type of situation is most commonly associated with islanding. Stability in this case is a question of whether or not each island will reach an acceptable state of operating equilibrium with minimal loss of load. It is determined by the overall response of the island as evidenced by its mean frequency, rather than relative motion of machines. Generally, frequency stability problems are associated with inadequacies in equipment responses, poor coordination of control and protection equipment, or insufficient generation reserve. Examples of such problems are reported by Kundur et al. (1985), Chow et al. (1989), and Kundur (1981).

Over the course of a frequency instability, the characteristic times of the processes and devices that are activated by the large shifts in frequency and other system variables will range from a matter of seconds, corresponding to the responses of devices such as generator controls and protections, to several minutes, corresponding to the responses of devices such as prime mover energy supply systems and load voltage regulators.

Although frequency stability is impacted by fast as well as slow dynamics, the overall time frame of interest extends to several minutes. Therefore, it is categorized as a long-term phenomenon in Figure 8.1.

8.2.5 Comments on Classification

The classification of stability has been based on several considerations so as to make it convenient for identification of the causes of instability, the application of suitable analysis tools, and the development of corrective measures appropriate for a specific stability problem. There clearly is some overlap between the various forms of instability, since as systems fail, more than one form of instability may ultimately emerge. However, a system event should be classified based primarily on the dominant initiating phenomenon, separated into those related primarily with voltage, rotor angle, or frequency.

While classification of power system stability is an effective and convenient means to deal with the complexities of the problem, the overall stability of the system should always be kept in mind. Solutions to stability problems of one category should not be at the expense of another. It is essential to look at all aspects of the stability phenomena, and at each aspect from more than one viewpoint.

8.3 Historical Review of Stability Problems

As electric power systems have evolved over the last century, different forms of instability have emerged as being important during different periods. The methods of analysis and resolution of stability problems were influenced by the prevailing developments in computational tools, stability theory, and power system control technology. A review of the history of the subject is useful for a better understanding of the electric power industry's practices with regard to system stability.

Power system stability was first recognized as an important problem in the 1920s (Steinmetz, 1920; Evans and Bergvall, 1924; Wilkins, 1926). The early stability problems were associated with remote power plants feeding load centers over long transmission lines. With slow exciters and noncontinuously acting voltage regulators, power transfer capability was often limited by steady-state as well as transient rotor angle instability due to insufficient synchronizing torque. To analyze system stability, graphical techniques such as the equal area criterion and power circle diagrams were developed. These methods were successfully applied to early systems that could be effectively represented as two-machine systems.

As the complexity of power systems increased, and interconnections were found to be economically attractive, the complexity of the stability problems also increased and systems could no longer be treated as two-machine systems. This led to the development in the 1930s of the network analyzer, which was capable of power flow analysis of multimachine systems. System dynamics, however, still had to be analyzed by solving the swing equations by hand using step-by-step numerical integration. Generators were represented by the classical "fixed voltage behind transient reactance" model. Loads were represented as constant impedances.

Improvements in system stability came about by way of faster fault clearing and fast-acting excitation systems. Small-signal aperiodic instability was virtually eliminated by the implementation of continuously acting voltage regulators. With increased dependence on controls, the emphasis of stability studies moved from transmission network problems to generator problems, and simulations with more detailed representations of synchronous machines and excitation systems were required.

The 1950s saw the development of the analog computer, with which simulations could be carried out to study in detail the dynamic characteristics of a generator and its controls rather than the overall behavior of multimachine systems. Later in the 1950s, the digital computer emerged as the ideal means to study the stability problems associated with large interconnected systems.

In the 1960s, most of the power systems in the United States and Canada were part of one of two large interconnected systems, one in the east and the other in the west. In 1967, low capacity HVDC ties were also established between the east and west systems. At present, the power systems in North America form virtually one large system. There were similar trends in the growth of interconnections in other countries. While interconnections result in operating economy and increased reliability through mutual assistance, they contribute to increased complexity of stability problems and increased consequences of instability. The Northeast Blackout of November 9, 1965, made this abundantly clear; it focused the attention of the public and of regulatory agencies, as well as of engineers, on the problem of stability and importance of power system reliability.

Until recently, most industry effort and interest has been concentrated on *transient* (*rotor angle*) *stability*. Powerful transient stability simulation programs have been developed that are capable of modeling large complex systems using detailed device models. Significant improvements in transient stability performance of power systems have been achieved through the use of high-speed fault clearing, high-response exciters, series capacitors, and special stability controls and protection schemes.

The increased use of high response exciters, coupled with decreasing strengths of transmission systems, has led to an increased focus on *small-signal* (*rotor angle*) *stability*. This type of angle instability is often seen as local plant modes of oscillation, or in the case of groups of machines interconnected by weak links, as interarea modes of oscillation. Small-signal stability problems have led to the development of special study techniques, such as modal analysis using eigenvalue techniques (Martins, 1986; Kundur et al., 1990). In addition, supplementary control of generator excitation systems, static var compensators, and HVDC converters is increasingly being used to solve system oscillation problems. There has also been a general interest in the application of power electronic based controllers referred to as FACTS (flexible AC transmission systems) controllers for damping of power system oscillations (IEEE PES Special Publication, 1996).

In the 1970s and 1980s, frequency stability problems experienced following major system upsets led to an investigation of the underlying causes of such problems and to the development of long-term dynamic simulation programs to assist in their analysis (Davidson et al., 1975; Converti et al., 1976; Ontario Hydro, 1989; Stubbe et al., 1989; Inoue et al., 1995). The focus of many of these investigations was on the performance of thermal power plants during system upsets (Kundur, 1981; Younkins and Johnson, 1981; Kundur et al., 1985; Chow et al., 1989). Guidelines were developed by an IEEE working group for enhancing power plant response during major frequency disturbances (IEEE Working Group, 1983). Analysis and modeling needs of power systems during major frequency disturbances were also addressed in a recent CIGRE Task Force report (1999).

Since the late 1970s, voltage instability has been the cause of several power system collapses worldwide (IEEE, 1990; Kundur, 1994; Taylor, 1994). Once associated primarily with weak radial distribution systems, voltage stability problems are now a source of significant concern in highly developed and mature networks as a result of heavier loadings and power transfers over long distances. Consequently, voltage stability is increasingly being addressed in system planning and operating studies. Powerful analytical tools are available for its analysis (Gao et al., 1992; Morison et al., 1993; Van Cutsem et al., 1995), and well-established criteria and study procedures have evolved.

Another trend that is having a significant impact on the dynamic characteristics of modern power systems is the increasing reliance on renewable energy resources, particularly wind energy. Wind generation technology has matured over the past several decades into an economically viable and environmentally favorable source of energy. Today wind generation has become a significant source of generation mix in many power systems around the world. This requires development and implementation of protection and control systems for wind farms so as to be able to satisfactorily ride through network disturbances and thereby contribute to satisfactory performance of the overall power system. Detailed studies with appropriate models should be carried out to ensure satisfactory integration of wind farms into power systems.

Present-day power systems, in addition to their changing dynamic characteristics, are being operated under increasingly stressed conditions due to the prevailing trend to make the most of existing facilities. Increased competition, open transmission access, and construction and environmental constraints are shaping the operation of electric power systems in new ways that present greater challenges for secure system operation. This is abundantly clear from the increasing number of major power-grid blackouts that have been experienced in recent years; for example, Brazil blackout of March 11, 1999; Northeast U.S.–Canada blackout of August 14, 2003; Southern Sweden and Eastern Denmark blackout of September 23, 2003; and Italian blackout of September 28, 2003. Planning and operation of today's power systems require a careful consideration of all forms of system instability. Significant advances have been made in recent years in providing the study engineers with a number of powerful tools and techniques. A coordinated set of complementary programs, such as the one described by Kundur et al. (1994) makes it convenient to carry out a comprehensive analysis of power system stability.

8.4 Consideration of Stability in Power System Design and Operation

For reliable service, a power system must remain intact and be capable of withstanding a wide variety of disturbances. Owing to economic and technical limitations, no power system can be stable for all possible disturbances or contingencies. In practice, power systems are designed and operated so as to be stable for a selected list of contingencies, normally referred to as "design contingencies" (Kundur, 1994). Experience dictates their selection. The contingencies are selected on the basis that they have a significant probability of occurrence and a sufficiently high degree of severity, given the large number of elements comprising the power system. The overall goal is to strike a balance between costs and benefits of achieving a selected level of system security.

While security is primarily a function of the physical system and its current attributes, secure operation is facilitated by

- Proper selection and deployment of preventive and emergency controls
- Assessing stability limits and operating the power system within these limits

Security assessment for establishing system operating limits has been historically conducted in an off-line operation planning environment in which stability for the near-term forecasted system conditions is exhaustively determined. The results of stability limits are loaded into lookup tables that are accessed by the operator to assess the security of a prevailing system operating condition.

In the new competitive utility environment, power systems can no longer be operated in a very structured and conservative manner; the possible types and combinations of power transfer transactions may grow enormously. The present trend is, therefore, to use online dynamic security assessment. This is feasible with today's computer hardware and stability analysis software (Morison et al., 2004).

In addition to online dynamic security assessment, a wide range of other new and emerging technologies could assist in significantly minimizing the occurrence and impact of widespread blackouts. These include

- Adaptive relaying
- Wide-area monitoring and control
- Flexible AC transmission (FACTS) devices
- Distributed generation technologies

Acknowledgments

The definition and classification of power system stability presented in this section is based on the report prepared by a joint IEEE/CIGRE Task Force on Power System Stability Terms, Classification, and Definitions. The membership of the Task Force comprised of Prabha Kundur (Convener), John Paserba (Secretary), Venkat Ajjarapu, Goran Andersson, Anjan Bose, Claudio Canizares, Nikos Hatziargyriou, David Hill, Alex Stankovic, Carson Taylor, Thierry Van Cutsem, and Vijay Vittal. The report has been published in the IEEE Transactions on Power Systems, August 2004 and as CIGRE Technical Brochure 231, June 2003.

References

Chow, Q.B., Kundur, P., Acchione, P.N., and Lautsch, B., Improving nuclear generating station response for electrical grid islanding, *IEEE Trans. Energy Convers.*, EC-4, 3, 406, 1989.

CIGRÉ Task Force 38.02.14 Report, Analysis and modelling needs of power systems under major frequency disturbances, 1999.

CIGRÉ Working Group 14.05 Report, Guide for planning DC links terminating at AC systems locations having short-circuit capacities, Part I: AC/DC Interaction Phenomena, CIGRÉ Guide No. 95, 1992.

Converti, V., Gelopulos, D.P., Housely, M., and Steinbrenner, G., Long-term stability solution of interconnected power systems, *IEEE Trans. Power App. Syst.*, PAS-95, 1, 96, 1976.

Davidson, D.R., Ewart, D.N., and Kirchmayer, L.K., Long term dynamic response of power systems—An analysis of major disturbances, *IEEE Trans. Power App. Syst.*, PAS-94, 819, 1975.

Diaz de Leon II, J.A., and Taylor, C.W., Understanding and solving short-term voltage stability problems, in *Proceedings of the IEEE/PES 2000 Summer Meeting*, Seattle, WA, 2000.

Evans, R.D. and Bergvall, R.C., Experimental analysis of stability and power limitations, *AIEE Trans.*, 43, 39–58, 1924.

Gao, B., Morison, G.K., and Kundur, P., Voltage stability evaluation using modal analysis, *IEEE Trans. Power Syst.*, PWRS-7, 4, 1529, 1992.

Gao, B., Morison, G.K., and Kundur, P., Towards the development of a systematic approach for voltage stability assessment of large scale power systems, *IEEE Trans. Power Syst.*, 11, 3, 1314, 1996.

IEEE PES Special Publication, FACTS Applications, Catalogue No. 96TP116-0, 1996.

IEEE Special Publication 90TH0358-2-PWR, Voltage stability of power systems: Concepts, analytical tools and industry experience, 1990.

IEEE Working Group, Guidelines for enhancing power plant response to partial load rejections, *IEEE Trans. Power App. Syst.*, PAS-102, 6, 1501, 1983.

Inoue, T., Ichikawa, T., Kundur, P., and Hirsch, P., Nuclear plant models for medium- to long-term power system stability studies, *IEEE Trans. Power Syst.*, 10, 141, 1995.

Kundur, P., A survey of utility experiences with power plant response during partial load rejections and system disturbances, *IEEE Trans. Power App. Syst.*, PAS-100, 5, 2471, 1981.

Kundur, P., *Power System Stability and Control*, McGraw-Hill, New York, 1994.

Kundur, P., Lee, D.C., Bayne, J.P., and Dandeno, P.L., Impact of turbine generator controls on unit performance under system disturbance conditions, *IEEE Trans. Power App. Syst.*, PAS-104, 1262, 1985.

Kundur, P. and Morison, G.K., A review of definitions and classification of stability problems in today's power systems, Paper presented at the *Panel Session on Stability Terms and Definitions, IEEE PES Winter Meeting*, New York, 1997.

Kundur, P., Morison, G.K., and Balu, N.J., A comprehensive approach to power system analysis, CIGRE Paper 38–106, presented at the 1994 Session, Paris, France.

Kundur, P., Rogers, G.J., Wong, D.Y., Wang, L., and Lauby, M.G., A comprehensive computer program package for small signal stability analysis of power systems, *IEEE Trans. Power Syst.*, 5, 1076, 1990.

Martins, N., Efficient eigenvalue and frequency response methods applied to power system small-signal stability studies, *IEEE Trans. Power Syst.*, PWRS-1, 217, 1986.

Morison, G.K., Gao, B., and Kundur, P., Voltage stability analysis using static and dynamic approaches, *IEEE Trans. Power Syst.*, 8, 3, 1159, 1993.

Morison, G.K., Wang, L., and Kundur, P., Power System Security Assessment, *IEEE Power & Energy Mag.*, 2(5), 30–39, September/October 2004.

Ontario Hydro, Long-term dynamics simulation: Modeling requirements, Final Report of Project 2473-22, EPRI Report EL-6627, 1989.

Steinmetz, C.P., Power control and stability of electric generating stations, *AIEE Trans.*, XXXIX, 1215, 1920.

Stubbe, M., Bihain, A., Deuse, J., and Baader, J.C., STAG a new unified software program for the study of dynamic behavior of electrical power systems, *IEEE Trans. Power Syst.*, 4, 1, 1989.

Taylor, C.W., *Power System Voltage Stability*, McGraw-Hill, New York, 1994.

Van Cutsem, T., Jacquemart, Y., Marquet, J.N., and Pruvot, P., A comprehensive analysis of mid-term, voltage stability, *IEEE Trans. Power Syst.*, 10, 1173, 1995.

Van Cutsem, T. and Mailhot, R., Validation of a fast voltage stability analysis method on the Hydro-Quebec system, *IEEE Trans. Power Syst.*, 12, 282, 1997.

Van Cutsem, T. and Vournas, C., *Voltage Stability of Electric Power Systems*, Kluwer Academic Publishers, Dordrecht, the Netherlands, 1998.

Wilkins, R., Practical aspects of system stability, *AIEE Trans.*, 41–50, 1926.

Younkins, T.D. and Johnson, L.H., Steam turbine overspeed control and behavior during system disturbances, *IEEE Trans. Power App. Syst.*, PAS-100, 5, 2504, 1981.

9
Transient Stability

Kip Morison
British Columbia Hydro and Power Authority

9.1 Introduction 9-1
9.2 Basic Theory of Transient Stability 9-1
   Swing Equation • Power–Angle Relationship •
   Equal Area Criterion
9.3 Methods of Analysis of Transient Stability 9-7
   Modeling • Analytical Methods • Simulation Studies
9.4 Factors Influencing Transient Stability 9-9
9.5 Transient Stability Considerations in System Design 9-10
9.6 Transient Stability Considerations in System Operation 9-11
References 9-12

9.1 Introduction

As discussed in Chapter 7, power system stability was recognized as a problem as far back as the 1920s, at which time the characteristic structure of systems consisted of remote power plants feeding load centers over long distances. These early stability problems, often a result of insufficient synchronizing torque, were the first emergence of transient instability. As defined in Chapter 8, *transient stability* is the ability of a power system to remain in synchronism when subjected to large transient disturbances. These disturbances may include faults on transmission elements, loss of load, loss of generation, or loss of system components such as transformers or transmission lines.

Although many different forms of power system stability have emerged and become problematic in recent years, transient stability still remains a basic and important consideration in power system design and operation. While it is true that the operation of many power systems is limited by phenomena such as voltage stability or small-signal stability, most systems are prone to transient instability under certain conditions or contingencies and hence the understanding and analysis of transient stability remain fundamental issues. Also, we shall see later in this chapter that transient instability can occur in a very short time frame (a few seconds) leaving no time for operator intervention to mitigate problems; it is therefore essential to deal with the problem in the design stage or severe operating restrictions may result.

In this chapter, we discuss the basic principles of transient stability, methods of analysis, control and enhancement, and practical aspects of its influence on power system design and operation.

9.2 Basic Theory of Transient Stability

Most power system engineers are familiar with plots of generator rotor angle (δ) versus time as shown in Figure 9.1. These "swing curves" plotted for a generator subjected to a particular system disturbance show whether a generator rotor angle recovers and oscillates around a new equilibrium point as in trace "a" or whether it increases aperiodically such as in trace "b." The former case is deemed to be transiently stable, and the latter case transiently unstable. What factors determine whether a machine will be stable

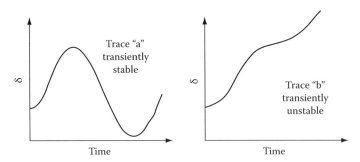

FIGURE 9.1 Plots showing the trajectory of generator rotor angle through time for transient stable and transiently unstable cases.

or unstable? How can the stability of large power systems be analyzed? If a case is unstable, what can be done to enhance its stability? These are some of the questions we seek to answer in this section.

Two concepts are essential in understanding transient stability: (1) the swing equation and (2) the power–angle relationship. These can be used together to describe the equal area criterion, a simple graphical approach to assessing transient stability [1–3].

9.2.1 Swing Equation

In a synchronous machine, the prime mover exerts a mechanical torque T_m on the shaft of the machine and the machine produces an electromagnetic torque T_e. If, as a result of a disturbance, the mechanical torque is greater than the electromagnetic torque, an accelerating torque T_a exists and is given by

$$T_a = T_m - T_e \tag{9.1}$$

This ignores the other torques caused by friction, core loss, and windage in the machine. T_a has the effect of accelerating the machine, which has an inertia J (kg · m²) made up of the inertia of the generator and the prime mover, and, therefore,

$$J \frac{d\omega_m}{dt} = T_a = T_m - T_e \tag{9.2}$$

where
 t is time in seconds
 ω_m is the angular velocity of the machine rotor in mechanical rad/s

It is common practice to express this equation in terms of the inertia constant H of the machine. If ω_{0m} is the rated angular velocity in mechanical rad/s, J can be written as

$$J = \frac{2H}{\omega_{0m}^2} VA_{base} \tag{9.3}$$

Therefore

$$\frac{2H}{\omega_{0m}^2} VA_{base} \frac{d\omega_m}{dt} = T_m - T_e \tag{9.4}$$

Transient Stability

And now, if ω_r denotes the angular velocity of the rotor (rad/s) and ω_0 its rated value, the equation can be written as

$$2H \frac{d\bar{\omega}_r}{dt} = \bar{T}_m - \bar{T}_e \tag{9.5}$$

Finally it can be shown that

$$\frac{d\bar{\omega}_r}{dt} = \frac{d^2\delta}{\omega_0 dt^2} \tag{9.6}$$

where δ is the angular position of the rotor (elec. rad/s) with respect to a synchronously rotating reference frame.

Combining Equations 9.5 and 9.6 results in the *swing equation* (Equation 9.7), so called because it describes the swings of the rotor angle δ during disturbances:

$$\frac{2H}{\omega_0} \frac{d^2\delta}{dt^2} = \bar{T}_m - \bar{T}_e \tag{9.7}$$

An additional term $(-K_D \Delta \bar{\omega}_r)$ may be added to the right-hand side of Equation 9.7 to account for a component of damping torque not included explicitly in T_e.

For a system to be *transiently stable* during a disturbance, it is necessary for the rotor angle (as its behavior is described by the swing equation) to oscillate around an equilibrium point. If the rotor angle increases indefinitely, the machine is said to be *transiently unstable* as the machine continues to accelerate and does not reach a new state of equilibrium. In multimachine systems, such a machine will "pull out of step" and lose synchronism with the rest of the machines.

9.2.2 Power–Angle Relationship

Consider a simple model of a single generator connected to an infinite bus through a transmission system as shown in Figure 9.2. The model can be reduced as shown by replacing the generator with a constant voltage behind a transient reactance (classical model). It is well known that there is a maximum power that can be transmitted to the infinite bus in such a network. The relationship between the electrical power of the generator P_e and the rotor angle of the machine δ is given by

$$P_e = \frac{E' E_B}{X_T} \sin\delta = P_{max} \sin\delta \tag{9.8}$$

where

$$P_{max} = \frac{E' E_B}{X_T} \tag{9.9}$$

Equation 9.8 can be shown graphically as Figure 9.3 from which it can be seen that as the power initially increases δ increases until reaching 90° when P_e reaches its maximum. Beyond $\delta = 90°$, the power decreases until at $\delta = 180°$, $P_e = 0$. This is the so-called power–angle relationship and describes the transmitted power as a function of rotor angle. It is clear from Equation 9.9 that the maximum power

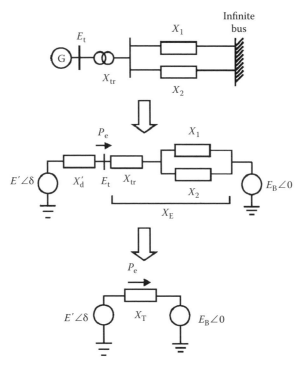

FIGURE 9.2 Simple model of a generator connected to an infinite bus.

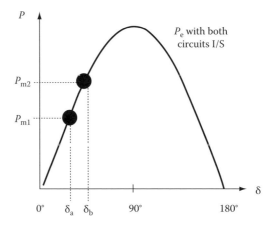

FIGURE 9.3 Power–angle relationship for case with both circuits in service.

is a function of the voltages of the generator and infinite bus and, more importantly, a function of the transmission system reactance; the larger the reactance (e.g., the longer or weaker the transmission circuits), the lower the maximum power.

Figure 9.3 shows that for a given input power to the generator P_{m1}, the electrical output power is P_e (equal to P_m) and the corresponding rotor angle is δ_a. As the mechanical power is increased to P_{m2}, the rotor angle advances to δ_b. Figure 9.4 shows the case with one of the transmission lines removed, causing an increase in X_T and a reduction in P_{max}. It can be seen that for the same mechanical input (P_{m1}), the situation with one line removed causes an increase in rotor angle to δ_c.

Transient Stability

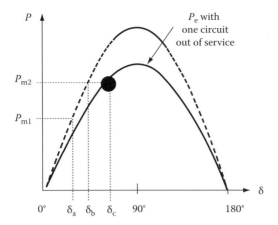

FIGURE 9.4 Power–angle relationship for case with one circuit out of service.

9.2.3 Equal Area Criterion

By combining the dynamic behavior of the generator as defined by the swing equation, with the power–angle relationship, it is possible to illustrate the concept of transient stability using the *equal area criterion*.

Consider Figure 9.5 in which a step change is applied to the mechanical input of the generator. At the initial power P_{m0}, $\delta = \delta_0$ and the system is at operating point "a." As the power is increased in a step to P_{m1} (accelerating power = $P_{m1} - P_e$), the rotor cannot accelerate instantaneously, but traces the curve up to point "b" at which time $P_e = P_{m1}$ and the accelerating power is zero. However, the rotor speed is greater than the synchronous speed and the angle continues to increase. Beyond b, $P_e > P_m$ and the rotor decelerates until reaching a maximum δ_{max} at which point the rotor angle starts to return toward b.

As we will see, for a single-machine infinite bus system, it is not necessary to plot the swing curve to determine if the rotor angle of the machine increases indefinitely, or if it oscillates around an equilibrium point. The equal area criterion allows stability to be determined using graphical means. While this method is not generally applicable to multimachine systems, it is a valuable learning aid.

Starting with the swing equation as given by Equation 9.7 and interchanging per unit power for torque,

$$\frac{d^2\delta}{dt^2} = \frac{\omega_0}{2H}(P_m - P_e) \tag{9.10}$$

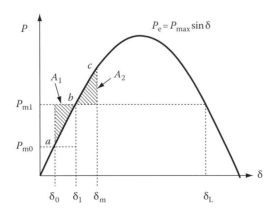

FIGURE 9.5 Power–angle curve showing the areas defined in the equal area criterion. Plot shows the result of a step change in mechanical power.

Multiplying both sides by $2\delta/dt$ and integrating gives

$$\left[\frac{d\delta}{dt}\right]^2 = \int_{\delta_0}^{\delta} \frac{\omega_0(P_m - P_e)}{H} d\delta \quad \text{or} \quad \frac{d\delta}{dt} = \sqrt{\int_{\delta_0}^{\delta} \frac{\omega_0(P_m - P_e)}{H} d\delta} \quad (9.11)$$

δ_0 represents the rotor angle when the machine is operating synchronously prior to any disturbance. It is clear that for the system to be stable, δ must increase, reach a maximum (δ_{max}), and then change direction as the rotor returns to complete an oscillation. This means that $d\delta/dt$ (which is initially zero) changes during the disturbance, but must, at a time corresponding to δ_{max}, become zero again. Therefore, as a stability criterion

$$\int_{\delta_0}^{\delta} \frac{\omega_0}{H}(P_m - P_e) d\delta = 0 \quad (9.12)$$

This implies that the area under the function $P_m - P_e$ plotted against δ must be zero for a stable system, which requires Area 1 to be equal to Area 2. Area 1 represents the energy gained by the rotor during acceleration and Area 2 represents energy lost during deceleration.

Figures 9.6 and 9.7 show the rotor response (defined by the swing equation) superimposed on the power–angle curve for a stable case and an unstable case, respectively. In both cases, a three-phase fault is applied to the system given in Figure 9.2. The only difference in the two cases is that the fault-clearing time has been increased for the unstable case. The arrows show the trace of the path followed by the rotor angle in terms of the swing equation and power–angle relationship. It can be seen that for the stable case, the energy gained during rotor acceleration is equal to the energy dissipated during deceleration ($A_1 = A_2$) and the rotor angle reaches a maximum and recovers. In the unstable case, however, it can be seen that the energy gained during acceleration is greater than that dissipated during deceleration (since the fault is applied for a longer duration) meaning that $A_1 > A_2$ and the rotor continues to advance and does not recover.

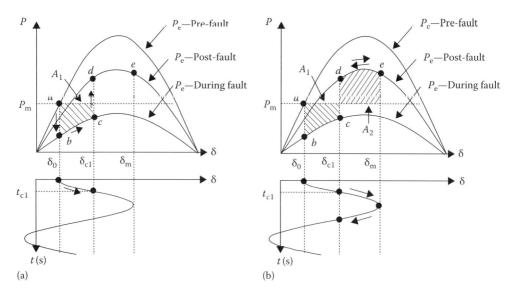

FIGURE 9.6 Rotor response (defined by the swing equation) superimposed on the power–angle curve for a stable case. (a) Shows the trajectory up until the fault is cleared at t_{cl}; moving from the origin, to point a to point b and to point c on the power-angle curves. (b) Shows the trajectory after the fault is cleared; moving from point c to point d and to point e on the power-angle curves.

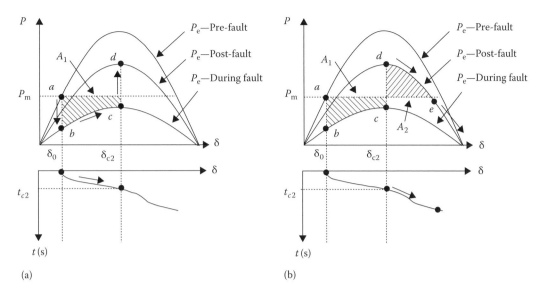

FIGURE 9.7 Rotor response (defined by the swing equation) superimposed on the power–angle curve for an unstable case. (a) Shows the trajectory up until the fault is cleared at t_{c1}; moving from the origin, to point a to point b and to point c on the power-angle curves. (b) Shows the trajectory after the fault is cleared; moving from point c to point d and to point e on the power-angle curves.

9.3 Methods of Analysis of Transient Stability

9.3.1 Modeling

The basic concepts of transient stability presented earlier are based on highly simplified models. Practical power systems consist of large numbers of generators, transmission circuits, and loads.

For stability assessment, the power system is normally represented using a positive sequence model. The network is represented by a traditional positive sequence power flow model, which defines the transmission topology, line reactances, connected loads and generation, and pre-disturbance voltage profile.

Generators can be represented with various levels of detail, selected based on such factors as length of simulation, severity of disturbance, and accuracy required. The most basic model for synchronous generators consists of a constant internal voltage behind a constant transient reactance, and the rotating inertia constant (H). This is the so-called classical representation that neglects a number of characteristics: the action of voltage regulators, variation of field flux linkage, the impact of the machine physical construction on the transient reactances for the direct and quadrature axis, the details of the prime mover or load, and saturation of the magnetic core iron. Historically, classical modeling was used to reduce computational burden associated with more detailed modeling, which is not generally a concern with today's simulation software and computer hardware. However, classical modeling still may be used for machines that are very remote from a disturbance (particularly in very large system models) and where more detailed model data are not available.

In general, synchronous machines are represented using detailed models, which capture the effects neglected in the classical model including the influence of generator construction (damper windings, saturation, etc.), generator controls (excitation systems including power system stabilizers, etc.), the prime mover dynamics, and the mechanical load. With the increasing penetration of wind generation in many power systems throughout the world, it is often necessary to use specific models to represent these types of machines. There is a variety of wind generator types with various control schemes [4] including conventional induction generators, doubly fed asynchronous generators, and fully converted generators. The unique dynamics of these devices should be taken into account in systems with significant wind generator penetrations. In situations where large wind farms exist, comprised of many small-sized generating units, it may be impractical to represent individual machines and, therefore, models representing the complete farms may be more appropriate.

Loads have always been a challenge to represent in stability simulations. Loads represented in most simulations are aggregates of a large number of component loads, such as all the individual loads connected to a substation for example. The individual loads may vary greatly in individual characteristics and may each vary uniquely with time; creating models that can accurately represent these mixed attributes is difficult. Aggregate loads have historically been represented by simplified models that capture the voltage dependency and, in some cases, frequency dependency of the load. However, recognizing that many loads include significant percentages of induction motors, modeling that can capture the dynamic effects of induction motors are now commonly used in composite load models such as those described in [5].

There are a myriad of other devices, such as HVDC lines and controls and static var devices, which may require detailed representation in transient stability studies. Finally, system protections are often represented including line protections (such as mho distance relays), out-of-step protections, loss of excitation protections, or special protection schemes.

Although power system models may be extremely large, representing thousands of generators and other devices producing systems with tens of thousands of system states, efficient numerical methods combined with modern computing power have made time-domain simulation software commonplace. It is also important to note that the time frame in which transient instability occurs is usually in the range of 1–5 s, so that simulation times need not be excessively long.

9.3.2 Analytical Methods

To accurately assess the system response following disturbances, detailed models are required for all critical elements. The complete mathematical model for the power system consists of a large number of algebraic and differential equations, including

- Generator stator algebraic equations
- Generator rotor circuit differential equations
- Swing equations
- Excitation system differential equations
- Prime mover and governing system differential equations
- Transmission network algebraic equations
- Load algebraic and differential equations

While considerable work has been done on *direct methods* of stability analysis in which stability is determined without explicitly solving the system differential equations (see Chapter 11), the most practical and flexible method of transient stability analysis is *time-domain simulation* using step-by-step numerical integration of the nonlinear differential equations. A variety of numerical integration methods are used, including *explicit* methods (such as Euler and Runge–Kutta methods) and *implicit* methods (such as the trapezoidal method). The selection of the method to be used depends largely on the stiffness of the system being analyzed. In systems in which time steps are limited by numerical stability rather than accuracy, implicit methods are generally better suited than the explicit methods.

9.3.3 Simulation Studies

Modern simulation tools offer sophisticated modeling capabilities and advanced numerical solution methods. Although each simulation tools differs somewhat, the basic requirements and functions are the same [6].

Transient Stability 9-9

9.3.3.1 Input Data

1. *Power flow*: This defines system topology and initial operating state.
2. *Dynamic data*: These include model types and associated parameters for generators, motors, line protections, special protections, and other dynamic devices and their controls.
3. *User-defined models*: In situations where standard models cannot adequately represent the dynamics of a device, user-defined models need to be provided. Most simulation programs provide the feature that permits the custom building of dynamic models by configuring a set of control blocks (and defined inputs) from a library.
4. *Program control data*: These specify such items as the type of numerical integration to use and time step.
5. *Switching data*: These include the details of the disturbance to be applied such as the time at which the fault is applied, where the fault is applied, the type of fault and its fault impedance if required, the duration of the fault, the elements lost as a result of the fault, and the total length of the simulation.
6. *System monitoring data*: These specify the quantities that are to be monitored (output) during the simulation. In general, it is not practical to monitor all quantities because system models are large, and recording all voltages, angles, flows, generator outputs, etc., at each integration time step would create an enormous volume. Therefore, it is a common practice to define a limited set of parameters to be recorded.

9.3.3.2 Output Data

1. *Simulation log*: This contains a listing of the actions that occurred during the simulation. It includes a recording of the actions taken to apply the disturbance, and reports on any operation of protections or controls, or any numerical difficulty encountered.
2. *Results output*: Most simulation tools provide sophisticated array of outputs from which the results of the simulations can be analyzed. These include scanning tools (to find largest deviations in specified variables, for example), multivariable tables and plots, animations, geographical visualization of results, and binary output finds for archiving results. These outputs are used in concert to determine if the system remained stable and to assess the details of the dynamic behavior of the system through time.

9.4 Factors Influencing Transient Stability

Many factors affect the transient stability of a generator in a practical power system. From the small system analyzed earlier, the following factors can be identified:

- The post-disturbance system reactance as seen from the generator. The weaker the post-disturbance system, the lower the P_{max} will be.
- The duration of the fault-clearing time. The longer the fault is applied, the longer the rotor will be accelerated and the more kinetic energy will be gained. The more energy that is gained during acceleration, the more difficult it is to dissipate it during deceleration.
- The inertia of the generator. The higher the inertia, the slower the rate of change of angle and the lesser the kinetic energy gained during the fault.
- The generator internal voltage (determined by excitation system) and infinite bus voltage (system voltage). The lower these voltages, the lower the P_{max} will be.
- The generator loading before the disturbance. The higher the loading, the closer the unit will be to P_{max}, which means that during acceleration, it is more likely to become unstable.
- The generator internal reactance. The lower the reactance, the higher the peak power and the lower the initial rotor angle.
- The generator output during the fault. This is a function of faults location and type of fault.

9.5 Transient Stability Considerations in System Design

As outlined in Section 9.1, transient stability is an important consideration that must be dealt with during the design of power systems. In the design process, time-domain simulations are conducted to assess the stability of the system under various conditions and when subjected to various disturbances. Since it is not practical to design a system to be stable under all possible disturbances, design criteria specify the disturbances for which the system must be designed to be stable. The criteria disturbances generally consist of the more statistically probable events, which could cause the loss of any system element and typically include three-phase faults cleared in normal time and line-to-ground faults with delayed clearing due to breaker failure. In most cases, stability is assessed for the loss of one element (such as a transformer or transmission circuit) with possibly one element out of service in the pre-disturbance system. In system design, therefore, a wide number of disturbances are assessed and if the system is found to be unstable (or marginally stable) a variety of actions can be taken to improve stability [1]. These include the following:

- *Reduction of transmission system reactance*: This can be achieved by adding additional parallel transmission circuits, providing series compensation on existing circuits, and by using transformers with lower leakage reactances.
- *High-speed fault clearing*: In general, two-cycle breakers are used in locations where faults must be removed quickly to maintain stability. As the speed of fault clearing decreases, so does the amount of kinetic energy gained by the generators during the fault.
- *Dynamic braking*: Shunt resistors can be switched in following a fault to provide an artificial electrical load. This increases the electrical output of the machines and reduces the rotor acceleration.
- *Regulate shunt compensation*: By maintaining system voltages around the power system, the flow of synchronizing power between generators is improved.
- *Reactor switching*: The internal voltages of generators, and therefore stability, can be increased by connected shunt reactors.
- *Single pole switching and reclosing*: Most power system faults are of the single-line-to-ground type. However, in most schemes, this type of fault will trip all three phases. If single pole switching is used, only the faulted phase is removed, and power can flow on the remaining two phases, thereby greatly reducing the impact of the disturbance. The single phase is reclosed after the fault is cleared and the fault medium is deionized.
- *Steam turbine fast valving*: Steam valves are rapidly closed and opened to reduce the generator accelerating power in response to a disturbance.
- *Generator tripping*: Perhaps one of the oldest and most common methods of improving transient stability, this approach disconnects selected generators in response to a disturbance that has the effect of reducing the power that is required to be transferred over critical transmission interfaces.
- *High-speed excitation systems*: As illustrated by the simple examples presented earlier, increasing the internal voltage of a generator has the effect of proving transient stability. This can be achieved by fast-acting excitation systems, which can rapidly boost field voltage in response to disturbances.
- *Special excitation system controls*: It is possible to design special excitation systems that can use discontinuous controls to provide special field boosting during the transient period, thereby improving stability.
- *Special control of HVDC links*: The DC power on HVDC links can be rapidly ramped up or down to assist in maintaining generation/load imbalances caused by disturbances. The effect is similar to generation or load tripping.

- *Controlled system separation and load shedding*: Generally considered a last resort, it is feasible to design system controls that can respond to separate, or island, a power system into areas with balanced generation and load. Some load shedding or generation tripping may also be required in selected islands. In the event of a disturbance, instability can be prevented from propagating and affecting large areas by partitioning the system in this manner. If instability primarily results in generation loss, load shedding alone may be sufficient to control the system.

9.6 Transient Stability Considerations in System Operation

While it is true that power systems are designed to be transiently stable, and many of the methods described previously may be used to achieve this goal, in actual practice, systems may be prone to being unstable. This is largely due to uncertainties related to assumptions made during the design process. These uncertainties result from a number of sources including the following:

- *Load and generation forecast*: The design process must use forecast information about the amount, distribution, and characteristics of the connected loads as well as the location and amount of connected generation. These all have a great deal of uncertainty. If the actual system load is higher than planned, the generation output will be higher, the system will be more stressed, and the transient stability limit may be significantly lower.
- *System topology*: Design studies generally assume all elements in service, or perhaps up to two elements out of service. In actual systems, there are usually many elements out of service at any one time due to forced outages (failures) or system maintenance. Clearly, these outages can seriously weaken the system and make it less transiently stable.
- *Dynamic modeling*: All models used for power system simulation, even the most advanced, contain approximations out of practical necessity.
- *Dynamic data*: The results of time-domain simulations depend heavily on the data used to represent the models for generators and the associated controls. In many cases, these data are not known (typical data are assumed) or are in error (either because they have not been derived from field measurements or due to changes that have been made in the actual system controls that have not been reflected in the data).
- *Device operation*: In the design process it is assumed that controls and protection will operate as designed. In the actual system, relays, breakers, and other controls may fail or operate improperly.

To deal with these uncertainties in actual system operation, safety margins are used. Operational (short-term) time-domain simulations are conducted using a system model, which is more accurate (by accounting for elements out on maintenance, improved short-term load forecast, etc.) than the design model. Transient *stability limits* are computed using these models. The limits are generally in terms of maximum flows allowable over critical interfaces, or maximum generation output allowable from critical generating sources. Safety *margins* are then applied to these computed limits. This means that actual system operation is restricted to levels (interface flows or generation) below the stability limit by an amount equal to a defined safety margin. In general, the margin is expressed in terms of a percentage of the critical flow or generation output. For example, an operation procedure might be to set the *operating limit* at a flow level 10% below the stability limit.

A growing trend in system operations is to perform transient stability assessment *online* in *near real time*. In this approach, the power flow defining the system topology and the initial operating state is derived, at regular intervals, from actual system measurements via the energy management system (EMS) using state-estimation methods. The derived power flow together with other data required for transient stability analysis is passed to transient stability software residing on dedicated computers and the computations required to assess all credible contingencies are performed within a specified cycle time. Using advanced analytical methods and high-end computer hardware, it is currently possible to assess the transient stability of very large systems, for a large number of contingencies, in cycle times typically ranging from 5 to 30 min. Since this online approach uses information derived directly from the actual power system, it eliminates a number of the uncertainties associated with load forecasting, generation forecasting, and prediction of system topology, thereby leading to more accurate and meaningful stability assessment.

References

1. Kundur, P., *Power System Stability and Control*, McGraw-Hill, Inc., New York, 1994.
2. Stevenson, W.D., *Elements of Power System Analysis*, 3rd edn., McGraw-Hill, New York, 1975.
3. Elgerd, O.I., *Electric Energy Systems Theory: An Introduction*, McGraw-Hill, New York, 1971.
4. Cigré working group WG C4.601 on Power System Security Assessment. Modeling and Dynamic Behavior of Wind Generation as It Relates to Power System Control and Dynamic Performance, CIGRE Technical Brochure, January 2007.
5. Kosterev, D. and A. Meklin, Load modeling in WECC, *Power Systems Conference and Exposition, 2006 (PSCE '06)*, Atlanta, GA, 2006.
6. IEEE Recommended Practice for Industrial and Commercial Power System Analysis, IEEE Std 399-1997, IEEE, New York, 1998.

10
Small-Signal Stability and Power System Oscillations

John Paserba
Mitsubishi Electric Power Products, Inc.

Juan Sanchez-Gasca
General Electric Energy

Lei Wang
Powertech Labs Inc.

Prabha S. Kundur
Kundur Power Systems Solutions, Inc.

Einar Larsen
General Electric Energy

Charles Concordia
Consultant

10.1 Nature of Power System Oscillations 10-1
 Historical Perspective • Power System Oscillations Classified by Interaction Characteristics • Summary on the Nature of Power System Oscillations
10.2 Criteria for Damping 10-7
10.3 Study Procedure 10-7
 Study Objectives • Performance Requirements • Modeling Requirements • System Condition Setup • Analysis and Verification
10.4 Mitigation of Power System Oscillations 10-10
 Siting • Control Objectives • Closed-Loop Control Design • Input-Signal Selection • Input-Signal Filtering • Control Algorithm • Gain Selection • Control Output Limits • Performance Evaluation • Adverse Side Effects • Power System Stabilizer Tuning Example
10.5 Higher-Order Terms for Small-Signal Analysis 10-18
10.6 Modal Identification 10-18
10.7 Summary 10-19
References 10-20

10.1 Nature of Power System Oscillations

10.1.1 Historical Perspective

Damping of oscillations has been recognized as important from the beginning. Before there were any power systems, oscillations in automatic speed controls (governors) initiated an analysis by J.C. Maxwell (speed controls were found necessary for the successful operation of the first steam engines).

Oscillations among generators appeared as soon as AC generators were operated in parallel. These oscillations were not unexpected, and in fact, were predicted from the concept of the power versus phase-angle curve gradient interacting with the electric generator rotary inertia, forming an equivalent mass-and-spring system. With a continually varying load and some slight differences in the design and loading of the generators, oscillations tended to be continually excited. In the case of hydrogenerators, in particular, there was very little damping, and so amortisseurs (damper windings) were installed, at first as an option. (There was concern about the increased short-circuit current and some people had to be persuaded to accept them [Crary and Duncan, 1941].) It is of interest to note that although the only significant source of actual negative damping here was the turbine speed governor (Concordia, 1969), the practical "cure" was found elsewhere. Two points were evident then and are still valid today. First, automatic control is practically the only source of negative damping, and, second, although it is obviously desirable to identify the sources of negative damping, the most effective and economical place to add damping may lie elsewhere.

After these experiences, oscillations seemed to disappear as a major problem. Although there were occasional cases of oscillations and evidently poor damping, the major analytical effort seemed to ignore damping entirely. All this changed rather suddenly in the 1960s, when the process of interconnection accelerated and more transmission and generation extended over large areas. Perhaps, the most important aspect was the wider recognition of the negative damping produced by the use of high-response generator voltage regulators in situations where the generator may be subject to relatively large angular swings, as may occur in extensive networks. (This possibility was already well known in the 1930s and 1940s but had not had much practical application then.) With the growth of extensive power systems, and especially with the interconnection of these systems by ties of limited capacity, oscillations reappeared. (Actually, they had never entirely disappeared but instead were simply not "seen.") There are several reasons for this reappearance:

1. For intersystem oscillations, the amortisseur is no longer effective, as the damping produced is reduced in approximately inverse proportion to the square of the effective external-impedance-plus-stator-impedance, and so it practically disappears.
2. The proliferation of automatic controls has increased the probability of adverse interactions among them. (Even without such interactions, the two basic controls—the speed governor and the generator voltage regulator—practically always produce negative damping for frequencies in the power system oscillation range: the governor effect, small and the automatic voltage regulator (AVR) effect, large.)
3. Even though automatic controls are practically the only devices that may produce negative damping, the damping of the uncontrolled system is itself very small and could easily allow the continually changing load and generation to result in unsatisfactory tie-line power oscillations.
4. A small oscillation in each generator that may be insignificant may add up to a tie-line oscillation that is very significant relative to its rating.
5. Higher tie-line loading increases both the tendency to oscillate and the importance of the oscillation.

To calculate the effect of damping on the system, the detail of system representation has to be considerably extended. The additional parameters required are usually much less well known than are the generator inertias and network impedances required for the "classical" studies. Further, the total damping of a power system is typically very small and is made up of both positive and negative components. Thus, if one wishes to get realistic results, one must include all the known sources. These sources include prime movers, speed governors, electrical loads, circuit resistance, generator amortisseurs, generator excitation, and, in fact, all controls that may be added for special purposes. In large networks, and particularly as they concern tie-line oscillations, the only two items that can be depended upon to produce positive damping are the electrical loads and (at least for steam-turbine-driven generators) the prime mover.

Although it is obvious that net damping must be positive for stable operation, why be concerned about its magnitude? More damping would reduce (but not eliminate) the tendency to oscillate and the magnitude of oscillations. As pointed out earlier, oscillations can never be eliminated, as even in the best-damped systems the damping is small, which is only a small fraction of the "critical damping." So the common concept of the power system as a system of masses and springs is still valid, and we have to accept some oscillations. The reasons why the power systems are often troublesome are various, depending on the nature of the system and the operating conditions. For example, when at first a few (or more) generators were paralleled in a rather closely connected system, oscillations were damped by the generator amortisseurs. If oscillations did occur, there was little variation in system voltage. In the simplest case of two generators paralleled on the same bus and equally loaded, oscillations between them would produce practically no voltage variation and what was produced would principally be at twice the oscillation frequency. Thus, the generator voltage regulators were not stimulated and did not participate in the activity. Moreover, the close coupling between the generators reduced the effective regulator gain considerably for the oscillation mode. Under these conditions, when voltage-regulator response was increased (e.g., to improve transient stability), there was little apparent decrease of system damping (in most cases), but appreciable improvement in transient stability. Instability through negative damping produced by increased voltage-regulator gain had already been demonstrated theoretically (Concordia, 1944).

Consider that the system just discussed is then connected to another similar system by a tie-line. This tie-line should be strong enough to survive the loss of any one generator but rather may be only a small fraction of system capacity. Now, the response of the system to tie-line oscillations is quite different from that just described. Because of the high external impedance seen by either system, not only is the positive damping by the generator amortisseurs largely lost, but also the generator terminal voltages become responsive to angular swings. This causes the generator voltage regulators to act, producing negative damping as an unwanted side effect. This sensitivity of voltage to angle increases as a strong function of initial angle and, thus, tie-line loading. Thus, in the absence of mitigating means, tie-line oscillations are very likely to occur, especially at heavy-line loading (and they have on numerous occasions as illustrated in Chapter 3 of CIGRE Technical Brochure No. 111, 1996). These tie-line oscillations are bothersome, especially as a restriction on the allowable power transfer, as relatively large oscillations are (quite properly) taken as a precursor to instability.

Next, as interconnection proceeds another system is added. If the two previously discussed systems are designated A and B, and a third system, C, is connected to B, then a chain A–B–C is formed. If power is flowing A → B → C or C → B → A, the principal (i.e., lowest frequency) oscillation mode is A against C, with B relatively quiescent. However, as already pointed out, the voltages of system B are varying. In effect, B is acting as a large synchronous condenser facilitating the transfer of power from A to C, and suffering voltage fluctuations as a consequence. This situation has occurred several times in the history of interconnected power systems and has been a serious impediment to progress. In this case, note that the problem is mostly in system B, while the solution (or at least mitigation) will be mostly in systems A and C. With any presently conceivable controlled voltage support, it would be practically impossible to maintain a satisfactory voltage solely in system B. On the other hand, without system B, for the same power transfer, the oscillations would be much more severe. In fact, the same power transfer might not be possible without, for example, a very high amount of series or shunt compensation. If the power transfer is A → B ← C or A ← B → C, the likelihood of severe oscillation (and the voltage variations produced by the oscillations) is much less. Further, both the trouble and the cure are shared by all three systems, so effective compensation is more easily achieved. For best results, all combinations of power transfers should be considered.

Aside from this abbreviated account of how oscillations grew in importance as interconnections grew in extent, it may be of interest to mention the specific case that seemed to precipitate the general acceptance of the major importance of improving system damping, as well as the general recognition of the generator voltage regulator as the major culprit in producing negative damping. This was the series of studies of the transient stability of the Pacific Intertie (AC and DC in parallel) on the west coast of the United States. In these studies, it was noted that for three-phase faults, instability was determined not by severe first swings of the generators but by oscillatory instability of the post-fault system, which had one of two parallel AC line sections removed and thus higher impedance. This showed that damping is important for transient as well as steady-state conditions and contributed to a worldwide rush to apply power system stabilizers (PSSs) to all generator-voltage regulators as a panacea for all oscillatory ills.

But the pressures of the continuing extension of electric networks and of increases in line loading have shown that the PSS alone is often not enough. When we push to the limit that limit is more often than not determined by lack of adequate damping. When we add voltage support at appropriate points in the network, we not only increase its "strength" (i.e., increased synchronizing power or smaller transfer impedance), but also improve its damping (if the generator voltage regulators have been producing negative damping) by relieving the generators of a good part of the work of voltage regulation and also reducing the regulator gain. This is so, whether or not reduced damping was an objective. However, the limit may still be determined by inadequate damping. How can it be improved? There are at least three options:

1. Add a signal (e.g., line current) to the voltage support device control.
2. Increase the output of the PSS (which is possible with the now stiffer system) or do both as found to be appropriate.
3. Add an entirely new device at an entirely new location. Thus, the proliferation of controls has to be carefully considered.

Oscillations of power system frequency as a whole can still occur in an isolated system, due to governor deadband or interaction with system frequency control, but is not likely to be a major problem in large interconnected systems. These oscillations are most likely to occur on intersystem ties among the constituent subsystems, especially if the ties are weak or heavily loaded. This is in a relative sense; an "adequate" tie planned for certain usual line loadings is nowadays very likely to be much more severely loaded and, thus, behave dynamically like a weak line as far as oscillations are concerned, quite aside from losing its emergency pickup capability. There has always been commercial pressure to utilize a line, perhaps originally planned to aid in maintaining reliability, for economical energy transfer simply because it is there. Now, however, there is also "open access" that may force a utility to use nearly every line for power transfer. This will certainly decrease reliability and may decrease damping, depending on the location of added generation.

10.1.2 Power System Oscillations Classified by Interaction Characteristics

Electric power utilities have experienced problems with the following types of subsynchronous frequency oscillations (Kundur, 1994):

- Local plant mode oscillations
- Interarea mode oscillations
- Torsional mode oscillations
- Control mode oscillations

Local plant mode oscillation problems are the most commonly encountered among the aforementioned types and are associated with units at a generating station oscillating with respect to the rest of the power system. Such problems are usually caused by the action of the AVRs of generating units operating at high output and feeding into weak-transmission networks; the problem is more pronounced with high-response excitation systems. The local plant oscillations typically have natural frequencies in the range of 1–2 Hz. Their characteristics are well understood and adequate damping can be readily achieved by using supplementary control of excitation systems in the form of PSSs.

Interarea modes are associated with machines in one part of the system oscillating against machines in other parts of the system. They are caused by two or more groups of closely coupled machines that are interconnected by weak ties. The natural frequency of these oscillations is typically in the range of 0.1–1 Hz. The characteristics of interarea modes of oscillation are complex and in some respects significantly differ from the characteristics of local plant modes (CIGRE Technical Brochure No. 111, 1996; Kundur, 1994; Rogers, 2000).

Torsional mode oscillations are associated with the turbine-generator rotational (mechanical) components. There have been several instances of torsional mode instability due to interactions with controls, including generating unit excitation and prime mover controls (Kundur, 1994):

- Torsional mode destabilization by excitation control was first observed in 1969 during the application of PSSs on a 555 MVA fossil-fired unit at the Lambton generating station in Ontario. The PSS, which used a stabilizing signal based on speed measured at the generator end of the shaft, was found to excite the lowest torsional (16 Hz) mode. The problem was solved by sensing speed between the two LP turbine sections and by using a torsional filter (Kundur et al., 1981; Watson and Coultes, 1973).

- Instability of torsional modes due to interaction with speed-governing systems was observed in 1983 during the commissioning of a 635 MVA unit at Pickering "B" nuclear generating station in Ontario. The problem was solved by providing an accurate linearization of steam valve characteristics and by using torsional filters (Lee et al., 1985).
- Control mode oscillations are associated with the controls of generating units and other equipment. Poorly tuned controls of excitation systems, prime movers, static var compensators (SVC), and high-voltage direct current (HVDC) converters are the usual causes of instability of control modes. Sometimes it is difficult to tune the controls so as to assure adequate damping of all modes. Kundur et al. (1981) describe the difficulty experienced in 1979 in tuning the PSSs at the Ontario Hydro's Nanticoke generating station. The stabilizers used shaft-speed signals with torsional filters. With the stabilizer gain high enough to stabilize the local plant mode oscillation, a control mode local to the excitation system and the generator field referred to as the "exciter mode" became unstable. The problem was solved by developing an alternative form of stabilizer that did not require a torsional filter (Lee and Kundur, 1986).

Although all of these categories of oscillations are related and can exist simultaneously, the primary focus of this section is on the electromechanical oscillations that affect interarea power flows.

A power system having multiple machines will act like a set of masses interconnected by a network of springs and will exhibit multiple modes of oscillation. In many systems, the damping of these electromechanical swing modes is a critical factor for operating the power system in a stable, thus secure manner (Kundur et al., 2004). The power transfer between such machines on the AC transmission system is a direct function of the angular separation between their internal voltage phasors. The torques that influence the machine oscillations can be conceptually split into synchronizing and damping components of torque (de Mello and Concordia, 1969). The synchronizing component "holds" the machines in the power system together and is important for system transient stability following large disturbances. For small disturbances, the synchronizing component of torque determines the frequency of an oscillation. Most stability texts present the synchronizing component in terms of the slope of the power–angle relationship, as illustrated in Figure 10.1, where K represents the

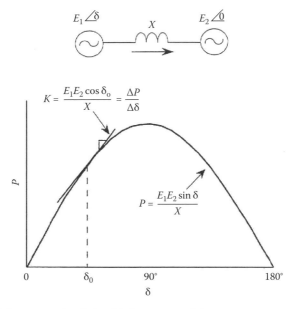

FIGURE 10.1 Simplified power–angle relationship between two AC systems.

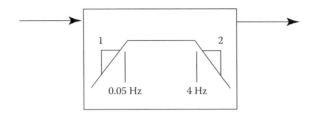

FIGURE 10.2 Conceptual block diagram of a power-swing mode.

amount of synchronizing torque. The damping component determines the decay of oscillations and is important for system stability following recovery from the initial swing. Damping is influenced by many system parameters, is usually small, and, as previously described, is sometimes negative in the presence of controls (which are practically the only "source" of negative damping). Negative damping can lead to spontaneous growth of oscillations until relays begin to trip system elements or a limit cycle is reached.

Figure 10.2 shows a conceptual block diagram of a power-swing mode, with inertial (M), damping (D), and synchronizing (K) effects identified. For a perturbation about a steady-state operating point, the modal accelerating torque ΔT_{ai} is equal to the modal electrical torque ΔT_{ei} (with the modal mechanical torque ΔT_{mi} considered to be 0). The effective inertia is a function of the total inertia of all machines participating in the swing; the synchronizing and damping terms are frequency dependent and are influenced by generator rotor circuits, excitation controls, and other system controls.

10.1.3 Summary on the Nature of Power System Oscillations

The preceding review leads to a number of important conclusions and observations concerning power system oscillations:

- Oscillations are due to natural modes of the system and therefore cannot be eliminated. However, their damping and frequency can be modified.
- As power systems evolve, the frequency and damping of existing modes change and new modes may emerge.
- The source of "negative" damping is power system controls, primarily excitation system automatic voltage regulators.
- Interarea oscillations are associated with weak transmission links and heavy power transfers.
- Interarea oscillations often involve more than one utility and may require the cooperation of all to arrive at the most effective and economical solution.
- PSSs are the most commonly used means of enhancing the damping of interarea modes.
- Continual study of the system is necessary to minimize the probability of poorly damped oscillations. Such "beforehand" studies may have avoided many of the problems experienced in power systems (see Chapter 3 of CIGRE Technical Brochure No. 111, 1996).

It must be clear that avoidance of oscillations is only one of many aspects that should be considered in the design of a power system and so must take its place in line along with economy, reliability, security, operational robustness, environmental effects, public acceptance, voltage and power quality, and certainly a few others that may need to be considered. Fortunately, it appears that many features designed to further some of these other aspects also have a strong mitigating effect in reducing oscillations. However, one overriding constraint is that the power system operating point must be stable with respect to oscillations.

10.2 Criteria for Damping

The rate of decay of the amplitude of oscillations is best expressed in terms of the damping ratio ζ. For an oscillatory mode represented by a complex eigenvalue $\sigma \pm j\omega$, the damping ratio is given by

$$\zeta = \frac{-\sigma}{\sqrt{\sigma^2 + \omega^2}} \tag{10.1}$$

The damping ratio ζ determines the rate of decay of the amplitude of the oscillation. The time constant of amplitude decay is $1/|\sigma|$. In other words, the amplitude decays to $1/e$ or 37% of the initial amplitude in $1/|\sigma|$ seconds or in $1/(2\pi\zeta)$ cycles of oscillation (Kundur, 1994). As oscillatory modes have a wide range of frequencies, the use of damping ratio rather than the time constant of decay is considered more appropriate for expressing the degree of damping. For example, a 5 s time constant represents amplitude decay to 37% of initial value in 110 cycles of oscillation for a 22 Hz torsional mode, in five cycles for a 1 Hz local plant mode, and in one-half cycle for a 0.1 Hz interarea mode of oscillation. On the other hand, a damping ratio of 0.032 represents the same degree of amplitude decay in five cycles, for example, for all modes.

A power system should be designed and operated so that the following criteria are satisfied for all expected system conditions, including post-fault conditions following design contingencies:

1. The damping ratio (ζ) of all system modes oscillation should exceed a specified value. The minimum acceptable damping ratio is system dependent and is based on operating experience and/or sensitivity studies; it is typically in the range 0.03–0.05.
2. The small-signal stability margin should exceed a specified value. The stability margin is measured as the difference between the given operating condition and the absolute stability limit ($\zeta = 0$) and should be specified in terms of a physical quantity, such as a power plant output, power transfer through a critical transmission interface, or system load level.

10.3 Study Procedure

Studies to investigate small-signal stability have become more and more important in power system planning and operation due to the following facts:

- Large-scale system interconnection
- Operation of system closer to transmission limits
- Integration of new generation techniques (such as wind and solar)
- Changing load characteristics
- Regulatory requirements for enforcing small-signal stability criteria

There is a general need for establishing study procedures with respect to power system oscillations. Such procedures should allow engineers to understand the study objectives and requirements, to prepare the appropriate models, to determine the existence of problems, to identify factors influencing the problem, and finally to provide information useful in developing control measures for mitigation.

10.3.1 Study Objectives

A small-signal stability study usually includes one or more of the following analysis objectives:

1. Review of overall small-signal stability of a power system under various conditions (Wang, 2005).
2. Identification of critical modes (local and interarea) and the associated modal characteristics to confirm existing problems or problems found in system planning (Kundur, 1994).

3. Assessment of the impact of generation and transmission expansion on small-signal stability (Arabi et al., 2000). This has become an important issue with the large-scale integration of renewable energy technologies (wind, solar, etc.) (Gautam et al., 2009).
4. Derivation of operation guidelines such as transfer limits subject to damping criteria (Chung et al., 2004).
5. Design and tuning of controls such as PSS (Bu et al., 2003).
6. Postmortem analysis of incidents involving system oscillations (EPRI, 1997; Kosterev et al., 2001).

10.3.2 Performance Requirements

Performance requirements for small-signal stability are mostly set by the minimum damping ratio of the worst oscillatory mode in a system, although some old criterion was set to measure the peak-to-peak decay rate of oscillations obtained from time-domain simulations. To maintain stable system operation, the minimum damping ratio of any mode in the system must be greater than zero. In studies, a reasonable margin is usually required by utilities for the minimum damping ratio (Midwest ISO, 2009; PJM, 2010). For example, a 3% minimum damping ratio is required for all modes in PJM transmission planning. Due to different nature and impact from different types of oscillatory modes, it is also fair to enforce different damping requirements for different types of modes; for example, 5% for local modes and 3% for interarea modes.

There are other small-signal performance standards. A notable example is the IEEE guide for the dynamic performance of excitation control systems (IEEE, 1990). Such standards are used to tune and validate control systems.

10.3.3 Modeling Requirements

Modeling requirements for small-signal stability analysis are similar to those for transient stability analysis. Some special considerations are as follows:

- System reduction or equivalencing is generally not recommended unless the study is entirely focused on local modes or the study is part of dynamic security assessment (DSA) performed for online conditions for which external system models are not available.
- Devices that have impact on oscillations should be included, for example, exciter/AVR/PSS, HVDC modulation controls, power oscillation damper (POD) on flexible AC transmission systems (FACTS) devices such as SVC and static compensator (STATCOM). This may even be a mandatory modeling requirement. For example, WECC has a policy that all PSS must be put in service to enhance system damping; this should be consistently represented in the system model (WECC, 2011).
- Devices not applicable for small-signal stability analysis should obviously not be included in the system model, for example, special protection systems.

It is well known that system oscillations (particularly local mode of oscillations) are sensitive to controls (such as AVR for generators) and their parameters. Inappropriate models and/or parameter values can lead to misleading or incorrect results from a study. A widely studied example is the Rush Island incident reported in (Shah et al., 1995). It is therefore important to validate the models prior to performing studies. Two complementary approaches can be used:

1. Measurement-based model validation. There are well-established procedures for model derivation and validation from field testing, for example, those set by WECC (WECC, 2010).
2. Simulation-based model validation. This refers to the use of simulation techniques to validate model performance. Typical types of simulations include exciter step responses and eigenvalue tests for isolated dynamic device models.

10.3.4 System Condition Setup

In a small-signal stability study, different system conditions often need to be included. These may range from

- Loading levels: These mostly affect the characteristics of interarea modes, which may exhibit significantly different damping from peak to light load conditions.
- Contingencies: Contingencies weaken transmission systems and thus would generally reduce the damping of oscillations for both local and interarea modes. Application of contingencies can follow the requirements from the applicable standards, for example, the NERC reliability standards for transmission planning (NERC, 2009).
- Transfers: In systems where the power transfer capacity on a transmission path may be limited due to small-signal stability, transfer limit assessment will need to include small-signal stability analysis with damping criteria. This will involve an iterative analysis process in which the transfer on the transmission path is adjusted in order to find the maximum transfer without violating the criteria. The transfer adjustments can be done with system dispatches set in study standards or with adaptive dispatches to increase the limit (Chung et al., 2004).

10.3.5 Analysis and Verification

Although a small-signal stability study may include different objectives, it usually has primary focus on identifying critical modes of oscillations in the system that might cause security concerns. This requires effective computation methods of determining oscillatory modes. The conventional nonlinear time-domain simulation method (the same method used for transient stability analysis) can be used for damping assessment of oscillations. This is often done with the help of additional tools to post-process simulation results, for example, by using the Prony analysis (Hauer, 1991). This method has the advantage that the system responses are faithfully produced with all modeling details (such as nonlinearities) included and the results available in the time-domain format are also easy to understand. On the other hand, the time-domain method may be deceptive in identifying power system oscillations as a mode of oscillations may not be excited in a specific simulation. Moreover, a simulation may be able to reveal only the basic information (mostly frequency and damping) for a mode of oscillations but not the information necessary to fully understand and to control the mode. These limitations are overcome by the modal analysis based on eigenvalue method, which computes the eigenvalues of the dynamic models of the system linearized about a specific operating point. The stability of each mode is clearly identified by the system's eigenvalues. Mode shapes and the relationships between different modes and system variables or parameters are identified using eigenvectors (Kundur, 1994). In addition, special eigenvalue computation algorithms are available to calculate local or interarea modes, as well as their associated modal information (Kundur, 1994; Kundur et al., 1990; Martins and Quintao, 2003; Martins et al., 1992, 1996; Semlyen and Wang, 1988; Wang and Semlyen, 1990). Powerful computer program packages incorporating the aforementioned computational features are also available for utility applications, thus providing comprehensive capabilities for analyses of power system oscillations (CIGRE, 1996, 2000; Wang et al., 2001).

In sum, modal analysis complemented by nonlinear time-domain simulations is the most effective procedure of studying power system oscillations. The following are the recommended steps for a systematic small-signal stability analysis:

1. Perform an eigenvalue scan for the types of modes of interest using a small-signal stability program. This will indicate the presence of poorly damped modes.
2. Perform a detailed eigenanalysis of the poorly damped modes. This will determine their characteristics and sources of the problem, and assist in developing mitigation measures. This will also identify the quantities to be monitored in time-domain simulations.
3. Perform time-domain simulations of the critical cases identified from the eigenanalysis. This is useful to confirm the results of small-signal analysis. In addition, it shows how system nonlinearities affect the oscillations.

In addition to eigenvalue analysis, other linear analysis techniques have been applied to small-signal stability analysis, in particular in designing remedial control measures. This includes use of transfer function zeros and residues, frequency responses, H_∞ analysis, etc. (Klein et al., 1995; Kundur, 1994; Martins and Quintao, 2003; Martins et al., 1992, 1996).

The IEEE Power Engineering Society Power System Dynamic Performance Committee has sponsored a series of panel sessions on small-signal stability and linear analysis techniques from 1998 to 2005, which can be found in Gibbard et al. (2001) and IEEE PES (2000, 2002, 2003, 2005). Further archival information can be found in IEEE PES (1995, 2006).

10.4 Mitigation of Power System Oscillations

In many power systems, equipment is installed to enhance various performance issues such as transient, oscillatory, or voltage stability (Kundur et al., 2004). In many instances, this equipment is power-electronic based, which generally means the device can be rapidly and continuously controlled. Examples of such equipment applied in the transmission system include an SVC, STATCOM, and thyristor-controlled series compensator (TCSC). To improve damping in a power system, a supplemental damping controller can be applied to the primary regulator of one of these transmission devices or to generator controls. The supplemental control action should modulate the output of a device in such a way as to affect power transfer such that damping is added to the power system swing modes of concern. This subsection provides an overview on some of the issues that affect the ability of damping controls to improve power system dynamic performance (CIGRE Technical Brochure No. 111, 1996; CIGRE Technical Brochure No. 116, 2000; Levine, 1995; Paserba et al., 1995).

10.4.1 Siting

Siting plays an important role in the ability of a device to stabilize a swing mode (Larsen et al., 1995; Martins and Lima, 1990; Pourbeik and Gibbard, 1996). Many controllable power system devices are sited based on issues unrelated to stabilizing the network (e.g., HVDC transmission and generators), and the only question is whether they can be utilized effectively as a stability aid. In other situations (e.g., SVC, STATCOM, TCSC, or other FACTS controllers), the equipment is installed primarily to help support the transmission system, and siting will be heavily influenced by its stabilizing potential. Device cost represents an important driving force in selecting a location. In general, there will be one location that makes optimum use of the controllability of a device. If the device is located at a different location, a device of larger size may be needed to achieve the desired stabilization objective. In some cases, overall costs may be minimized with nonoptimum locations of individual devices because other considerations must also be taken into account, such as land price and availability, environmental regulations, etc. (IEEE PES, 1996).

The inherent ability of a device to achieve a desired stabilization objective in a robust manner, while minimizing the risk of adverse interactions, is another consideration that can influence the siting decision. Most often, these other issues can be overcome by appropriate selection of input signals, signal filtering, and control design. This is not always possible, however, so these issues should be included in the decision-making process for choosing a site. For some applications, it will be desirable to apply the devices in a distributed manner. This approach helps maintain a more uniform voltage profile across the network, during both steady-state operation and after transient events. Greater security may also be possible with distributed devices because the overall system is more likely to tolerate the loss of one of the devices, but would likely come at a greater cost.

10.4.2 Control Objectives

Several aspects of control design and operation must be satisfied during both the transient and the steady-state operations of the power system, before and after a major disturbance. These aspects suggest that controls applied to the power system should

1. Survive the first few swings after a major system disturbance with some degree of safety. The safety factor is usually built into a Reliability Council's criteria (e.g., keeping voltages above some threshold during the swings).
2. Provide some minimum level of damping in the steady-state condition after a major disturbance (postcontingent operation). In addition to providing security for contingencies, some applications will require "ambient" damping to prevent spontaneous growth of oscillations in steady-state operation.
3. Minimize the potential for adverse side effects, which can be classified as follows:
    a. Interactions with high-frequency phenomena on the power system, such as turbine-generator torsional vibrations and resonances in the AC transmission network
    b. Local instabilities within the bandwidth of the desired control action
4. Be robust so that the control will meet its objectives for a wide range of operating conditions encountered in power system applications. The control should have minimal sensitivity to system operating conditions and component parameters since power systems operate over a wide range of operating conditions and there is often uncertainty in the simulation models used for evaluating performance. Also, the control should have minimum communication requirements.
5. Be highly dependable so that the control has a high probability of operating as expected when needed to help the power system. This suggests that the control should be testable in the field to ascertain that the device will act as expected should a contingency occur. This leads to the desire for the control response to be predictable. The security of system operations depends on knowing, with a reasonable certainty, what the various control elements will do in the event of a contingency.

10.4.3 Closed-Loop Control Design

Closed-loop control is utilized in many power system components. Voltage regulators, either continuous or discrete, are commonplace on generator excitation systems, capacitor and reactor banks, tap-changing transformers, and SVCs. Modulation controls to enhance power system stability have been applied extensively to generator exciters and to HVDC, SVC, and TCSC systems. A notable advantage of closed-loop control is that stabilization objectives can often be met with less equipment and impact on the steady-state power flows than is generally possible with open-loop controls. While the behavior of the power system and its components is usually predictable by simulation, its nonlinear character and vast size lead to challenging demands on system planners and operating engineers. The experience and intuition of these engineers is generally more important to the overall successful operation of the power system than the many available, elegant control design techniques (CIGRE Technical Brochure, 2000; Levine, 1995; Pal and Chaudhuri, 2005).

Typically, a closed-loop controller is always active. One benefit of such a closed-loop control is ease of testing for proper operation on a continuous basis. In addition, once a controller is designed for the worst-case contingency, the chance of a less-severe contingency causing a system breakup is lower than if only open-loop controls are applied. Disadvantages of closed-loop control involve primarily the potential for adverse interactions. Another possible drawback is the need for small step sizes, or vernier control in the equipment, which will have some impact on cost. If communication is needed, this could also be a challenge. However, experience suggests that adequate performance should be attainable using only locally measurable signals.

One of the most critical steps in control design is to select an appropriate input signal. The other issues are to determine the input filtering and control algorithm and to assure attainment of the stabilization objectives in a robust manner with minimal risk of adverse side effects. The following subsections discuss design approaches for closed-loop stability controls, so that the potential benefits can be realized on the power system.

10.4.4 Input-Signal Selection

The choice of using a local signal as an input to a stabilizing control function is based on several considerations:

1. The input signal must be sensitive to the swings on the machines and lines of interest. In other words, the swing modes of interest must be "observable" in the input signal selected. This is mandatory for the controller to provide a stabilizing influence.
2. The input signal should have as little sensitivity as possible to other swing modes on the power system. For example, for a transmission-line device, the control action will benefit only those modes that involve power swings on that particular line. If the input signal was also responsive to local swings within an area at one end of the line, then valuable control range would be wasted in responding to an oscillation that the damping device has little or no ability to control.
3. The input signal should have little or no sensitivity to its own output, in the absence of power swings. Similarly, there should be as little sensitivity to the action of other stabilizing controller outputs as possible. This decoupling minimizes the potential for local instabilities within the controller bandwidth (CIGRE Technical Brochure No. 116, 2000).

These considerations have been applied to a number of modulation control designs, which have eventually proven themselves in many actual applications (see Chapter 5 of CIGRE Technical Brochure No. 111, 1996). For example, the application of PSS controls on generator excitation systems was the first such study that reached the conclusion that speed or power is the best input signal, with frequency of the generator substation voltage being an acceptable choice as well (Kundur et al., 1989; Larsen and Swann, 1981). For SVCs, the conclusion was that the magnitude of line current flowing past the SVC is the best choice (Larsen and Chow, 1987). For torsional damping controllers on HVDC systems, it was found that using the frequency of a synthesized voltage close to the internal voltage of the nearby generator, calculated with locally measured voltages and currents, is best (Piwko and Larsen, 1982). In the case of a series device in a transmission line (such as a TCSC), the considerations listed earlier lead to the conclusion that using frequency of a synthesized remote voltage to estimate the center of inertia of an area involved in a swing mode is a good choice (Levine, 1995). This allows the series device to behave like a damper across the AC line.

10.4.5 Input-Signal Filtering

To prevent interactions with phenomena outside the desired control bandwidth, low-pass and high-pass filterings must be used for the input signal. In certain applications, notch filtering is needed to prevent interactions with certain lightly damped resonances. This has been the case with SVCs interacting with AC network resonances and modulation controls interacting with generator torsional vibrations. On the low-frequency end, the high-pass filter must have enough attenuation to prevent excessive response during slow ramps of power, or during the long-term settling following a loss of generation or load. This filtering must be considered while designing the overall control as it will strongly affect performance and the potential for local instabilities within the control bandwidth. However, finalizing such filtering usually must wait until the design for performance is completed, after which the attenuation needed at specific frequencies can be determined. During the control design work, a reasonable approximation of these filters needs to be included. Experience suggests that a high-pass break near 0.05 Hz (3 s washout

Small-Signal Stability and Power System Oscillations

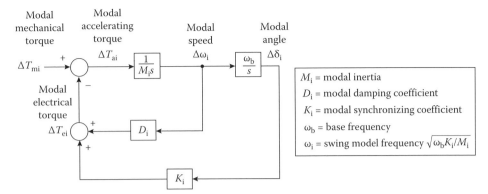

FIGURE 10.3 Initial input-signal filtering.

time constant) and a double low-pass break near 4 Hz (40 ms time constant), as shown in Figure 10.3, are suitable for a starting point. A control design that provides adequate stabilization of the power system with these settings for the input filtering has a high probability of being adequate after the input filtering parameters are finalized.

10.4.6 Control Algorithm

Levine (1995), CIGRE Technical Brochure No. 116 (2000), and Pal and Chaudhuri (2005) present many control design methods that can be utilized to design supplemental controls for power systems. Generally, the control algorithm for damping leads to a transfer function that relates an input signals to a device output. This statement is the starting point for understanding how deviations in the control algorithm affect system performance.

In general, the transfer function of the control (and input-signal filtering) is most readily discussed in terms of its gain and phase relationship versus frequency. A phase shift of 0° in the transfer function means that the output is proportional to the input and, for discussion purposes, is assumed to represent a pure damping effect on a lightly damped power swing mode. Phase lag in the transfer function (up to 90°) translates to a positive synchronizing effect, tending to increase the frequency of the swing mode when the control loop is closed. The damping effect will decrease with the sine of the phase lag. Beyond 90°, the damping effect will become negative. Conversely, phase lead is a desynchronizing influence and will decrease the frequency of the swing mode when the control loop is closed. Generally, the desynchronizing effect should be avoided. The preferred transfer function has between 0° and 45° of phase lag in the frequency range of the swing modes that the control is designed to damp.

10.4.7 Gain Selection

After the shape of the transfer function is designed to meet the desired control phase characteristics, the gain of the control is selected to obtain the desired level of damping. To maximize damping, the gain should be high enough to assure full utilization of the controlled device for the critical disturbances, but no higher, so that risks of adverse effects are minimized. Typically, the gain selection is done analytically with root-locus or Nyquist methods. However, the gain must ultimately be verified in the field (see Chapter 8 of CIGRE Technical Brochure No. 111, 1996).

10.4.8 Control Output Limits

The output of a damping control must be limited to prevent it from saturating the device being modulated. By saturating a controlled device, the purpose of the damping control would be defeated. As a general rule of thumb for damping, when a control is at its limits in the frequency range of interarea oscillations, the output of the controlled device should be just within its limits (Larsen and Swann, 1981).

10.4.9 Performance Evaluation

Good simulation tools are essential in applying damping controls to power transmission equipment for the purpose of system stabilization. The controls must be designed and tested for robustness with such tools. For many system operating conditions, the only feasible means of testing the system is by simulation, so confidence in the power system model is crucial. A typical large-scale power system model may contain up to 15,000 state variables or more. For design purposes, a reduced-order model of the power system is often desirable (Piwko et al., 1991; Wang et al., 1997). If the size of the study system is excessive, the large number of system variations and required parametric studies become tedious and prohibitively expensive for some linear analysis techniques and control design methods in general use today. A good understanding of the system performance can be obtained with a model that contains only the relevant dynamics for the problem under study. The key situations that establish the adequacy of controller performance and robustness can be identified from the reduced-order model, and then tested with the full-scale model. Note that CIGRE Technical Brochure No. 111 (1996), CIGRE Technical Brochure No. 116 (2000), and Kundur (1994), as well as Gibbard et al. (2001) and IEEE PES (2000, 2002, 2003, 2005) contain information on the application of linear analysis techniques for very large systems.

Field testing is also an essential part of applying supplemental controls to power systems. Testing needs to be performed with the controller open loop, comparing the measured response at its own input and the inputs of other planned controllers against the simulation models. Once these comparisons are acceptable, the system can be tested with the control loop closed. Again, the test results should have a reasonable correlation with the simulation program. Methods have been developed for performing such testing of the overall power system to provide benchmarks for validating the full-system model. Such testing can also be done on the simulation program to help arrive at the reduced-order models (Hauer, 1991; Kamwa et al., 1993) needed for the advanced control design methods (CIGRE Technical Brochure No. 116, 2000; Levine, 1995; Pal and Chaudhuri, 2005). Methods have also been developed to improve the modeling of individual components. These issues are discussed in great detail in Chapters 6 and 8 of CIGRE Technical Brochure No. 111 (1996).

10.4.10 Adverse Side Effects

Historically in the power industry, each major advance in improving system performance has created some adverse side effects. For example, the addition of high-speed excitation systems over 40 years ago caused the destabilization known as the "hunting" mode of the generators. The fix was PSSs, but it took over 10 years to learn how to tune them properly and there were some unpleasant surprises involving interactions with torsional vibrations on the turbine-generator shaft (Larsen and Swann, 1981).

The HVDC systems were also found to interact adversely with torsional vibrations (the subsynchronous torsional interaction [SSTI] problem), especially when augmented with supplemental modulation controls to damp power swings. Similar SSTI phenomena exist with SVCs, although to a lesser degree than with HVDC. Detailed study methods have since been established for designing systems with confidence that these effects will not cause trouble for normal operation (Bahrman et al., 1980; Piwko and Larsen, 1982). Another potential adverse side effect with SVC systems is that it can interact unfavorably with network resonances. This side effect caused a number of problems in the initial application of SVCs to transmission systems. Design methods now exist to deal with this phenomenon, and protective functions exist within SVC controls to prevent continuing exacerbation of an unstable condition (Larsen and Chow, 1987).

As the available technologies continue to evolve, such as the present industry focus on FACTS (IEEE PES, 1996), new opportunities arise for power system performance improvement. FACTS controllers introduce capabilities that may be an order of magnitude greater than existing equipment applied for stability improvement. Therefore, it follows that there may be much more serious consequences if they fail to operate properly. Robust operation and noninteraction of controls for these FACTS devices are critically important for stability of the power system (CIGRE Technical Brochure No. 116, 2000; Clark et al., 1995).

10.4.11 Power System Stabilizer Tuning Example

An example is presented in this section to illustrate the importance and method of PSS tuning. This is from a practical case for a large interconnected power system. An eigenvalue scan was performed for this system from which an unstable local mode was found for a generator at 1.97 Hz with −11.56% damping. Further investigation revealed that

- This generator has an AC exciter; see Figure 10.4.
- A PSS is already installed at this generator; see Figure 10.5. This PSS uses the generator speed as input and has three stages of phase compensation functions.
- The unstable mode can be clearly seen from the time-domain simulation when a step is applied to the reference point of the exciter of the generator; see Figure 10.6.
- The local mode becomes stable with a damping of 9.72% (at 1.43 Hz) when the PSS is disabled.

The aforementioned observation suggests that the parameters of the PSS are improperly tuned since the PSS apparently adds a large negative damping to the local mode. However, by inspection of both exciter/AVR and PSS parameters, no obvious problem can be found. It was necessary then to rely on a systematic analysis approach in order to

- Identify where the problem is
- Tune the PSS to improve the damping of the local mode

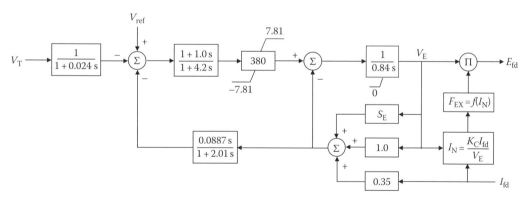

FIGURE 10.4 AC exciter and parameters.

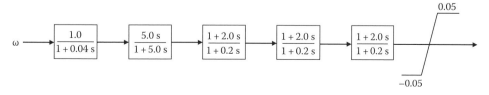

FIGURE 10.5 PSS and parameters.

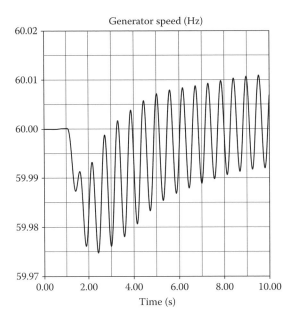

FIGURE 10.6 Unstable mode in time-domain simulation.

The analysis started with the calculation of the frequency response of the generator GEP transfer function. The phase characteristics of this transfer function are shown in Figure 10.7 together with the reversed phase characteristics of the PSS. For a properly tuned PSS, reversed phase characteristics of the PSS in Figure 10.7 should be a little below the generator-exciter-power system (GEP) phase curve for the frequency range of interest (in this case around 2 Hz), representing a small undercompensation by PSS. It is clear that the actual phase compensation provided by the PSS is far from what is required. Moreover, at the local mode frequency, the PSS undercompensates the GEP phase for about 130°. This actually indicates

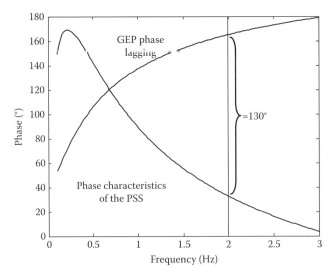

FIGURE 10.7 Phase characteristics of GEP and the original PSS.

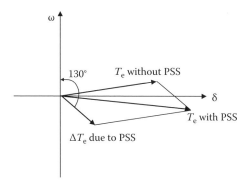

FIGURE 10.8 Phasor characteristics with the original PSS parameters.

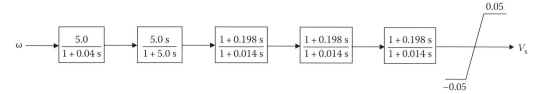

FIGURE 10.9 Tuned PSS parameters.

that the torque contributed by the PSS would include a negative damping component as shown in Figure 10.8. This is the reason that the PSS effectively reduces the local mode damping.

With the problem identified, the main focus is on the PSS phase tuning. Guided by the theory mentioned earlier, a lead/lag transfer function shown in Figure 10.9 was easily obtained, which provides a smooth phase compensation for the GEP transfer function with about 10°–20° undercompensation. Figure 10.10 shows the new phase compensation provided by the tuned PSS and the phasor diagram, indicating that the tuned PSS contributes to the improvement of both damping and synchronous torques.

Eigenvalue analysis was performed to confirm the effectiveness of the tuned PSS, and indeed the local mode damping is increased to 14%. This was further verified by the time-domain simulation performed with the same condition as Figure 10.6. The results in Figure 10.11 clearly show that the local mode is well damped.

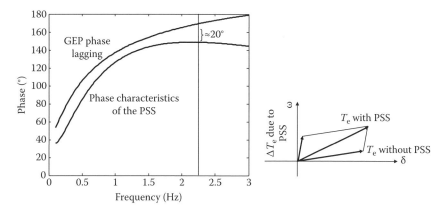

FIGURE 10.10 Phase characteristics of the tuned PSS and phasor diagram.

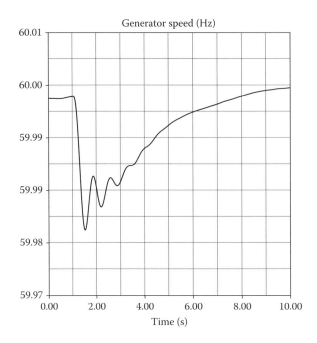

FIGURE 10.11 Local mode with tuned PSS.

10.5 Higher-Order Terms for Small-Signal Analysis

The implicit assumption in small-signal stability analysis is that the dynamic behavior of a power system in the neighborhood of an operating point of interest can be approximated by the response of a linear system.

In certain cases, such as when a power system is stressed, it has been suggested that linear analysis techniques might not provide an accurate picture of the system modal characteristics (Vittal et al., 1991). Under these circumstances, techniques that extend the domain of applicability of small-signal stability analysis become an attractive possibility for advancing the understanding of power system dynamics. Of particular interest is the study of modes and modal interactions that result from the combination of the individual system modes of the linearized system. These modes and their interactions are termed "higher-order modes" and "higher-order modal interactions," respectively.

The method of normal forms has been proposed as a means for studying higher-order modal interactions in power systems and several indices for quantifying higher-order modal characteristics have been introduced (Sanchez-Gasca et al., 2005 and references therein). In general, the method of normal forms consists of a sequence of coordinate transformations aimed at removing terms of increasing order from a Taylor series expansion (Guckenheimer and Holmes, 1983). For power system applications, due to the heavy computational burden associated with the computation of higher-order terms, work in this area has been focused on the Taylor series expansion evaluated up to second-order terms. The computational burden of the normal form analysis is large. Inclusion of even second-order terms for a large system represents a significant computational burden. Techniques need to be developed to reduce the computational burden. A related method also aimed at the study of higher-order modal interactions is described in Shanechi et al. (2003).

10.6 Modal Identification

Traditionally, the application of linear analysis techniques to study a power system is performed by first computing the system state matrices, followed by the application of a suitable eigenanalysis method to compute eigenvalues and mode shapes. However, the following practical issues may limit the applicability of this approach: (1) the program where the power system model resides does not include linearization capabilities and (2) only time-domain data are available. To circumvent these limitations, several techniques for calculating system modes and state space realizations from time-domain data have been proposed. These methods aim at computing the modal components present in the measured or simulated data and are known as modal identification methods. System modes that do not contribute to the time-domain data used in the modal identification are not computed.

In the context of power system analysis, the application of modal identification methods traces its roots to work done by Hauer and others using the Prony method (Hauer et al., 1990). The Prony method, which is based on fitting a weighted sum of exponential terms to a given signal, is perhaps the method for which more extensive results and applications have been reported in the power system literature. Since the inception of the Prony method as a power system analysis tool, other modal identification methods have been shown to be applicable to the analysis of lightly damped electromechanical oscillations in power systems. These methods include methods based on fast Fourier transform (FFT) analyses (Bounou et al., 1992; Lee and Poon, 1990), the eigensystem realization algorithm (ERA) (Kamwa et al., 1993), and the matrix pencil method (Crow and Singh, 2005). These methods are well developed and their application to the analysis and control of electromechanical oscillations has been extensively documented (e.g., Kamwa et al., 1996; Leirbukt et al., 1999; Sanchez-Gasca, 2001; Trudnowski et al., 1991).

The modal identification methods listed in the previous paragraph are typically applied to ringdown data that are obtained following the occurrence of a sudden transient event such as a line switching or the application of a step test signal. More recently, methods to compute system modes from data associated with the normal operation of a power system (ambient data) and from data generated as the result of a probing signal injected into the system for analysis purposes (probing data), have been proposed. There is a substantial amount of interest on these methods due to the practical implications associated with the ability to monitor the system modal characteristics while it operates in a normal state. These methods are designed to either process blocks of data or to compute the modes recursively (Trudnowski et al., 2008; Vanfretti et al., 2010; Zhou et al., 2008).

A more recent approach to modal identification is related to the application of time-frequency analysis techniques. These techniques aim at expanding modal identification to nonstationary nonlinear processes. The Hilbert–Huang Transform (HHT) is an analysis method that has been the focus of recent investigations (Messina, 2009).

10.7 Summary

This chapter on small-signal stability and power system oscillations shows that power systems contain many modes of oscillation due to a variety of interactions among components. Many of the oscillations are due to synchronous generator rotors swinging relative to one another. The electromechanical modes involving these masses usually occur in the frequency range of 0.1–2 Hz. Particularly troublesome are the interarea oscillations, which are typically in the frequency range of 0.1–1 Hz. The interarea modes are usually associated with groups of machines swinging relative to other groups across a relatively weak transmission path. The higher-frequency electromechanical modes (1–2 Hz) typically involve one or two generators swinging against the rest of the power system or electrically close machines swinging against each other.

These oscillatory dynamics can be aggravated and stimulated through a number of mechanisms. Heavy power transfers, in particular, can create interarea oscillation problems that constrain system operation. The oscillations themselves may be triggered through some event or disturbance on the power system or by shifting the system operating point across some steady-state stability boundary, where growing oscillations may be spontaneously created. Controller proliferation makes such boundaries increasingly difficult to anticipate. Once started, the oscillations often grow in magnitude over the span of many seconds. These oscillations may persist for many minutes and be limited in amplitude only by system nonlinearities. In some cases, they cause large generator groups to lose synchronism where a part of or the entire electrical network is lost. The same effect can be reached through slow-cascading outages when the oscillations are strong and persistent enough to cause uncoordinated automatic disconnection of key generators or loads. Sustained oscillations can disrupt the power system in other ways, even when they do not produce network separation or loss of resources. For example, power swings, which are not always troublesome in themselves, may have associated voltage or frequency swings that are unacceptable. Such concerns can limit power transfer even when oscillatory stability is not a direct concern.

Information presented in this chapter addressing power system oscillations included

- Nature of oscillations
- Criteria for damping
- Study procedure
- Mitigation of oscillations by control
- Higher-order terms for small-signal stability
- Modal identification

As to the priority of selecting devices and controls to be applied for the purpose of damping power system oscillations, the following summarizing remarks can be made:

1. Carefully tuned PSSs on the major generating units affected by the oscillations should be considered first. This is because of the effectiveness and relatively low cost of PSSs.
2. Supplemental controls added to devices installed for other reasons should be considered second. Examples include HVDC installed for the primary purpose of long-distance transmission or power exchange between asynchronous regions and SVC installed for the primary purpose of dynamic voltage support.
3. Augmentation of fixed or mechanically switched equipment with power electronics, including damping controls can be considered third. Examples include augmenting existing series capacitors with a TCSC.
4. The fourth priority for consideration is the addition of a new device in the power system for the primary purpose of damping.

References

Arabi, S., Kundur, P., Hassink, P., and Matthews, D., Small signal stability of a large power system as effected by new generation additions, in *Proceedings of the IEEE Power Engineering Society Summer Meeting*, Seattle, WA, July 16–20, 2000.

Bahrman, M.P., Larsen, E.V., Piwko, R.J., and Patel, H.S., Experience with HVDC turbine-generator torsional interaction at Square Butte, *IEEE Transactions on Power Apparatus and Systems*, 99, 966–975, 1980.

Bounou, M., Lefebvre, S., and Malhame, R.P., A spectral algorithm for extracting power system modes from time recordings, *IEEE Transactions on Power Systems*, 7(2), 665–672, May 1992.

Bu, L., Xu, W., Wang, L., Howell, F., and Kundur, P., A PSS tuning toolbox and its applications, in *Proceedings of the IEEE Power Engineering Society General Meeting*, Toronto, Ontario, Canada, July 13–17, 2003.

Chung, C.Y., Wang, L., Howell, F., and Kundur, P., Generation rescheduling methods to improve power transfer capability constrained by small-signal stability, *IEEE Transactions on Power Systems*, PWRS-19(1), 524–530, February 2004.

CIGRE Task Force 38.01.07 on Power System Oscillations, Analysis and control of power system oscillations, CIGRE Technical Brochure No. 111, December 1996, J. Paserba, Convenor.

CIGRE Task Force 38.02.16, Impact of the interaction among power system controllers, CIGRE Technical Brochure No. 116, 2000, N. Martins, Convenor.

Clark, K., Fardanesh, B., and Adapa, R., Thyristor controlled series compensation application study—Control interaction considerations, *IEEE Transactions on Power Delivery*, 10, 1031–1037, April 1995.

Concordia, C., Steady-state stability of synchronous machines as affected by voltage regulator characteristics, *AIEE Transactions*, 63, 215–220, 1944.

Concordia, C., Effect of prime-mover speed control characteristics on electric power system performance, *IEEE Transactions on Power Apparatus and Systems*, 88/5, 752–756, 1969.

Crary, S.B. and Duncan, W.E., Amortisseur windings for hydrogenerators, *Electrical World*, 115, 2204–2206, June 28, 1941.

Crow, M.L. and Singh, A., The matrix pencil for power system modal extraction, *IEEE Transactions on Power Systems*, 20(1), 501–502, February 2005.

de Mello, F.P. and Concordia, C., Concepts of synchronous machine stability as affected by excitation control, *IEEE Transactions on Power Apparatus and Systems*, 88, 316–329, 1969.

EPRI Report TR-108256, System disturbance stability studies for Western System Coordinating Council (WSCC), Prepared by Powertech Labs Inc., Surrey, British Columbia, Canada, September 1997.

Gautam, D., Vittal, V., and Harbour, T., Impact of increased penetration of DFIG-based wind turbine generators on transient and small signal stability of power systems, *IEEE Transactions on Power Systems*, PWRS-24(3), 1426–1434, August 2009.

Gibbard, M., Martins, N., Sanchez-Gasca, J.J., Uchida, N., and Vittal, V., Recent applications of linear analysis techniques, *IEEE Transactions on Power Systems*, 16(1), 154–162, February 2001. Summary of a 1998 Summer Power Meeting Panel Session on Recent Applications of Linear Analysis Techniques.

Guckenheimer, J. and Holmes, P., *Nonlinear Oscillations, Dynamical Systems, and Bifurcations of Vector Fields*, Springer-Verlag, New York, 1983.

Hauer, J.F., Application of Prony analysis to the determination of model content and equivalent models for measured power systems response, *IEEE Transactions on Power Systems*, 6, 1062–1068, August 1991.

Hauer, J.F., Demeure, C.J., and Scharf, L.L., Initial results in Prony analysis of power system response signals, *IEEE Transactions on Power Systems*, 5(1), 80–89, February 1990.

IEEE PES Special Publication 95-TP-101, Inter-area oscillations in power systems, 1995.

IEEE PES Special Publication 96-TP-116-0, FACTS applications, 1996.

IEEE PES panel session on recent applications of small signal stability analysis techniques, in *Proceedings of the IEEE Power Engineering Society Summer Meeting*, Seattle, WA, July 16–20, 2000.

IEEE PES panel session on recent applications of linear analysis techniques, in *Proceedings of the IEEE Power Engineering Society Winter Meeting*, New York, January 27–31, 2002.

IEEE PES panel session on recent applications of linear analysis techniques, in *Proceedings of the IEEE Power Engineering Society General Meeting*, Toronto, Ontario, Canada, July 13–17, 2003.

IEEE PES panel session on recent applications of linear analysis techniques, in *Proceedings of the IEEE Power Engineering Society General Meeting*, San Francisco, CA, June 12–16, 2005.

IEEE PES Special Publication 06TP177, Recent applications of linear analysis techniques for small signal stability and control, 2006.

IEEE Standard 421.2-1990, *IEEE Guide for Identification, Testing, and Evaluation of the Dynamic Performance of Excitation Control Systems*, IEEE, New York, 1990.

Kamwa, I., Grondin, R., Dickinson, J., and Fortin, S., A minimal realization approach to reduced-order modeling and modal analysis for power system response signals, *IEEE Transactions on Power Systems*, 8(3), 1020–1029, 1993.

Kamwa, I., Trudel, G., and Gerin-Lajoie, L., Low-order black-box models for control system design in large power systems, *IEEE Transactions on Power Systems*, 11(1), 303–311, February 1996.

Klein, M., Rogers, G.J., Farrokhpay, S., and Balu, N.J., H_∞ damping controller design in large power system, *IEEE Transactions on Power Systems*, PWRS-10(1), 158–165, February 1995.

Kosterev, G.N., Mittelstadt, W.A., Viles, M., Tuck, B., Burns, J., Kwok, M., Jardim, J., and Garnett, G., Model validation and analysis of WSCC system oscillations following Alberta separation on August 4, 2000, Final Report by Bonneville Power Administration and BC Hydro, January 2001.

Kundur, P., *Power System Stability and Control*, McGraw-Hill, New York, 1994.

Kundur, P., Klein, M., Rogers, G.J., and Zywno, M.S., Application of power system stabilizers for enhancement of overall system stability, *IEEE Transactions on Power Systems*, 4, 614–626, May 1989.

Kundur, P., Lee, D.C., and Zein El-Din, H.M., Power system stabilizers for thermal units: Analytical techniques and on-site validation, *IEEE Transactions on Power Apparatus and Systems*, 100, 81–85, January 1981.

Kundur, P., Paserba, J., Ajjarapu, V., Andersson, G., Bose, A., Canizares, C., Hatziargyriou, N. et al. (IEEE/CIGRE Joint Task Force on Stability Terms and Definitions), Definition and classification of power system stability, *IEEE Transactions on Power Systems*, 19, 1387–1401, August 2004.

Kundur, P., Rogers, G., Wong, D., Wang, L., and Lauby, M., A comprehensive computer program package for small signal stability analysis of power systems, *IEEE Transactions on Power Systems*, PWRS-5(4), 1076–1083, November 1990.

Larsen, E.V. and Chow, J.H., SVC control design concepts for system dynamic performance, Application of static var systems for system dynamic performance, IEEE Special Publication No. 87TH1087-5-PWR on Application of Static Var Systems for System Dynamic Performance, San Francisco, CA, pp. 36–53, 1987.

Larsen, E., Sanchez-Gasca, J., and Chow, J., Concepts for design of FACTS controllers to damp power swings, *IEEE Transactions on Power Systems*, 10(2), 948–956, May 1995.

Larsen, E.V. and Swann, D.A., Applying power system stabilizers, Parts I, II, and III, *IEEE Transactions on Power Apparatus and Systems*, 100, 3017–3046, 1981.

Lee, D.C., Beaulieu, R.E., and Rogers, G.J., Effects of governor characteristics on turbo-generator shaft torsionals, *IEEE Transactions on Power Apparatus and Systems*, 104, 1255–1261, June 1985.

Lee, D.C. and Kundur, P., Advanced excitation controls for power system stability enhancement, CIGRE Paper 38-01, Paris, France, 1986.

Lee, K.C. and Poon, K.P., Analysis of power system dynamic oscillations with beat phenomenon by Fourier transformation, *IEEE Transactions on Power Systems*, 5(1), 148–153, February 1990.

Leirbukt, A.B., Chow, J.H., Sanchez-Gasca, J.J., and Larsen, E.V., Damping control design based on time-domain identified models, *IEEE Transactions on Power Systems*, 14(1), 172–178, February 1999.

Levine, W.S., Ed., *The Control Handbook*, CRC Press, Boca Raton, FL, 1995.

Martins, N. and Lima, L., Determination of suitable locations for power system stabilizers and static var compensators for damping electromechanical oscillations in large scale power systems, *IEEE Transactions on Power Systems*, 5(4), 1455–1469, November 1990.

Martins, N., Lima, L.T.G., and Pinto, H.J.C.P., Computing dominant poles of power system transfer functions, *IEEE Transactions on Power Systems*, 11(1), 162–170, February 1996.

Martins, N., Pinto, H.J.C.P., and Lima, L.T.G., Efficient methods for finding transfer function zeros of power systems, *IEEE Transactions on Power Systems*, 7(3), 1350–1361, August 1992.

Martins, N. and Quintao, P.E.M., Computing dominant poles of power system multivariable transfer functions, *IEEE Transactions on Power Systems*, 18(1), 152–159, February 2003.

Messina, A.R., *Inter-Area Oscillations in Power Systems*, Springer-Verlag, New York, 2009.

Midwest ISO, Business practice manual for transmission planning, BPM-020-r1, July 8, 2009.

NERC, Reliability Standards for Transmission Planning, TPL-001 to TPL-006, available from www.nerc.com (accessed on May 18, 2009).

Pal, B. and Chaudhuri, B., *Robust Control in Power Systems*, Springer Science and Business Media Inc., New York, 2005.

Paserba, J.J., Larsen, E.V., Grund, C.E., and Murdoch, A., Mitigation of inter-area oscillations by control, IEEE PES Special Publication 95-TP-101 on Interarea Oscillations in Power Systems, 1995.

Piwko, R.J. and Larsen, E.V., HVDC System control for damping subsynchronous oscillations, *IEEE Transactions on Power Apparatus and Systems*, 101(7), 2203–2211, 1982.

Piwko, R., Othman, H., Alvarez, O., and Wu, C., Eigenvalue and frequency domain analysis of the inter-mountain power project and the WSCC network, *IEEE Transactions on Power Systems*, 6, 238–244, February 1991.

PJM Manual 14B, PJM region transmission planning process, available from www.pjm.com, November 18, 2010.

Pourbeik, P. and Gibbard, M., Damping and synchronizing torques induced on generators by FACTS stabilizers in multimachine power systems, *IEEE Transactions on Power Systems*, 11(4), 1920–1925, November 1996.

Rogers, G., *Power System Oscillations*, Kluwer Academic Publishers, Norwell, MA, 2000.

Sanchez-Gasca, J.J., Computation of turbine-generator subsynchronous torsional modes from measured data using the eigensystem realization algorithm, in *Proceedings of the IEEE PES Winter Meeting*, Columbus, OH, January 2001.

Sanchez-Gasca, J., Vittal, V., Gibbard, M., Messina, A., Vowles, D., Liu, S., and Annakkage, U., Inclusion of higher-order terms for small-signal (modal) analysis: Committee report—Task force on assessing the need to include higher-order terms for small-signal (modal) analysis, *IEEE Transactions on Power Systems*, 20(4), 1886–1904, November 2005.

Semlyen, A. and Wang, L., Sequential computation of the complete eigensystem for the study zone in small signal stability analysis of large power systems, *IEEE Transactions on Power Systems*, PWRS-3(2), 715–725, May 1988.

Shah, K.S., Berube, G.R., and Beaulieu, R.E., Testing and modelling of the Union Electric generator excitation systems, in *Missouri Valley Electric Association Engineering Conference*, Kansas City, MO, April 5–7, 1995.

Shanechi, H., Pariz, N., and Vaahedi, E., General nonlinear representation of large-scale power systems, *IEEE Transactions on Power Systems*, 18(3), 1103–1109, August 2003.

Trudnowski, D.J., Pierre, J.W., Zhou, N., Hauer, J.F., and Parashar, M., Performance of three mode-meter block-processing algorithms for automated dynamic stability assessment, *IEEE Transactions on Power Systems*, 23(2), 680–690, May 2008.

Trudnowski, D.J., Smith, J.R., Short, T.A., and Pierre, D.A., An application of Prony methods in PSS design for multimachine systems, *IEEE Transactions on Power Systems*, 6(1), 118–126, February 1991.

Vanfretti, L., Garcia-Valle, R., Uhlen, K., Johansson, E., Trudnowski, D., Pierre, J.W., Chow, J.H., Samuelsson, O., Østergaard, J., and Martin, K.E., Estimation of Eastern Denmark's electromechanical modes from ambient phasor measurement data, in *Proceedings of the IEEE PES General Meeting*, Minneapolis, MN, July 2010.

Vittal, V., Bhatia, N., and Fouad, A., Analysis of the inter-area mode phenomenon in power systems following large disturbances, *IEEE Transactions on Power Systems*, 6(4), 1515–1521, November 1991.

Wang, L., New England oscillation study, Final Report by Powertech Labs Inc. for ISO New England, May 2005.

Wang, L., Howell, F., Kundur, P., Chung, C.Y., and Xu, W., A tool for small-signal security assessment of power systems, in *Proceedings of the IEEE PES PICA 2001*, Sydney, Australia, May 2001.

Wang, L., Klein, M., Yirga, S., and Kundur, P., Dynamic reduction of large power systems for stability studies, *IEEE Transactions on Power Systems*, PWRS-12(2), 889–895, May 1997.

Wang, L. and Semlyen, A., Application of sparse eigenvalue techniques to the small signal stability analysis of large power systems, *IEEE Transactions on Power Systems*, PWRS-5(2), 635–642, May 1990.

Watson, W. and Coultes, M.E., Static exciter stabilizing signals on large generators—Mechanical problems, *IEEE Transactions on Power Apparatus and Systems*, 92, 205–212, January/February 1973.

WECC (2010), WECC generating unit model validation policy, available from www.wecc.biz (accessed on May 14, 2010).

WECC (2011), WECC standard VAR-501-WECC-1–Power system stabilizer, available from www.wecc.biz (accessed on July 1, 2011).

Zhou, N, Trudnowski, D.J., Pierre, J.W., and Mittelstadt, W.A., Electromechanical mode online estimation using regularized robust RLS methods, *IEEE Transactions on Power Systems*, 23(4), 1670–1680, November 2008.

11
Voltage Stability

Yakout Mansour
California Independent System Operator

Claudio Cañizares
University of Waterloo

11.1 Basic Concepts 11-1
  Generator-Load Example • Load Modeling • Effect of Load Dynamics on Voltage Stability
11.2 Analytical Framework 11-9
  Power Flow Analysis • Continuation Methods • Optimization or Direct Methods • Timescale Decomposition
11.3 Mitigation of Voltage Stability Problems 11-12
References 11-13

Voltage stability refers to "the ability of a power system to maintain steady voltages at all buses in the system after being subjected to a disturbance from a given initial operating condition" (IEEE-CIGRE, 2004). If voltage stability exists, the voltage and power of the system will be controllable at all times. In general, the inability of the system to supply the required demand leads to voltage instability (voltage collapse).

In general, power system dynamics are always associated with variations in voltage to various extents depending on the dynamic phenomena under consideration. For example, the classical angular stability phenomena can result in significant voltage drop or even collapse at points of the grid closer to the swing center for brief instants within the dynamic window. Other cases could be observed due to malfunctioning of voltage control equipment. It should be noted here that these are not the types of voltage dynamics being explained in this chapter and are explained in different chapters in this and other books in the literature.

11.1 Basic Concepts

Voltage instability of radial distribution systems has been well recognized and understood for decades (Venikov, 1970 and 1980) and was often referred to as load instability. Large interconnected power networks did not face the phenomenon until late 1970s and early 1980s.

Most of the early developments of the major high voltage (HV) and extra HV (EHV) networks and interties faced the classical machine angle stability problem. Innovations in both analytical techniques and stabilizing measures made it possible to maximize the power transfer capabilities of the transmission systems. The result was increasing transfers of power over long distances of transmission. As the power transfer increased, even when angle stability was not a limiting factor, many utilities have been facing a shortage of voltage support. The result ranged from postcontingency operation under reduced voltage profile to total voltage collapse. Major outages attributed to this problem have been experienced in the northeastern part of the United States, France, Sweden, Belgium, Japan, along with other localized cases of voltage collapse (Mansour, 1990; US–Canada, 2004). Accordingly, voltage stability has imposed itself as a governing factor in both planning and operating criteria of a number of utilities. Consequently, sound analytical procedures and quantitative measures of proximity to voltage instability have been developed for the past two decades.

11.1.1 Generator-Load Example

The simple generator-load model depicted in Figure 11.1 can be used to readily explain the basic concepts behind voltage stability phenomena. The power flow model of this system can be represented by the following equations:

$$0 = P_L - \frac{V_1 V_2}{X_L} \sin \delta$$

$$0 = k P_L - \frac{V_2^2}{X_L} - \frac{V_1 V_2}{X_L} \cos \delta$$

$$0 = Q_G - \frac{V_1^2}{X_L} + \frac{V_1 V_2}{X_L} \cos \delta$$

where
$\delta = \delta_2 - \delta_1$
$P_G = P_L$ (no losses)
$Q_L = k P_L$ (constant power factor load)

All solutions to these power flow equations, as the system load level P_L is increased, can be plotted to yield PV curves (bus voltage vs. active power load levels) or QV curves (bus voltage vs. reactive power load levels) for this system. For example, Figure 11.2 depicts the PV curves at the load bus obtained from these equations for $k = 0.25$ and $V_1 = 1$ pu when generator limits are neglected, and for two values of X_L to simulate a transmission system outage or contingency by increasing its value. Figure 11.3 depicts the power flow solution when reactive power limits are considered, for $Q_{Gmax} = 0.5$ and $Q_{Gmin} = -0.5$. Notice that these PV curves can be readily transformed into QV curves by properly scaling the horizontal axis.

In Figure 11.2, the maximum loading corresponds to a singularity of the Jacobian of the power flow equations, and may be associated with a *saddle-node bifurcation* of a dynamic model of this system (Cañizares, 2002). (A saddle-node bifurcation is defined in a power flow model of the power grid, which is considered a nonlinear system, as a point at which two power flow solutions merge and disappear as typically the load, which is a system parameter, is increased; the Jacobian of the power flow equations becomes singular at this "bifurcation" or "merging" point.) Observe that if the system were operating at a load level of $P_L = 0.7$ pu, the contingency would basically result in the disappearance of an operating point (power flow solution), thus leading to a voltage collapse.

Similarly, if there is an attempt to increase P_L (Q_L) beyond its maximum values in Figure 11.3, the result is a voltage collapse of the system, which is also observed if the contingency depicted in this figure occurs at the operating point associated with $P_L = 0.6$ pu. The maximum loading points correspond in this case to a maximum limit on the generator reactive power Q_G, with the Jacobian of the power flow

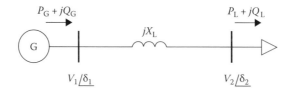

FIGURE 11.1 Generator-load example.

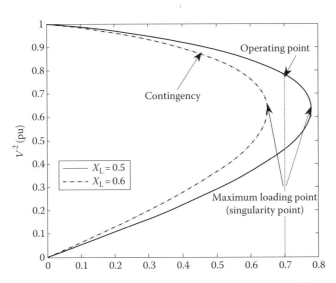

FIGURE 11.2 *PV* curve for generator-load example without generator reactive power limits.

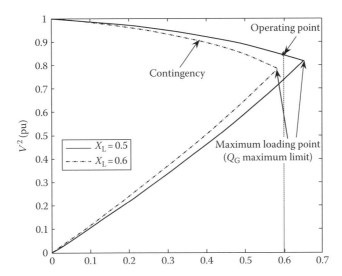

FIGURE 11.3 *PV* curve for generator-load example considering generator reactive power limits.

being nonsingular. This point may be associated with a *limit-induced bifurcation* of a dynamic model of this system (Cañizares, 2002). (A limit-induced bifurcation is defined in a power flow model of the nonlinear power grid as a point at which two power flow solutions merge as the load is increased; the Jacobian of the power flow equations at this point is not singular and corresponds to a power flow solution, where a system controller reaches a control limit, such as a voltage regulating generator reaching a maximum reactive power limit.)

For this simple generator-load example, different *PV* and *QV* curves can be computed depending on the system parameters chosen to plot these curves. For example, the family of curves shown in Figure 11.4 is produced by maintaining the sending end voltage constant, while the load at the receiving end is varied at a constant power factor and the receiving end voltage is calculated. Each curve is calculated at a specific power factor and shows the maximum power that can be transferred at this particular power factor, which is also referred to as the maximum system loadability. Note that the limit can be increased by providing more reactive support at the receiving end [limit (2) vs. limit (1)], which is

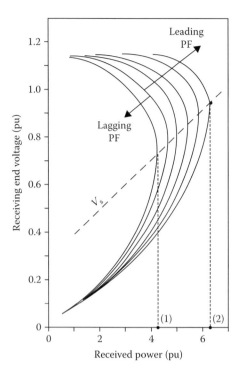

FIGURE 11.4 $P_L V_2$ characteristics.

effectively pushing the power factor of the load in the leading direction. It should also be noted that the points on the curves below the limit line V_s characterize unstable behavior of the system, where a drop in demand is associated with a drop in the receiving end voltage, leading to eventual collapse. Proximity to voltage instability is usually measured by the distance (in pu power) between the operating point on the PV curve and the limit of the same curve; this is usually referred to as the system loadability margin.

Another family of curves similar to that of Figure 11.5 can be produced by varying the reactive power demand (or injection) at the receiving end while maintaining the real power and the sending end voltage constant. The relation between the receiving end voltage and the reactive power injection at the receiving end is plotted to produce the so-called QV curves of Figure 11.5. The bottom of any given curve characterizes the voltage stability limit. Note that the behavior of the system on the right side of the limit is such that an increase in reactive power injection at the receiving end results in a receiving end voltage rise, while the opposite is true on the left side because of the substantial increase in current at the lower voltage, which, in turn, increases reactive losses in the network substantially. The proximity to voltage instability or voltage stability margin is measured as the difference between the reactive power injection corresponding to the operating point and the bottom of the curve. As the active power transfer increases (upward in Figure 11.5), the reactive power margin decreases, as does the receiving end voltage.

11.1.2 Load Modeling

Voltage instability is typically associated with relatively slow variations in network and load characteristics. Network response in this case is highly influenced by the slow-acting control devices such as transformer on-load tap changers (LTCs), automatic generation control, generator field current limiters, generator overload reactive capability, undervoltage load shedding relays, and switchable reactive devices. Load characteristics with respect to changing voltages play also a major role in voltage stability. The characteristics of such devices, as to how they influence the network response to voltage variations, are generally understood and well covered in the literature.

Voltage Stability

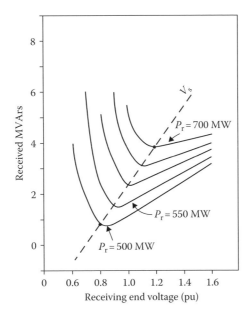

FIGURE 11.5 $Q_L V_2$ characteristics.

While it might be possible to identify the voltage response characteristics of a large variety of individual equipment of which a power network load is comprised, it is not practical or realistic to model network load by individual equipment models. Thus, the aggregate load model approach is much more realistic. However, load aggregation requires making certain assumptions, which might lead to significant differences between the observed and simulated system behavior. It is for these reasons that load modeling in voltage stability studies, as in any other kind of stability study, is a rather important and difficult issue.

Field test results as reported by Hill (1993) and Xu et al. (1997) indicate that typical response of an aggregate load to step-voltage changes is of the form shown in Figure 11.6. The response is a reflection of the collective effects of all downstream components ranging from LTCs to individual household loads. The time span for a load to recover to steady state is normally in the range of several seconds to minutes,

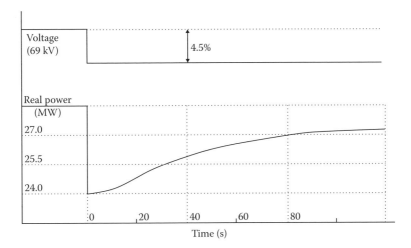

FIGURE 11.6 Aggregate load response to a step-voltage change.

depending on the load composition. Responses for real and reactive power are qualitatively similar. It can be seen that a sudden voltage change causes an instantaneous power demand change. This change defines the transient characteristics of the load and was used to derive static load models for angular stability studies. When the load response reaches steady state, the steady-state power demand is a function of the steady-state voltage. This function defines the steady-state load characteristics known as voltage-dependent load models in power flow studies.

The typical load–voltage response characteristics can be modeled by a generic dynamic load model proposed in Figure 11.7. In this model (Xu and Mansour, 1994), x is the state variable. $P_t(V)$ and $P_s(V)$ are the transient and steady-state load characteristics, respectively, and can be expressed as

$$P_t = V^a \quad \text{or} \quad P_t = C_2 V^2 + C_1 V + C_o$$
$$P_s = P_o V^a \quad \text{or} \quad P_s = P_o(d_2 V^2 + d_1 V + d_o)$$

where V is the pu magnitude of the voltage imposed on the load. It can be seen that, at steady state, the state variable x of the model is constant. The input to the integration block, $E = P_s - P$, must be zero and, as a result, the model output is determined by the steady-state characteristics $P = P_s$. For any sudden voltage change, x maintains its predisturbance value initially, because the integration block cannot change its output instantaneously. The transient output is then determined by the transient characteristics $P - xP_t$. The mismatch between the model output and the steady-state load demand is the error signal e. This signal is fed back to the integration block that gradually changes the state variable x. This process continues until a new steady state ($e = 0$) is reached. Analytical expressions of the load model, including real (P) and reactive (Q) power dynamics, are

$$T_p \frac{dx}{dt} = P_s(V) - P, \quad P = xP_t(V)$$

$$T_q \frac{dy}{dt} = Q_s(V) - Q, \quad Q = yQ_t(V)$$

$$P_t(V) = V^a, \quad P_s(V) = P_o V^a; \quad Q_t(V) = V^\beta, \quad Q_s(V) = Q_o V^\beta$$

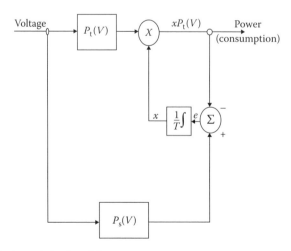

FIGURE 11.7 A generic dynamic load model.

11.1.3 Effect of Load Dynamics on Voltage Stability

As illustrated with the help of the aforementioned generator-load example, voltage stability may occur when a power system experiences a large disturbance, such as a transmission line outage. It may also occur if there is no major disturbance, but the system's operating point shifts slowly toward stability limits. Therefore, the voltage stability problem, as other stability problems, must be investigated from two perspectives, the large-disturbance stability and the small-signal stability.

Large-disturbance voltage stability is event oriented and addresses problems such as postcontingency margin requirement and response of reactive power support. Small-signal voltage stability investigates the stability of an operating point. It can provide such information as to the areas vulnerable to voltage collapse. In this section, the effect of load dynamics on large- and small-disturbance voltage stability is analyzed by examining the interaction of a load center with its supply network, and key parameters influencing voltage stability are identified. Since the real power dynamic behavior of an aggregate load is similar to its reactive power counterpart, the analysis is limited to reactive power only.

11.1.3.1 Large-Disturbance Voltage Stability

To facilitate the explanation, assume that the voltage dynamics in the supply network are fast as compared to the aggregate dynamics of the load center. The network can then be modeled by three quasi-steady-state VQ characteristics (QV curves), predisturbance, postdisturbance, and postdisturbance-with-reactive-support, as shown in Figure 11.8. The load center is represented by a generic dynamic load. This load-network system initially operates at the intersection of the steady-state load characteristics and the predisturbance network VQ curve, point a.

The network experiences an outage that reduces its reactive power supply capability to the postdisturbance VQ curve. The aggregate load responds (see Section 11.1.2) instantaneously with its transient characteristics ($\beta = 2$, constant impedance in this example) and the system operating point jumps to point b. Since, at point b, the network reactive power supply is less than load demand for the given voltage,

$$T_q \frac{dy}{dt} = Q_s(V) - Q(V) > 0$$

the load dynamics will try to draw more reactive power by increasing the state variable y. This is equivalent to increasing the load admittance if $\beta = 2$, or the load current if $\beta = 1$. It drives the operating point to a lower voltage. If the load demand and the network supply imbalance persist, the system will

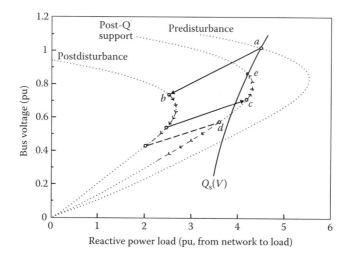

FIGURE 11.8 Voltage dynamics as viewed from VQ plane.

continuously operate on the intersection of the postdisturbance VQ curve and the drifting transient load curve with a monotonically decreasing voltage, leading to voltage collapse.

If reactive power support is initiated shortly after the outage, the network is switched to the third VQ curve. The load responds with its transient characteristics and a new operating point is formed. Depending on the switch time of reactive power support, the new operating point can be either c, for fast response, or d, for slow response. At point c, power supply is greater than load demand ($Q_s(V) - Q(V) < 0$); the load then draws less power by decreasing its state variable, and as a result, the operating voltage is increased. This dynamic process continues until the power imbalance is reduced to zero, namely a new steady-state operating point is reached (point e). On the other hand, for the case with slow response reactive support, the load demand is always greater than the network supply. A monotonic voltage collapse is the ultimate end.

A numerical solution technique can be used to simulate the aforementioned process. The equations for the simulation are

$$T_q \frac{dy}{dt} = Q_s(V) = Q(t); \quad Q(t) = yQ_t(V)$$

$$Q(t) = \text{Network}(V_s t)$$

where the function Network($V_s t$) consists of three polynomials, each representing one VQ curve. Figure 11.8 shows the simulation results in VQ coordinates. The load voltage as a function of time is plotted in Figure 11.9. The results demonstrate the importance of load dynamics for explaining the voltage stability problem.

A classic example of this phenomenon can be illustrated by a typical scenario associated with AC networks that involve dynamic of fast-acting devices such as induction motors and electronic controlled loads. In this case, a large disturbance happens at a time when the load has a significant volume of air-conditioning loads. When the disturbance occurs, the air-conditioning motors decelerate and, in the process, draw high current which in turn drives the voltage further down. Following fault clearing,

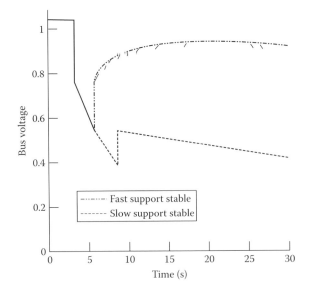

FIGURE 11.9 Simulation of voltage collapse.

motors draw high currents while attempting to restore their operating condition and may stall and aggravate the situation further, especially in cases where a weaker grid is established, by clearing the disturbance through tripping parts of the grid. The result would be a possible voltage collapse and massive loss of load. Several cases of voltage collapse around the world documented similar behavior to this example.

11.1.3.2 Small-Signal Voltage Stability

The voltage characteristics of a power system can be analyzed around an operating point by linearizing the power flow equations around the operating point and analyzing the resulting sensitivity matrices. Breakthroughs in computational algorithms have made these techniques efficient and helpful in analyzing large-scale systems, taking into account virtually all the important elements affecting the phenomenon. In particular, singular value decomposition and modal techniques should be of particular interest to the reader and are thoroughly described by Mansour (1993), Lof et al. (1992, 1993), Gao et al. (1992), and Cañizares (2002).

11.2 Analytical Framework

The slow nature of the network and load response associated with the phenomenon makes it possible to analyze the problem in two frameworks: (1) long-term dynamic framework, in which all slow-acting devices and aggregate bus loads are represented by their dynamic models (the analysis in this case is done through a dynamic quasi-dynamic simulation of the system response to contingencies or load variations) or (2) steady-state framework (e.g., power flow) to determine if the system can reach a stable operating point following a particular contingency. This operating point could be a final state or a midpoint following a step of a discrete control action (e.g., transformer tap change).

The proximity of a given system to voltage instability and the control actions that may be taken to avoid voltage collapse are typically assessed by various indices and sensitivities. The most widely used are (Cañizares, 2002)

- Loadability margins, that is, the "distance" in MW or MVA to a point of voltage collapse, and sensitivities of these margins with respect to a variety of parameters, such as active/reactive power load variations or reactive power levels at different sources
- Singular values of the system Jacobian or other matrices obtained from these Jacobians, and their sensitivities with respect to various system parameters
- Bus voltage profiles and their sensitivity to variations in active and reactive power of the load and generators, or other reactive power sources
- Availability of reactive power supplied by generators, synchronous condensers, and static var compensators and its sensitivity to variations in load bus active and/or reactive power

These indices and sensitivities, as well as their associated control actions, can be determined using a variety of the computational methods described in the following.

11.2.1 Power Flow Analysis

Partial *PV* and *QV* curves can be readily calculated using power flow programs. In this case, the demand of load center buses is increased in steps at a constant power factor while the generators' terminal voltages are held at their nominal value, as long as their reactive power outputs are within limits; if a generator's reactive power limit is reached, the corresponding generator bus is treated as another load bus. The *PV* relation can then be plotted by recording the MW demand level against a "central" load bus voltage at the load center. It should be noted that power flow solution algorithms diverge very close to or past the maximum loading point, and do not produce the unstable portion of the *PV* relation. The *QV* relation, however, can be produced in full by assuming a fictitious synchronous condenser at a central load bus in the load center (this is a "parameterization" technique also used in the continuation methods described in the following). The QV relation is then plotted for this particular bus as a representative of the load center by varying the voltage of the bus (now converted to a voltage control bus by the addition of the synchronous condenser) and recording its value against the reactive power injection of the synchronous condenser. If the limits on the reactive power capability of the synchronous condenser are made very high, the power flow solution algorithm will always converge at either side of the QV relation.

11.2.2 Continuation Methods

A popular and robust technique to obtain full PV and/or QV curves is the continuation method (Cañizares, 2002). This methodology basically consists of two power-flow-based steps: the predictor and the corrector, as illustrated in Figure 11.10. In the predictor step, an estimate of the power flow solution for a load P increase (point 2 in Figure 11.10) is determined based on the starting solution (point 1) and an estimate of the changes in the power flow variables (e.g., bus voltages and angles). This estimate may be computed using a linearization of the power flow equations, that is, determining the "tangent vector" to the manifold of power flow solutions. Thus, in the example depicted in Figure 11.10,

$$\Delta x = x_2 - x_1$$

$$= k J_{PF1}^{-1} \left. \frac{\partial f_{PF}}{\partial P} \right|_1 \Delta P$$

where
- J_{PF1} is the Jacobian of the power flow equations $f_{PF}(x) = 0$, evaluated at the operating point 1
- x is the vector of power flow variables (load bus voltages are part of x)
- $\partial f_{PF}/\partial P|_1$ is the partial derivative of the power flow equations with respect to the changing parameter P evaluated at the operating point 1
- k is a constant used to control the length of the step (typically $k = 1$), which is usually reduced by halves to guarantee a solution of the corrector step near the maximum loading point, and thus avoiding the need for a parameterization step

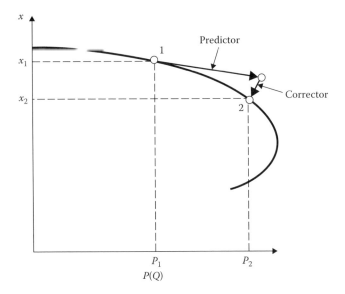

FIGURE 11.10 Continuation power flow.

Observe that the predictor step basically consists in determining the sensitivities of the power flow variables x with respect to changes in the loading level P.

The corrector step can be as simple as solving the power flow equations for $P = P_2$ to obtain the operating point 2 in Figure 11.10, using the estimated values of x yielded by the predictor as initial guesses. Other more sophisticated and computationally robust techniques, such as a "perpendicular intersection" method, may be used as well.

11.2.3 Optimization or Direct Methods

The maximum loading point can be directly computed using optimization-based methodologies (Rosehart et al., 2003), which yield the maximum loading margin to a voltage collapse point and a variety of sensitivities of the power flow variables with respect to any system parameter, including the loading levels (Milano et al., 2006). These methods basically solve the optimal power flow (OPF) problem:

$$\begin{aligned}
\text{Max.} \quad & P \\
\text{s.t.} \quad & f_{\text{PF}}(x, P) = 0 \rightarrow \text{power flow equations} \\
& x_{\min} \leq x \leq x_{\max} \rightarrow \text{limits}
\end{aligned}$$

where P represents the system loading level; the power flow equations f_{PF} and variable x should include the reactive power flow equations of the generators, so that the generator's reactive power limits can be considered in the computation. The Lagrange multipliers associated with the constraints are basically sensitivities that can be used for further analyses or control purposes. Well-known optimization techniques, such as interior point methods, can be used to obtain loadability margins and sensitivities by solving this particular OPF problem for real-sized systems.

Approaching voltage stability analysis from the optimization point of view has the advantage that certain variables, such as generator bus voltages or active power outputs, can be treated as optimization parameters. This allows treating the problem not only as a voltage stability margin computation, but also as a means to obtain an "optimal" dispatch to maximize the voltage stability margins.

11.2.4 Timescale Decomposition

The PV and QV relations produced results corresponding to an end state of the system where all tap changers and control actions have taken place in time and the load characteristics were restored to a constant power characteristic. It is always recommended and often common to analyze the system behavior in its transition following a disturbance to the end state. Apart from the full long-term time simulation, the system performance can be analyzed in a quasi-dynamic manner by breaking the system response down into several time windows, each of which is characterized by the states of the various controllers and the load recovery (Mansour, 1993). Each time window can be analyzed using power flow programs modified to reflect the various controllers' states and load characteristics. Those time windows (Figure 11.11) are primarily characterized by

1. Voltage excursion in the first second after a contingency as motors slow, generator voltage regulators respond, etc.
2. The period 1–20 s when the system is quiescent until excitation limiting occurs
3. The period 20–60 s when generator over excitation protection has operated
4. The period 1–10 min after the disturbance when LTCs restore customer load and further increase reactive demand on generators
5. The period beyond 10 min when automatic generation control (AGC), phase angle regulators, operators, etc., come into play

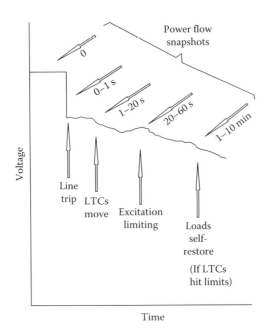

FIGURE 11.11 Breaking the system response down into time periods.

The sequential power flow analysis aforementioned can be extended further by properly representing in the simulation some of the slow system dynamics, such as the LTCs (Van Cutsem and Vournas, 1996).

11.3 Mitigation of Voltage Stability Problems

The following methods can be used to mitigate voltage stability problems:

Must-run generation: Operate uneconomic generators to change power flows or provide voltage support during emergencies or when new lines or transformers are delayed.

Series capacitors: Use series capacitors to effectively shorten long lines, thus, decreasing the net reactive loss. In addition, the line can deliver more reactive power from a strong system at one end to one experiencing a reactive shortage at the other end.

Shunt capacitors: Though the heavy use of shunt capacitors can be part of the voltage stability problem, sometimes additional capacitors can also solve the problem by freeing "spinning reactive reserve" in generators. In general, most of the required reactive power should be supplied locally, with generators supplying primarily active power.

Static compensators (SVCs and STATCOMs): Static compensators, the power electronics-based counterpart to the synchronous condenser, are effective in controlling voltage and preventing voltage collapse, but have very definite limitations that must be recognized. Voltage collapse is likely in systems heavily dependent on static compensators when a disturbance exceeding planning criteria takes these compensators to their ceiling.

Operate at higher voltages: Operating at higher voltage may not increase reactive reserves, but does decrease reactive demand. As such, it can help keep generators away from reactive power limits, and thus, help operators maintain control of voltage. The comparison of receiving end QV curves for two sending end voltages shows the value of higher voltages.

Secondary voltage regulation: Automatic voltage regulation of certain load buses, usually referred to as pilot buses, that coordinately controls the total reactive power capability of the reactive power sources in pilot buses' areas, has proven to be an effective way to improve voltage stability (Cañizares et al., 2005). These are basically hierarchical controls that directly vary the voltage set points of generators and static compensators on a pilot bus' control area, so that all controllable reactive power sources are coordinated to adequately manage the reactive power capability in the area, keeping some of these sources from reaching their limits at relatively low load levels.

Undervoltage load shedding: A small load reduction, even 5%–10%, can make the difference between collapse and survival. Manual load shedding is used today for this purpose (some utilities use distribution voltage reduction via system control and data acquisition [SCADA]), though it may be too slow to be effective in the case of a severe reactive shortage. Inverse time–undervoltage relays are not widely used, but can be very effective. In a radial load situation, load shedding should be based on primary side voltage. In a steady-state stability problem, the load shed in the receiving system will be most effective, even though voltages may be lowest near the electrical center (shedding load in the vicinity of the lowest voltage may be more easily accomplished, and still be helpful).

Lower power factor generators: Where new generation is close enough to reactive-short areas or areas that may occasionally demand large reactive reserves, a 0.80 or 0.85 power factor generator may sometimes be appropriate. However, shunt capacitors with a higher power factor generator having reactive overload capability may be more flexible and economic.

Use generator reactive overload capability: Generators should be used as effectively as possible. Overload capability of generators and exciters may be used to delay voltage collapse until operators can change dispatch or curtail load when reactive overloads are modest. To be most useful, reactive overload capability must be defined in advance, operators trained in its use, and protective devices set so as not to prevent its use.

References

Cañizares, C.A., ed., Voltage stability assessment: Concepts, practices and tools, IEEE-PES Power Systems Stability Subcommittee Special Publication, IEEE Catalog Number SP101PSS, August 2002.

Cañizares, C.A., Cavallo, C., Pozzi, M., and Corsi, S., Comparing secondary voltage regulation and shunt compensation for improving voltage stability and transfer capability in the Italian power system, *Electric Power Systems Research*, 73, 67–76, 2005.

Gao, B., Morison, G.K., and Kundur, P., Voltage stability evaluation using modal analysis, *IEEE Transactions on Power Systems*, 7, 1529–1542, 1992.

Hill, D.J., Nonlinear dynamic load models with recovery for voltage stability studies, *IEEE Transactions on Power Systems*, 8, 166–176, 1993.

IEEE-CIGRE Joint Task Force on Stability Terms and Definitions (Kundur, P., Paserba, J., Ajjarapu, V., Andersson, G., Bose, A., Cañizares, C., Hatziargyriou, N., Hill, D., Stankovic, A., Taylor, C., Van Cutsem, T., and Vittal, V.), Definition and classification of power system stability, *IEEE Transactions on Power Systems*, 19, 1387–1401, 2004.

Lof, P.-A., Andersson, G., and Hill, D.J., Voltage stability indices for stressed power systems, *IEEE Transactions on Power Systems*, 8, 326–335, 1993.

Lof, P.-A., Smed, T., Andersson, G., and Hill, D.J., Fast calculation of a voltage stability index, *IEEE Transactions on Power Systems*, 7, 54–64, 1992.

Mansour, Y., ed., Voltage stability of power systems: Concepts, analytical tools, and industry experience, IEEE PES Special Publication #90TH0358-2-PWR, 1990.

Mansour, Y., ed., Suggested techniques for voltage stability analysis, IEEE Special Publication #93TH0620-5-PWR, IEEE Power & Energy Society, New York, 1993.

Milano, F., Cañizares, C.A., and Conejo, A.J., Sensitivity-based security-constrained OPF market clearing model, *IEEE Transactions on Power Systems*, 20, 2051–2060, 2006.

Rosehart, W., Cañizares, C.A., and Quintana, V., Multi-objective optimal power flows to evaluate voltage security costs, *IEEE Transactions on Power Systems*, 18, 578–587, 2003.

US–Canada Power System Outage Task Force, Final Report on the August 14, 2003 Blackout in the United States and Canada: Causes and Recommendations, April 2004.

Van Cutsem, T. and Vournas, C.D., Voltage stability analysis in transient and midterm timescales, *IEEE Transactions on Power Systems*, 11, 146–154, 1996.

Venikov, V., *Transient Processes in Electrical Power Systems*, Mir Publishers, Moscow, Russia, 1970 and 1980.

Xu, W. and Mansour, Y., Voltage stability analysis using generic dynamic load models, *IEEE Transactions on Power Systems*, 9, 479–493, 1994.

Xu, W., Vaahedi, E., Mansour, Y., and Tamby, J., Voltage stability load parameter determination from field tests on B. C. Hydro's system, *IEEE Transactions on Power Systems*, 12, 1290–1297, 1997.

12
Direct Stability Methods

12.1	Review of Literature on Direct Methods.......................	12-2
12.2	The Power System Model..	12-4
	Review of Stability Theory	
12.3	The Transient Energy Function.....................................	12-7
12.4	Transient Stability Assessment......................................	12-9
12.5	Determination of the Controlling UEP	12-9
12.6	The Boundary Controlling UEP Method	12-10
12.7	Applications of the TEF Method and Modeling Enhancements..	12-11
	References..	12-11

Vijay Vittal
Arizona State University

Direct methods of stability analysis determine the transient stability (as defined in Chapter 7 and described in Chapter 8) of power systems without explicitly obtaining the solutions of the differential equations governing the dynamic behavior of the system. The basis for the method is Lyapunov's second method, also known as Lyapunov's direct method, to determine stability of systems governed by differential equations. The fundamental work of A.M. Lyapunov (1857–1918) on stability of motion was published in Russian in 1893, and was translated into French in 1907 (Lyapunov, 1907). This work received little attention and for a long time was forgotten. In the 1930s, Soviet mathematicians revived these investigations and showed that Lyapunov's method was applicable to several problems in physics and engineering. This revival of the subject matter has spawned several contributions that have led to the further development of the theory and application of the method to physical systems.

The following example motivates the direct methods and also provides a comparison with the conventional technique of simulating the differential equations governing the dynamics of the system. Figure 12.1 shows an illustration of the basic idea behind the use of the direct methods. A vehicle, initially at the bottom of a hill, is given a sudden push up the hill. Depending on the magnitude of the push, the vehicle will either go over the hill and tumble, in which case it is unstable, or the vehicle will climb only part of the way up the hill and return to a rest position (assuming that the vehicle's motion will be damped), i.e., it will be stable. In order to determine the outcome of disturbing the vehicle's equilibrium for a given set of conditions (mass of the vehicle, magnitude of the push, height of the hill, etc.), two different methods can be used:

1. Knowing the initial conditions, obtain a time solution of the equations describing the dynamics of the vehicle and track the position of the vehicle to determine how far up the hill the vehicle will travel. This approach is analogous to the traditional time domain approach of determining stability in dynamic systems.
2. The approach based on Lyapunov's direct method would consist of characterizing the motion of the dynamic system using a suitable Lyapunov function. The Lyapunov function should satisfy certain sign definiteness properties. These properties will be addressed later in this subsection. A natural

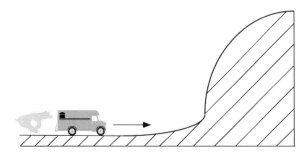

FIGURE 12.1 Illustration of idea behind direct methods.

choice for the Lyapunov function is the system energy. One would then compute the energy injected into the vehicle as a result of the sudden push, and compare it with the energy needed to climb the hill. In this method, there is no need to track the position of the vehicle as it moves up the hill.

These methods are simple to use if the calculations involve only one vehicle and one hill. The complexity increases if there are several vehicles involved as it becomes necessary to determine (a) which vehicles will be pushed the hardest, (b) how much of the energy is imparted to each vehicle, (c) which direction will they move, and (d) how high a hill must they climb before they will go over the top.

The simple example presented here is analogous to analyzing the stability of a one-machine-infinite-bus power system. The approach presented here is identical to the well-known equal area criterion (Kimbark, 1948; Anderson and Fouad, 1994) which is a direct method for determining transient stability for the one-machine-infinite-bus power system. For a more detailed discussion of the equal area criterion and its relationship to Lyapunov's direct method refer to Pai (1981), Chapter 4; Pai (1989), Chapter 1; Fouad and Vittal (1992), Chapter 3.

12.1 Review of Literature on Direct Methods

In the review presented here, we will deal only with work relating to the transient stability analysis of multimachine power systems. In this case the simple example presented above becomes quite complex. Several vehicles which correspond to the synchronous machines are now involved. It also becomes necessary to determine (a) which vehicles will be pushed the hardest, (b) what portion of the disturbance energy is distributed to each vehicle, (c) in which directions the vehicles move, and (d) how high a hill must the vehicles climb before they will go over.

Energy criteria for transient stability analysis were the earliest of all direct methods of multimachine power system transient stability assessment. These techniques were extensions of the equal area criterion to power systems with more than two generators represented by the classical model (Anderson and Fouad, 1994, Chapter 2). Researchers from the Soviet Union conducted early work in this area (1930s and 1940s). There were very few results on this topic in Western literature during the same period. In the 1960s the application of Lyapunov's direct method to power systems generated a great deal of activity in the academic community. In most of these investigations, the classical power system model was used. The early work on energy criteria dealt with two main issues: (a) characterization of the system energy and (b) the critical value of the energy.

Several excellent references that provide a detailed review of the development of the direct methods for transient stability exist. Ribbens-Pavella (1971a) and Fouad (1975) are early review papers and provide a comprehensive review of the work done in the period 1960–1975. Detailed reviews of more recent work are conducted in Bose (1984), Ribbens-Pavella and Evans (1985), Fouad and Vittal (1988), and Chiang et al. (1995). The following textbooks provide a comprehensive review and also present detailed descriptions of the various approaches related to direct stability methods: Pai (1981, 1989), Fouad and Vittal

(1992), Ribbens-Pavella (1971b), and Pavella and Murthy (1994). These references provide a thorough and detailed review of the evolution of the direct methods. In what follows, a brief review of the field and the evolutionary steps in the development of the approaches are presented.

Gorev (1971) first proposed an energy criteria based on the lowest saddle point or unstable equilibrium point (UEP). This work influenced the thinking of power system direct stability researchers for a long time. Magnusson (1947) presented an approach very similar to that of Gorev's and derived a potential energy function with respect to the (posttransient) equilibrium point of the system. Aylett (1958) studied the phase-plane trajectories of multimachine systems using the classical model. An important aspect of this work is the formulation of the system equations based on the intermachine movements. In the period that followed, several important publications dealing with the application of Lyapunov's method to power systems appeared. These works largely dealt with the aspects of obtaining better Lyapunov function, and determining the least conservative estimate of the domain of attraction. Gless (1966) applied Lyapunov's method to the one machine classical model system. El-Abiad and Nagappan (1966) developed a Lyapunov function for multimachine system and demonstrated the approach on a four machine system. The stability results obtained were conservative, and the work that followed this largely dealt with improving the Lyapunov function. A sampling of the work following this line of thought is presented in Willems (1968), Pai et al. (1970), and Ribbens-Pavella (1971a,b). These efforts were followed by the work of Tavora and Smith (1972a–c) dealing with the transient energy of a multimachine system represented by the classical model. They formulated the system equations in the Center of Inertia (COI) reference frame and also in the internode coordinates which is similar to the formulation used by Aylett (1958). Tavora and Smith obtained expressions for the total kinetic energy of the system and the transient kinetic energy, which the authors say determines stability. This was followed by work of Gupta and El-Abiad (1976), which recognized that the UEP of interest is not the one with the lowest energy, but rather the UEP closest to the system trajectory. Uyemura et al. (1972) made an important contribution by developing a technique to approximate the path-dependent terms in the Lyapunov functions by path-independent terms using approximations for the system trajectory.

The work by Athay, Podmore, and colleagues (Athay et al., 1979) is the basis for the transient energy function (TEF) method used today. This work investigated many issues dealing with the application of the TEF method to large power systems. These included

1. COI formulation and approximation of path-dependent terms
2. Search for the UEP in the direction of the faulted trajectory
3. Investigation of the potential energy boundary surface (PEBS)
4. Application of the technique to power systems of practical sizes
5. Preliminary investigation of higher-order models for synchronous generators

This work was followed by the work at Iowa State University by Fouad and colleagues (1981), which dealt with the determination of the correct UEP for stability assessment. This work also identified the appropriate energy for system separation and developed the concept of corrected kinetic energy. Details regarding this work are presented in Fouad and Vittal (1992).

The work that followed largely dealt with developing the TEF method into a more practical tool, and with improving its accuracy, modeling features, and speed. An important development in this area was the work of Bergen and Hill (1981). In this work the network structure was preserved for the classical model. As a result, fast techniques that incorporated network sparsity could be used to solve the problem. A concerted effort was also carried out to extend the applicability of the TEF method to realistic systems. This included improvements in modeling features, algorithms, and computational efficiency. Work related to the large-scale demonstration of the TEF method is found in Carvalho et al. (1986). The work dealing with extending the applicability of the TEF method is presented in Fouad et al. (1986). Significant contributions to this aspect of the TEF method can also be found in Padiyar and Sastry (1987), Padiyar and Ghosh (1989), and Abu-Elnaga et al. (1988).

In Chiang (1985) and Chiang et al. (1987, 1988), a significant contribution was made to provide an analytical justification for the stability region for multimachine power systems, and a systematic procedure to obtain the controlling UEP was also developed. Zaborszky et al. (1988) also provide a comprehensive analytical foundation for characterizing the region of stability for multimachine power systems.

With the development of a systematic procedure to determine and characterize the region of stability, a significant effort was directed toward the application of direct methods for online transient stability assessment. This work, reported in Waight et al. (1994) and Chadalavada et al. (1997), has resulted in an online tool which has been implemented and used to rank contingencies based on their severity. Another online approach implemented and being used at B.C. Hydro is presented in Mansour et al. (1995). A recent effort with regard to classifying and ranking contingencies quite similar to the one presented in Chadalavada et al. (1997) is described in Chiang et al. (1998).

Some recent efforts (Ni and Fouad, 1987; Hiskens et al., 1992; Jing et al., 1995) also deal with the inclusion of FACTS devices in the TEF analysis.

12.2 The Power System Model

The classical power system model will now be presented. It is the "simplest" power system model used in stability studies and is limited to the analysis of first swing transients. For more details regarding the model, the reader is referred to Anderson and Fouad (1994), Fouad and Vittal (1992), Kundur (1994), and Sauer and Pai (1998). The assumptions commonly made in deriving this model are

For the synchronous generators

1. Mechanical power input is constant.
2. Damping or asynchronous power is negligible.
3. The generator is represented by a constant EMF behind the direct axis transient (unsaturated) reactance.
4. The mechanical rotor angle of a synchronous generator can be represented by the angle of the voltage behind the transient reactance.

The load is usually represented by passive impedances (or admittances), determined from the predisturbance conditions. These impedances are held constant throughout the stability study. This assumption can be improved using nonlinear models. See Fouad and Vittal (1992), Kundur (1994), and Sauer and Pai (1998) for more details. With the loads represented as constant impedances, all the nodes except the internal generator nodes can be eliminated. The generator reactances and the constant impedance loads are included in the network bus admittance matrix. The generators' equations of motion are then given by

$$M_i \frac{d\omega_i}{dt} = P_i - P_{ei}$$

$$\frac{d\delta_i}{dt} = \omega_i \quad i = 1, 2, \ldots, n$$

(12.1)

where

$$P_{ei} = \sum_{\substack{j=1 \\ j \neq i}}^{n} \left[C_{ij} \sin(\delta_i - \delta_j) + D_{ij} \cos(\delta_i - \delta_j) \right]$$

(12.2)

$P_i = P_{mi} - E_i^2 G_{ii}$
$C_{ij} = E_i E_j B_{ij}, D_{ij} = E_i E_j G_{ij}$
P_{mi} = Mechanical power input
G_{ii} = Driving point conductance
E_i = Constant voltage behind the direct axis transient reactance
ω_i, δ_i = Generator rotor speed and angle deviations, respectively, with respect to a synchronously rotating reference frame
M_i = Inertia constant of generator
B_{ij} (G_{ij}) = Transfer susceptance (conductance) in the reduced bus admittance matrix

Equation 12.1 is written with respect to an arbitrary synchronous reference frame. Transformation of this equation to the inertial center coordinates not only offers physical insight into the transient stability problem formulation in general, but also removes the energy associated with the motion of the inertial center which does not contribute to the stability determination. Referring to Equation 12.1, define

$$M_T = \sum_{i=1}^{n} M_i$$

$$\delta_0 = \frac{1}{M_T} \sum_{i=1}^{n} M_i$$

then,

$$M_T \dot{\omega}_0 = \sum_{i=1}^{n} P_i - P_{ei} = \sum_{i=1}^{n} P_i - 2 \sum_{i=1}^{n-1} \sum_{j=i+1}^{n} D_{ij} \cos \delta_{ij}$$

$$\dot{\delta}_0 = \omega_0$$

(12.3)

The generators' angles and speeds with respect to the inertial center are given by

$$\theta_i = \delta_i - \delta_0$$
$$\tilde{\omega}_i = \omega_i - \omega_0 \quad i = 1, 2, \ldots, n$$

(12.4)

and in this coordinate system the equations of motion are given by

$$M_i \dot{\tilde{\omega}}_i = P_i - P_{mi} - \frac{M_i}{M_T} P_{COI}$$

$$\dot{\theta}_i = \tilde{\omega}_i \quad i = 1, 2, \ldots, n$$

(12.5)

12.2.1 Review of Stability Theory

A brief review of the stability theory applied to the TEF method will now be presented. This will include a few definitions, some important results, and an analytical outline of the stability assessment formulation.

The definitions and results that are presented are for differential equations of the type shown in Equations 12.1 and 12.5. These equations have the general structure given by

$$\dot{x}(t) = f(t, x(t))$$

(12.6)

The system described by Equation 12.6 is said to be *autonomous* if $f(t, x(t)) \equiv f(x)$, i.e., independent of t and is said to be nonautonomous otherwise.

A point $x_0 \in R^n$ is called an *equilibrium point* for the system (Equation 12.6) at time t_0 if $f(t, x_0) \equiv 0$ for all $t \geq t_0$.

An equilibrium point x_e of Equation 12.6 is said to be an isolated equilibrium point if there exists some neighborhood S of x_e which does not contain any other equilibrium point of Equation 12.6.

Some precise definitions of stability in the sense of Lyapunov will now be presented. In presenting these definitions, we consider systems of equations described by Equation 12.6, and also assume that Equation 12.6 possesses an isolated equilibrium point at the origin. Thus, $f(t, 0) = 0$ for all $t \geq 0$.

The equilibrium $x = 0$ of Equation 12.6 is said to be *stable* in the sense of Lyapunov, or simply stable if for every real number $\varepsilon > 0$ and initial time $t_0 > 0$ there exists a real number $\delta(\varepsilon, t_0) > 0$ such that for all initial conditions satisfying the inequality $\|x(t_0)\| = \|x_0\| < \delta$, the motion satisfies $\|x(t)\| < \varepsilon$ for all $t \geq t_0$.

The symbol $\|\cdot\|$ stands for a norm. Several norms can be defined on an n-dimensional vector space. Refer to Miller and Michel (1983) and Vidyasagar (1978) for more details. The definition of stability given above is unsatisfactory from an engineering viewpoint, where one is more interested in a stricter requirement of the system trajectory to eventually return to some equilibrium point. Keeping this requirement in mind, the following definition of asymptotic stability is presented.

The equilibrium $x = 0$ of Equation 12.6 is *asymptotically stable* at time t_0 if

1. $x = 0$ is stable at $t = t_0$
2. For every $t_0 \geq 0$, there exists an $\eta(t_0) > 0$ such that $\lim_{t \to \infty} \|x(t)\| \to 0$ whenever $\|x(t)\| < \eta$ (ATTRACTIVITY)

This definition combines the aspect of stability as well as attractivity of the equilibrium. The concept is local, because the region containing all the initial conditions that converge to the equilibrium is some portion of the state space. Having provided the definitions pertaining to stability, the formulation of the stability assessment procedure for power systems is now presented. The system is initially assumed to be at a predisturbance steady-state condition governed by the equations

$$\dot{x}(t) = f^p(x(t)) \quad -\infty < t \leq 0 \tag{12.7}$$

The superscript p indicates predisturbance. The system is at equilibrium, and the initial conditions are obtained from the power flow solution. At $t = 0$, the disturbance or the fault is initiated. This changes the structure of the right-hand sides of the differential equations, and the dynamics of the system are governed by

$$\dot{x}(t) = f^f(x(t)) \quad 0 < t \leq t_{cl} \tag{12.8}$$

where the superscript f indicates faulted conditions. The disturbance or the fault is removed or cleared by the protective equipment at time t_{cl}. As a result, the network undergoes a topology change and the right-hand sides of the differential equations are again altered. The dynamics in the postdisturbance or postfault period are governed by

$$\dot{x}(t) = f(x(t)) \quad t_{cl} < t \leq \infty \tag{12.9}$$

The stability analysis is done for the system in the postdisturbance period. The objective is to ascertain asymptotic stability of the postdisturbance equilibrium point of the system governed by Equation 12.9. This is done by obtaining the domain of attraction of the postdisturbance equilibrium and determining if the initial conditions of the postdisturbance period lie within this domain of attraction or outside it. The domain of attraction is characterized by the appropriately determined value of the TEF. In the literature survey presented previously, several approaches to characterize the domain of attraction were mentioned. In earlier approaches (El-Abiad and Nagappan, 1966; Tavora and Smith, 1972a–c), this was done by obtaining the UEP of the postdisturbance system and determining the one with the lowest level of potential energy with respect to the postdisturbance equilibrium. This value of potential energy then characterized the domain of attraction. In the work that followed, it was found that this approach provided very conservative results for power systems. In Gupta and El-Abiad (1976), it was recognized that the appropriate UEP was dependent on the fault location, and the concept of closest UEP was developed. An approach to determine the domain of attraction was also presented by Kakimoto et al. (1978a,b) and

Direct Stability Methods

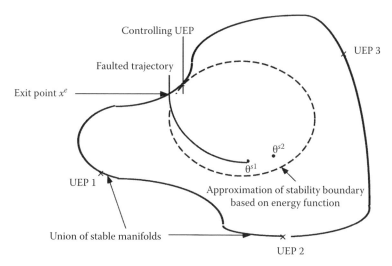

FIGURE 12.2 Conceptual framework of TEF approach.

Kakimoto and Hayashi (1981) based on the concept of the PEBS. For a given disturbance trajectory, the PEBS describes a "local" approximation of the stability boundary. The process of finding this local approximation is associated with the determination of the stability boundary of a lower dimensional system (see Fouad and Vittal [1992], Chapter 4 for details). It is formed by joining points of maximum potential energy along any direction originating from the postdisturbance stable equilibrium point. The PEBS constructed in this manner is orthogonal to the equipotential curves. In addition, along the direction orthogonal to the PEBS, the potential energy achieves a local maximum at the PEBS. In Athay et al. (1979), several simulations on realistic systems were conducted. These simulations, together with the synthesis of previous results in the area led to the development of a procedure to determine the correct UEP to characterize the domain of attraction. The results obtained were much improved, but in terms of practical applicability there was room for improvement. The work presented in Fouad et al. (1981) and Carvalho et al. (1986) made several important contributions to determining the correct UEP. The term *controlling UEP* was established, and a systematic procedure to determine the controlling UEP was developed. This will be described later. In Chiang (1985) and Chiang et al. (1987, 1988), a thorough analytical justification for the concept of the controlling UEP and the characterization of the domain of attraction was developed. This provides the analytical basis for the application of the TEF method to power systems. These analytical results in essence show that the stability boundary of the postdisturbance equilibrium point is made up of the union of the stable manifolds of those UEPs contained on the stability boundary. The boundary is then approximated locally using the energy function evaluated at the controlling UEP. The conceptual framework of the TEF approach is illustrated in Figure 12.2.

12.3 The Transient Energy Function

The TEF can be derived from Equation 12.5 using first principles. Details of the derivation can be found in Pai (1981, 1989), Fouad and Vittal (1992), Athay et al. (1979). For the power system model considered in Equation 12.5, the TEF is given by

$$V = \frac{1}{2}\sum_{i=1}^{n} M_i \tilde{\omega}_i^2 - \sum_{i=1}^{n} P_i\left(\theta_i - \theta_i^{s2}\right) - \sum_{i=1}^{n-1}\sum_{j=i+1}^{n}\left[C_{ij}\cos\left(\theta_{ij} - \theta_{ij}^{s2}\right) - \int_{\theta_i^{s2}+\theta_j^{s2}}^{\theta_i+\theta_j} D_{ij}\cos\theta_{ij}\,d(\theta_i + \theta_j)\right] \quad (12.10)$$

where $\theta_{ij} = \theta_i - \theta_j$.

The first term on the right-hand side of Equation 12.10 is the kinetic energy. The next three terms represent the potential energy. The last term is path dependent. It is usually approximated (Uyemura et al., 1972; Athay et al., 1979) using a straight line approximation for the system trajectory. The integral between two points θ^a and θ^b is then given by

$$I_{ij} = D_{ij} \frac{\theta_i^b - \theta_i^a + \theta_j^b - \theta_j^a}{\theta_{ij}^b - \theta_{ij}^a} \left(\sin \theta_{ij}^b - \sin \theta_{ij}^a \right). \tag{12.11}$$

In Fouad et al. (1981), a detailed analysis of the energy behavior along the time domain trajectory was conducted. It was observed that in all cases where the system was stable following the removal of a disturbance, a certain amount of the total kinetic energy in the system was not absorbed. This indicates that not all the kinetic energy created by the disturbance, contributes to the instability of the system. Some of the kinetic energy is responsible for the intermachine motion between the generators and does not contribute to the separation of the severely disturbed generators from the rest of the system. The kinetic energy associated with the gross motion of k machines having angular speeds $\tilde{\omega}_1, \tilde{\omega}_2, \ldots, \tilde{\omega}_k$ is the same as the kinetic energy of their inertial center. The speed of the inertial center of that group and its kinetic energy are given by

$$\tilde{\omega}_{cr} = \frac{\sum_{i=1}^{k} M_i \tilde{\omega}_i}{\sum_{i=1}^{k} M_i} \tag{12.12}$$

$$V_{KE_{cr}} = \frac{1}{2} \left[\sum_{i=1}^{k} M_i \right] (\tilde{\omega}_{cr})^2 \tag{12.13}$$

The disturbance splits the generators of the system into two groups: the critical machines and the rest of the generators. Their inertial centers have inertia constants and angular speeds $M_{cr}, \tilde{\omega}_{cr}$ and $M_{sys}, \tilde{\omega}_{sys}$, respectively. The kinetic energy causing the separation of the two groups is the same as that of an equivalent one-machine-infinite-bus system having inertia constant M_{eq} and angular speed $\tilde{\omega}_{eq}$ given by

$$M_{eq} = \frac{M_{cr} \times M_{sys}}{M_{eq} + M_{sys}}$$

$$\tilde{\omega}_{eq} = (\tilde{\omega}_{cr} - \tilde{\omega}_{sys}) \tag{12.14}$$

and the corresponding kinetic energy is given by

$$V_{KE_{corr}} = \frac{1}{2} M_{eq} (\tilde{\omega}_{eq})^2 \tag{12.15}$$

The kinetic energy term in Equation 12.10 is replaced by Equation 12.15.

12.4 Transient Stability Assessment

As described previously, the transient stability assessment using the TEF method is done for the final postdisturbance configuration. The stability assessment is done by comparing two values of the transient energy V. The value of V is computed at the end of the disturbance. If the disturbance is a simple fault, the value of V at fault clearing V_{cl} is evaluated.

The other value of V that largely determines the accuracy of the stability assessment is the critical value of V, V_{cr}, which is the potential energy at the controlling UEP for the particular disturbance being investigated.

If $V_{cl} < V_{cr}$, the system is stable, and if $V_{cl} > V_{cr}$, the system is unstable. The assessment is made by computing the energy margin ΔV given by

$$\Delta V = V_{cr} - V_{cl} \tag{12.16}$$

Substituting for V_{cr} and V_{cl} from Equation 12.10 and invoking the linear path assumption for the path dependent integral between the conditions at the end of the disturbance and the controlling UEP, we have

$$\Delta V = -\frac{1}{2} M_{eq} \tilde{\omega}_{eq}^{cl\,2} - \sum_{i=1}^{n} P_i \left(\theta_i^u - \theta_i^{cl} \right)$$

$$- \sum_{i=1}^{n-1} \sum_{j=i+1}^{n} \left[C_{ij} \left(\cos\theta_{ij}^u - \cos\theta_{ij}^{cl} \right) \right] - D_{ij} \frac{\theta_i^u - \theta_i^{cl} + \theta_j^u - \theta_j^{cl}}{\left(\theta_{ij}^u - \theta_{ij}^{cl} \right)} \left(\sin\theta_{ij}^u - \sin\theta_{ij}^{cl} \right) \tag{12.17}$$

where $(\theta^{cl}, \tilde{\omega}^{cl})$ are the conditions at the end of the disturbance and $(\theta^u, \mathbf{0})$ represents the controlling UEP. If ΔV is greater than zero the system is stable, and if ΔV is less than zero, the system is unstable. A qualitative measure of the degree of stability (or instability) can be obtained if ΔV is normalized with respect to the corrected kinetic energy at the end of the disturbance (Fouad et al., 1981).

$$\Delta V_n = \frac{\Delta V}{V_{KE_{corr}}} \tag{12.18}$$

For a detailed description of the computational steps involved in the TEF analysis, refer to Fouad and Vittal (1992), Chapter 6.

12.5 Determination of the Controlling UEP

A detailed description of the rationale in developing the concept of the controlling UEP is provided in Fouad and Vittal (1992), Section 5.4. A criterion to determine the controlling UEP based on the normalized energy margin is also presented. The criterion is stated as follows. The postdisturbance trajectory approaches (if the disturbance is large enough) the controlling UEP. This is the UEP with the lowest normalized potential energy margin. The determination of the controlling UEP involves the following key steps:

1. Identifying the correct UEP
2. Obtaining a starting point for the UEP solution close to the exact UEP
3. Calculation of the exact UEP

Identifying the correct UEP involves determining the advanced generators for the controlling UEP. This is referred to as the mode of disturbance (MOD). These generators generally are the most severely disturbed generators due to the disturbance. The generators in the MOD are not necessarily those that lose synchronism. The computational details of the procedure to identify the correct UEP and obtain a starting point for the exact UEP solution are provided in Fouad and Vittal (1992), Section 6.6. An outline of the procedure is provided below:

1. Candidate modes to be tested by the MOD test depend on how the disturbance affects the system. The selection of the candidate modes is based on several disturbance severity measures obtained at the end of the disturbance. These severity measures include kinetic energy and acceleration. A ranked list of machines is obtained using the severity measures. From this ranked list, the machines or group of machines at the bottom of the list are included in the group forming the rest of the system and $V_{KE_{corr}}$ is calculated. In a sequential manner, machines are successively added to the group forming the rest of the system and $V_{KE_{corr}}$ is calculated and stored.
2. The list of $V_{KE_{corr}}$ calculated above is sorted in descending order and only those groups within 10% of the maximum $V_{KE_{corr}}$ in the list are retained.
3. Corresponding to the MOD for each of the retained groups of machines in step 2, an approximation to the UEP corresponding to that mode is constructed using the postdisturbance stable equilibrium point. For a given candidate mode, where machines i and j are contained in the critical group, an estimate of the approximation to the UEP for an n-machine system is given by $\left[\hat{\theta}_{ij}^u\right]^T = \left[\theta_1^{s2}, \theta_2^{s2}, \ldots, [\pi - \theta_i^{s2}], \ldots, [\pi - \theta_j^{s2}], \ldots, \theta_n^{s2}\right]$. This estimate can be further improved by accounting for the motion of the COI, and using the concept of the PEBS to maximize the potential energy along the ray drawn from the estimate and the postdisturbance stable equilibrium point θ^{s2}.
4. The normalized potential energy margin for each of the candidate modes is evaluated at the approximation to the exact UEP, and the mode corresponding to the lowest normalized potential energy margin is then selected as the mode of the controlling UEP.
5. Using the approximation to the controlling UEP as a starting point, the exact UEP is obtained by solving the nonlinear algebraic equation given by

$$f_i = P_i - P_{mi} - \frac{M_i}{M_T} P_{COI} = 0 \quad i = 1, 2, \ldots, n \tag{12.19}$$

The solution of these equations is a computationally intensive task for realistic power systems. Several investigators have made significant contributions to determining an effective solution. A detailed description of the numerical issues and algorithms to determine the exact UEP solution are beyond the scope of this handbook. Several excellent references that detail these approaches are available. These efforts are described in Fouad and Vittal (1992), Section 6.8.

12.6 The Boundary Controlling UEP Method

The Boundary Controlling UEP (BCU) method (Chiang, 1985; Chiang et al., 1987, 1988) provides a systematic procedure to determine a suitable starting point for the controlling UEP solution. The main steps in the procedure are as follows:

1. Obtain the faulted trajectory by integrating the equations

$$M_i \dot{\tilde{\omega}}_i = P_i^f - P_{ei}^f - \frac{M_i}{M_T} P_{COI}^f$$

$$\dot{\theta}_i = \tilde{\omega}_i, \quad i = 1, 2, \ldots, n \tag{12.20}$$

Values of θ obtained from Equation 12.20 are substituted in the postfault mismatch equation given by Equation 12.19. The exit point x^e is then obtained by satisfying the condition $\sum_{i=1}^{n} -f_i \tilde{\omega}_i = 0$.

2. Using θ^e as the starting point, integrate the associated gradient system equations given by

$$\dot{\theta}_i = P_i - P_{ei} - \frac{M_i}{M_T} P_{COI}, \quad i = 1, 2, \ldots, n-1$$

$$\theta_n = -\frac{\sum_{i=1}^{n-1} M_i \theta_i}{M_n} \tag{12.21}$$

At each step of the integration, evaluate $\sum_{i=1}^{n} |f_i| = F$ and determine the first minimum of F along the gradient surface. Let θ^* be the vector of rotor angles at this point.

3. Using θ^* as a starting point in Equation 12.19, obtain the exact solution for the controlling UEP.

12.7 Applications of the TEF Method and Modeling Enhancements

The preceding subsections have provided the important steps in the application of the TEF method to analyze the transient stability of multimachine power systems. In this subsection, a brief mention of the applications of the technique and enhancements in terms of modeling detail and application to realistic power systems is provided. Inclusion of detailed generator models and excitation systems in the TEF method are presented in Athay et al. (1979), Fouad et al. (1986), and Waight et al. (1994). The sparse formulation of the system to obtain more efficient solution techniques is developed in Bergen and Hill (1981), Abu-Elnaga et al. (1988), and Waight et al. (1994). The application of the TEF method for a wide range of problems including dynamic security assessment are discussed in Fouad and Vittal (1992), Chapters 9 and 10; Chadalavada et al. (1997); and Mansour et al. (1995). The availability of a qualitative measure of the degree of stability or instability in terms of the energy margin makes the direct methods an attractive tool for a wide range of problems. The modeling enhancements that have taken place and the continued development in terms of computational efficiency and computer hardware, make direct methods a viable candidate for online transient stability assessment (Waight et al., 1994; Mansour et al., 1995; Chadalavada et al., 1997). This feature is particularly effective in the competitive market environment to calculate operating limits with changing conditions. There are several efforts underway dealing with the development of direct methods and a combination of time simulation techniques for online transient stability assessment. These approaches take advantage of the superior modeling capability available in the time simulation engines, and use the qualitative measure provided by the direct methods to derive preventive and corrective control actions and estimate limits. This line of investigation has great potential and could become a vital component of energy control centers in the near future.

References

Abu-Elnaga, M.M., El-Kady, M.A., and Findlay, R.D., Sparse formulation of the transient energy function method for applications to large-scale power systems, *IEEE Trans. Power Syst.*, PWRS-3(4), 1648–1654, November 1988.

Anderson, P.M. and Fouad, A.A., *Power System Control and Stability*, IEEE Press, New York, 1994.

Athay, T., Sherkat, V.R., Podmore, R., Virmani, S., and Puech, C., Transient energy stability analysis, in *Systems Engineering for Power: Emergency Operation State Control—Section IV*, Davos, Switzerland, U.S. Department of Energy Publication No. Conf.-790904-PL, 1979.

Aylett, P.D., The energy-integral criterion of transient stability limits of power systems, in *Proceedings of Institution of Electrical Engineers*, 105C(8), London, U.K., pp. 527–536, September 1958.

Bergen, A.R. and Hill, D.J., A structure preserving model for power system stability analysis, *IEEE Trans. Power App. Syst.*, PAS-100(1), 25–35, January 1981.

Bose, A., Chair, IEEE Committee Report, Application of direct methods to transient stability analysis of power systems, *IEEE Trans. Power App. Syst.*, PAS-103(7), 1629–1630, July 1984.

Carvalho, V.F., El-Kady, M.A., Vaahedi, E., Kundur, P., Tang, C.K., Rogers, G., Libaque, J., Wong, D., Fouad, A.A., Vittal, V., and Rajagopal, S., Demonstration of large scale direct analysis of power system transient stability, Electric Power Research Institute Report EL-4980, December 1986.

Chadalavada, V. et al., An on-line contingency filtering scheme for dynamic security assessment, *IEEE Trans. Power Syst.*, 12(1), 153–161, February 1997.

Chiang, H.-D., A theory-based controlling UEP method for direct analysis of power system transient stability, in *Proceedings of the* 1989 *International Symposium on Circuits and Systems*, 3, Portland, OR, pp. 65–69, 1985.

Chiang, H.-D., Chiu, C.C., and Cauley, G., Direct stability analysis of electric power systems using energy functions: Theory, application, and perspective, *IEEE Proceedings*, 83(11), 1497–1529, November 1995.

Chiang, H.D., Wang, C.S., and Li, H., Development of BCU classifiers for on-line dynamic contingency screening of electric power systems, Paper No. PE-349, in *IEEE Power Engineering Society Summer Power Meeting*, San Diego, CA, July 1998.

Chiang, H.-D., Wu, F.F., and Varaiya, P.P., Foundations of the direct methods for power system transient stability analysis, *IEEE Trans. Circuits Syst.*, 34, 160–173, February 1987.

Chiang, H.-D., Wu, F.F., and Varaiya, P.P., Foundations of the potential energy boundary surface method for power system transient stability analysis, *IEEE Trans. Circuits Syst.*, 35(6), 712–728, June 1988.

El-Abiad, A.H. and Nagappan, K., Transient stability regions of multi-machine power systems, *IEEE Trans. Power App. Syst.*, PAS-85(2), 169–178, February 1966.

Fouad, A.A., Stability theory-criteria for transient stability, in *Proceedings of the Engineering Foundation Conference on System Engineering for Power, Status and Prospects*, Henniker, NH, NIT Publication No. Conf.-750867, August 1975.

Fouad, A.A., Kruempel, K.C., Mamandur, K.R.C., Stanton, S.E., Pai, M.A., and Vittal, V., Transient stability margin as a tool for dynamic security assessment, EPRI Report EL-1755, March 1981.

Fouad, A.A. and Vittal, V., The transient energy function method, *Int. J. Electr. Power Energy Syst.*, 10(4), 233–246, October 1988.

Fouad, A.A. and Vittal, V., *Power System Transient Stability Analysis Using the Transient Energy Function Method*, Prentice-Hall, Inc., Upper Saddle River, NJ, 1992.

Fouad, A.A., Vittal, V., Ni, Y.X., Pota, H.R., Nodehi, K., and Oh, T.K., Extending application of the transient energy function method, Report EL-4980, Palo Alto, CA, EPRI, 1986.

Gless, G.E., Direct method of Liapunov applied to transient power system stability, *IEEE Trans. Power App. Syst.*, PAS-85(2), 159–168, February 1966.

Gorev, A.A., *Criteria of Stability of Electric Power Systems*. Electric Technology and Electric Power Series. The All Union Institute of Scientific and Technological Information and the Academy of Sciences of the USSR (in Russia). Moscow, Russia, 1971 (in Russian).

Gupta, C.L. and El-Abiad, A.H., Determination of the closest unstable equilibrium state for Lyapunov's method in transient stability studies, *IEEE Trans. Power App. Syst.*, PAS-95, 1699–1712, September/October 1976.

Hiskens, I.A. et al., Incorporation of SVC into energy function method, *IEEE Trans. Power Syst.*, PWRS-7, 133–140, February 1992.

Jing, C. et al., Incorporation of HVDC and SVC models in the Northern States Power Co. (NSP) network for on-line implementation of direct transient stability assessment, *IEEE Trans. Power Syst.*, 10(2), 898–906, May 1995.

Kakimoto, N. and Hayashi, M., Transient stability analysis of multimachine power systems by Lyapunov's direct method, in *Proceedings of 20th Conference on Decision and Control*, San Diego, CA, 1981.

Kakimoto, N., Ohsawa, Y., and Hayashi, M., Transient stability analysis of electric power system via Lure-Type Lyapunov function, Parts I and II, *Trans. IEE Japan*, 98, 516, 1978a.

Kakimoto, N., Ohsawa, Y., and Hayashi, M., Transient stability analysis of large-scale power systems by Lyapunov's direct method, *IEEE Trans. Power App. Syst.*, 103(1), 160–167, January 1978b.

Kimbark, E.W., *Power System Stability*, I, John Wiley & Sons, New York, 1948.

Kundur, P., *Power System Stability and Control*, McGraw-Hill, New York, 1994.

Lyapunov, M.A., Problème Général de la Stabilité du Mouvement, *Ann. Fac. Sci. Toulouse*, 9, 203–474, 1907 (French, translation of the original paper published in 1893 in *Comm. Soc. Math. Kharkow*; reprinted as Vol. 17 in *Annals of Mathematical Studies*, Princeton, NJ, 1949).

Magnusson, P.C., Transient energy method of calculating stability, *AIEE Trans.*, 66, 747–755, 1947.

Mansour, Y., Vaahedi, E., Chang, A.Y., Corns, B.R., Garrett, B.W., Demaree, K., Athay, T., and Cheung, K., B.C. Hydro's on-line transient stability assessment (TSA): Model development, analysis, and post-processing, *IEEE Trans. Power Syst.*, 10(1), 241–253, February 1995.

Miller, R.K. and Michel, A.N., *Ordinary Differential Equations*, Academic Press, New York, 1983.

Ni, Y.-X. and Fouad, A.A., A simplified two terminal HVDC model and its use in direct transient stability assessment, *IEEE Trans. Power Syst.*, PWRS-2(4), 1006–1013, November 1987.

Padiyar, K.R. and Ghosh, K.K., Direct stability evaluation of power systems with detailed generator models using structure preserving energy functions, *Int. J. Electr. Power Energy Syst.*, 11(1), 47–56, January 1989.

Padiyar, K.R. and Sastry, H.S.Y., Topological energy function analysis of stability of power systems, *Int. J. Electr. Power Energy Syst.*, 9(1), 9–16, January 1987.

Pai, M.A., *Power System Stability*, North-Holland Publishing Co., Amsterdam, the Netherlands, 1981.

Pai, M.A., *Energy Function Analysis for Power System Stability*, Kluwer Academic Publishers, Boston, MA, 1989.

Pai, M.A., Mohan, A., and Rao, J.G., Power system transient stability regions using Popov's method, *IEEE Trans. Power App. Syst.*, PAS-89(5), 788–794, May/June 1970.

Pavella, M. and Murthy, P.G., *Transient Stability of Power Systems: Theory and Practice*, John Wiley & Sons, Inc., New York, 1994.

Ribbens-Pavella, M., Critical survey of transient stability studies of multi-machine power systems by Lyapunov's direct method, in *Proceedings of 9th Annual Allerton Conference on Circuits and System Theory*, Monticello, IL, October 1971a.

Ribbens-Pavella, M., Transient stability of multi-machine power systems by Lyapunov's direct method, in *IEEE Winter Power Meeting Conference Paper*, New York, 1971b.

Ribbens-Pavella, M. and Evans, F.J., Direct methods for studying of the dynamics of large scale electric power systems—A survey, *Automatica*, 21(1), 1–21, 1985.

Sauer, P.W. and Pai, M.A., *Power System Dynamics and Stability*, Prentice Hall, New York, 1998.

Tavora, C.J. and Smith, O.J.M., Characterization of equilibrium and stability in power systems, *IEEE Trans. Power App. Syst.*, PAS-72, 1127–1130, May/June 1972a.

Tavora, C.J. and Smith, O.J.M., Equilibrium analysis of power systems, *IEEE Trans. Power App. Syst.*, PAS-72, 1131–1137, May/June 1972b.

Tavora, C.J. and Smith, O.J.M., Stability analysis of power systems, *IEEE Trans. Power App. Syst.*, PAS-72, 1138–1144, May/June 1972c.

Uyemura, K., Matsuki, J., Yamada, I., and Tsuji, T., Approximation of an energy function in transient stability analysis of power systems, *Electr. Eng. Jpn.*, 92(6), 96–100, November/December 1972.

Vidyasagar, M., *Nonlinear Systems Analysis*, Prentice-Hall, Englewood Cliffs, NJ, 1978.

Waight, J.G. et al., Analytical methods for contingency selection and ranking for dynamic security analysis, Report TR-104352, Palo Alto, CA, September 1994.

Willems, J.L., Improved Lyapunov function for transient power-system stability, in *Proceedings of the Institution of Electrical Engineers*, 115(9), London, U.K., pp. 1315–1317, September 1968.

Zaborszky, J., Huang, G., Zheng, B., and Leung, T.-C., On the phase-portrait of a class of large nonlinear dynamic systems such as the power system, *IEEE Trans. Autom. Control*, 32, 4–15, January 1988.

13
Power System Stability Controls

13.1	Review of Power System Synchronous Stability Basics	13-2
13.2	Concepts of Power System Stability Controls	13-5
	Feedback Controls • Feedforward Controls • Synchronizing and Damping Torques • Effectiveness and Robustness • Actuators • Reliability Criteria	
13.3	Types of Power System Stability Controls and Possibilities for Advanced Control	13-7
	Excitation Control • Prime Mover Control Including Fast Valving • Generator Tripping • Fast Fault Clearing, High-Speed Reclosing, and Single-Pole Switching • Dynamic Braking • Load Tripping and Modulation • Reactive Power Compensation Switching or Modulation • Current Injection by Voltage Sourced Inverters • Fast Voltage Phase Angle Control • HVDC Link Supplementary Controls • Adjustable Speed (Doubly Fed) Synchronous Machines • Controlled Separation and Underfrequency Load Shedding	
13.4	Dynamic Security Assessment	13-14
13.5	"Intelligent" Controls	13-14
13.6	Wide-Area Stability Controls	13-15
13.7	Effect of Industry Restructuring on Stability Controls	13-16
13.8	Experience from Recent Power Failures	13-16
13.9	Summary	13-16
	References	13-16

Carson W. Taylor
(retired)
Bonneville Power Administration

Power system synchronous or angle instability phenomenon limits power transfer, especially where transmission distances are long. This is well recognized and many methods have been developed to improve stability and increase allowable power transfers.

The synchronous stability problem has been fairly well solved by fast fault clearing, thyristor exciters, power system stabilizers (PSSs), and a variety of other stability controls such as generator tripping. Fault clearing of severe short circuits can be less than three cycles (50 ms for 60 Hz frequency) and the effect of the faulted line outage on generator acceleration and stability may be greater than that of the fault itself. The severe multiphase short circuits are infrequent on extra high voltage (EHV) transmission networks.

Nevertheless, more intensive use of available generation and transmission, more onerous load characteristics, greater variation in power schedules, and other negative aspects of industry restructuring pose new concerns. Recent large-scale cascading power failures have heightened the concerns.

In this chapter we describe the state-of-the-art of power system angle stability controls. Controls for voltage stability are described in another chapter and in other literature [1–5].

We emphasize controls employing relatively new technologies that have actually been implemented by electric power companies, or that are seriously being considered for implementation. The technologies include applied control theory, power electronics, microprocessors, signal processing, transducers, and communications.

Power system stability controls must be effective and robust. Effective in an engineering sense means "cost-effective." Control robustness is the capability to operate appropriately for a wide range of power system operating and disturbance conditions.

13.1 Review of Power System Synchronous Stability Basics

Many publications, for example Refs. [6–9,83], describe the basics—which we briefly review here. Power generation is largely by synchronous generators, which are interconnected over thousands of kilometers in very large power systems. Thousands of generators must operate in synchronism during normal and disturbance conditions. Loss of synchronism of a generator or group of generators with respect to another group of generators is *instability* and could result in expensive widespread power blackouts.

The essence of synchronous stability is the balance of individual generator electrical and mechanical torques as described by Newton's second law applied to rotation:

$$J\frac{d\omega}{dt} = T_m - T_e$$

where
  J is moment of inertia of the generator and prime mover
  ω is speed
  T_m is mechanical prime mover torque
  T_e is electrical torque related to generator electric power output

The generator speed determines the generator rotor angle changes relative to other generators. Figure 13.1 shows the basic "swing equation" block diagram relationship for a generator connected to a power system.

The conventional equation form and notation are used. The block diagram is explained as follows:

- The inertia constant, H, is proportional to the moment of inertia and is the kinetic energy at rated speed divided by the generator MVA rating. Units are MW-s/MVA, or s.
- T_m is mechanical torque in per unit. As a first approximation it is assumed to be constant. It is, however, influenced by speed controls (governors) and prime mover and energy supply system dynamics.

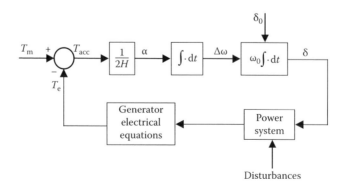

FIGURE 13.1 Block diagram of generator electromechanical dynamics.

- ω_0 is rated frequency in radians/second.
- δ_0 is predisturbance rotor angle in radians relative to a reference generator.
- The power system block comprises the transmission network, loads, power electronic devices, and other generators, prime movers, and energy supply systems with their controls. The transmission network is generally represented by algebraic equations. Loads and generators are represented by algebraic and differential equations.
- Disturbances include short circuits, and line and generator outages. A severe disturbance is a three-phase short circuit near the generator. This causes electric power and torque to be zero, with accelerating torque equal to T_m. (Although generator current is very high for the short circuit, the power factor, and active current and active power are close to zero.) Other switching (discrete) events for stabilization such as line reclosing may be included as disturbances to the differential–algebraic equation model (hybrid DAE math model).
- The generator electrical equations block represents the internal generator dynamics.

Figure 13.2 shows a simple conceptual model: a remote generator connected to a large power system by two parallel transmission lines with an intermediate switching station. With some approximations adequate for a second of time or so following a disturbance, Figure 13.3 block diagram is realized. The basic relationship between power and torque is $P = T\omega$. Since speed changes are quite small, power is considered equal to torque in per unit. The generator representation is a constant voltage, E', behind a reactance. The transformer and transmission lines are represented by inductive reactances. Using the relation $S = E'I^*$, the generator electrical power is the well-known relation:

$$P_e = \frac{E'V}{X}\sin\delta$$

where
 V is the large system (infinite bus) voltage
 X is the total reactance from the generator internal voltage to the large system

The above equation approximates characteristics of a detailed, large-scale model, and illustrates that the power system is fundamentally a highly nonlinear system for large disturbances.

FIGURE 13.2 Remote power plant to large system. Short circuit location is shown.

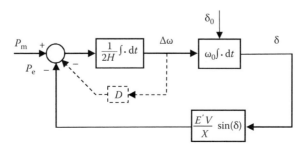

FIGURE 13.3 Simplified block diagram of generator electromechanical dynamics.

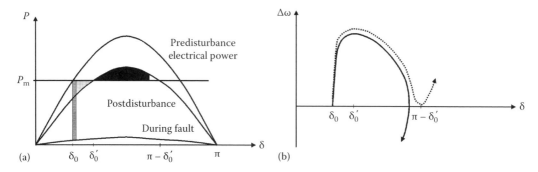

FIGURE 13.4 (a) Power–angle curve and equal area criterion. Dark shading for acceleration energy during fault. Light shading for additional acceleration energy because of line outage. Black shading for deceleration energy. (b) Angle–speed phase plane. Dotted trajectory is for unstable case.

Figure 13.4a shows the relation graphically. The predisturbance operating point is at the intersection of the load or mechanical power characteristic and the electrical power characteristic. Normal stable operation is at δ_0. For example, a small increase in mechanical power input causes an accelerating power that increases δ to increase P_e until accelerating power returns to zero. The opposite is true for the unstable operating point at $\pi - \delta_0$. δ_0 is normally less than 45°.

During normal operation, mechanical and electrical torques are equal and a generator runs at close to 50 or 60 Hz rated frequency. If, however, a short circuit occurs (usually with removal of a transmission line), the electric power output will be momentarily partially blocked from reaching loads and the generator (or group of generators) will accelerate, with increase in generator speed and angle. If the acceleration relative to other generators is too great, synchronism will be lost. Loss of synchronism is an unstable, runaway situation with large variations of voltages and currents that will normally cause protective separation of a generator or a group of generators. Following short circuit removal, the electrical torque and power developed as angle increases will decelerate the generator. If deceleration reverses angle swing prior to $\pi - \delta_0'$, stability is maintained at the new operating point δ_0' (Figure 13.4). If the swing is beyond $\pi - \delta_0'$, accelerating power or torque again becomes positive, resulting in runaway increase in angle and speed, and instability.

Figure 13.4a illustrates the equal area stability criterion for "first swing" stability. If the decelerating area (energy) above the mechanical power load line is greater than the accelerating area below the load line, stability is maintained.

Stability controls increase stability by decreasing the accelerating area or increasing the decelerating area. This may be done by either increasing the electrical power–angle relation, or by decreasing the mechanical power input.

For small disturbances the block diagram, Figure 13.3, can be linearized. The block diagram would then be that of a second-order differential equation oscillator. For a remote generator connected to a large system the oscillation frequency is 0.8–1.1 Hz.

Figure 13.3 also shows a damping path (dashed, damping power or torque in-phase with speed deviation) that represents mechanical or electrical damping mechanisms in the generator, turbine, loads, and other devices. Mechanical damping arises from the turbine torque–speed characteristic, friction and windage, and components of prime mover control in-phase with speed. At an oscillation frequency, the electrical power can be resolved into a component in-phase with angle (synchronizing power) and a component in quadrature (90° leading) in-phase with speed (damping power). Controls, notably generator automatic voltage regulators with high gain, can introduce negative damping at some oscillation frequencies. (In any feedback control system, high gain combined with time delays can cause positive feedback and instability.) For stability, the net damping must be positive for both normal conditions and for large disturbances with outages. Stability controls may also be added to improve damping. In some cases, stability controls are designed to improve both synchronizing and damping torques of generators.

The above analysis can be generalized to large systems. For first swing stability, synchronous stability between two critical groups of generators is of concern. For damping, many oscillation modes are present, all of which require positive damping. The low frequency modes (0.1–0.8 Hz) are most difficult to damp. These modes represent interarea oscillations between large portions of a power system.

13.2 Concepts of Power System Stability Controls

Figure 13.5 shows the general structure for analysis of power system stability and for development of power system stability controls. The feedback controls are mostly local, continuous controls at power plants. The feedforward controls are discontinuous, and may be local at power plants and substations or wide area.

Stability problems typically involve disturbances such as short circuits, with subsequent removal of faulted elements. Generation or load may be lost, resulting in generation–load imbalance and frequency excursions. These disturbances stimulate power system electromechanical dynamics. Improperly designed or tuned controls may contribute to stability problems; as mentioned, one example is negative damping torques caused by generator automatic voltage regulators.

Because of power system synchronizing and damping forces (including the feedback controls shown in Figure 13.5), stability is maintained for most disturbances and operating conditions.

13.2.1 Feedback Controls

The most important feedback (closed-loop) controls are the generator excitation controls (automatic voltage regulator often including PSS). Other feedback controls include prime mover controls, controls for reactive power compensation such as static var systems, and special controls for HVDC links. These controls are generally linear, continuously active, and based on local measurements.

There are, however, interesting possibilities for very effective discontinuous feedback controls, with microprocessors facilitating implementation. Discontinuous controls have certain advantages over continuous controls. Continuous feedback controls are potentially unstable. In complex power systems, continuously controlled equipment may cause adverse modal interactions [10]. Modern digital controls, however, can be discontinuous, and take no action until variables are out-of-range. This is analogous to biological systems (which have evolved over millions of years) that operate on the basis of excitatory stimuli [11].

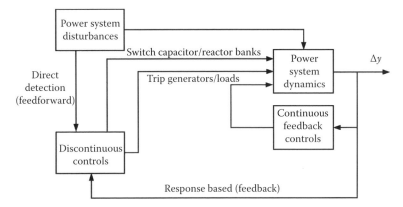

FIGURE 13.5 General power system structure showing local and wide-area, continuous and discontinuous stability controls. (From Taylor, C.W. et al., *Proc. IEEE Special Issue on Energy Infrastructure Defense Systems*, 93, 892, 2005. With permission.)

Bang–bang discontinuous control can operate several times to control large amplitude oscillations, providing time for linear continuous controls to become effective. If stability is a problem, generator excitation control including PSSs should be high performance.

13.2.2 Feedforward Controls

Also shown in Figure 13.5 are specialized feedforward (open-loop) controls that are powerful stabilizing forces for severe disturbances and for highly stressed operating conditions. Short circuit or outage events can be directly detected to initiate preplanned actions such as generator or load tripping, or reactive power compensation switching. These controls are rule-based, with rules developed from simulations (i.e., pattern recognition). These "event-based" controls are very effective since rapid control action prevents electromechanical dynamics from becoming stability threatening.

"Response-based" or feedback discontinuous controls are also possible. These controls initiate stabilizing actions for arbitrary disturbances that cause significant "swing" of measured variables.

Controls such as generator or load tripping can ensure a postdisturbance equilibrium with sufficient region of attraction. With fast control action the region of attraction can be small compared to requirements with only feedback controls.

Discontinuous controls have been termed discrete supplementary controls [8], special stability controls [12], special protection systems, remedial action schemes, and emergency controls [13]. Discontinuous controls are very powerful. Although the reliability of emergency controls is often an issue [14], adequate reliability can be obtained by design. Generally, controls are required to be as reliable as primary protective relaying. Duplicated or multiple sensors, redundant communications, and duplicated or voting logic are common [15].

Response-based discontinuous controls are often less expensive than event-based controls because fewer sensors and communications paths are needed. These controls are often "one-shot" controls, initiating a single set of switching actions. For slow dynamics, however, the controls can initiate a discontinuous action, observe response, and then initiate additional discontinuous action if necessary.

Undesired operation by some feedforward controls is relatively benign, and controls can be "trigger happy." For example, infrequent misoperation or unnecessary operation of HVDC fast power change, reactive power compensation switching, and transient excitation boosting (TEB) may not be very disruptive. Misoperation of generator tripping (especially of steam-turbine generators), fast valving, load tripping, or controlled separation, however, are disruptive and costly.

13.2.3 Synchronizing and Damping Torques

Power system electromechanical stability means that synchronous generators and motors must remain in synchronism following disturbances—with positive damping of rotor angle oscillations (swings). For very severe disturbances and operating conditions, loss of synchronism (instability) occurs on the first forward swing within about 1 s. For less severe disturbances and operating conditions, instability may occur on the second or subsequent swings because of a combination of insufficient synchronizing and damping torques at synchronous machines.

13.2.4 Effectiveness and Robustness

Power systems have many electromechanical oscillation modes, and each mode can potentially become unstable. Lower frequency interarea modes are the most difficult to stabilize. Controls must be designed to be effective for one or more modes, and must not cause adverse interactions for other modes.

There are recent advances in robust control theory, especially for linear systems. For real nonlinear systems, emphasis should be on knowing uncertainty bounds and on sensitivity analysis using detailed nonlinear, large-scale simulation. For example, the sensitivity of controls to different operating conditions and load characteristics must be studied. On-line simulation using actual operating conditions reduces uncertainty, and can be used for control adaptation.

13.2.5 Actuators

Actuators may be mechanical or power electronic. There are tradeoffs between cost and performance. Mechanical actuators (circuit breakers, turbine valves) are lower cost, and are usually sufficiently fast for electromechanical stability (e.g., two-cycle opening time, five-cycle closing time circuit breakers). They have restricted operating frequency and are generally used for feedforward controls.

Circuit breaker technology and reliability have improved in recent years [16,17]. Bang–bang control (up to perhaps five operations) for interarea oscillations with periods of 2 s or longer is feasible [18]. Traditional controls for mechanical switching have been simple relays, but advanced controls can approach the sophistication of controls of, for example, thyristor-switched capacitor banks.

Power electronic phase control or switching using thyristors has been widely used in generator exciters, HVDC, and static var compensators. Newer devices, such as insulated gate bipolar transistor (IGBT) and gate commutated thyristor (GCT/IGCT), now have voltage and current ratings sufficient for high power transmission applications. Advantages of power electronic actuators are very fast control, unrestricted switching frequency, and minimal transients.

For economy, existing actuators should be used to the extent possible. These include generator excitation and prime mover equipment, HVDC equipment, and circuit breakers. For example, infrequent generator tripping may be cost-effective compared to new power electronic actuated equipment.

13.2.6 Reliability Criteria

Experience shows that instability incidents are usually not caused by three-phase faults near large generating plants that are typically specified in deterministic reliability criteria. Rather they are the result of a combination of unusual failures and circumstances. The three-phase fault reliability criterion is often considered an *umbrella* criterion for less predictable disturbances involving multiple failures such as single-phase short circuits with "sympathetic" tripping of unfaulted lines. Of main concern are multiple *related* failures involving lines on the same right-of-way or with common terminations.

13.3 Types of Power System Stability Controls and Possibilities for Advanced Control

Stability controls are of many types including

- Generator excitation controls
- Prime mover controls including fast valving
- Generator tripping
- Fast fault clearing
- High-speed reclosing and single-pole switching
- Dynamic braking
- Load tripping and modulation
- Reactive power compensation switching or modulation (series and shunt)
- Current injection by voltage source inverter devices (STATCOM, UPFC, SMES, battery storage)
- Fast phase angle control

- HVDC link supplementary controls
- Adjustable speed (doubly fed) synchronous machines
- Controlled separation and underfrequency load shedding

We will summarize these controls. Chapter 17 of Ref. [7] provides considerable additional information. Reference [19] describes use of many of these controls in Japan.

13.3.1 Excitation Control

Generator excitation controls are a basic stability control. Thyristor exciters with high ceiling voltage provide powerful and economical means to ensure stability for large disturbances. Modern automatic voltage regulators and PSSs are digital, facilitating additional capabilities such as adaptive control and special logic [20–23].

Excitation control is almost always based on local measurements. Therefore full effectiveness may not be obtained for interarea stability problems where the normal local measurements are not sufficient. Line drop compensation [24,25] is one method to increase the effectiveness (sensitivity) of excitation control, and improve coordination with static var compensators that normally control transmission voltage with small droop.

Several forms of discontinuous control have been applied to keep field voltage near ceiling levels during the first forward interarea swing [7,26,27]. The control described in Refs. [7,26] computes change in rotor angle locally from the PSS speed change signal. The control described in Ref. [27] is a feedforward control that injects a decaying pulse into the voltage regulators at a large power plant following remote direct detection of a large disturbance. Figure 13.6 shows simulation results using this TEB.

13.3.2 Prime Mover Control Including Fast Valving

Fast power reduction (fast valving) at accelerating sending-end generators is an effective means of stability improvement. Use has been limited, however, because of the coordination required between characteristics of the electrical power system, the prime mover and prime mover controls, and the energy supply system (boiler).

Digital prime mover controls facilitate addition of special features for stability enhancement. Digital boiler controls, often retrofitted on existing equipment, may improve the feasibility of fast valving.

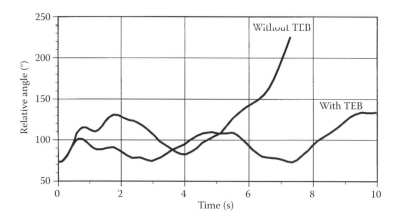

FIGURE 13.6 Rotor angle swing of Grand Coulee Unit 19 in Pacific Northwest relative to the San Onofre nuclear plant in Southern California. The effect of transient excitation boosting (TEB) at the Grand Coulee Third Power Plant following bipolar outage of the Pacific HVDC Intertie (3100 MW) is shown. (From Taylor, C.W. et al., *IEEE Trans. Power Syst.*, 8, 1291, 1993.)

Fast valving is potentially lower cost than tripping of turbo-generators. References [7,28] describe concepts, investigations, and recent implementations of fast valving. Two methods of steam-turbine fast valving are used: momentary and sustained. In momentary fast valving, the reheat turbine intercept valves are rapidly closed and then reopened after a short time delay. In sustained fast valving, the intercept values are also rapidly opened and reclosed, but with the control valves partially closed for sustained power reduction. Sustained fast valving may be necessary for a stable post-disturbance equilibrium.

13.3.3 Generator Tripping

Generator tripping is an effective (cost-effective) control especially if hydro units are used. Tripping of fossil units, especially gas- or oil-fired units, may be attractive if tripping to house load is possible and reliable. Gas turbine and combined-cycle plants constitute a large percentage of the new generation. Occasional tripping of these units is feasible and can become an attractive stability control in the future.

Most generator tripping controls are event-based (based on outage of generating plant out going lines or outage of tie lines). Several advanced response-based generator tripping controls, however, have been implemented.

The automatic trend relay (ATR) is implemented at the Colstrip generating plant in eastern Montana [29]. The plant consists of two 330-MW units and two 700-MW units. The microprocessor-based controller measures rotor speed and generator power and computes acceleration and angle. Tripping of 16%–100% of plant generation is based on 11 trip algorithms involving acceleration, speed, and angle changes. Because of the long distance to Pacific Northwest load centers, the ATR has operated many times, both desirably and undesirably. There are proposals to use voltage angle measurement information (Colstrip 500-kV voltage angle relative to Grand Coulee and other Northwest locations) to adaptively adjust ATR settings, or as additional information for trip algorithms. Another possibility is to provide speed or frequency measurements from Grand Coulee and other locations to base algorithms on speed difference rather than only Colstrip speed [30].

A Tokyo Electric Power Company stabilizing control predicts generator angle changes and decides the minimum number of generators to trip [31]. Local generator electric power, voltage, and current measurements are used to estimate angles. The control has worked correctly for several actual disturbances.

The Tokyo Electric Power Company (TEPCO) is also developing an emergency control system, which uses a predictive prevention method for step-out of pumped storage generators [32,33]. In the new method, the generators in TEPCO's network that swing against their local pumped storage generators after serious faults are treated as an external power system. The parameters in the external system, such as angle and moment of inertia, are estimated using local on-line information, and the behavior of local pumped storage generators is predicted based on equations of motion. Control actions (the number of generators to be tripped) are determined based on the prediction.

Reference [34] describes response-based generator tripping using a phase-plane controller. The controller is based on the apparent resistance–rate of change of apparent resistance (R–Rdot) phase plane, which is closely related to an angle difference–speed difference phase plane between two areas. The primary use of the controller is for controlled separation of the Pacific AC Intertie. Figure 13.7 shows simulation results where 600 MW of generator tripping reduces the likelihood of controlled separation.

13.3.4 Fast Fault Clearing, High-Speed Reclosing, and Single-Pole Switching

Clearing time of close-in faults can be less than three cycles using conventional protective relays and circuit breakers. Typical EHV circuit breakers have two-cycle opening time. One-cycle breakers have been developed [35], but special breakers are seldom justified. High magnitude short circuits may be detected as fast as one-fourth cycle by nondirectional overcurrent relays. Ultrahigh speed traveling wave relays are also available [36]. With such short clearing times, and considering that most EHV faults

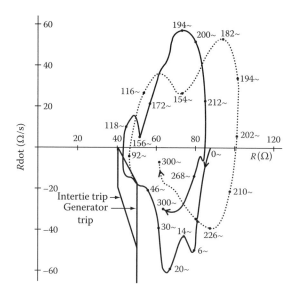

FIGURE 13.7 R–Rdot phase plane for loss of Pacific HVDC Intertie (2000 MW). Solid trajectory is without additional generator tripping. Dashed trajectory is with additional 600 MW of generator tripping initiated by the R–Rdot controller generator trip switching line. (From Haner, J.M. et al., *IEEE Trans. Power Delivery*, PWRD-1, 35, 1986.)

are single-phase, the removed transmission lines or other elements may be the major contributor to generator acceleration. This is especially true if non-faulted equipment is also removed by sympathetic relaying.

High-speed reclosing is an effective method of improving stability and reliability. Reclosing is before the maximum of the first forward angular swing, but after 30–40 cycle time for arc extinction. During a lightning storm, high-speed reclosing keeps the maximum number of lines in service. High-speed reclosing is effective when unfaulted lines trip because of relay misoperations.

Unsuccessful high-speed reclosing into a permanent fault can cause instability, and can also compound the torsional duty imposed on turbine-generator shafts. Solutions include reclosing only for single-phase faults, and reclosing from the weaker remote end with hot-line checking prior to reclosing at the generator end. Communication signals from the weaker end indicating successful reclosing can also be used to enable reclosing at the generator end [37].

Single-pole switching is a practical means to improve stability and reliability in extra high voltage networks where most circuit breakers have independent pole operation [38,39]. Several methods are used to ensure secondary arc extinction. For short lines, no special methods are needed. For long lines, the four-reactor scheme [40,41] is most commonly used. High-speed grounding switches may be used [42]. A hybrid reclosing method used successfully by Bonneville Power Administration (BPA) on many lines over many years employs single-pole tripping, but with three-pole tripping on the backswing followed by rapid three-pole reclosure; the three-pole tripping ensures secondary arc extinction [38]. Single-pole switching may necessitate positive sequence filtering in stability control input signals.

For advanced stability control, signal processing and pattern recognition techniques may be developed to detect secondary arc extinction [43,44]. Reclosing into a fault is avoided and single-pole reclosing success is improved.

High-speed reclosing or single-pole switching may not allow increased power transfers because deterministic reliability criteria generally specify permanent faults. Nevertheless, fast reclosing provides "defense-in-depth" for frequently occurring single-phase temporary faults and false operation of protective relays. The probability of power failures because of multiple line outages is greatly reduced.

13.3.5 Dynamic Braking

Shunt dynamic brakes using mechanical switching have been used infrequently [7]. Normally the insertion time of a few hundred milliseconds is fixed. One attractive method not requiring switching is neutral-to-ground resistors in generator step-up transformers; braking automatically results for ground faults—which are most common. Often, generator tripping, which helps ensure a postdisturbance equilibrium, is a better solution.

Thyristor switching of dynamic brakes has been proposed. Thyristor switching or phase control minimizes generator torsional duty [45], and can also be a subsynchronous resonance countermeasure [46].

13.3.6 Load Tripping and Modulation

Load tripping is similar in concept to generator tripping but is at the receiving end to reduce deceleration of receiving-end generation. Interruptible industrial load is commonly used. For example, Ref. [47] describes tripping of up to 3000 MW of industrial load following outages during power import conditions.

Rather than tripping large blocks of industrial load, it may be possible to trip low priority commercial and residential load such as space and water heaters, or air conditioners. This is less disruptive and the consumer may not even notice brief interruptions. The feasibility of this control depends on implementation of direct load control as part of demand side management and on the installation of high-speed communication links to consumers with high-speed actuators at load devices. Although unlikely because of economics, appliances such as heaters could be designed to provide frequency sensitivity by local measurements.

Load tripping is also used for voltage stability. Here the communication and actuator speeds are generally not as critical. It is also possible to modulate loads such as heaters to damp oscillations [48–50]. Clearly load tripping or modulation of small loads will depend on the economics, and the development of fast communications and actuators.

13.3.7 Reactive Power Compensation Switching or Modulation

Controlled series or shunt compensation improves stability, with series compensation generally being the most powerful. For switched compensation, either mechanical or power electronic switches may be used. For continuous modulation, thyristor phase control of a reactor (TCR) is used. Mechanical switching has the advantage of lower cost. The operating times of circuit breakers are usually adequate, especially for interarea oscillations. Mechanical switching is generally single insertion of compensation for synchronizing support. In addition to previously mentioned advantages, power electronic control has advantages in subsynchronous resonance performance.

For synchronizing support, high-speed series capacitor switching has been used effectively on the North American Pacific AC Intertie for over 25 years [51]. The main application is for full or partial outages of the parallel Pacific HVDC Intertie (event-driven control using transfer trip over microwave radio). Series capacitors are inserted by circuit breaker opening; operators bypass the series capacitors some minutes after the event. Response-based control using an impedance relay was also used for some years, and new response-based controls are being investigated.

Thyristor-based series compensation switching or modulation has been developed with several installations in service or planned [32,52,53]. Thyristor-controlled series compensation (TCSC) allows significant time–current dependent increase in series reactance over nominal reactance. With appropriate controls, this increase in reactance can be a powerful stabilizing force.

Thyristor-controlled series compensation was chosen for the 1020-km, 500-kV intertie between the Brazilian North–Northeast networks and the Southeast network [54]. The TCSCs at each end of the intertie are modulated using line power measurements to damp low frequency (0.12 Hz) oscillations. Figure 13.8, from commissioning field tests [55], shows the powerful stabilizing benefits of TCSCs.

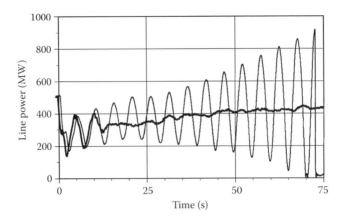

FIGURE 13.8 Effect of TCSCs for trip of a 300-MW generator in the North–Northeast Brazilian network. Results are from commissioning field tests in March 1999. The thin line without TCSC power oscillation damping shows interconnection separation after 70 s. The thick line with TCSC power oscillation damping shows rapid oscillation damping.

Reference [56] describes a TCSC application in China for integration of a remote power plant using two parallel 500-kV transmission lines (1300 km). Transient stability simulations indicate that 25% thyristor-controlled compensation is more effective than 45% fixed compensation. Several advanced TCSC control techniques are promising. The state-of-the-art is to provide both transient stability and damping control modes. Reference [57] surveys TCSC stability controls, providing 85 references.

For synchronizing support, high-speed switching of shunt capacitor banks is also effective. Again on the Pacific AC Intertie, four 200-MVAr shunt banks are switched for HVDC and 500-kV AC line outages [18]. These banks plus other 500-kV shunt capacitor/reactor banks and series capacitors are also switched for severe voltage swings.

High-speed mechanical switching of shunt banks as part of a static var system is common. For example, the Forbes SVS near Duluth, Minnesota, includes two 300-MVAr 500-kV shunt capacitor banks [58]. Generally it is effective to augment power electronic controlled compensation with fixed or mechanically switched compensation.

Static var compensators are applied along interconnections to improve synchronizing and damping support. Voltage support at intermediate points allows operation at angles above 90°. SVCs are modulated to improve oscillation damping. A seminal study [6,59] showed line current magnitude to be the most effective input signal. Synchronous condensers can provide similar benefits, but nowadays are not competitive with power electronic control. Available SVCs in load areas may be used to indirectly modulate load to provide synchronizing or damping forces.

Digital control facilitates new strategies. Adaptive control—gain supervision and optimization—is common. For series or shunt power electronic devices, control mode selection allows bang–bang control, synchronizing versus damping control, and other nonlinear and adaptive strategies.

13.3.8 Current Injection by Voltage Sourced Inverters

Advanced power electronic controlled equipment employs gate turn-off thyristors, IGCTs, or IGBTs. Reference [6] describes use of these devices for oscillation damping. As with conventional thyristor-based equipment, it is often effective for voltage source inverter control to also direct mechanical switching.

Voltage sourced inverters may also be used for real power series or shunt injection. Superconducting magnetic energy storage (SMES) or battery storage is the most common. For angle stability control, injection of real power is more effective than reactive power. For transient stability improvement, SMES can be of smaller MVA size and lower cost than a STATCOM. SMES is less location dependent than a STATCOM.

13.3.9 Fast Voltage Phase Angle Control

Voltage phase angles and thereby rotor angles can be directly and rapidly controlled by voltage sourced inverter series injection or by power electronic controlled phase shifting transformers. This provides powerful stability control. Although one type of thyristor-controlled phase shifting transformer was developed over 20 years ago [60], high cost has presumably prevented installations. Reference [61] describes an application study.

As modular devices, multiple voltage sourced converters can be combined in several shunt and series arrangements, and as back-to-back HVDC links. Reactive power injection devices include the shunt static compensator (STATCOM), static synchronous series compensator (SSSC), unified power flow controller (UPFC), and interline power flow controller (IPFC). The convertible static compensator (CSC) allows multiple configurations with one installation in service. These devices tend to be quite expensive and special purpose.

The UPFC combines shunt and series voltage sourced converters with common DC capacitor and controls, and provides shunt compensation, series compensation, and phase shifting transformer functions. At least one installation (not a transient stability application) is in service [62], along with a CSC installation [9].

One concept employs power electronic series or phase shifting equipment to control angles across an interconnection within a small range [63]. On a power–angle curve, this can be visualized as keeping high synchronizing coefficient (slope of power–angle curve) during disturbances.

BPA developed a novel method for transient stability by high-speed 120° phase rotation of transmission lines between networks losing synchronism [64]. This technique is very powerful (perhaps too powerful) and raises reliability and robustness issues especially in the usual case where several lines form the interconnection. It has not been implemented.

13.3.10 HVDC Link Supplementary Controls

HVDC links are installed for power transfer reasons. In contrast to the above power electronic devices, the available HVDC converters provide the actuators so that stability control is inexpensive. For long distance HVDC links within a synchronous network, HVDC modulation can provide powerful stabilization, with active and reactive power injections at each converter. Control robustness, however, is a concern [6,10].

References [6,65–67] describe HVDC link stability controls. The Pacific HVDC Intertie modulation control implemented in 1976 is unique in that a remote (wide-area) input signal from the parallel Pacific AC Intertie was used [66,67]. Figure 13.9 shows commissioning test results.

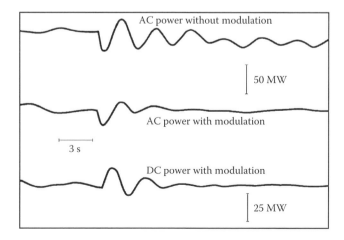

FIGURE 13.9 System response to Pacific AC Intertie series capacitor bypass with and without DC modulation. (From Cresap, R.L. et al., *IEEE Trans. Power App. Syst.*, PAS-98, 1053, 1978.)

13.3.11 Adjustable Speed (Doubly Fed) Synchronous Machines

Reference [68] summarizes stability benefits of adjustable speed synchronous machines that have been developed for pumped storage applications in Japan. Fast digital control of excitation frequency enables direct control of rotor angle.

13.3.12 Controlled Separation and Underfrequency Load Shedding

For very severe disturbances and failures, maintaining synchronism may not be possible or cost-effective. Controlled separation based on out-of-step detection or parallel path outages mitigates the effects of instability. Stable islands are formed, but underfrequency load shedding may be required in islands that were importing power.

References [34,69–71] describe advanced controlled separation schemes. Recent proposals advocate use of voltage phase angle measurements for controlled separation.

13.4 Dynamic Security Assessment

Control design and settings, along with transfer limits, are usually based on off-line simulation (time and frequency domain) and on field tests. Controls must then operate appropriately for a variety of operating conditions and disturbances.

Recently, however, on-line dynamic (or transient) stability and security assessment software have been developed. State estimation and on-line power flow provide the base operating conditions. Simulation of potential disturbances is then based on actual operating conditions, reducing uncertainty of the control environment. Dynamic security assessment is presently used to determine arming levels for generator tripping controls [72,73].

With today's computer capabilities, hundreds or thousands of large-scale simulations may be run each day to provide an organized database of system stability properties. Security assessment is made efficient by techniques such as fast screening and contingency selection, and smart termination of strongly stable or unstable cases. Parallel computation is straightforward using multiple workstations for different simulation cases; common initiation may be used for the different contingencies.

In the future, dynamic security assessment may be used for control adaptation to current operating conditions. Another possibility is stability control based on neural network or decision-tree pattern recognition. Dynamic security assessment provides the database for pattern recognition techniques. Pattern recognition may be considered data compression of security assessment results.

Industry restructuring requiring near real-time power transfer capability determination may accelerate the implementation of dynamic security assessment, facilitating advanced stability controls.

13.5 "Intelligent" Controls

Mention has already been made of rule-based controls and pattern recognition based controls. As a possibility, Ref. [74] describes a sophisticated self-organizing neural fuzzy controller (SONFC) based on the speed–acceleration phase plane. Compared to the angle–speed phase plane, control tends to be faster and both final states are zero (using angle, the postdisturbance equilibrium angle is not known in advance). The controllers are located at generator plants. Acceleration and speed can be easily measured or computed using, for example, the techniques developed for PSSs.

The SONFC could be expanded to incorporate remote measurements. Dynamic security assessment simulations could be used for updating or retraining of the neural network fuzzy controller. The SONFC is suitable for generator tripping, series or shunt capacitor switching, HVDC control, etc.

13.6 Wide-Area Stability Controls

The development of synchronized phasor measurements, fiber optic communications, digital controllers, and other IT advances have spurred development of wide-area controls. Wide-area controls offer increased observability and controllability, and as mentioned above, may be either continuous or discontinuous. They may augment local controls, or provide supervisory or adaptive functions rather than primary control. In particular, voltage phase angles, related to generator rotor angles, are often advocated as input signals.

The additional time delays because of communications are a concern, and increase the potential for adverse dynamic interactions. Figure 13.10, however, shows that latency for fiber optic communications (SONET) can be less than 25 ms, which is adequate for interarea stability.

Wide-area continuous controls include PSSs applied to generator automatic voltage regulators, and to static var compensators and other power electronic devices. For some power systems, wide-area controls are technically more effective than local controls [75,76].

Referring to Figure 13.5, discontinuous controls are often wide-area. Control inputs can be from multiple locations and control output actions can be taken at multiple locations. Most wide-area discontinuous controls directly detect fault or outage events (feedforward control). These controls generally involve preplanned binary logic rules and employ programmable logic controllers. For example, if line A and line B trip, then disconnect sending-end generators at power plants C and D. These schemes can be quite complex—BPA's remedial action scheme for the Pacific AC Intertie comprises around 1000 AND/OR decisions, with fault tolerant logic computers at two control centers.

BPA is developing a feedback wide-area stability and voltage control system (WACS) employing discontinuous control actions [77]. Inputs are from phasor measurements at eight locations, with generator tripping and capacitor or reactor switching actions available at many locations via existing remedial action scheme circuits. The WACS controller has two algorithms that cater to both angle and voltage stability problems.

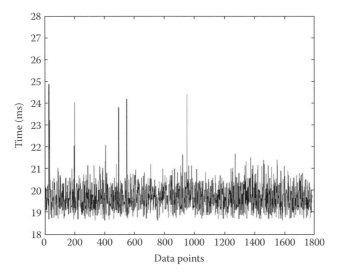

FIGURE 13.10 Fiber optic communications latency over 1 min. Bonneville Power Administration phasor measurement unit at Slatt Substation to BPA control center. (From Taylor, C.W. et al., *Proc. IEEE Special Issue on Energy Infrastructure Defense Systems*, 93, 892, 2005. With permission.)

13.7 Effect of Industry Restructuring on Stability Controls

Industry restructuring has many impacts on power system stability. Frequently changing power transfer patterns cause new stability problems. Most stability and transfer capability problems must be solved by new controls and new substation equipment, rather than by new transmission lines.

Different ownership of generation, transmission, and distribution makes the necessary power system engineering more difficult. New power industry reliability standards along with ancillary services mechanisms are being developed. Generator or load tripping, fast valving, higher than standard exciter ceilings, and PSSs may be ancillary services. In large interconnections, independent grid operators or reliability coordination centers may facilitate dynamic security assessment and centralized stability controls.

13.8 Experience from Recent Power Failures

Recent cascading power outages demonstrated the impact of control and protection failures, the need for "defense-in-depth," and the need for advanced stability controls.

The July 2, 1996 and August 10, 1996 power failures [78–81] in western North America, the August 14, 2003 failure in northeastern North America [82], and other failures demonstrate need for improvements and innovations in stability control areas such as

- Fast insertion of reactive power compensation, and fast generator tripping using response-based controls
- Special HVDC and SVC control
- PSS design and tuning
- Controlled separation
- Power system modeling and data validation for control design
- Adaptation of controls to actual operating conditions
- Local or wide-area automatic load shedding
- Prioritized upgrade of control and protection equipment including generator excitation equipment

13.9 Summary

Power system angle stability can be improved by a wide variety of controls. Some methods have been used effectively for many years, both at generating plants and in transmission networks. New control techniques and actuating equipment are promising.

We provide a broad survey of available stability control techniques with emphasis on implemented controls, and on new and emerging technology.

References

1. CIGRÉ TF 38.02.12, Criteria and Countermeasures for Voltage Collapse, CIGRÉ Brochure No. 101, Summary in *Electra*, October 1995.
2. CIGRÉ WG 34.08, Protection against Voltage Collapse, CIGRÉ Brochure No. 128, 1998. Summary in *Electra*, 179, 111–126, August 1998.
3. IEEE Power System Relaying Committee WG K12, System Protection and Voltage Stability, 93 THO 596-7 PWR, 1993.
4. Taylor, C.W., *Power System Voltage Stability*, McGraw-Hill, New York, 1994.
5. Van Cutsem, T. and Vournas, C., *Voltage Stability of Electric Power Systems*, Kluwer Academic, Dordrecht, the Netherlands, 1998.
6. CIGRÉ TF 38.01.07, Analysis and Control of Power System Oscillations, Brochure No. 111, December 1996.

7. Kundur, P., *Power System Stability and Control*, McGraw-Hill, New York, 1994.
8. IEEE Discrete Supplementary Control Task Force, A description of discrete supplementary controls for stability, *IEEE Transactions on Power Apparatus and Systems*, PAS-97, 149–165, January/February 1978.
9. Arabi, S., Hamadanizadeh, H., and Fardanesh, B., Convertible static compensator performance studies on the NY state transmission system, *IEEE Transactions on Power Systems*, 17(3), 701–706, August 2002.
10. Hauer, J.F., Robust damping controls for large power systems, *IEEE Control Systems Magazine*, 9, 12–19, January 1989.
11. Studt, T., Computer scientists search for ties to biological intelligence, *R&D Magazine*, 40, 77–78, October 1998.
12. IEEE Special Stability Controls Working Group, Annotated bibliography on power system stability controls: 1986–1994, *IEEE Transactions on Power Systems*, 11(2), 794–800, August 1996.
13. Djakov, A.F., Bondarenko, A., Portnoi, M.G., Semenov, V.A., Gluskin, I.Z., Kovalev, V.D., Berdnikov, V.I., and Stroev, V.A., The operation of integrated power systems close to operating limits with the help of emergency control systems, *CIGRÉ*, Paper 39–109, 1998.
14. Anderson, P.M. and LeReverend, B.K. (IEEE/CIGRÉ Committee Report), Industry experience with special protection schemes, *IEEE Transactions on Power Systems*, 11(3), 1166–1179, August 1996.
15. Dodge, D., Doel, W., and Smith, S., Power system stability control using fault tolerant technology, ISA instrumentation in power industry, *Proceedings of the 33rd Power Instrumentation Symposium*, Vol. 33, paper 90–1323, May 21–23, Toronto, Ontario, Canada, 1990.
16. CIGRÉ TF 13.00.1, Controlled switching—A state-of-the-art survey, *Electra*, 163, 65–97, December 1995.
17. Brunke, J.H., Esztergalyos, J.H., Khan, A.H., and Johnson, D.S., Benefits of microprocessor-based circuit breaker control, CIGRÉ, Paper 23/13-10, 1994.
18. Furumasu, B.C. and Hasibar, R.M., Design and installation of 500-kV back-to-back shunt capacitor banks, *IEEE Transactions on Power Delivery*, 7(2), 539–545, April 1992.
19. Torizuka, T. and Tanaka, H., An outline of power system technologies in Japan, *Electric Power Systems Research*, 44, 1–5, 1998.
20. IEEE Digital Excitation Applications Task Force, Digital excitation technology—A review of features, functions and benefits, *IEEE Transactions on Energy Conversion*, 12(3), 255–258, September 1997.
21. Bollinger, K.E., Nettleton, L., Greenwood-Madsen, T., and Salyzyn, M., Experience with digital power system stabilizers at steam and hydro generating stations, *IEEE Transactions on Energy Conversion*, 8(2), 172–177, June 1993.
22. Hajagos, L.M. and Gerube, G.R., Utility experience with digital excitation systems, *IEEE Transactions on Power Systems*, 13(1), 165–170, February 1998.
23. Arcidiancone, V., Corsi, S., Ottaviani, G., Togno, S., Baroffio, G., Raffaelli, C., and Rosa, E., The ENEL's experience on the evolution of excitation control systems through microprocessor technology, *IEEE Transactions on Energy Conversion*, 13(3), 292–299, September 1998.
24. Rubenstein, A.S. and Walkley, W.W., Control of reactive KVA with modern amplidyne voltage regulators, *AIEE Transactions*, Part III, 1957(12), 961–970, December 1957.
25. Dehdashti, A.S., Luini, J.F., and Peng, Z., Dynamic voltage control by remote voltage regulation for pumped storage plants, *IEEE Transactions on Power Systems*, 3(3), 1188–1192, August 1988.
26. Lee, D.C. and Kundur, P., Advanced excitation controls for power system stability enhancement, CIGRÉ, Paper 38–01, 1986.
27. Taylor, C.W., Mechenbier, J.R., and Matthews, C.E., Transient excitation boosting at grand coulee third power plant, *IEEE Transactions on Power Systems*, 8(3), 1291–1298, August 1993.
28. Bhatt, N.B., Field experience with momentary fast turbine valving and other special stability controls employed at AEP's Rockport Plant, *IEEE Transactions on Power Systems*, 11(1), 155–161, February 1996.

29. Stigers, C.A., Woods, C.S., Smith, J.R., and Setterstrom, R.D., The acceleration trend relay for generator stabilization at Colstrip, *IEEE Transactions on Power Delivery*, 12(3), 1074–1081, July 1997.
30. Kosterev, D.N., Esztergalyos, J., and Stigers, C.A., Feasibility study of using synchronized phasor measurements for generator dropping controls in the Colstrip system, *IEEE Transactions on Power Systems*, 13(3), 755–762, August 1998.
31. Matsuzawa, K., Yanagihashi, K., Tsukita, J., Sato, M., Nakamura, T., and Takeuchi, A., Stabilizing control system preventing loss of synchronism from extension and its actual operating experience, *IEEE Transactions on Power Systems*, 10(3), 1606–1613, August 1995.
32. Kojima, Y., Taoka, H., Oshida, H., and Goda, T., On-line modeling for emergency control systems, *IFAC/CIGRÉ Symposium on Control of Power Systems and Power Plant*, Beijing, China, pp. 627–632, 1997.
33. Imai, S., Syoji, T., Yanagihashi, K., Kojima, Y., Kowada, Y., Oshida, H., and Goda, T., Development of predictive prevention method for mid-term stability problem using only local information, *Transactions of IEE Japan*, 118-B(9), 1998.
34. Haner, J.M., Laughlin, T.D., and Taylor, C.W., Experience with the R–Rdot out-of-step relay, *IEEE Transactions on Power Delivery*, PWRD-1(2), 35–39, April 1986.
35. Berglund, R.O., Mittelstadt, W.A., Shelton, M.L., Barkan, P., Dewey, C.G., and Skreiner, K.M., One-cycle fault interruption at 500 kV: System benefits and breaker designs, *IEEE Transactions on Power Apparatus and Systems*, PAS-93, 1240–1251, September/October 1974.
36. Esztergalyos, J.H., Yee, M.T., Chamia, M., and Lieberman, S., The development and operation of an ultra high speed relaying system for EHV transmission lines, CIGRÉ, Paper 34–04, 1978.
37. Behrendt, K.C., Relay-to-relay digital logic communication for line protection, monitoring, and control, *Proceedings of the 23rd Annual Western Protective Relay Conference*, Spokane, WA, October 1996.
38. IEEE Committee Report, Single-pole switching for stability and reliability, *IEEE Transactions on Power Systems*, PWRS-1, 25–36, May 1986.
39. Belotelov, A.K., Dyakov, A.F., Fokin, G.G., Ilynichnin, V.V., Leviush, A.I., and Strelkov, V.M., Application of automatic reclosing in high voltage networks of the UPG of Russia under new conditions, CIGRÉ, Paper 34–203, 1998.
40. Knutsen, N., Single-phase switching of transmission lines using reactors for extinction of the secondary arc, CIGRÉ, Paper 310, 1962.
41. Kimbark, E.W., Suppression of ground-fault arcs on single-pole switched lines by shunt reactors, *IEEE Transactions on Power Apparatus and Systems*, PAS-83(3), 285–290, March 1964.
42. Hasibar, R.M., Legate, A.C., Brunke, J.H., and Peterson, W.G., The application of high-speed grounding switches for single-pole reclosing on 500-kV power systems, *IEEE Transactions on Power Apparatus and Systems*, PAS-100(4), 1512–1515, April 1981.
43. Fitton, D.S., Dunn, R.W., Aggarwal, R.K., Johns, A.T., and Bennett, A., Design and implementation of an adaptive single pole autoreclosure technique for transmission lines using artificial neural networks, *IEEE Transactions on Power Delivery*, 11(2), 748–756, April 1996.
44. Djuric, M.B. and Terzija, V.V., A new approach to the arcing faults detection for fast autoreclosure in transmission systems, *IEEE Transactions on Power Delivery*, 10(4), 1793–1798, October 1995.
45. Bayer, W., Habur, K., Povh, D., Jacobson, D.A., Guedes, J.M.G., and Marshall, D.A., Long distance transmission with parallel ac/dc link from Cahora Bassa (Mozambique) to South Africa and Zimbabwe, CIGRÉ, Paper 14–306, 1996.
46. Donnelly, M.K., Smith, J.R., Johnson, R.M., Hauer, J.F., Brush, R.W., and Adapa, R., Control of a dynamic brake to reduce turbine-generator shaft transient torques, *IEEE Transactions on Power Systems*, 8(1), 67–73, February 1993.
47. Taylor, C.W., Nassief, F.R., and Cresap, R.L., Northwest power pool transient stability and load shedding controls for generation–load imbalances, *IEEE Transactions on Power Apparatus and Systems*, PAS-100(7), 3486–3495, July 1981.

48. Samuelsson, O. and Eliasson, B., Damping of electro-mechanical oscillations in a multimachine system by direct load control, *IEEE Transactions on Power Systems*, 12(4), 4, 1604–1609, November 1997.
49. Kamwa, I., Grondin, R., Asber, D., Gingras, J.P., and Trudel, G., Active power stabilizers for multimachine power systems: Challenges and prospects, *IEEE Transactions on Power Systems*, 13(4), 1352–1358, November 1998.
50. Dagle, J., Distributed-FACTS: End-use load control for power system dynamic stability enhancement, EPRI Conference, *The Future of Power Delivery in the 21st Century*, La Jolla, San Diego, CA, pp. 18–20, November 1997.
51. Kimbark, E.W., Improvement of system stability by switched series capacitors, *IEEE Transactions on Power Apparatus and Systems*, PAS-85(2), 180–188, February 1966.
52. Christl, N., Sadek, K., Hedin, R., Lützelberger, P., Krause, P.E., Montoya, A.H., McKenna, S.M., and Torgerson, D., Advanced series compensation (ASC) with thyristor controlled impedance, CIGRÉ, Paper 14/37/38-05, 1992.
53. Piwko, R.J., Wegner, C.A., Furumasu, B.C., Damsky, B.L., and Eden, J.D., The slatt thyristor-controlled series capacitor project—Design, installation, commissioning and system testing, CIGRÉ, Paper 14-104, 1994.
54. Gama, C., Leoni, R.L., Gribel, J., Fraga, R., Eiras, M.J., Ping, W., Ricardo, A., Cavalcanti, J., and Tenório, R., Brazilian North–South interconnection—Application of thyristor controlled series compensation (TCSC) to damp inter-area oscillation mode, CIGRÉ, Paper 14–101, 1998.
55. Gama, C., Brazilian North–South interconnection—Control application and operating experience with a TCSC, *Proceedings of 1999 IEEE/PES Summer Meeting*, Edmonton, Alberta, Canada, July 1999, pp. 1103–1108.
56. Zhou, X. et al., Analysis and control of Yimin–Fentun 500 kV TCSC system, *Electric Power Systems Research*, 46, 157–168, 1998.
57. Zhou, X. and Liang, J., Overview of control schemes for TCSC to enhance the stability of power systems, *IEE Proceedings—Generation, Transmission and Distribution*, 146(2), 125–130, March 1999.
58. Sybille, G., Giroux, P., Dellwo, S., Mazur, R., and Sweezy, G., Simulator and field testing of Forbes SVC, *IEEE Transactions on Power Delivery*, 11(3), 1507–1514, July 1996.
59. Larsen, E.V. and Chow, J.H., SVC control design concepts For system dynamic performance, Application of Static Var Systems for System Dynamic Performance, IEEE Special Publication 87TH1087-5-PWR, 36–53, 1987.
60. Stemmler, H. and Güth, G., The thyristor-controlled static phase shifter—A new tool for power flow control in ac transmission systems, *Brown Boveri Review*, 69(3), 73–78, March 1982.
61. Fang, Y.J. and Macdonald, D.C., Dynamic quadrature booster as an aid to system stability, *IEE Proceedings—Generation, Transmission and Distribution*, 145(1), 41–47, January 1998.
62. Rahman, M., Ahmed, M., Gutman, R., O'Keefe, R.J., Nelson, R.J., and Bian, J., UPFC application on the AEP system: Planning considerations, *IEEE Transactions on Power Systems*, 12(4), 1695–1701, November 1997.
63. Christensen, J.F., New control strategies for utilizing power system network more effectively, *Electra*, 173, 5–16, August 1997.
64. Cresap, R.L., Taylor, C.W., and Kreipe, M.J., Transient stability enhancement by 120-degree phase rotation, *IEEE Transactions on Power Apparatus and Systems*, PAS-100, 745–753, February 1981.
65. IEEE Committee Report, HVDC controls for system dynamic performance, *IEEE Transactions on Power Systems*, 6(2), 743–752, May 1991.
66. Cresap, R.L., Scott, D.N., Mittelstadt, W.A., and Taylor, C.W., Operating experience with modulation of the Pacific HVDC Intertie, *IEEE Transactions on Power Apparatus and Systems*, PAS-98, 1053–1059, July/August 1978.
67. Cresap, R.L., Scott, D.N., Mittelstadt, W.A., and Taylor, C.W., Damping of Pacific AC Intertie oscillations via modulation of the parallel Pacific HVDC Intertie, CIGRÉ, Paper 14–05, 1978.

68. CIGRÉ TF 38.02.17, Advanced Angle Stability Controls, CIGRÉ Brochure No. 155, April 2000.
69. Ohura, Y., Suzuki, M., Yanagihashi, K., Yamaura, M., Omata, K., Nakamura, T., Mitamura, S., and Watanabe, H., A predictive out-of-step protection system based on observation of the phase difference between substations, *IEEE Transactions on Power Delivery*, 5(4), 1695–1704, November 1990.
70. Taylor, C.W., Haner, J.M., Hill, L.A., Mittelstadt, W.A., and Cresap, R.L., A new out-of-step relay with rate of change of apparent resistance augmentation, *IEEE Transactions on Power Apparatus and Systems*, PAS-102(3), 631–639, March 1983.
71. Centeno, V., Phadke, A.G., Edris, A., Benton, J., Gaudi, M., and Michel, G., An adaptive out of step relay, *IEEE Transactions on Power Delivery*, 12(1), 61–71, January 1997.
72. Mansour, Y., Vaahedi, E., Chang, A.Y., Corns, B.R., Garrett, B.W., Demaree, K., Athay, T., and Cheung, K., B.C. Hydro's on-line transient stability assessment (TSA) model development, analysis, and post-processing, *IEEE Transactions on Power Systems*, 10(1), 241–253, February 1995.
73. Ota, H., Kitayama, Y., Ito, H., Fukushima, N., Omata, K., Morita, K., and Kokai, Y., Development of transient stability control system (TSC system) based on on-line stability calculation, *IEEE Transactions on Power Systems*, 11(3), 1463–1472, August 1996.
74. Chang, H.-C. and Wang, M.-H., Neural network-based self-organizing fuzzy controller for transient stability of multimachine power systems, *IEEE Transactions on Energy Conversion*, 10(2), 339–347, June 1995.
75. Kamwa, I., Grondin, R., and Hebert, Y., Wide-area measurement based stabilizing control of large power systems—A decentralized/hierarchical approach, *IEEE Transactions on Power Systems*, 16(1), 136–153, February 2001.
76. Kamwa, I., Heniche, A., Trudel, G., Dobrescu, M., Grondin, R., and Lefebvre, D., Assessing the technical value of FACTS-based wide-area damping control loops, *Proceedings of IEEE/PES 2005 General Meeting*, San Francisco, CA, June 2005, Vol. 2, pp. 1734–1743.
77. Taylor, C.W., Erickson, D.C., Martin, K.E., Wilson, R.E., and Venkatasubramanian, V., WACS—Wide-area stability and voltage control system: R&D and on-line demonstration, *Proceedings of the IEEE Special Issue on Energy Infrastructure Defense Systems*, 93(5), 892–906, May 2005.
78. WSCC reports on July 2, 1996 and August 10, 1996 outages—Available at www.wscc.com
79. Taylor, C.W. and Erickson, D.C., Recording and analyzing the July 2 cascading outage, *IEEE Computer Applications in Power*, 10(1), 26–30, January 1997.
80. Hauer, J., Trudnowski, D., Rogers, G., Mittelstadt, W., Litzenberger, W., and Johnson, J., Keeping an eye on power system dynamics, *IEEE Computer Applications in Power*, 10(1), 26–30, January 1997.
81. Kosterev, D.N., Taylor, C.W., and Mittelstadt, W.A., Model validation for the August 10, 1996 WSCC system outage, *IEEE Transactions on Power Systems*, 14(3), 967–979, August 1999.
82. US-Canada Power System Outage Task Force, Final Report on the August 14, 2003 Blackout in the United States and Canada: Causes and Recommendations, April 2004.
83. IEEE/CIGRÉ Joint Task Force on Stability Terms and Definitions, Definition and classification of power system stability, *IEEE Transactions on Power Systems*, 19(2), 1387–1401, August 2004. Also published as CIGRÉ Technical Brochure No. 231, with summary in *Electra*, June 2003.

14
Power System Dynamic Modeling

	14.1 Modeling Requirements	**14**-1
	14.2 Generator Modeling	**14**-2
	Rotor Mechanical Model • Generator Electrical Model • Saturation Modeling	
	14.3 Excitation System Modeling	**14**-5
	14.4 Prime Mover Modeling	**14**-6
	Wind-Turbine-Generator Systems	
William W. Price	14.5 Load Modeling	**14**-8
Consultant	14.6 Transmission Device Models	**14**-10
	Static VAr Systems	
Juan Sanchez-Gasca	14.7 Dynamic Equivalents	**14**-12
General Electric Energy	References	**14**-13

14.1 Modeling Requirements

Analysis of power system dynamic performance requires the use of computational models representing the nonlinear differential–algebraic equations of the various system components. While scale models or analog models are sometimes used for this purpose, most power system dynamic analysis is performed with digital computers using specialized programs. These programs include a variety of models for generators, excitation systems, governor-turbine systems, loads, and other components. The user is, therefore, concerned with selecting the appropriate models for the problem at hand and determining the data to represent the specific equipment on his or her system. The focus of this article is on these concerns.

The choice of appropriate models depends heavily on the timescale of the problem being analyzed. Figure 14.1 shows the principal power system dynamic performance areas displayed on a logarithmic timescale ranging from microseconds to days. The lower end of the band for a particular item indicates the smallest time constants that need to be included for adequate modeling. The upper end indicates the approximate length of time that must be analyzed. It is possible to build a power system simulation model that includes all dynamic effects from very fast network inductance/capacitance effects to very slow economic dispatch of generation. However, for efficiency and ease of analysis, normal engineering practice dictates that only models incorporating the dynamic effects relevant to the particular performance area of concern be used.

This section focuses on the modeling required for analysis of power system stability, including transient stability, oscillatory stability, voltage stability, and frequency stability. For this purpose, it is normally adequate to represent the electrical network elements (transmission lines and transformers) by

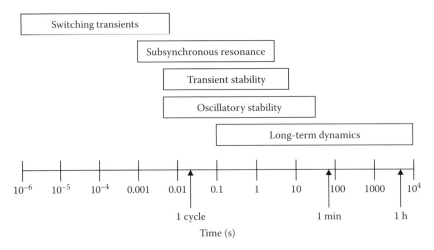

FIGURE 14.1 Timescale of power system dynamic phenomena.

algebraic equations. The effect of frequency changes on the inductive and capacitive reactances is sometimes included, but is usually neglected, since for most stability analysis, the frequency changes are small. The modeling of the various system components for stability analysis purposes is discussed in the remainder of this section. For greater detail, the reader is referred to Kundur (1994a) and the other references cited hereafter.

14.2 Generator Modeling

Most conventional generating plants use synchronous machines as electrical generators. The primary exceptions are wind-turbine-generator (WTG) systems and solar photovoltaic (PV) systems. WTG system modeling is discussed in Section 14.4.1. Solar PV systems involve DC to AC converters, which produce constant power (for constant irradiance) and either constant power factor, or voltage regulation using converter firing angle control. Standard models of both WTG and PV systems are starting to become available.

The model of a synchronous generator consists of two parts: the acceleration equations of the turbine-generator rotor and the generator electrical flux dynamics.

14.2.1 Rotor Mechanical Model

The acceleration equations are simply Newton's second law of motion applied to the rotating mass of the turbine-generator rotor, as shown in block diagram form in Figure 14.2. The following points should be noted:

1. The inertia constant (H) represents the stored energy in the rotor in MW-s, normalized to the MVA rating of the generator. Typical values are in the range of 3–15, depending on the type and size of the turbine generator. If the inertia (J) of the rotor is given in kg-m/s, H is computed as follows:

$$H = 5.48 \times 10^{-9} \frac{J(\text{RPM})^2}{\text{MVA rating}} \text{ MW-s/MVA}.$$

2. Sometimes, mechanical power and electrical power are used in this model instead of the corresponding torques. Since power equals torque multiplied by rotor speed, the difference is small for operation close to nominal speed. However, there will be some effect on the damping of oscillations (IEEE Transactions, February 1999).

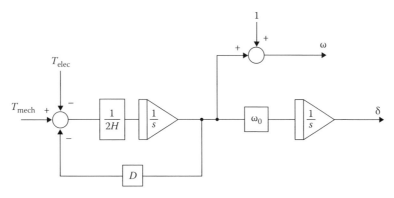

FIGURE 14.2 Generator rotor mechanical model.

3. Most models include the damping factor (D), shown in Figure 14.2. It is used to model oscillation damping effects that are not explicitly represented elsewhere in the system model. The selection of a value for this parameter has been the subject of much debate (IEEE Transactions, February 1999). Values from 0 to 4 or higher are sometimes used. The recommended practice is to avoid the use of this parameter by including sources of damping in other models, for example, generator amortisseur and eddy current effects, load frequency sensitivity, etc.

14.2.2 Generator Electrical Model

The equivalent circuit of a three-phase synchronous generator is usually rendered as shown in Figure 14.3. The three phases are transformed into a two-axis equivalent, with the direct (d) axis in phase with the rotor field winding and the quadrature (q) axis 90 electrical degrees ahead. For a more complete discussion of this transformation and of generator modeling, see IEEE Standard 1110-1991. In this equivalent circuit, r_a and L_l represent the resistance and leakage inductance of the generator stator, L_{ad} and L_{aq} represent the mutual inductance between stator and rotor, and the remaining elements represent rotor windings or equivalent windings. This equivalent circuit assumes that the mutual coupling between

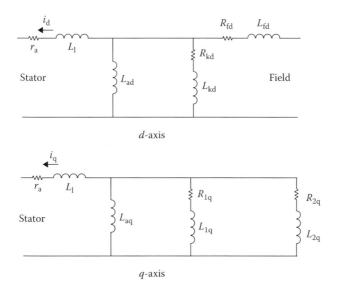

FIGURE 14.3 Generator equivalent circuit.

TABLE 14.1 Generator Parameter Relationships

	d-Axis	q-Axis
Synchronous inductance	$L_d = L_l + L_{ad}$	$L_q = L_l + L_{aq}$
Transient inductance	$L'_d = L_l + \dfrac{L_{ad}L_{fd}}{L_{ad}+L_{fd}}$	$L'_q = L_l + \dfrac{L_{aq}L_{1q}}{L_{aq}+L_{1q}}$
Subtransient inductance	$L''_d = L_l + \dfrac{L_{ad}L_{fd}L_{kd}}{L_{ad}L_{fd}+L_{ad}L_{kd}+L_{fd}L_{kd}}$	$L''_q = L_l + \dfrac{L_{aq}L_{1q}L_{2q}}{L_{aq}L_{1q}+L_{aq}L_{2d}+L_{1q}L_{2q}}$
Transient open circuit time constant	$T'_{do} = \dfrac{L_{ad}+L_{fd}}{\omega_0 R_{fd}}$	$T'_{qo} = \dfrac{L_{aq}+L_{1q}}{\omega_0 R_{1q}}$
Subtransient open circuit time constant	$T''_{do} = \dfrac{L_{ad}L_{fd}+L_{ad}L_{kd}+L_{fd}L_{kd}}{\omega_0 R_{kd}(L_{ad}+L_{fd})}$	$T''_{qo} = \dfrac{L_{aq}L_{1q}+L_{aq}L_{2q}+L_{1q}L_{2q}}{\omega_0 R_{2q}(L_{aq}+L_{1q})}$

the rotor windings and between the rotor and stator windings is the same. Additional elements can be added (IEEE Standard 1110-1991) to account for unequal mutual coupling, but most models do not include this, since the data are difficult to obtain and the effect is small.

The rotor circuit elements may represent either physical windings on the rotor or eddy currents flowing in the rotor body. For solid-iron rotor generators, such as steam-turbine generators, the field winding to which the DC excitation voltage is applied is normally the only physical winding. However, additional equivalent windings are required to represent the effects of eddy currents induced in the body of the rotor. Salient-pole generators, typically used for hydro-turbine generators, have laminated rotors with lower eddy currents. However, these rotors often have additional amortisseur (damper) windings embedded in the rotor.

Data for generator modeling are usually supplied as synchronous, transient, and subtransient inductances and open circuit time constants. The approximate relationships between these parameters and the equivalent network elements are shown in Table 14.1. Note that the inductance values are often referred to as reactances. At nominal frequency, the per-unit inductance and reactance values are the same. However, as used in the generator model, they are really inductances, which do not change with changing frequency.

These parameters are normally supplied by the manufacturer. Two values are often given for some of the inductance values, a saturated (rated voltage) and unsaturated (rated current) value. The unsaturated values should be used, since saturation is usually accounted for separately, as discussed in Section 14.2.3.

For salient-pole generators, one or more of the time constants and inductances may be absent from the data, since fewer equivalent circuits are required. Depending on the program, either separate models are provided for this case or the same model is used with certain parameters set to zero or equal to each other.

14.2.3 Saturation Modeling

Magnetic saturation effects may be incorporated into the generator electrical model in various ways. The data required from the manufacturer are the open circuit saturation curve, showing generator terminal voltage versus field current. If the field current is given in amperes, it can be converted to per unit by dividing by the field current at rated terminal voltage on the air gap (no saturation) line. (This value of field current is sometimes referred to as AFAG or IFAG.) Often the saturation data for a generator model are input as only two points on the saturation curve, for example, at rated voltage and 120% of rated voltage. The model then automatically fits a curve to these points.

The open circuit saturation curve characterizes saturation in the *d*-axis only. Ideally, saturation of the *q*-axis should also be represented, but the data for this are difficult to determine and are usually not provided. Some models provide an approximate representation of *q*-axis saturation based on the *d*-axis saturation data (IEEE Standard 1110-1991).

14.3 Excitation System Modeling

The excitation system provides the DC voltage to the field winding of the generator and modulates this voltage for control purposes. There are many different configurations and designs of excitation systems. Stability programs usually include a variety of models capable of representing most systems. These models normally include the IEEE standard excitation system models, described in IEEE Standard 421.5 (2005). Reference should be made to that document for a description of the various models and typical data for commonly used excitation system designs. This standard is periodically updated to include new excitation system designs.

The excitation system consists of several subsystems, as shown in Figure 14.4. The excitation power source provides the DC voltage and current at the levels required by the generator field. The excitation power may be provided by a rotating exciter, either a DC generator or an AC generator (alternator) and rectifier combination, or by controlled rectifiers supplied from the generator

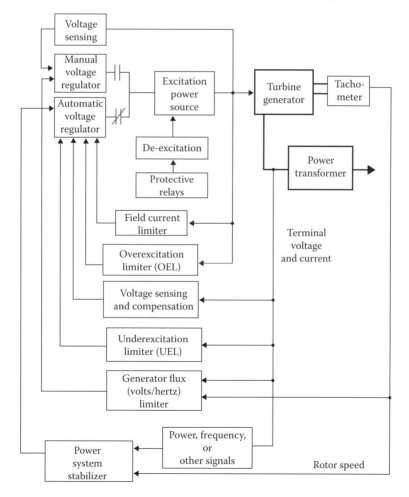

FIGURE 14.4 Excitation system model structure.

terminals (or other AC source). Excitation systems with these power sources are often classified as "DC," "AC," and "static," respectively. The maximum (ceiling) field voltage available from the excitation power source is an important parameter. Depending on the type of system, this ceiling voltage may be affected by the magnitude of the field current or the generator terminal voltage, and this dependency must be modeled since these values may change significantly during a disturbance.

The automatic voltage regulator (AVR) provides for control of the terminal voltage of the generator by changing the generator field voltage. There are a variety of designs for the AVR, including various means of ensuring stable response to transient changes in terminal voltage. The speed with which the field voltage can be changed is an important characteristic of the system. For the DC and most of the AC excitation systems, the AVR controls the field of the exciter. Therefore, the speed of response is limited by the exciter's time constant. The speed of response of excitation systems is characterized according to IEEE Standard 421.2 (1990).

A power system stabilizer (PSS) is frequently, but not always, included in an excitation system. It is designed to modulate the AVR input to increase damping of intermachine oscillations. The input to the PSS may be generator rotor speed, electrical power, or other signals. The PSS usually is designed with linear transfer functions whose parameters are tuned to produce positive damping for the range of oscillation frequencies of concern. It is important that reasonably correct values be used for these parameters. The output of the PSS is limited, usually to ±5% of rated generator terminal voltage, and this limit value must be included in the model.

The excitation system includes several other subsystems designed to protect the generator and excitation system from excessive duty under abnormal operating conditions. Normally, these limiters and protective modules do not come into play for analysis of transient and oscillatory stability. However, for longer-term simulations, particularly related to voltage instability, overexcitation limiters (OEL) and underexcitation limiters (UEL) may need to be modeled. While there are many designs for these limiters, typical systems are described in IEEE Transactions (December and September 1995).

14.4 Prime Mover Modeling

The system that drives the generator rotor is often referred to as the prime mover. The prime mover system includes the turbine (or other engine) driving the shaft, the speed control system, and the energy supply system for the turbine. The following are the most common prime mover systems:

1. Steam turbine
    a. Fossil fuel (coal, gas, or oil) boiler
    b. Nuclear reactor
2. Hydro turbine
3. Combustion turbine (gas turbine)
4. Combined cycle (gas turbine and steam turbine)
5. Wind turbine

Other less common and generally smaller prime movers include geothermal steam turbine, solar thermal turbine or reciprocating engine, and diesel engine.

For analysis of transient and oscillatory stability, greatly simplified models of the prime mover are sufficient since, with some exceptions, the response times of the prime movers to system disturbances are slow compared with the time duration of interest, typically 10–20 s or less. For simple transient stability analysis of only a few seconds duration, the prime mover model may be omitted altogether by assuming that the mechanical power output of the turbine remains constant. An exception is for a steam-turbine system equipped with "fast valving" or "early valve actuation" (EVA). These systems are designed to reduce turbine power output rapidly for nearby faults by quickly closing the intercept valves between the high-pressure and low-pressure turbine sections (Younkins et al., 1987).

For analysis of disturbances involving significant frequency excursions, the turbine and speed control (governor) systems must be modeled. Simplified models for steam and hydro-turbine-governor systems are given in IEEE Transactions (December 1973, February 1992) and these models are available in most stability programs. Models for gas turbines and combined cycle plants are less standard, but typical models have been described in several references (Rowen, 1983; Hannett and Khan, 1993; IEEE Transactions, August 1994).

For long-term simulations involving system islanding and large frequency excursions, more detailed modeling of the energy supply systems may be necessary. There are a great many configurations and designs for these systems. Models for typical systems have been published (IEEE Transactions, May 1991). However, detailed modeling is often less important than incorporating key factors that affect the plant response, such as whether the governor is in service and where the output limits are set.

For a fossil fuel steam plant, the coordination between the speed control and steam pressure control systems has an important impact on the speed with which the plant will respond to frequency excursions. If the governor directly controls the turbine valves (boiler-follow mode), the power output of the plant will respond quite rapidly, but may not be sustained due to reduction in steam pressure. If the governor controls fuel input to the boiler (turbine-follow mode), the response will be much slower but can be sustained. Modern coordinated controls will result in an intermediate response to these two extremes. The plant response will also be slowed by the use of "sliding pressure" control, in which valves are kept wide open and power output is adjusted by changing the steam pressure.

Hydro plants can respond quite rapidly to frequency changes if the governors are active. Some reduction in transient governor response is often required to avoid instability due to the "nonminimum phase" response characteristic of hydro turbines, which causes the initial response of power output to be in the opposite of the expected direction. This characteristic can be modeled approximately by the simple transfer function: $(1 - sT_w)/(1 + sT_w/2)$. The parameter T_w is called the water starting time and is a function of the length of the penstock and other physical dimensions. For high-head hydro plants with long penstocks and surge tanks, more detailed models of the hydraulic system may be necessary.

Gas (combustion) turbines can be controlled very rapidly, but are often operated at maximum output (base load), as determined by the exhaust temperature control system, in which case they cannot respond in the upward direction. However, if operated below base load, they may be able to provide output in excess of the base load value for a short period following a disturbance, until the exhaust temperature increases to its limit. Typical models for gas turbines and their controls are found in Rowen (1983), IEEE Transactions (February 1993), and Yee et al. (2008).

Combined cycle plants come in a great variety of configurations, which makes representation by a typical model difficult (IEEE Transactions, 1994). The steam turbine is supplied from a heat recovery steam generator (HRSG). Steam is generated by the exhaust from the gas turbines, sometimes with supplementary firing. Often the power output of the steam turbine is not directly controlled by the governor, but simply follows the changes in gas-turbine output as the exhaust heat changes. Since the time constants of the HRSG are very long (several minutes), the output of the steam turbine can be considered constant for most studies.

14.4.1 Wind-Turbine-Generator Systems

As large clusters of WTGs become more widely installed on power systems, they must be included in system dynamic performance studies. This requires special modeling because the generation technologies used for WTGs differ significantly from the directly connected synchronous generators that are universally used for all of the other types of generation discussed previously. There are four principal generation technologies in use for WTGs:

1. Induction generator—a "squirrel-cage" induction machine operating at essentially constant speed as determined by the power available in the wind. Terminal voltage cannot be controlled.
2. Induction generator with controlled field resistance—a wound-rotor induction machine with external rotor resistance controlled electronically to permit some variation, for example, ±10%, in rotor speed. Terminal voltage cannot be controlled.
3. Doubly fed asynchronous generator—a wound-rotor induction machine with its three-phase field voltage supplied by a power electronic (PE) converter connected to the machine terminals. The field voltage magnitude and frequency are controlled to regulate terminal voltage and to vary the machine speed over a wide, for example, ±30%, range.
4. Full converter system—a generator connected to the system through a PE converter. The generator speed is decoupled from system frequency and can be controlled as desired, while the converter is used to regulate voltage and supply reactive power.

Computational models have been developed for each of these technologies, plus the electrical controls required by the latter three (Kazachkov et al., 2003; Koessler et al., 2003; Miller et al., 2003; Pourbeik et al., 2003). Most large WTGs also have blade pitch control systems that regulate shaft speed in response to wind fluctuations and electrical system disturbances. Several industry groups are working toward the development of standard models for each of these technologies (Ellis et al., 2011).

For most studies, it is not necessary to represent the individual WTGs in a wind farm (cluster, park). One or a few aggregate machines can be used to represent the wind farm by the following procedures:

1. Aggregate WTG model same as individual but with MVA rating equal n times individual WTG rating
2. Aggregate generator step-up transformer same as individual but with MVA rating equal to n times individual transformer rating
3. Interconnection substation modeled as is
4. Aggregate collector system modeled as a single line with charging capacitance equal to total of the individual collector lines/cables and with series R and X adjusted to give approximately the same P and Q output at the interconnection substation as the full system with all WTGs at rated output

Stability analysis programs generally have simplified standard models for all of the WTG technologies, and may have more detailed models of some designs.

14.5 Load Modeling

For dynamic performance analysis, the transient and steady-state variation of the load P and Q with changes in bus voltage and frequency must be modeled. Accurate load modeling is difficult due to the complex and changing nature of the load and the difficulty in obtaining accurate data on its characteristics. Therefore, sensitivity studies are recommended to determine the impact of the load characteristics on the study results of interest. This will help to guide the selection of a conservative load model or focus attention on where load-modeling improvements should be sought.

For most power system analysis purposes, "load" refers to the real and reactive power supplied to lower voltage subtransmission or distribution systems at buses represented in the network model. In addition to the variety of actual load devices connected to the system, the "load" includes the intervening distribution feeders, transformers, shunt capacitors, etc., and may include voltage control devices, including automatic tap-changing transformers, induction voltage regulators, automatically switched capacitors, etc.

For transient and oscillatory stability analysis, several levels of detail can be used, depending on the availability of information and the sensitivity of the results to the load modeling detail. IEEE Transactions (May 1993, August 1995) discuss recommended load-modeling procedures. A brief discussion is given here:

1. *Static load model*—The simplest model is to represent the active and reactive load components at each bus by a combination of constant impedance, constant current, and constant power components, with a simple frequency sensitivity factor, as shown in the following formula:

$$P = P_0\left[P_1\left(\frac{V}{V_0}\right)^2 + P_2\left(\frac{V}{V_0}\right) + P_3\right](1 + L_{DP}\Delta f)$$

$$Q = Q_0\left[Q_1\left(\frac{V}{V_0}\right)^2 + Q_2\left(\frac{V}{V_0}\right) + Q_3\right](1 + L_{DP}\Delta f)$$

If nothing is known about the characteristics of the load, it is recommended that constant current be used for the real power and constant impedance for the reactive power, with frequency factors of 1 and 2, respectively. This is based on the assumption that typical loads are about equally divided between motor loads and resistive (heating) loads.

Most stability programs provide for this type of load model, often called a ZIP model. Sometimes an exponential function of voltage is used instead of the three separate voltage terms. An exponent of 0 corresponds to constant power, 1 to constant current, and 2 to constant impedance. Intermediate values or larger values can be used if available data so indicate. The following, more general model, permitting greater modeling flexibility, is recommended in IEEE Transactions (August 1995):

$$P = P_0\left[K_{PZ}\left(\frac{V}{V_0}\right)^2 + K_{PI}\left(\frac{V}{V_0}\right) + K_{PC} + K_{P1}\left(\frac{V}{V_0}\right)^{npV1}(1 + n_{PF1}\Delta f) + K_{P2}\left(\frac{V}{V_0}\right)^{npV2}(1 + n_{PF2}\Delta f)\right]$$

$$Q = Q_0\left[K_{QZ}\left(\frac{V}{V_0}\right)^2 + K_{QI}\left(\frac{V}{V_0}\right) + K_{QC} + K_{Q1}\left(\frac{V}{V_0}\right)^{nQV1}(1 + n_{QF1}\Delta f) + K_{Q2}\left(\frac{V}{V_0}\right)^{nQV2}(1 + n_{QF2}\Delta f)\right]$$

2. *Induction motor dynamic model*—For loads subjected to large fluctuations in voltage and/or frequency, the dynamic characteristics of the motor loads become important. Induction motor models are usually available in stability programs. Except in the case of studies of large motors in an industrial plant, individual motors are not represented. But one or two motor models representing the aggregation of all of the motors supplied from a bus can be used to give the approximate effect of the motor dynamics (Nozari et al., 1987). Typical motor data are given in the General Electric Company Load Modeling Reference Manual (1987). For analysis of voltage instability and other low voltage conditions, motor load modeling must include the effects of motor stalling and low-voltage tripping by protective devices.
3. *Composite load model*—For particular studies, more accurate modeling of certain loads may be necessary. This may include representation of the approximate average feeder and transformer impedance as a series element between the network bus and the bus where the load models are connected. For long-term analysis, the automatic adjustment of transformer taps may be represented by simplified models. Several load components with different characteristics may be connected to the load bus to represent the composition of the load. The Western Electricity Coordinating Council (WECC) has developed a Composite Load Model with the structure shown in Figure 14.5 and described in Kosterev et al. (2008). Special attention is focused on the modeling of air conditioners, both three phase and single phase.

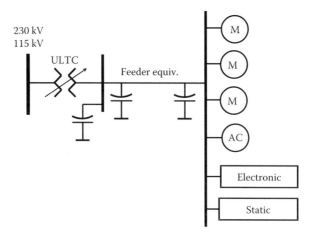

FIGURE 14.5 Composite load model structure.

Load-modeling data can be acquired in several ways, none of which is entirely satisfactory, but contribute to the knowledge of the load characteristics:

1. *Staged testing of load feeders*—Measurements can be made of changes in real and reactive power on distribution feeders when intentional changes are made in the voltage at the feeder, for example, by changing transformer taps or switching a shunt capacitor. The latter has the advantage of providing an abrupt change that may provide some information on the dynamic response of the load as well as the steady-state characteristics. This approach has limitations in that only a small range of voltage can be applied, and the results are only valid for the conditions (time of day, season, temperature, etc.) when the tests were conducted. This type of test is most useful to verify a load model determined by other means.
2. *System disturbance monitoring*—Measurements can be made of power, voltage, and frequency at various points in the system during system disturbances, which may produce larger voltage (and possibly frequency) changes than can be achieved during staged testing. This requires installation and maintenance of monitors throughout the system, but this is becoming common practice on many systems for other purposes. Again, the data obtained will only be valid for the conditions at the time of the disturbance, but over time many data points can be collected and correlated.
3. *Composition-based modeling*—Load models can also be developed by obtaining information on the composition of the load in particular areas of the system. Residential, commercial, and various types of industrial loads are composed of various proportions of specific load devices. The characteristics of many specific devices are well known (General Electric Company, 1987). The mix of devices can be determined from load surveys, customer SIC classifications, and typical compositions of different types of loads (General Electric Company, 1987).

14.6 Transmission Device Models

For the most part, the elements of the transmission system, including overhead lines, underground cables, and transformers, can be represented by the same algebraic models used for steady-state (power flow) analysis. Lines and cables are normally represented by a pi-equivalent with lumped values for the series resistance and inductance and the shunt capacitance. Transformers are normally represented by their leakage inductance, resistance, and tap ratios. Transformer magnetizing inductance and eddy current (no-load) losses are sometimes included.

Other transmission devices that require special modeling include high-voltage direct current (HVDC) systems, static VAr systems (SVSs), and other PE devices. HVDC systems vary widely in their controls and often require specialized models. Kundur (1994b) provides a good discussion of HVDC converters and controls. Modeling of SVS is discussed in Section 14.6.1. Other PE devices include a number of devices (TCSC, UPFC, etc.) generally referred to as flexible AC transmission systems (FACTS) devices. Due to the developmental nature of many of these technologies and specialized designs that are implemented, the modeling usually must be customized to the particular device.

14.6.1 Static VAr Systems

SVSs are shunt-connected devices whose primary application is rapid voltage control. When equipped with a damping controller, SVSs can also be used to damp electromechanical oscillations. An SVS consists of various combinations of continuous and discretely switched elements operated in a coordinated fashion by an automated control system (Pourbeik et al., 2010). Descriptions of SVS functionalities, components, and structures can be found in several publications (CIGRE, 1993; IEEE Transactions, February 1994; Kundur, 1994a; CIGRE, 2000).

Three SVS models have been developed within a Task Force organized by the WECC for implementation in the main commercial transient stability simulation programs in use in North America. These models reflect current technology and functionalities associated with these devices. The models are the following (Pourbeik et al., 2010):

1. Thyristor-controlled reactor-based SVS (static VAr compensator)—includes a thyristor-controlled reactor (TCR) together with a thyristor-switched capacitor (TSC) and filter banks (FB). The control action is continuous.
2. TSC/TSR-based SVS—includes only switched elements; the control action is discrete.
3. Voltage source converter-based SVS (STATCOM)—the fundamental PE device is a voltage source converter (VSC).

The three models listed can be used in conjunction with mechanically switched shunts (MSSs) and include commonly used control functions:

- Automatic voltage regulator
- Coordinated switching operation of MSSs
- Slow susceptance regulator (SVC)
- Slow current regulator (STATCOM)
- Deadband control
- Linear and nonlinear droop

The SVS models also include protective functions for anomalous voltage and current conditions. The model structures of the TCR-based SVS, TSC/TSR-based SVS, and STATCOM are very similar. Figure 14.6 shows the block diagram for the TCR-based SVS. The input and output of this model are a bus voltage and the SVS susceptance, respectively. The main voltage control path includes a PI controller (K_{pv}, K_{iv}) and a lead-lag block (T_{c2}, T_{b2}) for transient gain reduction; the time constant T_2 is typically small and reflects inherent delays in the hardware. The input labeled Vsig allows for the model to be connected with a damping controller.

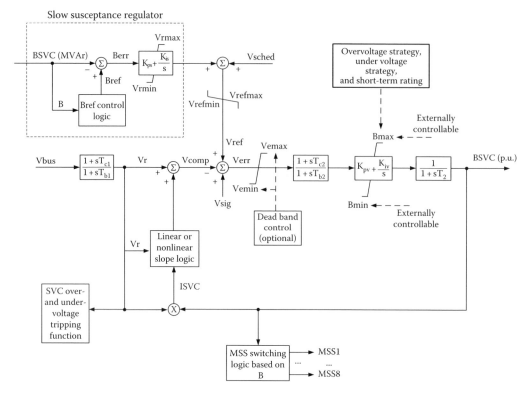

FIGURE 14.6 TCR-based SVS.

14.7 Dynamic Equivalents

It is often not feasible or necessary to include the entire interconnected power system in the model being used for a dynamic performance study. A certain portion of the system that is the focus of the study, the "study system," is represented in detail. The remainder of the system, the "external system," is represented by a simplified model that is called a dynamic equivalent. The requirements for the equivalent depend on the objective of the study and the characteristics of the system. Several types of equivalents are discussed as follows:

1. *Infinite bus*—If the external system is very large and stiff, compared with the study system, it may be adequate to represent it by an infinite bus, that is, a generator with very large inertia and very small impedance. This is often done for studies of industrial plant power systems or distribution systems that are connected to higher voltage transmission systems.
2. *Lumped inertia equivalent*—If the external system is not infinite with respect to the study system but is connected at a single point to the study system, a simple equivalent consisting of a single equivalent generator model may be used. The inertia of the generator is set approximately equal to the total inertia of all of the generators in the external area. The internal impedance of the equivalent generator should be set equal to the short-circuit (driving point) impedance of the external system viewed from the boundary bus.
3. *Coherent machine equivalent*—For more complex systems, especially when interarea oscillations are of interest, some form of coherent machine equivalent should be used. In this case, groups of generators in the external system are combined into single lumped inertia equivalents if these groups oscillate together for interarea modes of oscillation. Determination of such equivalents requires specialized calculations for which software is available (Price et al., 1996, 1998).

References

CIGRE Working Group 38-05, Analysis and Optimization of SVC Use in Transmission Systems, 1993.

CIGRE Working Group 14.19, Static Synchronous Compensator (STATCOM), 2000.

Damping representation for power system stability analysis, *IEEE Transactions*, IEEE Committee Report, PWRS-14, February 1999, 151–157.

Dynamic models for combined cycle plants in power systems, *IEEE Transactions*, IEEE Committee Report, PWRS-9, August 1994, 1698–1708.

Dynamic models for fossil fueled steam units in power system studies, *IEEE Transactions*, IEEE Committee Report, PWRS-6, May 1991, 753–761.

Dynamic models for steam and hydro turbines in power system studies, *IEEE Transactions*, IEEE Committee Report, PAS-92, December 1973, 1904–1915.

Ellis, A., Muljadi, E., Sanchez-Gasca, J., and Kazachkov, Y., Generic models for simulation of wind power plants in bulk system planning studies, *IEEE PES General Meeting*, July 2011.

General Electric Company, Load modeling for power flow and transient stability computer studies—Load modeling reference manual, EPRI Final Report EL-5003, Vol. 2, January 1987.

Hannett, L.N. and Khan, A., Combustion turbine dynamic model validation from tests, *IEEE Transactions*, PWRS-8, February 1993, 152–158.

Hydraulic turbine and turbine control models for system dynamic studies, *IEEE Transactions*, PWRS-7, February 1992, 167–179.

IEEE Standard 1110-1991, IEEE Guide for Synchronous Generator Modeling Practices in Stability Analysis, 1991.

IEEE Standard 421.2-1990, IEEE Guide for Identification, Testing, and Evaluation of the Dynamic Performance of Excitation Control Systems, 1990.

IEEE Standard 421.5-2005, IEEE Recommended Practice for Excitation System Models for Power System Stability Studies, 2005.

Kazachkov, Y.A., Feltes, J.W., and Zavadil, R., Modeling wind farms for power system stability studies, *PES General Meeting*, Toronto, Ontario, Canada, July 2003.

Koessler, R.J., Pillutla, S., Trinh, L.H., and Dickmander, D.L., Integration of large wind farms into utility grids, Part I, *PES General Meeting*, Toronto, Ontario, Canada, July 2003.

Kosterev, D., Meklin, A., Undrill, J., Lesieutre, B., Price, W., Chassin, D., Bravo, R., and Yang, S., Load modeling in power system studies: WECC Progress Update, *IEEE PES General Meeting*, Pittsburgh, PA, 2008.

Kundur, P., *Power System Stability and Control*, McGraw-Hill, New York, 1994a.

Kundur, P., *Power System Stability and Control*, Section 10.9, Modelling of HVDC systems, McGraw-Hill, New York, 1994b.

Load representation for dynamic performance analysis, *IEEE Transactions*, IEEE Committee Report, PWRS-8, May 1993, 472–482.

Miller, N.W., Sanchez-Gasca, J.J., Price, W.W., and Delmerico, R.W., Dynamic modeling of GE 1.5 and 3.6 MW wind turbine-generators for stability simulations, *PES General Meeting*, Toronto, Ontario, Canada, July 2003.

Nozari, F., Kankam, M.D., and Price, W.W., Aggregation of induction motors for transient stability load modeling, *IEEE Transactions*, PWRS-2, November 1987, 1096–1103.

Pourbeik, P., Koessler, R.J., Dickmander, D.L., and Wong, W., Integration of large wind farms into utility grids, Part 2, *PES General Meeting*, Toronto, Ontario, Canada, July 2003.

Pourbeik, P., Sullivan, D., Boström, A., Sanchez-Gasca, J., Kazachkov, Y., Kowalski, J., Salazar, A., and Sudduth, B., Developing generic static VAr system models—A WECC Task Force effort, *Proceedings of 2010 IEEE PES Transmission and Distribution Conference and Exposition*, New Orleans, LA, April 19–22, 2010.

Price, W.W., Hargrave, A.W., Hurysz, B.J., Chow, J.H., and Hirsch, P.M., Large-scale system testing of a power system dynamic equivalencing program, *IEEE Transactions*, PWRS-13, August 1998, 768–774.

Price, W.W., Hurysz, B.J., Chow, J.H., and Hargrave, A.W., Advances in power system dynamic equivalencing, *Proceedings of the Fifth Symposium of Specialists in Electric Operational and Expansion Planning (V SEPOPE)*, Recife, Brazil, May 1996, pp. 155–169.

Recommended models for overexcitation limiting devices, *IEEE Transactions*, IEEE Committee Report, EC-10, December 1995, 706–713.

Rowen, W.I., Simplified mathematical representations of heavy-duty gas turbines, *ASME Transactions (Journal of Engineering for Power)*, 105(1), October 1983, 865–869.

Standard load models for power flow and dynamic performance simulation, *IEEE Transactions*, IEEE Committee Report, PWRS-10, August 1995, 1302–1313.

Static var compensator models for power flow and dynamic performance simulation, *IEEE Transactions*, IEEE Committee Report, PWRS-9, February 1994, 229–240.

Underexcitation limiter models for power system stability studies, *IEEE Transactions*, IEEE Committee Report, EC-10, September 1995, 524–531.

Yee, S.K., Milanovic, J.V., and Hughes, F.M., Overview and comparative analysis of gas turbine models for system stability studies, *IEEE Transactions on Power Systems*, 23(1), February 2008, 108–118.

Younkins, T.D., Kure-Jensen, J. et al., Fast valving with reheat and straight condensing steam turbines, *IEEE Transactions*, PWRS-2, May 1987, 397–404.

15
Wide-Area Monitoring and Situational Awareness

Manu Parashar
ALSTOM Grid, Inc.

Jay C. Giri
ALSTOM Grid, Inc.

Reynaldo Nuqui
Asea Brown Boveri

Dmitry Kosterev
Bonneville Power Administration

R. Matthew Gardner
Dominion Virginia Power

Mark Adamiak
General Electric

Dan Trudnowski
Montana Tech

Aranya Chakrabortty
North Carolina State University

Rui Menezes de Moraes
Universidade Federal Fluminense

Vahid Madani
Pacific Gas & Electric

Jeff Dagle
Pacific Northwest National Laboratory

Walter Sattinger
Swiss Grid

Damir Novosel
Quanta Technology

15.1 Introduction 15-2
Drivers for Wide-Area Monitoring and Situational Awareness • What Is Situation Awareness? • Situation Awareness for Power Grid Operations • Grid Operator Visualization Advancements

15.2 WAMS Infrastructure 15-8
Phasor Measurement Unit • Phasor Data Concentrator • Phasor Gateway and NASPInet • Emerging Protocols and Standards

Mevludin Glavic
Quanta Technology

Yi Hu
Quanta Technology

Ian Dobson
Iowa State University

Arun Phadke
Virginia Tech

James S. Thorp
Virginia Tech

15.3 WAMS Monitoring Applications 15-15
   Angle Monitoring and Alarming • Small-Signal Stability
   Monitoring • Voltage Stability Monitoring • Transient Stability
   Monitoring • Improved State Estimation
15.4 WAMS in North America 15-35
   North American SynchroPhasor Initiative
15.5 WAMS Worldwide 15-36
   WAMS Applications in Europe • WAMS Applications in Brazil
15.6 WAMS Deployment Roadmap 15-40
References 15-41

15.1 Introduction

The power grid operating conditions are continually changing—every second, every minute, and every hour of the day. This is because changes in electricity demand dictate immediate, instantaneous changes in electricity production; consequentially, voltages, currents, and power flows are dynamically changing, at all times, all across the vast electricity delivery network called the power grid.

The challenge is to ensure that these changing power-system operating conditions always stay within safe limits for the present instance in time, as well as for a set of postulated, potential contingencies that might occur; if these safe limits are violated, equipment may be tripped by protection systems, which would further compromise the ability to deliver power across the grid. Maintaining the integrity of the grid means to ensure that operating conditions are always safe, while successfully supplying generation to meet the ever-changing customer demand.

Grid conditions need to be monitored in a timely periodic manner in order to immediately detect any adverse conditions, as soon as they arise, so that corrective actions could be implemented to mitigate potentially harmful conditions that could lead to a widespread grid collapse. Since there is a tremendous volume of constantly changing conditions across the grid, the challenge is to sift through these data to identify conditions that are potential imminent problems that need operator attention. The challenge is to convert vast amounts of data into useful information.

As early as in 1965, visualization of the grid was stated to be a high priority to ensure the integrity of grid operations. In the aftermath of the 1965 blackout of the Northeast United States and Canada, the findings from the blackout report included the following: "control centers should be equipped with display and recording equipment which provide operators with as clear a picture of system conditions as possible." Since then many more blackouts have occurred, small and large, around the world, and in almost all cases, improvements in visibility of grid conditions were identified as one of the primary recommendations.

15.1.1 Drivers for Wide-Area Monitoring and Situational Awareness

On August 14, 2003, the largest blackout in the history of the North American power grid occurred. Subsequently, numerous experts from across the industry were brought together to create a blackout investigation team. A primary objective of this team was to perform in-depth post-event analyses to identify the root causes of the event and, more importantly, to make recommendations on what could be done to prevent future occurrences of such events. The report [1] identified four root causes: inadequate system understanding, *inadequate situational awareness*, inadequate tree trimming, and inadequate reliability coordinator diagnostic support. This report gave a sudden new prominence to the term "situation awareness" or "situational awareness." Interviews that were conducted subsequent to the August 14, 2003, blackout identified the following gaps in operator situation awareness:

- Inadequate information sharing and communication between neighboring system operators.
- Information is available and not always used. Displays and visualization tools need to increase the availability and utility of real-time information.
- Information about grid operations decisions need to be shared along with real-time information.
- Need to redefine what "Normal Operations" means.

Control centers have traditionally relied on SCADA (supervisory control and data acquisition) measurements to monitor real-time grid conditions. SCADA measurements include flows in the transmission lines and voltages at substations and are used with a steady-state, positive-sequence model of the electrical network to calculate (using a state estimation application) the network conditions at all modeled substations. These are displayed to the operator at the control center. The limitation is that the operator is able to monitor conditions in only the portion of the electrical network that is being measured by one's SCADA system or the grid that is within one's own jurisdictional boundary. Wide-area monitoring system (WAMS) addresses this limitation. A WAMS objective is to provide real-time grid monitoring capabilities for a much larger expanse of the entire electrical interconnection; in other words, not just one's own SCADA system, but of neighboring systems as well!

WAMS is a recent technological evolution at control centers with a primary objective being to help improve situational awareness of the grid interconnection. WAMS is an evolving infrastructure that consists of measurements from across the grid, innovative analysis capabilities, and advanced visualization. The primary objective of WAMS is to provide grid operators with an enhanced view of the current dynamic grid conditions of the electric interconnection in a comprehensive intuitive manner and to facilitate prompt confident decision making to ensure improved grid reliability.

Trends in the electric utility industry that are driving the rapid deployment of WAMS include

- The aging electrical infrastructure—resulting in more equipment malfunctions
- The aging of the workforce—resulting in loss of grid operations expertise
- The growth of renewable energy resources—which provides less predictable, variable energy outputs that impact system reliability
- The growth of distributed generation and demand response—which results in greater uncertainty in generation/load balance at the low-voltage grid system, which impacts system reliability
- The growth of synchrophasor measurement devices—which provide fast, sub-second, time-tagged data from across the entire interconnection, which provide better grid visibility
- Advances in more sophisticated visualization technologies—which provide a more holistic view of grid conditions from multiple, diverse data sources from across the interconnection

WAMS are making their way into electric utility systems driven by requirements to improving system economics, efficiency, and reliability. Time-synchronized phasor data from WAMS sampled at a sub-second rate from locations all across the grid provide a more representative state of the power system and offer improved solutions to handle the operational challenges of modern power systems.

Figure 15.1 is a typical depiction of a WAMS. It consists of phasor measurement units (PMUs) dispersed across a wide geographical region to provide sub-second, time-tagged measurements of voltages and currents to several monitoring centers where software applications process the phasor data and convert it to information for the operator for improved monitoring and decision making to ensure improved operation of the power system.

- *Improving system economics*: Environmental constraints in some cases have put a hold on the construction of new transmission lines resulting in transmission bottlenecks; this results in existing facilities being loaded closer to their operational limits, which in turn increases operation risk.

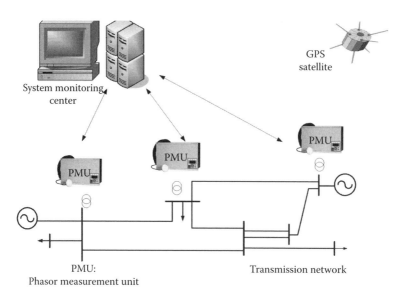

FIGURE 15.1 Wide-area monitoring systems (WAMS).

Bottlenecks impose dispatch limits on generating units and often result in noneconomic operation of the power system. Such limits are currently estimated a priori from the models of the power system at hand. The problem is that inaccurate models result in the calculation of inaccurate limits; also, the models are typically not updated frequently enough, so that the calculated limits may not reflect current conditions. Time-synchronized phasor data from WAMS are useful for calculating transmission limits in real time. For example, WAMS can estimate conductor temperatures and transmission line sags, parameters that reflect the current state of conductor capacity. On a wider scale, synchrophasor voltage angles can be used in conjunction with existing asynchronous SCADA measurements to significantly improve the estimation of the state of the system. Good state estimates increase the accuracy and lessen the uncertainties of line loadings and stability margins. With less model uncertainties, capacity margins reserved by operators for contingencies can be decreased, thereby releasing reserved transmission capacity to potentially lower cost generating units. Other dispatch strategies that are not economical, such as not using generating capacity in order to guard against oscillatory instabilities can be relaxed when wide-area monitoring and control systems are installed to guard against such instabilities.

- *Improving system reliability*: Driven by load growth and delays in transmission reinforcement, modern power systems are being operated closer to their stability limits and have an increased risk of angle and voltage instability and thermal overloads. Additionally, the variability and unpredictability of unconventional sources, such as from renewable energy resources, carry new operational challenges. Maintaining system reliability under these conditions require additional closed loop controls and improved operator tools for managing the operation of the whole network. With improved tools and visualization, operators can make more responsive and accurate dispatch and switching decisions to improve system reliability. WAMS can help operators increase their situational awareness by increasing observability of their control area and neighboring systems. For example, evolving situations from outside the control area can be viewed more promptly in conjunction with geographical information systems (GIS). With WAMS, information on oscillatory disturbances can be provided quickly to grid operators. Power systems typically leave traces of evidence in voltages and currents that can be used for identifying the nature of a disturbance. Wide-area time-synchronized frequency information can help operators locate remote generator, line or load outages that could potentially impact system integrity. In some

installations, WAMS have been deployed to monitor the risk of voltage instabilities of congested transmission corridors; such dedicated systems are more accurate and enhance operators' awareness of how close the system is to a voltage collapse.
- *Meeting smart grid challenges*: Smart grid initiatives are imposing new reliability and economic requirements that will impact how transmission systems will be monitored, protected, and controlled in the future. Large renewable energy resources will be deployed across the power system; these include transmission-level renewable energy "farms" (wind or solar generation) and distribution-level distributed generation resources. These deployments introduce higher variability in generation–load balance, which subsequently impacts system reliability. The variability also imposes new challenges to maintain power quality and frequency and voltage stability.

New wide-area control systems (WACS), based on remote measurements from multiple locations and acting on a range of controllers are being designed and implemented to handle such disturbances. In coordination with controllable devices such as flexible AC transmission systems (FACTS), HVDC, SVCs, TCSCs, storage, and generator excitation systems, WACS are designed to issue controls quickly to mitigate power oscillations or stability problems. WAMS can be extended into the lower voltage distribution network to allow operators to monitor evolving problem situations from the low voltage distribution system such as instability of distributed generators that may propagate into the transmission network

15.1.2 What Is Situation Awareness?

From Contemporary Psychology: "Situation awareness is, simply put, understanding the situation in which one is operating." Situation awareness is more comprehensively defined as "…the perception of the elements in the environment within a volume of time and space, the comprehension of their meaning, and the projection of their status in the near future" [2,3]. It transcends the more traditional human factors/visualization studies centered on transactions between the operator and the computer, and needs to focus on factors and interaction/tool requirements that address issues of sense making and information sharing.

Situation awareness is based on getting real-time information from the system being monitored. These inputs are typically asynchronous and from diverse, different parts of the system. The objective of situation awareness is to assimilate the real-time data, to assess the vulnerability of the current state, to make short-term projections based on personal experience or analytical tools, and to identify and issue corrective actions, if necessary. Hence, situation awareness consists of the following three stages: *perception, comprehension, and projection*.

These first three stages form the critical input to—but are separate from—implementation stages (if necessary), which include *decision making and action*.

15.1.2.1 Perception

Stage 1 involves perceiving the status, attributes, and dynamics of relevant elements in the environment. Within the power grid operations context, this equates to the operator being aware of the current state of the power grid—power flows across critical flowgates, congested paths, acceptable voltage profiles, adequate reactive reserves, available generation, frequency trends, line and equipment status, as well as weather patterns, winds, and lightning information. It is shown that 76% of situational awareness errors in pilots were related to not perceiving the needed information [4].

15.1.2.2 Comprehension

Comprehension of the situation is based on a synthesis of disjointed Stage 1 elements. Stage 2 goes beyond simply being aware of the elements that are present, by putting them together to form patterns with other elements, and developing a holistic picture of the environment including a comprehension of the significance of information and events.

15.1.2.3 Projection

It is the ability to utilize the status and dynamics of the elements and a comprehension of the situation (both Stage 1 and Stage 2), to project future actions of the elements in the environment.

15.1.2.4 Decision Making

This stage involves using all the knowledge gained in the first three stages to identify the best recourse to mitigate or eliminate the perceived problem. Here analytical tools and on-the-job experience are used to come up with a suitable plan of action.

15.1.2.5 Action

This is the final stage of implementing the decision that has been made. Here operator control displays are used to precisely locate the means by which the action can be implemented and then to actually issue a command action to the system. Once the action is issued the operator needs to verify successful implementation of the action and to ensure that the system is subsequently stabilized and does not need any additional analysis or control.

15.1.3 Situation Awareness for Power Grid Operations

For most of the time, the grid is in the Normal or Alert state and the electricity supply chain functions quite well with built-in automatic controls (e.g., AGC, protection schemes), and does not require operator intervention. It is only when a sudden disturbance occurs that an operator needs to be involved. When this happens, the typical operator's thought process sequence is as follows:

1. Just received a new problem alert!
    a. Is it a valid alert or a false alarm?
        i. Has any limit been violated?
            A. If so, how serious is this violation; can it be ignored?
2. Where is the problem located?
    a. What is the likely root cause?
3. Is there any corrective or mitigative action that could be taken now?
    a. What is the action?
        i. Must it be implemented now; or can I just wait and monitor the situation?
4. Has the problem been resolved?
    a. Is there any follow-up action I need to take?

15.1.4 Grid Operator Visualization Advancements

As the saying goes, "a picture is worth a thousand words." More importantly, the *correct picture is worth a million words!* What this means is that providing the operator with a concise pictorial depiction of voluminous grid data is meaningful; whereas, providing a pictorial depiction of voluminous grid data that *need immediate operator attention* is immensely more meaningful! This is the objective of an advanced, intelligent situation awareness; to provide timely information that needs prompt action, for current system conditions.

To successfully deploy situation awareness in power-system operations, the key is to capture data from a large number of different sources and to present the data in a manner that helps the operators' understanding of evolving events in complex, dynamic situations. Hence, while WAMS plays an important role in situation awareness, an advanced operator visualization framework is also required to be able to present real-time conditions in a timely, prompt manner; as well as to be able to navigate and drill down to discover additional information, such as the specific location of the problem; and more importantly to be able to identify and implement corrective actions to mitigate the potential risk of grid operations failure.

Power-system situation awareness visualization falls along the following perspectives, axes, or dimensions [5]:

1. *Spatial, geographical*: Situation awareness in power systems is highly spatial in nature. An operator may be responsible for a small region of the grid interconnection (one's own control area), or for multiple control areas, or for a large region of the interconnected grid. So, for example, a weather front moving across the region has different priorities based on the operator's regional responsibilities. Wide-area grid responsibilities are where GIS and other geo-spatial visualization technologies can be effectively applied.
2. *Voltage levels*: Grid operations and control is typically separated into transmission (higher voltages) and distribution (lower voltages). So, for example, a distribution operator may be more concerned about maintaining a uniform voltage profile, while the transmission operator is concerned about wide-area interconnection stability, market system interfaces, and area transfer capacity issues.
3. *Temporal*: Timescales for operators are typically in the seconds range. While real-time data refreshes from the traditional SCADA systems are typically received every 2–4 s from the field, the WAMS data are captured at a much higher sub-second rate. Regardless of the data source, operators have to assimilate data, make decisions, and issue action in the tens of seconds or minutes time frame. It is also important to correlate past history of grid conditions with current conditions to make an educated guess as to what future conditions are likely to be. Operator decisions are predicated on imminent future grid conditions.
4. *Functional*: Operators within a control center are typically assigned different roles and responsibilities. At a large utility, different operators typically work on different functional aspects of grid operations. These different functions include
   a. Voltage control
   b. Transmission dispatch—switching and outages
   c. Generation dispatch, AGC
   d. Reliability coordination, contingency analysis—what-if scenarios
   e. Oversight of remedial action or special protection schemes
   f. Supervisory oversight of all control center operators
   g. Market system operation, etc.

Situation awareness, therefore, consists of looking at the system from multiple different perspectives in a holistic manner. For an accurate assessment of the state of any complex, multidimensional system, the system needs to be "viewed" at from various different angles, perspectives, and potential what-if scenarios; these different views are then intelligently combined to synthesize a "true" assessment. Local regions are viewed microscopically and the entire system is viewed macroscopically. This intelligent synthesis of information from various diverse perspectives will improve the operator's capabilities and confidence to make prompt, "correct" decisions. This forms the very foundation of what is called an advanced visualization framework.

Synchrophasors and WAMS technology is a smart-grid-enabling technology that not only complements existing energy management systems, but also provides Situational Awareness tools with additional information to quickly assess the current grid conditions. Figure 15.2 illustrates how a PMU-based WAMS and network-model-based EMS hybrid solution within an advanced visualization framework can offer true situation awareness to grid operators.

Measurement-based WAMS techniques may be applied to quickly and accurately *assess* current grid conditions over a wide-area basis, such as monitoring phase angular separation as a measure of grid steady-state stress, detecting rapid changes in phase angle or frequency measurements—indicative of sudden weakening of power grid due to line outages or generator trips or to identity potential voltage or oscillatory stability problems.

Where the real-time network-model-based dynamic security assessment such as voltage stability, small-signal stability, and transient stability assessment fit in is providing the much needed *predictive*

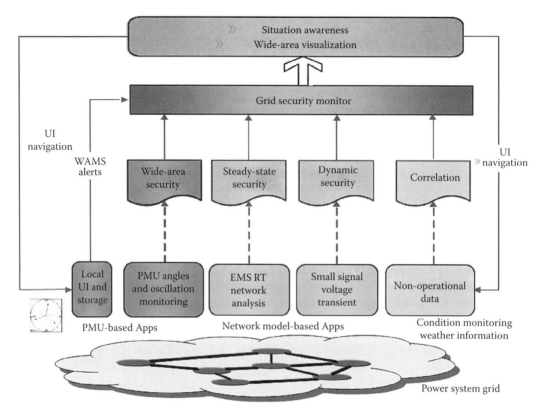

FIGURE 15.2 Integration of WAMS within an advanced visualization framework.

element to help the control center operator's decision-making process. Once the operators have made an assessment of the current state and its vulnerability, operators will need to rely on "What-if" analytical tools to be able to make decisions that will prevent adverse conditions if a specific contingency or disturbance were to occur and make recommendations on corrective actions. Thus, the focus shifts from "Problem Analysis" (reactive) to "Decision Making" (preventive).

15.2 WAMS Infrastructure

15.2.1 Phasor Measurement Unit

PMUs had their origin in a computer relay known as symmetrical component distance relay [6] for transmission line protection. This relaying algorithm calculated positive, negative, and zero sequence components of voltages and currents in order to facilitate execution of an efficient relaying algorithm. Starting in about 1982, the measurement algorithms were separated into a stand-alone function that became the basis of the modern PMU. In recent years, there has been a marked increase in PMU installations on power transmission networks as part of a WAMS. Applications of WAMS in power-system postmortem analysis, real-time monitoring, improved protection, and control have been—and are being—developed so that at present there are major undertakings to implement these applications on most large power systems around the world [7].

15.2.1.1 Phasors and Synchrophasors

A phasor is a complex number representing a sinusoidal function of time. A sinusoid given by $x(t) = X_m \cos(\omega t + \phi)$ has a phasor representation $\mathbf{X} \cong (X_m/\sqrt{2})\varepsilon^{j\phi}$. The PMUs use sampled data obtained from the sinusoid using a sampling clock with a sampling frequency ω_s, which is generally an integral multiple of the nominal signal frequency $\tilde{\omega}$. In order to avoid aliasing errors in phasor estimation from sampled data, it is necessary to use an anti-aliasing filter to attenuate frequencies greater than $\omega_s/2$. The sampled data version of the phasor with N samples per fundamental frequency period is given by the discrete Fourier transform

$$\mathbf{X} = \frac{\sqrt{2}}{N} \sum_{n=0}^{N-1} x(n\Delta T)\varepsilon^{-j\left(\frac{2n\pi}{N}\right)} \quad (15.1)$$

where ΔT is the time interval between samples.

A *synchrophasor* is a phasor representation that uses a common time signal (UTC) to define the instant when the measurement is made. By using a common timing signal, it becomes possible to combine phasors obtained from different locations on a common phasor diagram. The timing pulses provided by the global positioning system satellites are commonly used to furnish UTC reference in synchrophasor measurement systems (SMS).

15.2.1.2 Generic Phasor Measurement Unit

A generic PMU is represented in Figure 15.3. The GPS receiver is often an integral part of the PMU. Analog input signals are first filtered to remove extraneous interfering signals present in a power-system substation before applying the anti-aliasing filters. The timing pulse provided by the GPS receiver is used to produce a phase-locked oscillator at the required sampling rate. Synchrophasors are continuously computed with each arriving data sample, and the measured phasor with the time stamp based on the UTC provided by the GPS receiver is output as a continuous data stream to various applications. In practice a hierarchical system of PMUs and phasor data concentrators (PDCs) is employed to collect data over a wide area [8].

15.2.1.3 Positive Sequence Measurements

Most PMUs provide individual phase voltages and currents, as well as positive sequence quantities. The positive sequence phasor ($\mathbf{X}_1$) is calculated from phase quantities ($\mathbf{X}_a, \mathbf{X}_b, \mathbf{X}_c$) with the familiar equation: $\mathbf{X}_1 = (1/3)(\mathbf{X}_a + \mathbf{X}_b e^{j2\pi/3} + \mathbf{X}_c \varepsilon^{-j2\pi/3})$.

15.2.1.4 Transients and Off-Nominal Frequency Signals

Most PMUs use a fixed frequency sampling clock, which is keyed to the nominal frequency of the power system. However, the power-system frequency does vary constantly, and the PMU measurement system must take the prevailing frequency into account and apply required corrections to the estimated phasor.

In addition, there are transient phenomena due to faults and switching operations, harmonics etc., which must be reckoned with. As the PMU calculates synchrophasors with every sample of the input signal, there are phasor estimates that include effects of the transients. Care must be exercised

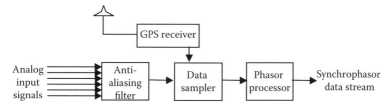

FIGURE 15.3 Building blocks of a generic PMU.

when using these phasors. Most PMU applications assume quasi-steady conditions over the measurement window. Phasors with transients in the measurement windows must be discarded for such applications.

15.2.1.5 IEEE Standards

IEEE standard C37.118 and its forthcoming variants C37.118.1 and C37.118.2 dictate the requirements for synchrophasors. PMUs compliant with these standards will assure interoperability of units of different manufacture.

15.2.2 Phasor Data Concentrator

A key attribute of PMUs is their ability to include a precise GPS timestamp with each measurement of when the measurement was taken. For applications that rely on data from multiple PMUs, it is vital that the measurements taken from these different devices be time aligned based on their original time tag to create a system-wide snapshot of synchronized measurements as a collective group before they can be useful to these applications. The PDC fulfills precisely this function. A PDC collects phasor data from multiple PMUs or other PDCs, aligns the data by time tag to create a time-synchronized dataset, and streams this combined dataset in real time to the applications. To accommodate the varying latencies in data delivery from individual PMUs and to ensure that these delayed data packets are not missed, PDCs typically buffer the input data streams and have a certain "wait time" before outputting the aggregated data stream.

Since PMUs may utilize various data formats (e.g., IEEE 1344, IEEE C37.118, BPA Stream), data rates, and communication protocols (e.g., TCP, UDP) for streaming data to the PDC, the PDC must, therefore, not only support these different formats on its input side, it should be able to down-sample (or up-sample) the input streams to a standard reporting rate, and massage the various datasets into a common format output stream. Appropriate anti-aliasing filters should be used whenever the data are down-sampled. Furthermore, as there may be multiple users of the data, the PDC should also be able to distribute received data to multiple users simultaneously, each of which may have different data requirements that are application specific.

The functions of a PDC can vary depending on its role or its location between the source PMUs and the higher-level applications. Broadly speaking, there are three levels of PDCs (Figure 15.4):

- *Local or substation PDC*: A local PDC is generally located at the substation for managing the collection and communication of time-synchronized data from multiple PMUs within the substation or neighboring substations, and sending this time-synchronized aggregated dataset to higher level concentrators at the control center. Since the local PDC is close to the PMU source, it is

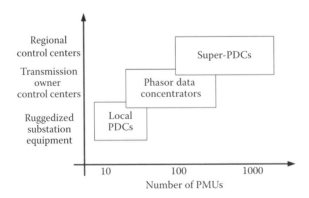

FIGURE 15.4 The three levels of PDCs: (1) local or substation PDCs, (2) control center PDCs, and (3) regional substation PDCs.

Samples (kbits/s)	Number of PMUs			
	2	10	40	100
30	57	220	836	2085
60	114	440	1672	4170
120	229	881	3345	8340

FIGURE 15.5 Data produced by different numbers of PMUs and sampling rates (kbits/s).

typically configured for minimal latency (i.e., short wait times). It is also most commonly utilized for all local control operations in order to avoid time delay from passing information and control decisions up through the communications and analysis system. Local PDCs may include a short-term storage to prevent against network failure. A local PDC is generally a hardware-based device that should require limited maintenance and can operate independently if it loses communications with the rest of the synchrophasor network.
- *Control center PDC*: This PDC operates within a control center environment and aggregates data from multiple PMUs and substation PDCs primarily located within a utility's footprint. It must be capable of simultaneously sending multiple output streams to different applications, such as visualization and alarming software, databases, and energy management systems, each with its own data requirements. Control center PDC architecture must be able to parallelize in order to handle expected future loads with minimal overhead penalty, satisfy the high-availability needs of a production system, perform its duties regardless of vendor and device type, and should use a hardware abstraction layer to protect the end user or data consumer. PDCs must be adaptable to new protocols and output formats as well as interfaces with data-using applications.
- *Super-PDC*: A super-PDC operates on a regional scale and is not only responsible for collecting and correlating phasor measurements from hundreds of PMUs and multiple PDCs spanning several utilities; it may also be responsible for brokering the data exchange between these utilities. In addition to supporting applications such as wide-area monitoring and visualization software, energy management systems and SCADA applications, it should be capable of archiving vast amounts of data (several TB/day) that are being gathered from the large number of PMU/PDC devices. Super-PDCs are, therefore, typically enterprise level software systems running on clustered server hardware to accommodate scalability to meet the growing PMU deployment and utility needs.

While some PMUs provide the ability to store many days of phasor data in the event of a communications failure, phasor data are typically stored at the PDC and/or super-PDC levels of the synchrophasor data system. Depending on the number of PMUs and the data rates, the volume of phasor data can quickly add up. This is illustrated in Figure 15.5, which shows the amount of data produced (in kbits/s) by differing numbers of PMUs at different sampling rates (assuming there are 20 measurements per PMU). Given the large volume of phasor data, data storage requirements can become significant. As an example, the Super-PDC operated by Tennessee Valley Authority (TVA) is presently networked to around 120 PMUs and is archiving approximately 36 GB/day.

15.2.3 Phasor Gateway and NASPInet

Although PMUs of various WAMS are typically deployed and owned by the respective asset owners, such as transmission owners, there are many situations in which the synchrophasor measurement data, as well as other types of data, messages, and control commands, will need to be exchanged between

WAMS of different entities, such as among asset owners, between asset owners and ISOs/RTOs/RCs, between asset owners and research communities, and so on.

The data exchange needs among various types of WAMS of different entities are typically driven by the real-time applications of these WAMS. However, other types of data exchange needs also exist. One example is exchanging stored synchrophasor data for many off-line post-event analyses and archiving applications. Many WAMS are also part of an integrated wide-area monitoring, protection, and control system (WAMPACS). For such systems, data exchange needs among various WAMS/WAMPACS of different entities will also be driven by wide-area protection and control applications.

Depending on the types of the applications, the quality of service (QoS) requirements, such as latency, reliability, etc., for data exchange could be very different from each other.

One major concern for an entity to exchange data with other entities is cyber security. Data should be provided to those that the data owner agreed to provide the data, and the entities that receive the data should also be assured that they are receiving the data from the data sources that they requested the data. The exchanged data could be used in mission critical applications, such as WAMS applications for assisting system operators in real-time decision making, or WACS/WAPS applications to perform real-time control and protection actions. Breach of security will expose the power grid to malicious attacks and illegal data access for unauthorized use.

Gateways can be used to shield entities from unauthorized access of their networks and provide a means to facilitate a secure and QoS assured data exchange among various entities.

The North American power industry has long recognized that existing inter-entity data exchange infrastructures, such as NERCnet for ICCP EMS/SCADA data exchange, are not capable of meeting the previously discussed data exchange needs. North American SynchroPhasor Initiative (NASPI) was created to facilitate "a collaborative effort between the U.S. Department of Energy, the North American Electric Reliability Corporation, and North American electric utilities, vendors, consultants, federal and private researchers and academics." NASPI's mission is "to improve power system reliability and visibility through wide area measurement and control." NASPI's ultimate objective "is to decentralize, expand, and standardize the current synchrophasor infrastructure through the introduction of a wide-area data exchange network across the entire North American continent (NASPInet)."

The data and network management task team (DNMTT) of NASPI had undertaken the task to determine the overall requirements of the NASPInet and its general architecture. Its work had led to an envisioned NASPInet consisting of two major components, phasor gateways (PG) and a data bus (DB). The NASPInet DB includes a NASPInet wide-area network (WAN) and associated services to provide basic connectivity, QoS management, performance monitoring, information security management, and access policy enforcement over different service classes of data exchanged through NASPInet. A NASPInet PG is the sole access point to NASPInet for an entity, which connects an entity to the NASPInet WAN and manages, in conjunction with NASPInet DB services, the connected devices (PMUs, PDCs, Applications, etc.,) on the entity's own network side, manages QoS, administers cyber security and access rights, performs necessary data conversions, and interfaces with connected entity's own network.

Based on the extensive preliminary work of DNMTT of NASPI, the U.S. Department of Energy (DOE) funded a project to develop the NASPInet specification for PG and DB, respectively.

NASPInet, as it is specified, will be a decentralized publish/subscribe-based data exchange network. There will be no single centralized authority to manage the data publishing and subscribing. Instead, all subscriptions to any published data object (hereafter called "signal") will be managed by the entity that publishes the data. The NASPInet DB services facilitate and data owners manage the data subscriptions of their published data through NASPInet PGs owned and operated by the data owners. Hence, we are slowly transitioning from the hierarchical "hub-n-spoke" type of an architectural design that comprised of the PMU → substation PDC → control center PDC → super-PDC, to a more distributed architecture consisting of the NASPInet PG and DB that integrate with the PDCs or directly with the PMUs in the field.

Though not explicitly defined in the specifications, the term "signal" for NASPInet published/subscribed data objects is not restricted to synchrophasor data points only. A "signal" of NASPInet could be any data object that will be exchanged across the NASPInet to support various wide-area monitoring, protection, and control applications, which could include but is not limited to synchrophasor data, analog measurement data, digital data, event notification, control command, and so on.

To support the data publishing and subscription, each published signal must be uniquely identifiable across the entire NASPInet with all associated information of the signal. The associated information of a signal includes type of the signal, where it is generated, relationships with other relevant signals, ownership information, where data are published, and so on. For example, a synchrophasor data point generated by a PDC through downsampling of a received synchrophasor input data point will need to provide such associated information as the signal type description (a synchrophasor; voltage or current measurement; positive sequence or single-phase value; accuracy class; and reporting rate; etc.), the PDC ID to indicate where it is generated and to link it to its ownership information contained in the associated information of the PDC, the input signal's ID to show its relationship with the input signal, and the publishing PG's ID to indicate where the signal can be subscribed.

NASPInet requires a NASPInet signal to be published to go through a signal registration process before making it available for subscription and uniquely identifiable. The registration process is accomplished by the relevant PG that publishes it and the Name & Directory Service (NDS) provided by NASPInet's data bus. The signal registration process includes steps of publishing PG provides all associated information of the signal to NDS, NDS registering the signal by validating and storing all associated information in its metadata database, NDS assigning a 128 bit unique identification number to the signal, and NDS providing the unique signal ID to publishing PG for future reference to the signal.

There is a diverse range of wide-area monitoring, protection and control applications that can and will benefit from using wide-area measurement data exchanged across the NASPInet. Different applications could have very different QoS requirements on the data in terms of accuracy, latency, availability, etc. In NASPInet specifications, five general classes of data defined initially by DNMTT were described with desired QoS requirements, such as latency, availability, etc., specified. It is anticipated that the number of data classes, the definition of each data class, and the QoS requirements for each data class may change over time as NASPInet grows. However, NASPInet is required to ensure that QoS of each data subscription is satisfied when there are hundreds and thousands of concurrent active subscriptions for different data classes.

NASPInet is required to implement a comprehensive resources management mechanism to ensure the QoS of each subscription of any data class, which includes resource condition monitoring, resource usage monitoring, QoS performance monitoring, QoS provisioning, and traffic management.

NASPInet resource condition monitoring function monitors the conditions of each resource both in real time and for historical archiving, including logging, reporting, and alarming on any failure or out-of-service condition for any of NASPInet resources.

NASPInet resource usage monitoring and tracking functions must be able to track all resources that are involved in the data delivery chain from data entering NASPInet to data leaving NASPInet. The logged information enables detailed resource usage to be archived, which can be used to determine instant, peak, and average loading level for each resource and so on.

NASPInet QoS performance monitoring function monitors the end-to-end QoS performance of each subscription from the publishing PG to subscribing PG. NASPInet QoS performance monitoring function achieves this through real-time QoS performance measurement, logging, reporting, and alarming for each subscription. The logged information enables historical QoS performance information for each subscription to be archived and analyzed, and the aggregated QoS performance information for overall NASPInet to be derived.

Because resources usage may not be evenly distributed and failure/outage may occur from time to time, NASPInet resources management mechanism is required to implement a traffic management function for real-time traffic management during both normal and abnormal conditions based on the assigned priority level of each subscription.

To safeguard the reliable operation and secure the data exchange, NASPInet is required to implement a comprehensive security framework for identification and authentication, access control, information assurance, and security monitoring and auditing.

Only authorized devices/equipment can be connected to NASPInet and use NASPInet's resources. Users must also be authorized to use the NASPInet. Unauthorized connections or users should be detected and reported.

NASPInet must be able to securely authorize, assign, and authenticate each and every device/equipment connected to the NASPInet. For example, Data Bus must be able to securely authorize, assign, and authenticate each and every PG connected to it, and a PG must be able to securely authorize, assign, and authenticate each and every data generating/consuming device (e.g., PMU/application) that is connected on entity's own network.

NASPInet is required to be able to set and enforce the proper levels of access privileges and rights to each and every device/equipment connected to the NASPInet on an individual device/equipment basis, as well as to each and every user on an individual user basis. For example, a data publishing only PG shall not be able to subscribe to any data published by other publishing PGs, and a user authorized to access historical data only shall not be able to subscribe to any real-time data.

The NASPInet is required to implement information assurance functions to enable NASPInet guarantee the confidentiality and integrity of the data exchanged through NASPInet that include secure subscription setup, subscription-based data and control flow security, key management, and information integrity assurance. These functions serve two main objectives: keep data and control flows from any unauthorized access and at the same time ensure that data and control flows have not been tampered with or degraded when traveling across the NASPInet.

One of the major challenges in NASPInet's implementation is to balance the requirement of using system resources more efficiently and simplifying the implementation and data handling under a one-to-many publish–subscribe scenario. It is anticipated that a real-time synchrophasor data stream published by a PG may be subscribed by many subscribers through many PGs. Under this condition, data multicast could be a very efficient mechanism to distribute data across the NASPInet with minimal bandwidth needs. However, in a generalized scenario, each subscriber may only subscribe to a subset of data contained in the data packet of the published data stream, while any two subscribers do not have exactly the same subset of data. Under this scenario, each subscription must have its own subscription key in order to ensure that other subscribers will not be able to know which data subset that it is subscribing.

15.2.4 Emerging Protocols and Standards

As communication requirements evolve, the standards community has been ready to update communication profiles to meet these requirements. One example of this is the recently developed IEC 90-5 report on routed GOOSE and Sample Value (SV) datasets that is an upgrade to the IEC 61850 GOOSE and SV profiles, respectively.

The delivery of synchrophasor measurements is usually required at multiple locations including multiple utilities. The implementation of this requirement can be met through a publisher–subscriber architecture. The IEC GOOSE and SV datasets are such mechanisms. The Generic Object Oriented Substation Event (GOOSE) is a user-defined dataset that is sent primarily on detection of a change of any value in the dataset. SV is also a user-defined dataset; however, the data is streamed at a user-defined rate (e.g., the reporting rate of the synchrophasors). These implementations achieve this functionality through the use of a multicast Ethernet data frame that contains no IP address and no transport protocol. As such, when a GOOSE or SV message reaches a router, the packet is dropped.

In order to address this issue, a new IEC 61850 routed profile has been defined. This profile takes either a GOOSE or SV dataset and wraps it in a UDP/multicast IP wrapper. The multicast IP address enables the router to "route" the message to multiple other locations. As a multicast message, the tendency is for the message to be sent everywhere in the network (which is not desirable). To address "where" the data are delivered, the involved routers must implement the Internet Gateway Management Protocol (IGMP) version 3. This protocol is initiated by the subscribers in the system and it enables the routers to "learn" the paths to which the multicast message is to be routed.

Security is a key element in any communication system. To this end, the 90-5 profile implements authentication of a transmitted message and optional encryption of the same message. Authentication is achieved through the addition of a secure hash algorithm (SHA) at the end of a message. The hash code can only be decoded through the use of the proper key. The 90-5 profile defines a key exchange mechanism that is responsible for delivering keys to all registered subscribers. Encryption of the "dataset" of the message is to be implemented using the advanced encryption algorithm (AES). The same key that unlocks the hash code is used to decode the encrypted message. The key manager is responsible for periodically updating the keys in the subscribers. If a subscriber is removed from the approved receive list, it is no longer updated with a new key.

15.3 WAMS Monitoring Applications

Some of the applications that are significantly improved or enabled by using PMUs include Angle/Frequency Monitoring and Alarming, Small-Signal Stability and Oscillations Monitoring, Event and Performance Analysis, Dynamic Model Validation, Transient and Voltage Stability, Advanced System Integrity Protection Schemes (SIPS) or Remedial Actions Schemes (RAS), and Planned Power System Separation, Dynamic State Estimation Linear State Measurements. Some of them are "low-hanging fruit" applications with lower deployment challenge (e.g., Angle/Frequency Monitoring and Alarming, Small-Signal Stability and Oscillations Monitoring). Furthermore, to realize the benefits of the aforementioned applications, it is necessary to place PMUs at optimal locations based on maximizing benefits for multiple applications. Those locations may also depend on leveraging existing or planned infrastructure, PMU placement in neighboring systems, and other factors such as upgradeability, maintenance, redundancy, security, and communication requirements. Several of these applications are discussed in the following sections.

15.3.1 Angle Monitoring and Alarming

This section explains how synchrophasor angle measurements can be combined to indicate grid stress. It is convenient to assume a DC load flow model of the power grid. That is, the lossless grid has unity voltage magnitudes, and the voltage phasor bus angles θ are linearly related to the real power injections at buses. Transients in the synchrophasor measurements are allowed to settle before they are combined in order to indicate the steady-state stress.

15.3.1.1 Simple Case of a Double Circuit Line

Consider two equal transmission lines joining bus A to bus B. Suppose that synchrophasor measurements of the voltage phasor angles θ_A and θ_B are available. Then the stress on the lines can be measured by the angle difference $\theta_{AB} = \theta_A - \theta_B$. The angle difference θ_{AB} is proportional to the combined power flow through the two lines, but there is a distinction between monitoring the combined power flow and the angle difference. If one of the lines trips, the power flow from A to B remains the same, but the equivalent inductance doubles, the equivalent susceptance halves, and the angle θ_{AB} doubles. The doubling of θ_{AB} correctly indicates that the single line is more stressed than the two lines. Although this case is simple, it contains the essential idea of stress angle monitoring, and it can be generalized in two ways. The first way considers an angle difference between any pair of buses in the grid [9–11]. The second way generalizes to an angle difference across an area of the grid [12–14].

15.3.1.2 Angles between Pair of Buses

If A and B are any two buses in the grid with synchrophasor measurements, the angle difference $\theta_{AB} = \theta_A - \theta_B$ can be a measure of grid stress. The general pattern is that large angle differences correspond to higher stress holds. The angle between two buses generally responds to changes throughout the grid. This makes it harder to interpret changes in the angle or to set thresholds for undue stress. This difficulty can be reduced by monitoring multiple angle differences between multiple pairs of buses. The data in time series of multiple angle differences can be examined for regularities such as typical ranges of values so that unusual stresses can be detected [15].

15.3.1.3 Angles across Areas

Consider an area of the grid in which there are synchrophasor measurements at all the buses at the border of the area (i.e., all the area tie lines have a bus with synchrophasor measurements). The idea is to combine together the voltage phasor angles at the border buses to form a single angle across the area. The border buses are divided into A buses and B buses and the angle θ_{AB} across the area from the A buses to B buses is formed as a weighted linear combination of the border bus voltage angles [14]. For example, if the A buses are all north of all the B buses, then the area θ_{AB} from the A buses to the B buses measures the north-to-south stress across the area. The weights for combining together the border bus angles to calculate θ_{AB} are not arbitrary; they are specific weights calculated from a DC load flow model of the area. The area angle θ_{AB} behaves according to circuit laws, and this makes the area angle behave in accordance with engineering intuition. For example, the power flow north to south through the area is proportional to θ_{AB}, and the constant of proportionality is the area susceptance.

15.3.1.4 Internal and External Area Stress

The area stress and the area angle θ_{AB} are influenced both by power flows into the area through tie lines from other areas and powers injected inside the area, including the border buses. In fact [14], the area angle is the sum of an external stress angle due to the tie-line flows and an internal stress angle due to powers injected inside the area:

$$\theta_{AB} = \theta_{AB}^{into} + \theta_{AB}^{area} \qquad (15.2)$$

θ_{AB} can be obtained from the border bus angles, and, if the currents through the area tie lines are also measured by synchrophasors, then the external stress angle θ_{AB}^{into} can also be obtained. Equation 15.2 shows how the internal stress angle θ_{AB}^{area} can be obtained from measurements.

The internal stress angle θ_{AB}^{area} is useful because it only responds to changes inside the area. For example, if lines trip or power is redispatched outside the area, the internal stress angle θ_{AB}^{area} does not change. But the internal stress angle θ_{AB}^{area} changes if lines trip or power is redispatched inside the area. Therefore, changes in θ_{AB}^{area} indicate changes in stress that can be localized to causes within the area. The localization makes θ_{AB}^{area} easier to interpret. The localization also holds for θ_{AB} in the special case that the area is chosen to stretch all the way across the power grid [12,13]. In this special case, the area is called a "cutset area."

In summary, there are two ways to combine together synchrophasor measurements to monitor grid stress. The angle differences between multiple pairs of buses are simply obtained and these data can be mined to detect undue stress or unusual events. Computing the angle across an area is straightforward, but there are the additional requirements of synchrophasor measurements at all the buses along the border of the area and a DC load flow model of the area. The area angle obeys circuit laws and can give more specific information about events inside the area.

15.3.2 Small-Signal Stability Monitoring

Time-synchronized measurements provide rich information for near-real-time situational awareness of a power-system's small-signal dynamic properties via visualization and advanced signal processing. This information is becoming critical for the improved operational reliability of interconnected grids. A given mode's properties are described by its frequency, damping, and shape (i.e., the mode's magnitude and phase characterizing the observability of the mode at a particular location). Modal frequencies and damping are useful indicators of power-system stress, usually declining with increased load or reduced grid capacity. Mode shape provides critical information for operational control actions. Over the past two decades, many signal-processing techniques have been developed and tested to conduct modal analysis using only time-synchronized actual-system measurements [16]. Some techniques are appropriate for transient signals while others are for ambient signal conditions.

Near-real-time operational knowledge of a power system's modal properties may provide critical information for control decisions and, thus, enable reliable grid operation at higher loading levels. For example, modal shape may someday be used to optimally determine generator and/or load-tripping schemes to improve the damping of a dangerously low-damped mode. The optimization involves minimizing load shedding and maximizing improved damping. The two enabling technologies for such real-time applications are a reliable real-time synchronized measurement system and accurate modal analysis signal-processing algorithms.

This section strives to provide an overview of the state of the art and points the reader to more detailed literature. A starting point for the novice is Ref. [16].

15.3.2.1 Actual System Examples from the Western North American Power System

A power system experiences a wide variety of power oscillations:

- Slow power oscillations related to mistuned AGC bias (typically 20–50 s period, or 0.02–0.05 Hz)
- Inter-area electro-mechanical power oscillations (typically 1–5 s period, or 0.2–1.0 Hz)
- Power plant and individual generator oscillations
- Wind turbine torsional oscillations (usually about 1.5–2.5 Hz)
- Control issues in generator, SVC, and HVDC systems
- Steam generator torsional oscillations (about 5, 9–11 Hz, etc.)

The oscillations are always there. For example, switching lights and starting/stopping electric motors cause momentary imbalance between electric demand and supply. If the system is stable, the power imbalances are settled in oscillatory way and are normally well dampened. The concern is when the oscillations are growing in amplitude. The growing oscillations can result from power-system stress, unusual operating conditions, or failed controllers (PSS, excitation, etc.). The system can either transient into the oscillation because of an outage/failure or slowly build into the oscillation because of an increased system stress or forced oscillation. A forced oscillation is a condition where an apparatus such as a generator controller has failed and is operating in a limit cycle. Growing power oscillations can cause line opening and generator tripping, and in the worst case, lead to cascading power blackouts. Another risk is that the interactions between oscillations can lead to major equipment damage. The most known phenomenon is sub-synchronous resonance when a generator torsional oscillation resonate with LC circuits of series compensated lines, have the risk of breaking generator shaft or causing large overvoltages on the transmission system. Recently this phenomenon was observed with wind generators. In any case, it is very important to detect sustained or growing oscillations and to take mitigation actions.

15.3.2.1.1 Unstable Oscillation in the U.S. Western Interconnection on August 10, 1996

On August 10, 1996, a major outage occurred in the Western Interconnection, breaking the interconnection into four islands, resulting in loss of 30,390 MW of load and affecting 7.49 million customers. A combination of line outages, heavy transfers, large angular separation across the system, low reactive

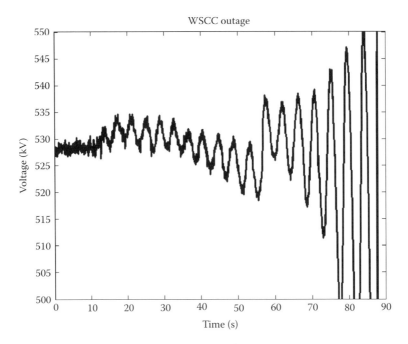

FIGURE 15.6 Malin 500 kV bus voltage, unstable oscillation during August 10, 1996, outage.

support, and equipment control issues put the system at the edge of collapse. The instability manifested in collapsing system voltages and growing North–South inter-area power oscillations [17]. Figure 15.6 shows a 500 kV bus voltage at the California–Oregon Intertie.

15.3.2.1.2 "Close Call" in the U.S. Western Interconnection on August 4, 2000

On August 4, 2000, a poorly dampened oscillation was observed across the Western Interconnection following a loss of a 400 MW tie line between British Columbia and Alberta. The oscillation lasted about 60 s. Although the power flow on the California–Oregon Intertie was well within the operating limits, power exports from Canada to United States were relatively high, and the relative phase angles between North and South were close to historic highs. The damping issues are more pronounced when the relative phase angles across the system are large. The event's response is shown in Figure 15.7.

15.3.2.1.3 Forced Inter-Area Power Oscillation on November 29, 2005

A forced oscillation occurred in the Western Interconnection on November 29, 2005. The oscillation was driven by steam supply instability at the Nova Joffre plant in Alberta. The plant has two gas turbines and a heat-recovery steam turbine. A fraction of the steam is typically sent to a process plant. An extractor control valve is set to keep the constant flow of steam to the process plant. Due to problems with a relief valve within the process plant, the extractor control valve began oscillating in a limit cycle at about 0.25 cycles per second. This resulted in power peak-to-peak oscillations of 15–20 MW observed at steam-turbine terminals. The power plant oscillation resonated with the North–South inter-area mode, causing peak-to-peak power oscillations on California–Oregon Intertie of 175 MW. This is shown in Figure 15.8.

15.3.2.1.4 Boundary Power Plant Oscillation on September 29, 2004

Boundary hydro-power plant is located in the northeastern part of the state of Washington. The plant has six hydro-power units with the total capacity of about 1050 MW. The plant is connected to Canada by a 230 kV line. There are three 230 kV lines connecting Boundary plant with the Spokane, WA area,

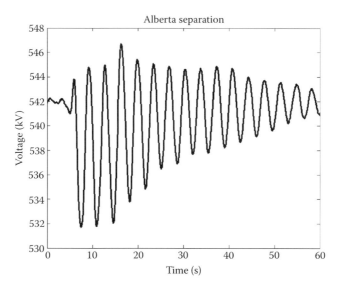

FIGURE 15.7 Malin 500 kV bus voltage, August 4, 2000, event.

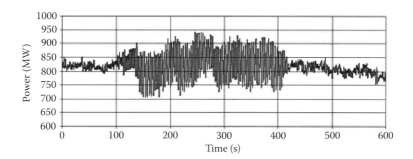

FIGURE 15.8 Forced oscillations on California–Oregon intertie on November 29, 2005.

referenced as Boundary–Bell 230 kV lines. The Boundary–Bell lines are about 90 miles long and have tap points feeding local area 115 kV subtransmission along the way.

On September 29, 2004, a tie line with Canada was out of service for maintenance, and Boundary power plant was radially connected to the Spokane area. A section of Boundary–Bell #3 line was also taken out of service for maintenance. Five out of six generating units were online. As the Boundary generation was ramping up to about 750 MW, the power oscillation developed at 0.8 Hz frequency. The oscillation was noticeable in Boundary 230 kV bus voltage, power plant output, and as far as Bell 230 kV bus voltage.

This Boundary plant oscillation was a classical example of a local electro-mechanical instability. The oscillation was the result of a weak transmission system due to several lines out of service. The appropriate operator actions were to reduce the power output until oscillation dampened and to operate at the reduced power output until the transmission lines are restored. This event is shown in Figure 15.9.

15.3.2.1.5 Pacific HVDC Intertie Oscillation on January 26, 2008

On January 26, 2008, a sustained oscillation occurred at Pacific HVDC intertie (PDCI). Transformer outages created a weak configuration on the inverter side, starting a high-frequency 4 Hz controller oscillation.

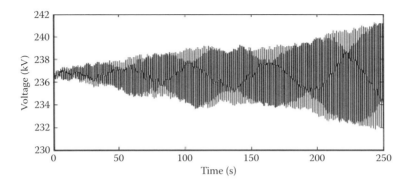

FIGURE 15.9 Growing oscillations at boundary on September 29, 2004.

15.3.2.2 Response Types

Analyzing and estimating power-system electromechanical dynamic effects are a challenging problem because the system is nonlinear, high order, and time varying; contains many electromechanical modes of oscillation close in frequency; and is primarily stochastic in nature. Design of signal-processing algorithms requires that one address each of these issues. Fortunately, the system behaves relatively linear when at a steady-state operating point [17].

We classify the response of a power system as one of two types: transient (sometimes termed a ringdown) and ambient. The basic assumption for the ambient case is that the system is excited by low-amplitude random variations (such as random load changes). This results in a response that is colored by the system dynamics. A transient response is typically larger in amplitude and is caused by a sudden switching action, or a sudden step or pulse input. The resulting time-domain response is a multimodal oscillation superimposed with the underlying ambient response.

The different types of responses are shown in Figure 15.10, which shows a widely published plot of the real power flowing on a major transmission line during the Western Interconnection event in 1996. Prior to the transient at the 400 s point, the system is in an ambient condition. After the ringdown at the 400 s point, the system returns to an ambient condition. The next event in the system causes an unstable oscillation.

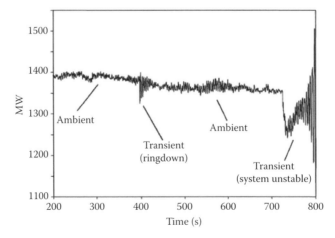

FIGURE 15.10 Real power flowing on a major transmission line during the Western North American power system breakup of 1996.

In terms of application, we classify modal frequency and damping estimation algorithms into two categories: (1) ringdown analyzers and (2) mode meters. A ringdown analysis tool operates specifically on the ringdown portion of the response; typically the first several cycles of the oscillation (5–20 s). Alternatively, a mode meter is applied to any portion of the response: ambient, transient, or combined ambient/transient. Ultimately, a mode meter is an automated tool that estimates modal properties continuously, and without reference to any exogenous system input.

15.3.2.3 Signal-Processing Methods for Estimating Modes

Many parametric methods have been applied to estimate power-system electromechanical modes. As stated previously, we classify these methods into two categories: ringdown analyzers and mode meters.

Ringdown analysis for power-system modal analysis is a relatively mature science. The underlying assumed signal model for these algorithms is a sum of damped sinusoids. The most widely studied ringdown analysis algorithm is termed Prony analysis. The pioneering paper by Hauer, Demeure, and Scharf [18] was the first to establish Prony analysis as a tool for power-system ringdown analysis. Many expansions and other algorithms have been researched since; the reader is referred to reference [16].

Ambient analysis of power-system data estimates the modes when the primary excitation to the system is random load variation, which results in a low-amplitude stochastic time series (ambient noise). A good place to begin ambient analysis is with non-parametric spectral-estimation methods, which are very robust since they make very few assumptions. The most widely used non-parametric method is the power spectral density [16], which provides an estimate of a signal's strength as a function of frequency. Thus, usually the dominate modes are clearly visible as peaks in the spectral estimate. While robust and insightful, non-parametric methods do not provide direct numerical estimates of a mode's damping and frequency. Therefore, to obtain further information parametric methods are applied.

The first application of signal-processing techniques for estimating modal frequency and damping terms from ambient data is contained in [19], where a Yule–Walker algorithm is employed. Many extensions and new algorithms have been explored since [19]; see [16] for an overview. This includes the regularized robust recursive least squares (R3LS) method [20].

An important component of a mode meter is the automated application of the algorithm. With all algorithms, several modes are estimated and many of them are "numerical artifacts." Typically, "modal energy" methods are used to determine which of the modes in the frequency range of the inter-area modes have the largest energy in the signal [21]. It is then assumed that this is the mode of most interest.

It is absolutely imperative to understand that because of the stochastic nature of the system, the accuracy of any mode estimation is limited. It is possible to significantly improve the estimation by exciting the system with a probing signal. A signal may be injected into the power system using a number of different actuators such as resistive brakes, generator excitation, or modulation of DC intertie signals. For example, operators of the western North American power system use both the 1400 MW Chief Joseph dynamic brake and modulation of the PDCI to routinely inject known probing signals into the system [16].

15.3.2.4 Mode Estimation Example

As mentioned previously, operators of the Western Interconnection periodically conduct extensive dynamic tests. These tests typically involve 0.5 s insertion of the Chief Joseph 1400 MW braking resistor in Washington state and probing of the power reference of the PDCI. The resulting system response provides rich data for testing mode-estimation algorithms. This section presents a few of these results.

Figure 15.11 shows the system response from a brake insertion along with several minutes of ambient data. The signal shown is the detrended real power flowing on a major transmission line.

Two recursive mode-estimation algorithms are applied to the data: the RLS and RRLS [20] algorithms. The resulting mode estimates are shown in Figures 15.12 and 15.13. The damping estimates for the 0.39 Hz mode are shown as this is the most lightly damped dominant mode. The results are

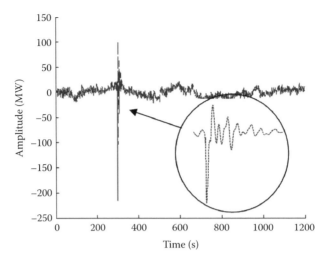

FIGURE 15.11 Brake response of U.S. Western Interconnection. Brake inserted at the 300 s point. Combined ambient and ringdown data from field measurements. Detrended power flowing on a major transmission line.

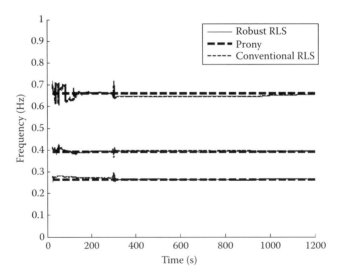

FIGURE 15.12 Frequency estimation of the major modes using the RRLS algorithm.

compared to a Prony analysis of the ringdown. More detailed results are shown in [20]. The RRLS algorithm provides a more accurate mode-damping estimate and the accuracy improves after the ringdown.

15.3.2.5 Estimating Mode Shape

Similar to the modal damping and frequency information, near-real-time operational knowledge of a power system's mode-shape properties may provide critical information for control decisions. For example, modal shape may someday be used to optimally determine generator and/or load tripping schemes to improve the damping of a dangerously low-damped mode. The optimization involves

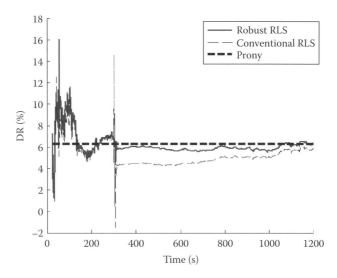

FIGURE 15.13 Damping ratio estimation of the major mode around 0.39 Hz.

minimizing load shedding and maximizing improved damping. Mode shape can be estimated from time-synchronized measurements.

The first published approach for estimating mode shape from time-synchronized measurements is contained in [22]. Follow-on methods are contained in [23–26]. An overview is provided in [16].

Figure 15.14 shows a control center implementation of small-signal stability monitoring displays.

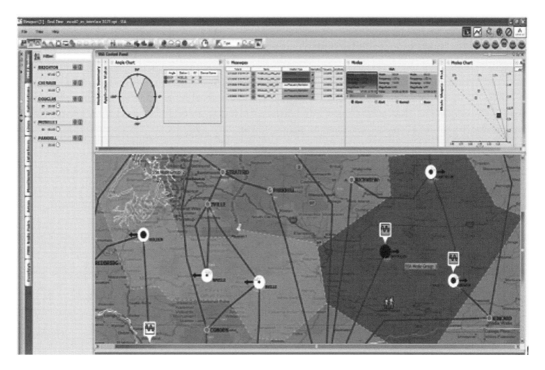

FIGURE 15.14 Synchrophasor-based small-signal stability monitoring within **e-terra**vision.

15.3.3 Voltage Stability Monitoring

Voltage stability refers to the ability of a power system to maintain steady voltages at all buses in the system after being subjected to a disturbance from a given initial operating condition [27]. Principal causes of voltage instability are [27–31]: heavy load system operation conditions, long distances between generation and load, low source voltages, and insufficient reactive power compensation. It is considered as a major threat for secure power-system operation in many power systems throughout the world. Voltage instability, resulting in voltage collapse, was reported as either main cause or being an important part of the problem in many partial or complete system blackouts. Some incidents are partly documented in [29]. Table 15.1 lists some recent incidents, not summarized in the literature so far (together with time frames and total load interruption) [32–34].

Wide-area voltage stability monitoring, detection, and control schemes built around PMUs supported by adequate communication infrastructure basically offer two major advantages with respect to existing voltage stability monitoring, instability detection, and control schemes:

1. PMU is GPS time-synchronized instrument [36,37] able to provide measurements at much higher rate of 10–120 samples/s and transmits them through a fast communication infrastructure to data concentrators. High sampling rates allow
   a. More accurate computation and better tracking of system stability degree
   b. Better and more robust identification of parameters associated with stability degree computation
   c. Anticipation (prediction) of short-term and long-term evolutions of chosen index
2. PMUs are time synchronized with the precision of less than 1 μs. Highly accurate time synchronization allows these devices deliver coherent (not average, due to skew time, as in traditional SCADA) picture of the full or partial system state. Consequently, all voltage stability indices computed from the knowledge of full or partial system state benefit for this advantage. However, averaging if deemed useful is still possible with these devices (e.g., for voltage short-term and long-term trending)

15.3.3.1 Description of Voltage Stability

Voltage instability essentially results from the inability of the combined transmission and generation system to deliver the power requested by loads [29] and is related to the maximum power that can be delivered by the transmission and generation system to the system loads. In order to establish power–voltage relationships and introduce the notion of maximum deliverable power, a simple two-bus (generator-transmission line-load) system is considered (shown in Figure 15.15).

TABLE 15.1 Voltage Collapse Incidents

Date	Location	Time Frame	Interrupted Load (MW)	Remark
06/08/95	Israel	19 min	~3,140	—
May 1997	Chile	30 min	2,000	—
08/14/03	USA–Canada	39 min	63,000	Estimated cost: $4–10 billion People affected: 50 million
09/23/03	South Sweden and East Denmark		6,550	People affected: 4 million
07/12/04	Southern Greece	30 min	5,000	—
11/11/09	Brazil–Paraguay	68 s	24,436	Voltage collapse in part of the system (68 s after initial event)

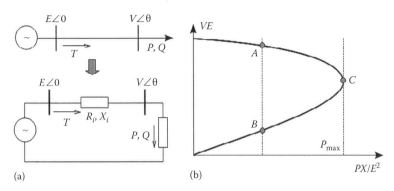

FIGURE 15.15 (a) Simple two-bus system and (b) power-voltage characteristics.

Assuming that the load behaves as an impedance with constant power factor ($X = R \tan \theta$) and using basic circuit theory equations for this system, active power consumed by the load can be expressed as

$$P = -\frac{RE^2}{(R_l + R)^2 + (X_l + R \tan \theta)^2} \tag{15.3}$$

Taking derivative of active power with respect to R and equalizing it with zero gives that at the extreme conditions (maximum deliverable power),

$$|\bar{Z}_l| = |\bar{Z}| \tag{15.4}$$

or maximum power that can be delivered to load is achieved when the load impedance is equal in magnitude to the transmission impedance [28–30].

If no assumption is made about the load (impedance behavior), the maximum deliverable power can be derived from power flow equations for simple two-bus system. Active and reactive powers consumed by the load can be expressed as

$$P = -\frac{EV}{X} \sin \theta \tag{15.5}$$

$$Q = -\frac{V^2}{X} + \frac{EV}{X} \cos \theta \tag{15.6}$$

Based on these two equations, the following power–voltage relationships can be established:

$$V = \sqrt{\frac{E^2}{2} - QX \pm \sqrt{\frac{E^4}{4} - X^2 P^2 - XE^2 Q}} \tag{15.7}$$

Assuming again constant load power factor, increase in active load power, and expressing load voltage magnitude as a function of this power results in well-known PV curve illustrated in Figure 15.15. PV curve gives relationship between voltage magnitude and active power of combined generation and transmission system. The system equilibrium is at the intersection of the PV curve and load

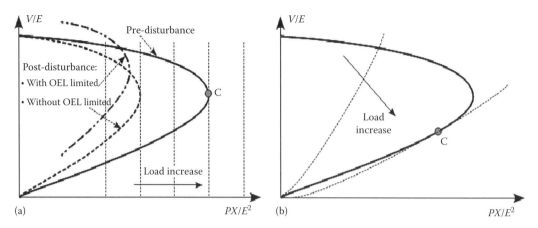

FIGURE 15.16 Voltage instability mechanisms. (a) Constant power load type, and (b) nonconstant power load type.

characteristic. As shown in the figure, for each load active power there are two operating points (A and B in Figure 15.15). Point A, characterized by high voltage magnitude and low current is normal operating point while point B, characterized by low voltage magnitude and high current for the same load power is generally not acceptable. Point C corresponds to the maximum deliverable power where two operating points coalesce. An attempt to operate the system beyond maximum deliverable power will generally result in voltage instability. This happens for two reasons:

- Due to smooth parameter (system load) changes
- Due to disturbances that decrease the maximum deliverable power

This is illustrated in Figure 15.16. If the load is of constant power type, with the increase in load active power, the system reaches maximum deliverable power (point C in Figure 15.16) and this point corresponds to the voltage instability point (often referred as critical point) [28,29,31]. Beyond the critical point the system equilibrium does not exist. In the same figure, dashed PV curve corresponds to post-disturbance system conditions (without a generator overexcitation limit [OEL]), depicting a decrease in maximum deliverable power. Further decrease in the maximum deliverable power (at higher voltage magnitude) is experienced if a generator OEL is activated (dash-dotted line in Figure 15.16).

If the load is not of constant power type (Figure 15.16), the critical point does not coincide with the maximum deliverable power and the system can operate at a part of lower portion of PV curve. However, the system operation at the lower portion of PV curve is generally not acceptable since the load would draw much higher current for the same power and for practical purposes voltage stability is associated with the maximum deliverable power.

15.3.3.2 Voltage Stability Monitoring and Instability Detection

Voltage stability monitoring is a process of continuous computation of the system stability degree, which can be seen as monitoring chosen stability index. The voltage stability index needs to reflect dominant phenomena linked to voltage instability in a particular power system and at the same time to be simple and practical for deployment. A wide variety of voltage stability indices have been proposed so far [28,29,38–48]. These indices serve as a measure of the proximity to the voltage instability (degree of system stability) by mapping current system state into a single (usually scalar) value. They are defined as a smooth, computationally inexpensive scalar with predictable shape that can be monitored as system operating conditions and parameters change [42].

In principle, any stability index could be used within voltage stability monitoring scheme but the following show the best promises to be used to this purpose:

- *Voltage magnitudes at critical locations* (*key load center and bulk transmission buses*): This is the simplest approach and consists of monitoring voltage magnitudes at critical locations and their comparison with predetermined thresholds. Voltage magnitude is not a good indicator of the security margin available at an operating point. On the other hand, when the system enters an emergency situation, low voltage of the affected buses is the first indication of an approaching collapse [28,29]. Short-term (about 1 min using PMUs) and long-term voltage trending plots [49] are near-term applications of synchrophasor technology, easy to deploy for voltage stability monitoring and detection.
- *Voltage stability indices derived from Thevenin impedance matching condition* [39–45]: Essentially, these indices measure stability degree of individual load buses (or a transmission corridor) by monitoring the equivalent Thevenin impedance of the system and equivalent impedance of local load (magnitude of these values are equal at the voltage instability point). Furthermore, stability degree can be expressed in terms of local voltage magnitude and voltage drop across the transmission path as well as in terms of power margin (MW or MVA). Computation of these indices does not require system model.
- *Loading margin of an operating point computed as the amount of load increase in a specific pattern that would cause voltage instability*: This index is based on physical quantities (usually MW) and as such easy to interpret and practical. Load margin computation requires system model (power flow model) and computation is performed using repetitive power flows, continuation power flows [50] specifying the load (at one load bus, region, or the system) as continuation parameter or direct method [51] solving equations describing the system model at the critical point. Sensitivity of computed margin with respect to any system parameter and control is easy to compute [52]. However, the computational costs are considered as the main disadvantage of this index [45].
- *Singular values and eigenvalues*: The focus is on monitoring the smallest singular value or eigenvalue of the system Jacobian matrices (usually power flow Jacobian is satisfactory for this purpose). These values become zero at the voltage instability point. Involved computations require system model and is often associated with higher computational costs [28,30,44].
- *Sensitivity-based voltage stability indices*: These indices relate changes in some system quantities to the changes in others. Different sensitivity factors can be used to this purpose [28,31,38]. However, some studies suggest the sensitivities of the total reactive power generation to individual load reactive powers as the best choice since these sensitivities are directly related to the smallest eigenvalue of Jacobian matrix and are computationally inexpensive. Computation of these indices requires system model [28,44].
- *Reactive power reserves*: Considerable decrease in reactive power reserves of system's key generators is a good indicator of system stress. Computation of reactive power reserves requires placement of measurement devices at several locations, does not require system model and, in principle, can make use of both SCADA and PMU-based measurements [46,47].
- *Singular value decomposition (SVD) applied to a measurement matrix*: The focus is on computing and tracking the largest singular value of the matrix. Measurement matrix is constructed from PMU measurements such that each column is a stacked vector of the available PMU measurements over a time window (two to three times the number of available PMUs) [48]. This matrix is updated as soon as new vector of measurements is acquired. Involved computations do not require system model [48].

Computation of chosen voltage stability index can be complemented with the stored results of off-line studies and observations. These results provide thresholds for chosen index. Another way to monitor voltage stability would be to use off-line observations (without computation of a voltage stability index) in order to build, periodically updated to account for changing system conditions, statistical model of the system and use it together with machine learning techniques such as decision trees (DTs), neural networks, and expert systems. The simplicity of DTs and easy interpretation of the decisions made, offer

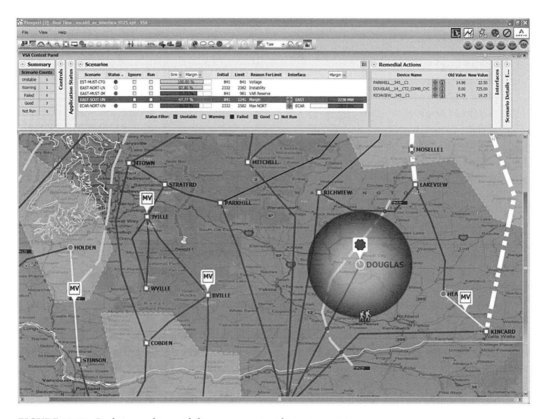

FIGURE 15.17 Real-time voltage stability assessment within **e-terra***vision*.

it as an attractive alternative for voltage stability monitoring [53]. DTs are automatically built off-line on the basis of learning set and a list of candidate attributes are further used in real time to assess quickly any new operating state, in terms of the values of its test attributes. In principle, DTs do not require synchronized measurement. SCADA measurements are enough since time skew should not be critical. However, they certainly can take advantage of these advanced measurements [53].

On-line voltage security assessment (VSA) tools can be used at the control center to measure the distance to voltage instability at any specific point in time (Figure 15.17). In this case, real-time measurements provide the base case and permit computation of the stability degree for base case and any postulated scenario. Commercially available VSA tools are still to be adapted in order to take advantages offered by PMUs [28,31,54,55].

Voltage instability detection is usually based on simple comparison of computed values of chosen index with its predefined thresholds. These thresholds are usually set pessimistic with respect to the theoretical values to allow timely detection of developing instability. On the other hand, some indices do not require any threshold but rely on the change in sign (most of sensitivity-based indices). Theoretical criteria for instability detection of the aforementioned indices are listed in Table 15.2.

Voltage trending application is strongly related to the use of voltage magnitudes as voltage stability index and when properly tuned this application could provide an early detection of developing instability. In addition, this application could be complemented by the computation of the sensitivities of voltage to active and reactive load powers (model-free sensitivities) that could be also used to measure the system stress and detect approaching instability [49]. Several voltage stability indices can be combined to define an efficient monitoring and detection scheme [43].

Figure 15.18 shows two control center implementations of voltage stability monitoring displays [56,57].

TABLE 15.2 Theoretical Values of Indices Threshold

Index	Threshold/Detection Criterion	Remark
Voltage magnitude	—	System dependent, no general criteria
Thevenin impedance matching condition	Either 1 or 0 (for power margins)	For practical purposes less than 1 or bigger than 0
Reactive power reserve	0 or 100	For practical purposes bigger than 0 in terms of Mvar reserve. 100 is threshold value of derived index [19]
Singular value	0	For practical purposes bigger than 0
Eigenvalue	0	For practical purposes bigger than 0
Sensitivities	Change in sign (positive to negative)	Does not require any tuning
SVD	Big change in two consecutively computed largest singular values	System dependent, no general criteria

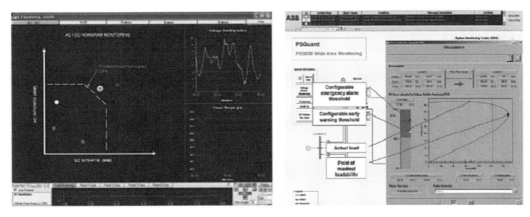

FIGURE 15.18 Control center voltage stability monitoring displays. (From Consortium for Electric Reliability Technology Solutions (CERTS), Nomogram validation application for CAISO utilizing phasor technology: Functional specification, Prepared for California Energy Commission, Berkeley, CA: CERTS Publication, 2006; ABB Ltd., Voltage stability monitoring: A PSGuard wide area monitoring system application, Zurich, Switzerland: ABB Switzerland Ltd., 2003. http://www.abb.com/poweroutage)

15.3.4 Transient Stability Monitoring

This section describes how synchrophasors can be used for transient and damping stability assessment of multi-machine power systems following large, nonlinear disturbances. Broadly speaking, transient stability is defined as the ability of the synchronous generators in a power system to synchronize with each other asymptotically over time from any arbitrary asynchronized state after being perturbed by some major disturbance. In the 1970s, the concept of energy functions was developed as a useful tool for such transient or synchronous stability analysis using passivity theory, the first two seminal papers being [16,58] followed by detailed methods for construction of energy functions in [59]. Viewing a power system equivalently as a network of coupled nonlinear oscillators, energy function is generically defined as the sum of its transient kinetic and potential energies capturing its cumulative oscillatory behavior, and can be used as a very useful metric to quantify the system's dynamic performance following a fault. In this section, we address the problem of constructing such energy functions using synchrophasor measurements with a particular focus on systems that are defined by predominantly two oscillating areas, or equivalently one dominant *inter-area* mode, as in Figure 15.19a. The validity of this approach is based on dominant power transfer paths being able to be separately modeled as interconnections of

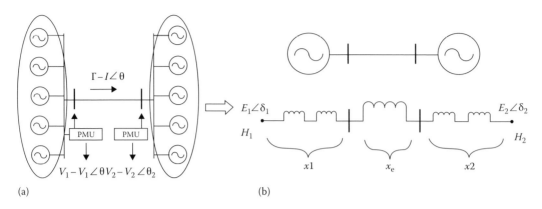

FIGURE 15.19 Two-machine dynamic equivalent of a two-area power system. (a) Two-area power system. (b) Two machine equivalent.

two machines or two groups of slow-coherent machines. Under such ideal conditions, the voltage and current phasor data from disturbances are used to estimate the swing energy associated with the disturbance and the quasi-steady state of the angular separation across the transfer path. The idea is illustrated on actual data recorded during a disturbance event in the WECC system.

15.3.4.1 Transient Stability Monitoring via Energy Functions

The study of energy functions stems from the fundamental swing dynamics defining the electromechanical motion of synchronous machines in any *n*-machine power system, given by Newton's laws of angular motion as

$$2H_i\Omega\ddot{\delta}_i = -d_i\omega_i + \underbrace{\sum_{j\neq i} E_i E_j B_{ij}(\sin(\delta_i - \delta_j) - \sin(\delta_{ij}^*))}_{u_i} \tag{15.8}$$

for, $i = 1, 2, \ldots, n$

where

$H_i, \ddot{\delta}_i, \omega_i, E_i, d_i, u_i$ denote the inertia, rotor angle, rotor speed, machine internal voltage, damping coefficient and driving input for the *i*th machine, respectively

$\Omega = 120\pi$ is a conversion factor from per unit to rad/s

B_{ij} is the admittance between *i*th and *j*th machine

δ_{ij}^* is the equilibrium angle between them before the disturbance

Considering every pair of machine, the energy function for the entire system can then be written as [60]

$$S = \underbrace{\sum_{i=1}^{n} \Omega H_i \omega_i^2}_{KE} + \underbrace{\sum_{k=1}^{n(n-1)/2} E_i E_j B_{ij} \int_{\delta_{ij}^*}^{\delta_k^*} (\sin(\delta_i - \delta_j) - \sin(\delta_{ij}^*))d\sigma}_{PE} \tag{15.9}$$

where
  KE stands for kinetic energy
  PE for potential energy

For the two-machine radial system shown in Figure 15.19a, substituting $n = 2$, the equivalent expression in Equation 15.9 reduces to

$$S = H\Omega\omega^2 + \frac{E_1 E_2}{x_e}\left(\cos(\delta_{op}) - \cos(\delta_{op}) + \sin(\delta_{op})(\delta_{op} - \delta)\right) \tag{15.10}$$

where $H = \dfrac{H1\,H2}{H1 + H2}$ is the equivalent inertia for the single-machine infinite bus representation of the two-machine system, jx_e is the equivalent reactance of the transmission line connecting the two equivalent machines, and $\delta = \delta_1 - \delta_2$. However, we must remember that each of the two equivalent machines in Figure 15.19b represents the slow coherent representation of several local machines inside each area as a result of which the bus voltages contain high-frequency local modes as well as slower inter-area modes. These fast and slow components need to be separated, before using the voltages to construct the *inter-area* energy function that can serve as a performance metric for monitoring the wide-area transient stability of this two-area system. We call the filtered slow component of the voltages as the quasi-steady states $\bar{V}_1$ and $\bar{V}_2$. In real time, the post-fault equilibrium angle δ_{op} or θ_{op} is not fixed either, but rather time varying, due to turbine-generator governing and other generation and load changes. Thus we can write

$$\delta = \hat{\delta} + \delta_{qss} \tag{15.11}$$

where $\hat{\delta}$ and δ_{qss} are, respectively, the swing component and the quasi-steady-state component of δ. We need to extract the quasi-steady-state value in order to approximate the post-disturbance equilibrium angle δ_{op} used in the energy function (15.10).

Based on the previous discussion, the following transient *inter-area* swing energy function

$$\bar{S} = H\Omega\omega(t)^2 + \frac{E_1 E_2}{x_e}\left(\cos(\delta_{op}) - \cos(\delta(t)) + \sin(\delta_{qss})(\delta_{qss} - \delta(t))\right) \tag{15.12}$$

can be proposed to model the energy excited in the system due to the dominant inter-area mode. Here, δ_{qss} is obtained by bandpass filtering of the measured response of $\delta(t)$ from PMUs. This, in turn, can serve as an effective performance metric for monitoring whether the two areas will synchronize with each other, in a *wide-area* sense, following any major disturbance.

15.3.4.2 Applications to U.S. Western North American Power System

We next use synchrophasor data from a disturbance event in the U.S. Western Interconnection to illustrate the construction of energy functions for a radial transfer path in this system. The variations of the bus angular separation and the bus frequency difference over time are shown in Figure 15.20a and b. The machine speed difference is mostly monomodal, but the angle difference θ shows a distinct quasi-steady-state variation. Bandpass filtering is used to separate the oscillation and the quasi-steady-state components of δ(t). The swing component is shown in Figure 15.20c. For the post-disturbance case, we get $x_e = 0.077$ pu from the least squares fitting, and the equivalent machine inertia is estimated to be $H = 119$ pu. Figure 15.20d–f show the kinetic energy V_{KE}, potential energy V_{PE} and the total swing energy V_E, which is the sum of the two. Note that oscillations are clearly visible in V_{KE} and V_{PE} and yet they literally disappear when they are

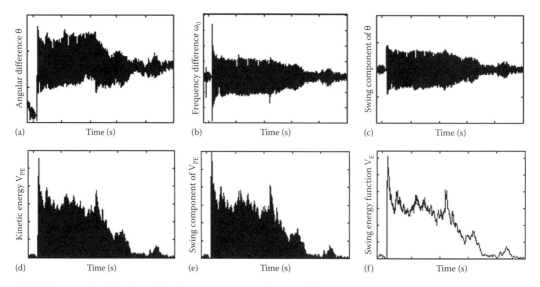

FIGURE 15.20 Synchrophasor-based transient energy functions for a Western Interconnection disturbance event swing. (a) Angle difference δ. (b) Frequency difference ω. (c) Swing angle component. (d) Swing kinetic energy. (e) Swing kinetic energy. (f) Swing energy function.

added together to form V_E. The oscillation is small-signal stable, although the damping is very low, and V_E eventually decays to a level commensurate with random perturbations on the system. If the system were negatively damped, V_E would grow to instability. The quasi-steady-state angle δ_{qss} indicates that the sending end and receiving end of the transfer path remain synchronized, and, therefore, is transiently stable. A sudden increase in δ_{qss} indicates the loss of a portion of the transmission system or the loss of generation at the load bus, both of which would stress the transfer path. If the disturbance had caused a separation of the transfer path, δ_{qss} would grow as synchronism would be lost.

15.3.5 Improved State Estimation

Conventional static state estimators enhanced with a few strategically placed synchronized phasor measurements have offered some improvements in state estimation. However, state estimators based entirely on synchrophasor measurements will produce a fundamental change in state estimation and its applications. For example, the EHV lines and buses in many systems can be considered to be a network with injections from the lower voltage portions of the system. If a sufficient number of bus voltages and line currents are measured (in rectangular form), then there is a linear relationship between the bus voltages (the state) and the measurements of the form

$$z = \begin{bmatrix} II \\ YA + Y_s \end{bmatrix} E + \varepsilon \quad (15.13)$$

Typically, more PMUs are installed for practical reasons but the minimum number of substations with measurements of both bus voltages and line currents required to observe all the bus voltages is approximately one-third the total number of EHV buses. The quantities in (15.13) are complex; II is a unit matrix with rows missing where there are no PMUs. The matrices II and A are real (containing only ones and zeros) but z, E, Y, and Y_s are complex. The Y matrix represents line admittances and Y_s represents the shunt elements. With $Y = G + jB$, $Y_s = G_s + jB_s$, $E = E_r + jE_x$, $z = z_r + jz_x$ in real and imaginary form, (15.13) takes the form of (15.14) and the estimate is given by (15.16):

$$\begin{bmatrix} z_r \\ z_x \end{bmatrix} = \begin{bmatrix} \begin{bmatrix} II \\ GA+G_s \\ 0 \\ BA+B_s \end{bmatrix} & \begin{bmatrix} 0 \\ -BA-B_s \\ II \\ GA+G_s \end{bmatrix} \end{bmatrix} \begin{bmatrix} E_r \\ E_x \end{bmatrix} + \varepsilon \qquad (15.14)$$

$$\mathbf{z} = \mathbf{Hx} + \boldsymbol{\varepsilon} \qquad (15.15)$$

$$\hat{\mathbf{x}} = (\mathbf{H}^T\mathbf{W}^{-1}\mathbf{H})^{-1}\mathbf{B}^T\mathbf{H}^{-1}\mathbf{z} = \mathbf{Mz} \qquad (15.16)$$

In (15.15), $\mathbf{z}$ is a vector of measurements including both voltages and currents, $\mathbf{x}$ is the state of the EHV system (the bus voltages in rectangular form), and $\boldsymbol{\varepsilon}$ is the vector of measurement errors with covariance matrix $\mathbf{W}$. The matrix $\mathbf{W}$ is typically assumed to be diagonal ($w_{ii} = \sigma_i^2$, $w_{ij} = 0\ i \neq j$), that is, the measurement errors are independent. The $\mathbf{W}^{-1}$ in (15.16) weights the difference between the actual measurements $\mathbf{z}$ and the estimated measurements $\hat{\mathbf{z}} = \mathbf{H}\hat{\mathbf{x}}$ with the variances giving more weight to measurements with small variances, that is, the estimate minimizes $\sum (z_i - \hat{z}_i)^2/\sigma_i^2$.

The matrix inverse in (15.16) is only symbolic. The Q–R algorithm or something equivalent is used to solve the set of over-defined equations in (15.15). The $\mathbf{H}$ matrix, and hence $\mathbf{M}$, are constant unless there are topology changes in the system. A topology processor can use breaker status if available from a "dual-use" line relay/PMU device or use line current measurements, if necessary, to track breaker status. The resulting estimator is linear (no iteration is involved). The estimator can run as frequently as once a cycle or once every two cycles. The estimator is truly dynamic, no static assumptions are involved, and the data are time tagged and organized by PDCs. A measurement vector corresponding to measurements with the same time tag is multiplied by a constant matrix (which changes with topology) to produce an estimate of the EHV voltages.

A second variation made possible with synchrophasor measurements is a three-phase estimator. Rather than combining the individual phase quantities to form a positive sequence voltage or current, the individual phase measurements can be time tagged and communicated to the control center, where an estimate of the individual phase voltages are computed with a matrix similar to $\mathbf{M}$ in (15.15) and (15.16) but with three times as many rows and columns. The three-phase estimator gives real-time dynamic information about the origins and the magnitudes of imbalances in the network.

The three-phase estimator also has the ability to provide calibration of the PTs and CTs in the network. The concept of calibrating measurements is not new but there is not actually a positive sequence CT to be calibrated. In the three-phase formulation, there are three actual CTs and PTs that have ratio errors.

With the notation of (15.15), the calibration problem is given in (15.17):

$$\mathbf{z} = \mathbf{KHx} + \boldsymbol{\varepsilon} \qquad (15.17)$$

where **K** is a diagonal matrix of ratio correction factors. It is clear that with a k for every measurement there are too many unknowns, since if **K** and **x** are solutions then $\alpha\mathbf{K}$ and $\mathbf{x}/\alpha$ are also solutions. One ratio correction factor is taken as 1 (the voltage measurement is taken as being correct) and that k is removed from the set of unknowns. A precision PT or a new high quality CVT can be used for this purpose.

Each PT and CT has a ratio error of the form, Measurement = k * TRUE, where k is the ratio correction factor, $k = |k| \angle \theta$. The estimation of all the xs and all the ks requires data over an extended time period in which the state of the system changes. The calibration is a batch solution for all states over the period along with the ratio correction factors. It is also assumed that the system model is known and accurate, the ratio correction factors are constant, and that one PT is essentially perfect.

PMU-based, three-phase state estimation can be accomplished using a multitude of monitoring solutions. The improved state estimation methodology described in this section is being implemented as shown in Figure 15.21. This diagram serves as a basic framework for substation PMU data. Existing serial communications, line voltages, line currents, etc., are not displayed.

The scale of the extra-high-voltage (EHV) state estimation problem is well suited for the deployment of synchrophasor technology. Generally speaking, while electrical interconnections can be vast on any scale, the patchwork of EHV networks composing the interconnection are generally rather sparse when compared to their lower-voltage transmission and subtransmission counterparts. For any arbitrary moderate-to-large utility within an expansive interconnection, there might exist hundreds of electric transmission substations at lower voltages (<345 kV). For that same utility, however, there may only exist scores if not dozens of EHV stations. Tersely expressed, the PMU-based EHV state estimation problem can be addressed with a modest number of PMUs.

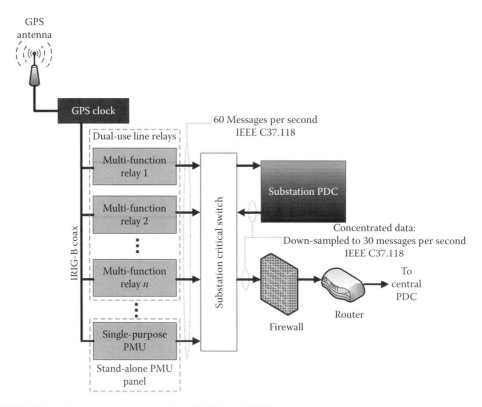

FIGURE 15.21 Substation architecture for a PMU-based EHV state estimator.

After identifying EHV stations to be monitored (resulting from a study to establish complete observability), two principles guide the selection of parameters to be monitored (Figure 15.21). The following must be captured for each selected station:

1. Three-phase voltage measurements for every EHV bus element that can be isolated.
2. Three-phase current measurements for all electric transmission lines and transformers connected at the EHV voltage level. All injections into the EHV network must be captured.

In many cases, digital line relays can serve as "dual-use" line relay/PMU devices with minimal settings changes and/or upgrades. Older digital relays may require a "cradle swap" to install new relay hardware. In any case, a stand-alone PMU solution can be used to augment any monitoring solution chosen.

15.4 WAMS in North America

15.4.1 North American SynchroPhasor Initiative

NASPI is a joint effort between the U.S. Department of Energy (DOE) and the North American Electric Reliability Corporation (NERC). The goal is to improve power system reliability through wide-area measurement, monitoring, and control. This shall be achieved by facilitating a robust, widely available, and secure synchronized data measurement infrastructure for the interconnected North American electric power system. It also includes associated analysis and monitoring tools for better planning and operation, and improved reliability.

To facilitate the development of synchrophasor technology, particularly to foster an environment of information exchange between utilities, the U.S. DOE initiated the Eastern Interconnection Phasor Project (EIPP) in October 2002, building on over a decade of experience in the Western Interconnection. In 2007, NERC formally joined DOE in the effort, and expanded it to include all interconnections within North America. At this time, the EIPP was renamed NASPI [61]. Updates of the NASPI program have been recently provided [62–64].

NASPI is structured as a working group made up of voluntary members from electric power operating organizations, reliability coordinators, suppliers of monitoring and communications network hardware and software, and researchers from industry, universities, and national laboratories. The working group is composed of five task teams who focus on various aspects of developing and deploying synchrophasor measurement technology. DOE, through the Consortium for Electric Reliability Technology Solutions (CERTS) and in collaboration with NERC, provides technical support to the task team activities. The task team leaders, together with the DOE program manager and representatives from NERC and CERTS, make up a leadership committee, whose role is to plan and coordinate the working group activities. An executive steering group provides oversight to the working group and engages the power industry at a senior management level to spread the word about the benefits of system-wide measurements and enlists support for the program. Some of the more recent achievements of the NASPI initiative include

- *NASPInet concept*: Under the leadership of the NASPI DNMTT, the concept of a distributed architecture linking the providers of the data (publishers) with applications (subscribers) using a publish-and-subscribe middleware and data bus concept is under development. Currently, the NASPInet architecture is at a conceptual design phase, and a detailed specification is under development. The vision is that this specification can be used by hardware or software vendors to provide an interface to NASPInet, either as a publisher or as a subscriber. The unifying concept that will provide this interface is a phasor data gateway. The next stage of development will be pilot demonstration projects to further refine and modify this specification based on lessons learned from interconnecting multiple vendors and a spectrum of applications in a common architectural framework.

- *System baselining*: Both the planning and operations task teams are currently involved in baselining activities or, in other words, determining "normal" phase angles so that abnormal conditions can be better defined and alarmed. The task teams are taking complimentary approaches to determining these "normal" phase angle separations. The operations task team is looking at observed angle separations for key phasor measurement locations, and evaluating these data over historical time frames. The planning task team is performing model-based studies to assess the phase angle separation under known heavily stressed conditions, and correlating these angles with specific changes to the operating conditions in the base case model study. Between these two approaches, the goal is to develop a more rigorous methodology for determining the thresholds at which the real-time monitoring tools should be alarmed based on the observed phase angle separation between monitoring locations.

In the United States, in 2010, DOE has also provided major stimulus with Smart Grid Investment Grant (SGIG) projects to speed up the deployment of PMU systems:

- WECC WISP (250 new PMUs)—$108 M (including PG&E, BPA, SCE, SRP)
- PJM (90 new PMUs)—$40 M
- NYISO (35 new PMUs)—$76 M
- Midwest ISO (150 new PMUs)—$35 M
- ISO New England (30 new PMUs)—$9 M
- Duke Energy Carolina (45 new PMUs)—$8 M
- Entergy (18 new PMUs)—$10 M
- American Transmission Company (5 new PMUs)—$28 M
- Midwest Energy (1 substation)—$1.5 M

15.5 WAMS Worldwide

15.5.1 WAMS Applications in Europe

Due to the challenging system operation conditions for European transmission system operators (TSOs) within the last few years as well as to modern technology and related software currently available at reasonable prices, substantial progress regarding dynamic system analysis tools has been reached. While PMU-based WAMS is a key component of these applications, as a result of different application requirements and strict rules concerning the security of data exchange, two main categories of applications have been set up independently [65–74]:

1. Research and development or demonstration projects
2. Industrial and TSO applications

15.5.1.1 Research and Development Projects

Such systems are mainly used by universities, which exchange their measurements via the public Internet. The measurement equipment and the required software for data acquisition and analysis are either based on development work of their own or comprise a combination of standard software and WAMS from leading manufacturers. As the related measurements are mainly from the distribution system, only frequency, voltage, and voltage phase angle are used for analysis and research.

15.5.1.2 TSO Applications

In the European TSO community, WAMS based on different technologies have been installed. On the one hand, there are stand-alone transient recorders, the measurements of which require remote collection and subsequent manual synchronization with measurements from other substations; on the other hand, there are systems where PMU and PDC technology provide for automatic and online data synchronization and analysis.

Common applications for both solutions are

- Dynamic model validation based on postmortem dynamic system analysis
- Monitoring of dynamic system performance

It is obvious that the second technology offers a wider range of applications that are in use today:

- Voltage phase angle difference monitoring
- Line thermal monitoring (medium value between two substations)
- Voltage stability monitoring (online P–V curves)
- Online monitoring of system damping (online modal analysis with online parameter estimation
- Intelligent alarming if predefined critical levels are exceeded
- Online monitoring of system loading

TSOs have already started to include some of the measurements, information, and alarms output by the WAM system within their SCADA systems, too.

Although the TSOs currently focus on using their WAMS mainly for their own system operation purposes, a few TSOs have already started to exchange PMU measurements between their PDCs. Swissgrid, being one of the driving forces for the application of the WAMS technology, has established links to eight European TSOs (see Figure 15.22).

Based on this system, a continuous dynamic monitoring of the Continental European system is performed.

15.5.2 WAMS Applications in Brazil

Brazil is the biggest country in the South American continent and its main electric energy source comes from hydroelectric plants (more than 90% of all produced electric energy in 2010). The hydro generation park is formed by power plants in cascade along 12 major hydrographic basins spread all over the Brazilian territory and many of these hydro plants are distant from the main load centers, which are situated in the southeast region. Due to the extension of the Brazilian territory, the rainfall profiles are complementary among geographic regions and variable over the year, as well as between dry and wet years. Thus, one of the main challenges for the Brazilian power-system operation is to optimize the available hydro resources, considering each river with their cascade power plants, the diverse energy production in each region, the existing transmission restrictions, and complementing the energy production with other energy resources (thermal, nuclear, wind, biomass, etc.), in order to obtain the minimum production cost while maintaining the system reliability. Some of the biggest hydroelectric plants in Brazil (like Itaipu and Tucurui) are located far from the load centers, resulting in bulk power transfers over long distances. Even in the near future, this picture will not change so much, as the most relevant power plants under development are located in the Amazon region, far from the main load centers by approximately 3000 km.

As in all systems of this proportion, disturbances due to significant generation and load unbalances may cause excessive frequency variations, voltage collapse situations, and even the splitting of certain parts of the network, with loss of important load centers.

The sole independent power system operator in Brazil (ONS) has been investigating the effective use of synchrophasor technology in power-system operation and is leading a major industry effort to deploy a large-scale SMS for the Brazilian interconnected power system [75]. In this effort, ONS conducted the first round of certification tests on commercial PMU models. These tests were performed to guarantee smooth system integration and the global performance of the SMS, considering that it will be a multi-owner system and will need to use PMUs from diverse manufacturers. The results from these tests have shown that nowadays synchrophasor technology is ready to be applied for wide-area monitoring and situation awareness applications, but the technology needs to evolve in order to allow more reliable real-time applications and wide-area protection and control applications [76]. The ongoing revision of

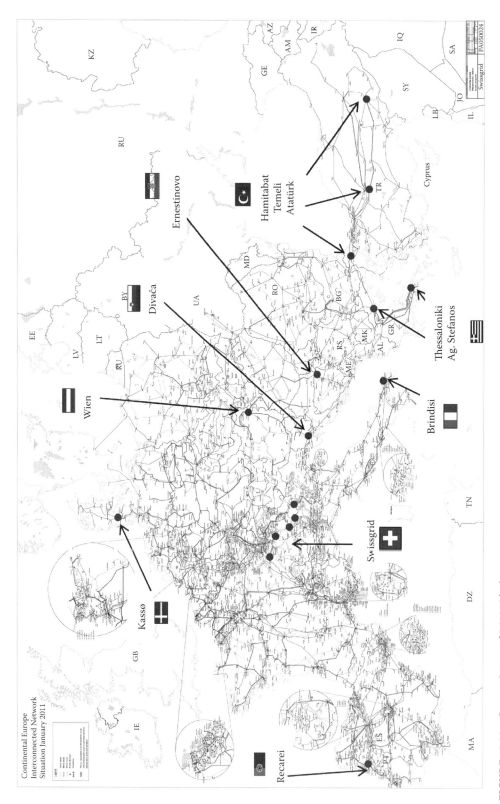

FIGURE 15.22 Current Swissgrid WAMS links.

the IEEE C37.118 standard surely will solve most of the current issues and will provide more adequate characteristics suitable for protection and control applications.

Considering the present technological status, the Brazilian WAMS will be used initially as a long-term system dynamics and event recording system for postmortem analysis, model validation, and mitigation solution investigation, but it is also expected to add synchrophasor applications to support the real-time power-system operation.

As a result of a research project, ONS has identified a number of candidate synchrophasor applications, and selected four applications for a proof-of-concept pilot implementation:

- *System stress monitoring (StressMon)*: During times of normal or abnormal operation the angle difference between two locations in the transmission system could be used to measure how much margin there is between the current operating conditions and an operating condition that would be impacted by either predetermined pre- or post-contingency stability violations. Phase angle differences between a limited number of preselected PMU locations would provide a measure of the overall condition of the power system. StressMon is a tool to monitor angle difference between pairs of locations or regions in the power system to detect proximity of predefined stability limits. Monitoring should consider trespassing of limits and deviation from forecasted reference values. Results can be used for decision support in real-time or for off-line auditing.
- *Closing a breaker in a loop in the transmission network (LoopAssist)*: Closing the breaker may cause an overload, when the phase angle difference across the breaker is too large. Also system stability may be affected or damage may occur to power system equipment. When closing a parallel connection in a transmission network (closing a loop) the phase angle across the circuit breaker provides a good measure for the impact of the control action. LoopAssist is a tool to monitor voltage magnitude and angle difference across circuit breakers involved in closing transmission loops in the power system. This function can be useful for providing a means to guide the operator on conditioning the power system for a valid reconnection situation avoiding overloads or to provide a measure of the impact incurred in closing a loop in the system. Results can be used for decision support in real time or for off-line auditing.
- *Closing a connection between two electrical islands (SynchAssist)*: When closing a tie line between two electrical islands, a synchronism check relay may block the control action when the conditions for synchronization are not satisfied. Typically, the relay will verify the synchronization conditions before allowing breaker closure to take place, based on frequency deviation, phase angle difference, and voltage magnitude difference. Showing the variation of the phase angle between the two islands in a time trend together with numerical values for the aforementioned criteria together with some other numerical values (e.g., actual generation and access synchronized generation capacity in both electrical islands) provides useful information for the operator to make an informed decision when to issue the breaker control action. When no synchronism check relay is available at the substation where the breaker is located, monitoring the periodic oscillation of the phase angle difference will provide an indicator when to issue a control action from the SCADA system. SynchAssist is a tool to monitor voltage magnitude difference, angle difference, and frequency deviation across transmission equipment involved in reconnecting electrical islands in the power system. This tool can be useful for providing information to guide the operator on conditioning the power system for a valid reconnection situation, by avoiding unstable situations, cascading events, or severe overloads. Results can be used for decision support in real time or for off-line auditing.
- *System oscillations monitoring (DampMon)*: Synchrophasor measurements can be used to monitor oscillations in power system quantities. These quantities can be raw or filtered phasor measurements or quantities calculated from phasor measurements such as line flows or corridor flows. Power system oscillations are usually initiated by sudden changes in the power system such as fault clearing, line switching, or generator tripping. These events cause generator shaft

oscillations that are usually damped within a very short period of time (seconds); however, when a system is heavily loaded, these oscillations can become poorly damped. In addition, even without these events it is possible that in heavily loaded systems oscillations occur. This application would calculate the amplitude of the oscillation using phasor measurement in real time on a sample-by-sample basis. It would further calculate the characteristic frequency and damping factor of the power oscillations. All three values would be displayed in real time, both as a trend display or as a numerical value in some format (e.g., bar chart). The results may be used to show the oscillations of the relevant phasor measurements, phase angle differences, or calculated flows. DampMon is a tool to monitor oscillations in power system quantities. These quantities can be raw or filtered phasor measurements or quantities calculated from phasor measurements such as line flows or corridor flows.

To confirm the adequacy of these applications, they were implemented on an application test platform based mainly on the control center existing EMS system.

Another important synchrophasor application initiative in Brazil comes from the Santa Catarina Federal University (UFSC) [77]. The project started in 2001 as a research carried out jointly by UFSC and a Brazilian industry. In 2003, the project got financial support from the Brazilian government, which allowed the deployment of a prototype SMS. This first system measures the distribution low voltage at nine of the university's laboratories, communicating with a PDC at UFSC over the public Internet. This system allowed recording the Brazilian interconnected power system's dynamic performance during recent major power system disturbances. Currently, another project from UFSC has installed PMUs on three 500 kV substations in the south of Brazil.

15.6 WAMS Deployment Roadmap

As discussed, PMU applications offer large reliability and financial benefits for customers/society and the electrical grid when implemented across the interconnected grid. Synchrophasors enable a better indication of grid stress and can be used to trigger corrective actions to maintain reliability. As measurements are reported 10–120 times per second, PMUs are well suited to track grid dynamics in real time. In general, this technology is instrumental for improving wide-area monitoring, protection, and control. Considering a large number of existing and potential applications, benefits are grouped in four categories:

1. Data analysis and visualization—significant benefits have already been achieved
2. Outage reduction and blackout prevention to improve system reliability, including real time control and protection—huge societal benefits
3. System operations and planning, including modeling and restoration—enable a paradigm shift with tracking grid dynamics and system measurements vs. estimation
4. Market operations and congestion management—large potential financial benefit as it enables utilization of accurate and optimal margins for power transfer (vs. worst-case scenario presently used in practice)

Given the nature of PMU implementation requiring participation of a broad base of users, the "overall industry roadmap" is an important step in designing and deploying large-scale PMU systems. NASPI has developed such a roadmap and is shown in Figure 15.23. This roadmap, which is based on applications, business needs, commercial availability, and cost, and complexity with deploying those applications, was developed through an interview process with industry experts and users [78,79]. The details of this roadmap are described as follows.

First, industry needs are identified (critical, moderate, unknown) regardless of the technology. Second, the value of the PMU technology, for each identified application, has been mapped related to importance in serving industry. This approach resulted in four categories: *necessary and critical*, *critical*

Wide-Area Monitoring and Situational Awareness

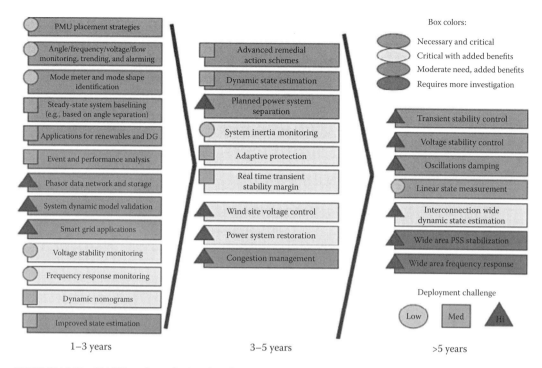

FIGURE 15.23 NASPI roadmap for SynchroPhasor applications.

with added benefits, moderate need with added benefits, and *requires more investigation.* Third, deployment challenges have been mapped for each application (low, medium, and high). The deployment challenges are defined based on technology (communications and hardware and software requirements) and application status (commercially available, pilot installation, research, and not developed).

The applications and infrastructure aspects (e.g., PMU locations, network, and data storage) are grouped into near term (1–3 years), medium term (3–5 years), or long term (more than 5 years). This roadmap focuses on business and reliability needs to commercialize and deploy PMU technology and applications addressing implementation risks. Applications in the near-term group reflect the immediate needs and deployment possibilities. Applications in the medium-term group largely reflect that even though the benefits are there, the commercial deployment is still further away due to deployment challenges and application commercialization. Applications in the long-term group indicate a combination of distant commercial status, extensive infrastructure requirements (and thus costs), and/or lengthy field trials.

References

1. U.S. Department of Energy and Natural Resources Canada, Final report of the U.S.-Canada power system outage task force, April 2004. https://reports.energy.gov/BlackoutFinal-Web.pdf (accessed January 2011).
2. Endsley, M.R., Design and evaluation of situation awareness enhancement, *Proceedings of the Human Factors Society 32nd Annual Meeting*, Anaheim, CA, Vol. 1, pp. 97–101, 1988. Santa Monica, CA: Human Factors Society.
3. Endsley, M.R., Toward a theory of situation awareness in a dynamic system, *Human Factors*, 37(1), 32–64, 1995.
4. Endsley, M.R., Farley, T.C., Jones, W.M., Midkiff, A.H., and Hansman, R.J., Situation awareness information requirements for commercial airline pilots, Technical report ICAT-98-1 to NASA Ames under Grant NAG, 1998.

5. Guttromson, R., Greitzer, F.L., Paget, M.L., and Schur, A., *Human Factors for Situation Assessment in Power Grid Operations*, Pacific Northwest National Laboratory, Richland, WA, 2007.
6. Phadke, A.G., Hlibka, T., and Ibrahim, M., Fundamental basis for distance relaying with symmetrical components, *IEEE Transactions on Power Apparatus and Systems*, 96(3), 635–646, March/April 1977.
7. Phadke, A.G. (guest editor), System of choice: Phasor measurements for real-time applications, *IEEE Power & Energy* magazine, special issue, September/October 2008.
8. Phadke, A.G. and Thorp, J.S., *Synchronized Phasor Measurements and Their Applications*, New York: Springer, 2008, Chapter 5.
9. Bhargava, B. and Salazar, A., Synchronized phasor measurement system (SPMS) for monitoring transmission system at SCE, Presented at *NASPI Meeting*, Carson, CA, May 2007.
10. Cummings, R.W., Predicting cascading failures, Presented at *NSF/EPRI Workshop on Understanding and Preventing Cascading Failures in Power Systems*, Westminster, CO, October 2005.
11. Parashar, M., Das A., and Carter, C., WECC phase angle baselining: Steady state analysis, Presented at *NASPI Meeting*, Chattanooga, TN, October 2009. http://www.naspi.org/resources/pitt/wecc_voltage%20angle_baselining_manu.pdf (accessed January 2011).
12. Venkatasubramanian, V., Yue, Y.X., Liu, G. et al., Wide-area monitoring and control algorithms for large power systems using synchrophasors, *IEEE Power Systems Conference and Exposition*, Seattle, WA, March 2009.
13. Dobson, I., Parashar, M., and Carter, C., Combining phasor measurements to monitor cutset angles, *Proceedings of the 2010 43rd Hawaii International Conference on System Sciences*, Kauai, HI, January 2010.
14. Dobson, I. and Parashar, M., A cutset area concept for phasor monitoring, *IEEE Power and Energy Society General Meeting*, Minneapolis, MN, July 2010.
15. Dobson, I., New angles for monitoring areas, *International Institute for Research and Education in Power Systems (IREP) Symposium*, Bulk Power System Dynamics and Control-VIII, Rio de Janeiro, Brazil, August 2010.
16. Trudnowski, D. and Pierre, J., Signal processing methods for estimating small-signal dynamic properties from measured responses, *Inter-Area Oscillations in Power Systems: A Nonlinear and Nonstationary Perspective*, ISBN 978 0 387 89529-1, New York: Springer, 2009, Chapter 1.
17. Kosterev, D.N., Taylor, C.W., and Mittelstadt, W.A., Model validation for the August 10, 1996 WSCC system outage, *IEEE Transactions on Power Systems*, 14(3), 967–979, August 1999.
18. Hauer, J.F., Demeure, C.J., and Scharf, L.L., Initial results in prony analysis of power system response signals, *IEEE Transactions on Power Systems*, 5(1), 80–89, February 1990.
19. Pierre, J.W., Trudnowski, D.J., and Donnelly, M.K., Initial results in electromechanical mode identification from ambient data, *IEEE Transactions on Power Systems*, 12(3), 1245–1251, August 1997.
20. Zhou, N., Trudnowski, D., Pierre, J., and Mittelstadt, W., Electromechanical mode on-line estimation using regularized robust RLS methods, *IEEE Transactions on Power Systems*, 23(4), 1670–1680, November 2008.
21. Trudnowski, D., Pierre, J., Zhou, N., Hauer, J., and Parashar, M., Performance of three mode-meter block-processing algorithms for automated dynamic stability assessment, *IEEE Transactions on Power Systems*, 23(2), 680–690, May 2008.
22. Trudnowski, D., Estimating electromechanical mode shape from synchrophasor measurements, *IEEE Transactions on Power Systems*, 23(3), 1188–1195, August 2008.
23. Dosiek, L., Trudnowski, D., and Pierre, J., New algorithms for mode shape estimation using measured data, *IEEE Power & Energy Society General Meeting*, Pittsburgh, PA, Paper no. PESGM2008-001014, July 2008.
24. Tuffner, F.K., Dosiek, L., Pierre, J.W., and Trudnowski, D., Weighted update method for spectral mode shape estimation from PMU measurements, *Proceedings of the IEEE Power Engineering Society General Meeting*, Piscataway, NJ, July 2010.

25. Dosiek, L., Pierre, J., Trudnowski, D., and Zhou, N., A channel matching approach for estimating electromechanical mode shape and coherence, *IEEE Power & Energy Society General Meeting*, Calgary, Alberta, Canada, Paper no. 09GM0255, July 26–30, 2009.
26. Zhou, N., Huang, Z., Dosiek, L., Trudnowski, D., and Pierre, J., Electromechanical mode shape estimation based on transfer function identification using PMU measurements, *IEEE Power & Energy Society General Meeting*, Calgary, Alberta, Canada, Paper no. 09GM0342, July 26–30, 2009.
27. Kundur, P., Paserba, J., Ajjarapu, V., Andersson, G., Bose, A., Canizares, C., Hatziargyriou, N., Hill, D., Stankovic, A., Taylor, C., Van Cutsem, T., and Vittal, V., Definition and classification of power system stability, *IEEE Transactions on Power Systems*, 19(2), 1387–1401, 2004.
28. Van Cutsem, T. and Vournas, C., *Voltage Stability of Electric Power Systems*, Boston, MA: Kluwer Academic Publisher, March 1998.
29. Taylor, C.W., *Power System Voltage Stability*, EPRI Power System Engineering Series, New York: McGraw-Hill, September 1994.
30. Kundur, P., *Power System Stability and Control*, EPRI Power System Engineering Series, New York: McGraw-Hill, 1994.
31. Ajjarapu, V., *Computational Techniques for Voltage Stability Assessment and Control*, New York: Springer, 2006.
32. Hain, Y. and Schweitzer, I., Analysis of the power blackout on June 8, 1995, in the Israel electric corporation, *IEEE Transactions on Power Systems*, 12(4): 1752–1758, 1997.
33. Vournas, C.D., Nikolaidis, V.C., and Tassoulis, A.A., Postmortem Analysis and Data Validation in the Wake of the 2004 Athens Blackout, *IEEE Transactions on Power Systems*, 21(3), 1331–1339, 2006.
34. Vargas, L.D., Quintana, V.H., and Miranda, R.D., Voltage collapse scenario in the Chilean interconnected system, *IEEE Transactions on Power Systems*, 14(4), 1415–1421, 1999.
35. Filho, J.M.O., Brazilian Blackout 2009, Blackout Watch, 2010. http://www.pacw.org/fileadmin/doc/MarchIssue2010/Brazilian_Blackout_march_2010.pdf (accessed January 2011).
36. Novosel, D., Madani, V., Bhargava, B., Vu, K., and Cole, J., Dawn of grid Synchronization: Benefits, practical applications, and deployment strategies for wide area monitoring, protection, and control, *IEEE Power and Energy Magazine*, 6(1), 49–60, 2008.
37. Phadke, A.G. and Thorp, J.S., *Synchronized Phasor Measurements and Their Applications*, New York: Springer, 2008.
38. Glavic, M. and Van Cutsem, T., Wide-area detection of voltage instability from synchronized phasor measurements. Part I: Principle, *IEEE Transactions on Power Systems*, 24(3), 1408–1416, 2009.
39. Vu, K., Begovic, M.M., Novosel, D., and Saha, M.M., Use of local measurements to estimate voltage stability margin, *IEEE Transactions on Power Systems*, 14(3), 1029–1035, 1999.
40. Corsi, S. and Taranto, G.N., A real-time voltage instability identification algorithm based on local phasor measurements, *IEEE Transactions on Power Systems*, 23(3), 1271–1279, 2008.
41. Larsson, M., Rehtanz, C., and Bertsch, J., Real-time voltage stability assessment of transmission corridors, *Proceedings of the IFAC Symposium on Power Plants and Power Systems*, Seoul, South Korea, 2003.
42. Parniani, M., Chow, J.H., Vanfretti, L., Bhargava, B., and Salazar, A., Voltage stability analysis of a multiple-infeed load center using phasor measurement data, *Proceedings of the 2006 IEEE Power System Conference and Exposition*, Atlanta, GA, 2006.
43. Milosevic, B. and Begovic, M., Voltage stability protection and control using a wide-area network of phasor measurements, *IEEE Transactions Power Systems*, 18(1), 121–127, 2003.
44. Gao, B., Morison, G.K., and Kundur, P., Voltage stability evaluation using modal analysis, *IEEE Transactions on Power Systems*, 8(3), 1159–1171, 1993.
45. Canizares, C. (editor/coordinator), *Voltage Stability Assessment: Concepts, Practices and Tools*, IEEE PES Publication, New York: Power System Stability Subcommittee, ISBN 0780379695, 2002.
46. Taylor, C.W. and Ramanathan, R., BPA reactive power monitoring and control following the August 10, 1996 power failure, *Proceedings of VI Symposium of Specialists in Electric Operational and Expansion Planning*, Salvador, Brazil, 1998.

47. Bao, L., Huang, Z., and Xu, W., Online voltage stability monitoring using var reserves, *IEEE Transactions on Power Systems*, 18(4), 1461–1469, 2003.
48. Overbye, T., Sauer, P., DeMarco, C., Lesieutre, B., and Venkatasubramanian, M., Using PMU data to increase situational awareness, PSERC Report 10-16, 2010.
49. NASPI Report, Real-time application of synchrophasors for improving reliability, 2010. www.naspi.org/rapir_final_draft_20101017.pdf (accessed January 2011).
50. Ajjarapu, V. and Christy, C., The continuation power flow: A tool for steady state voltage stability analysis, *IEEE Transactions on Power Systems*, 7(1), 416–423, 1992.
51. Canizares, C.A., Alvarado, F.L., DeMarco, C.L., Dobson, I., and Long, W.F., Point of collapse methods applied to AC/DC power systems, *IEEE Transactions on Power Systems*, 7(2), 673–683, 1992.
52. Greene, S., Dobson, I., and Alvarado, F.L., Sensitivity of the loading margin to voltage collapse with respect to arbitrary parameters, *IEEE Transactions on Power Systems*, 12(1), 262–272, 1997.
53. Diao, R., Sun, K., Vittal, V., O'Keefe, R.J., Richardson, M.R., Bhatt, N., Stradford, D., and Sarawgi, S.K., Decision tree-based online voltage security assessment using PMU measurements, *IEEE Transactions on Power Systems*, 24(2), 832–839, 2009.
54. Van Cutsem, T., An approach to corrective control of voltage instability using simulation and sensitivity, *IEEE Transactions on Power Systems*, 7(4), 1529–1542, 1993.
55. Ajjarapu, V. and Sakis, M.A.P., Preventing voltage collapse with protection systems that incorporate optimal reactive power control, PSERC Report 08–20, 2008.
56. Consortium for Electric Reliability Technology Solutions (CERTS), Nomogram validation application for CAISO utilizing phasor technology: Functional specification, Prepared for California Energy Commission, Berkeley, CA: CERTS Publication, 2006.
57. ABB Ltd., Voltage stability monitoring: A PSGuard wide area monitoring system application, Zurich, Switzerland: ABB Switzerland Ltd., 2003. http://www.abb.com/poweroutage (accessed January 2011).
58. Fouad, A.A. and Stanton, S.E., Transient stability of a multimachine power system, part I: Investigation of system trajectories, *IEEE Transactions on Power Systems*, PAS-100, 3408–3416, 1981.
59. Michel, A., Fouad, A., and Vittal, V., Power system transient stability using individual machine energy functions, *IEEE Transactions on Circuits and Systems*, 30(5), 266–276, May 1983.
60. Chow, J.H., Chakrabortty, A., Arcak, M., Bhargava, B., and Salazar, A., Synchronized phasor data based energy function analysis of dominant power transfer paths in large power systems, *IEEE Transactions on Power Systems*, 22(2), 727–734, May 2007.
61. Dagle, J.E., North American synchrophasor initiative, *Hawaii International Conference on System Sciences, HICSS-41*, Waikoloa, Big Island, HI, January 2008, Piscataway, NJ: IEEE Computer Society.
62. Dagle, J.E., North American synchrophasor initiative: An update of progress, *Hawaii International Conference on System Sciences, HICSS-42*, Waikoloa Village, HI, January 2009. Piscataway, NJ: IEEE Computer Society.
63. Dagle, J.E., The North American synchrophasor initiative (NASPI), invited panelist at the *IEEE Power & Energy Society General Meeting*, Minneapolis, MN, July 2010.
64. Dagle, J.E., North American synchrophasor initiative: An update of progress, *Hawaii International Conference on System Sciences*, HICSS-44, Koloa, Kauai, HI, January 2011. Piscataway, NJ: IEEE Computer Society.
65. Breulmann, H., Grebe, E., Lösing, M. et al., Analysis and damping of inter-area oscillations in the UCTE/CENTREL power system, *CIGRE Session 2000*, Paris, France.
66. ICOEUR–Intelligent coordination of operation and emergency control of EU and Russian power grids, http://icoeur.eu/other/downloads/deliveralbles/ICOEUR_D1_1_final.pdf (accessed January 2011).
67. Sattinger, W., WAMs initiatives in continental Europe, *IEEE Power & Energy Magazine*, 6(5), 58–59, September/October 2008.

68. Babnik, T., Gabrijel, U., Mahkovec, B., Perko, M., and Sitar, G., Wide area measurement system in action, *Proceedings of the IEEE Power Tech 2007*, Lausanne, Switzerland, pp. 232–[237], [COBISS. SI-ID 28917509], 2007.
69. Zdeslav, Č., Ivan, Š., Renata, M., and Veselin, S., Synchrophasor applications in the Croatian power system, *Western Protective Relay Conference*, Spokane, WA, October 20–22, 2009.
70. Reinhardt, P., Carnal, C., and Sattinger, W., Reconnecting Europe, *Power Engineering International*, pp. 23–25, January 2005.
71. Sattinger, W., Reinhard, P., and Bertsch, J., Operational experience with wide area monitoring systems, *CIGRE 2006 Session*, Paris, France, pp. B5–B216.
72. Sattinger, W., Baumann, R., and Rothermann, P., A new dimension in grid monitoring, *Transmission & Distribution World*, pp. 54–60, February 2007.
73. Sattinger, W., Awareness system based on synchronized phasor measurements, *IEEE Power Energy Society General Meeting*, Calgary, Alberta, Canada, 2009.
74. Grebe, E., Kabouris, J., Lopez, B.S., Sattinger, W., and Winter, W., Low frequency oscillations in the interconnected system of continental Europe, *IEEE Power Energy Society General Meeting*, Minneapolis, MN, 2010.
75. Moraes, R.M., Volskis, H.A.R, and Hu, Y., Deploying a large-scale PMU system for the Brazilian interconnected power system, *IEEE Third International Conference on Electric Utility Deregulation and Restructuring and Power Technologies*, Nanjing, China, April, 2008.
76. Moraes, R.M., Volskis, H.A.R., Hu, Y., Martin, K., Phadke, A.G., Centeno, V., and Stenbakken, G., PMU performance certification test process for WAMPAC systems, *CIGRÉ SC-B5 Annual Meeting & Colloquium*, Jeju, Korea, October 2009.
77. Decker, I.C., Silva, A.S., Agostini, M.N., Priote, F.B., Mayer, B.T., and Dotta, D., Experience and applications of phasor measurements to the Brazilian interconnected power system, *European Transactions on Electrical Power*, DOI: 10.1002/etep.537, 2010. Published online in Wiley Online Library http://www.wileyonlinelibrary.com (accessed January 2011).
78. Beard, L. and Chow, J., NASPI RITT report outNASPI, October 2010. http://www.naspi.org/meetings/workgroup/2010_october/presentations/taskteams/taskteam_report_ritt_beard_20101006.pdf (accessed January 2011).
79. Novosel, D., Madani, V., Bhargava, B., Vu, K., and Cole, J., Dawn of the grid synchronization, *IEEE Power and Energy Magazine*, 6, 49–60, January/February 2008.

16
Assessment of Power System Stability and Dynamic Security Performance

16.1	Definitions and Historical Perspective..	16-1
16.2	Phenomena of Interest	16-3
16.3	Security Criteria...	16-3
16.4	Modeling	16-5
	Power System Network • Generators • Loads • Advanced Transmission Technologies • Protective Devices • Model Validation	
16.5	Analysis Methods	16-8
	Power Flow Analysis • P–V Analysis and Continuation Power Flow Methods • Time-Domain Simulations • Eigenvalue Analysis • Direct Methods • Other Methods	
16.6	Control and Enhancements	16-11
16.7	Off-Line DSA	16-11
16.8	Online DSA..	16-12
	Monitor System Security • Determine Stability Limits • Recommend Preventative and Corrective Control Actions • Handling Distributed and Variable Generation • Verify Special Protection Systems • Settle Transactions in Power Market • Determine Active and Reactive Power Reserves • Help in Scheduling Equipment Maintenance • Calibrate and Validate Power System Models • Prepare Models for System Studies • Perform System Restoration • Perform Postmortem Analysis of Incidents	
16.9	Status and Summary	16-16
	References..	16-17

Lei Wang
Powertech Labs Inc.

Pouyan Pourbeik
Electric Power Research Institute

16.1 Definitions and Historical Perspective

Power system security, in the context of this chapter, is concerned with the degree of risk in a power system as it pertains to its ability to survive a disturbance without interruption to customer service (IEEE/CIGRE, 2004). It relates to robustness of the system to disturbances and, hence, depends on the system operating condition prior to the disturbance as well as the contingent probability of the disturbance. Note that the concept of security should not be confused with the concept of stability. Stability refers to the ability of a power system, for a given operating condition, to regain a state of operating equilibrium after being subjected to a disturbance, with most system variables bounded so that practically the entire system remains intact. A stable system condition may not necessarily be secure. For example, under certain conditions a disturbance may lead to unintentional loss of load (due to the action of protective systems), which may eventually render the system response stable. However, a secure system condition must be stable.

When disturbances occur, the various components of the power system respond and hopefully reach a new equilibrium condition that is acceptable according to some criteria. The process of performing mathematical analysis of these responses and determining the new equilibrium condition is called security assessment. In terms of the dominant physical performance concerned, security assessment used to be classified into static security assessment (SSA) and dynamic security assessment (DSA). Due to the requirements of ensuring overall security of a system in practical security assessment, the current trend is not to make such a distinction by generally referring to both as DSA. DSA has been formally defined by IEEE as follows:

> Dynamic Security Assessment is an evaluation of the ability of a certain power system to withstand a defined set of contingencies and to survive the transition to an acceptable steady-state condition.

By a contingency is meant the loss of one or more power system elements, such as a transmission line, power transformer, etc. Early power systems were often separate and isolated regions of generators and loads. As systems became larger and more interconnected, the possibility of disturbances propagating long distances increased. The northeast blackout of November 1965 started a major emphasis on the reliability and security of electric power systems. The benchmark paper by Tom Dy Liacco introduced the concept of the normal, emergency, and restorative operating states and their associated controls (Dy Liacco, 1967). The normal state is the state wherein the system is stable with all components within operating constraints. The emergency state arises when the system begins to lose stability or when component operating constraints are violated. The restorative state is when service to some customers has been lost—usually due to progression through the emergency state and the operation of protective devices. Two additional states, *Alert* and *In Extremis*, were later added as shown in Figure 16.1 (Kundur, 1994).

DSA is traditionally performed off-line using power system models assembled for operational studies. Transmission planning (i.e., looking at future generation and load scenarios) is performed using planning cases where preliminary data are used for planned future generation and transmission equipment. However, off-line DSA for current (or imminent) system conditions is performed using models of the current system configuration, where models of currently in-service equipment have preferably been validated. Recently, online DSA performed using real-time system data has become more attractive due to advances in computation algorithm development, real-time data acquisition and analysis, and

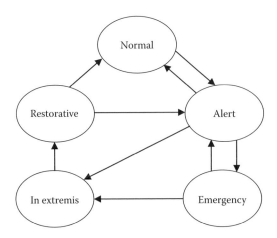

FIGURE 16.1 Operation states of a power system.

computer technologies. Another fundamental difference between online and off-line DSA analysis is that in off-line studies the engineer attempts to envisage possible onerous conditions that may lead to operating challenges and thus develop operating strategies to address them, whereas in online DSA the tool is working in real time to address actual present system operating conditions.

DSA contains three basic elements that are described in this chapter:

1. Setting the security criteria including the contingencies to be applied and the system performance expected
2. Building a set of system models necessary for the assessment
3. Performing the analysis using appropriate methods

Extensive literature on DSA practices is available; for example, Fouad (1988) and CIGRE (2007) give good review and summary on this subject.

16.2 Phenomena of Interest

While there are many phenomena that are of interest in the response of a power system following a contingency, DSA focuses primarily on two types of phenomena—static and dynamic. The static phenomena are the characteristics after the system settles to a new equilibrium, for example, power flow in circuits and voltages at major substations. Sometimes, the concern could be the ability of the system to settle into a new equilibrium, for example, the slow system collapse phenomenon due to voltage instability (Taylor, 1994). The dynamic phenomena are more related to the transient behavior of the system before settling into a new equilibrium, with the most basic concept concerning the issue of maintaining synchronous operation of the AC generators. This is usually referred to as transient stability (discussed in a previous chapter of this book).

The operation of the modern power system is becoming more and more complex. This is due to many factors, such as highly interconnected systems, use of advanced and fast controls, large-scale integration of asynchronous (or even nonrotary) generators such as wind turbines and solar photovoltaic (PV) arrays, changing load characteristics (e.g., the introduction of power electronic based loads such as variable frequency drives), unique considerations from power market operation, the looming prospect of mass deployment of electric vehicles, distributed generation, and many other factors. These factors have significant impact on the phenomena of interest for DSA. In addition to the traditional issues described earlier, proper emphasis needs now to be laid on other types of power system responses. These include electromechanical oscillations, transient voltages, and transient frequency. Studies for DSA should thus address such phenomena to ensure the overall security of the system.

16.3 Security Criteria

In general, to perform DSA one needs first to define two sets of criteria:

1. Contingencies: These are the disturbances that may occur in a system.
2. System performance: These criteria define the desired physical responses of a system following a contingency.

The NERC transmission planning standards (NERC, 2009) define the various system conditions. The category A condition is the no-contingency or pre-contingency condition during which all facilities in a system are in service. In terms of severity, contingencies may be further classified into

- Category B: event resulting in the loss of a single element, for example, the loss of a single generator, transmission line, power transformer, or single HVDC pole, with or without a fault
- Category C: event resulting in the loss of two elements with or without fault
- Category D: an extreme event resulting in the loss of two or more elements removed of cascading out of service

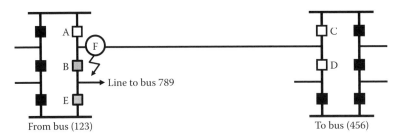

Event Sequence	
Time (Cycles)	Event
0	Single-line-to-ground fault at one end of a transmission line (from bus 123 to bus 456)
4	The line with fault is tripped after breakers A, C, and D are opened; however, the fault still stays assuming a stuck breaker B (which does not open)
10	A second line from bus 123 to bus 789 is tripped as a result of opening breaker E by the backup protection scheme
10	Fault is cleared

FIGURE 16.2 Sample N-2 contingency.

The event shown in Figure 16.2 illustrates a sample category C event, that is, the loss of two elements, resulting from a stuck breaker for a breaker and a half substation configuration.

System performance may be interpreted in two levels. First, the performance of a power system may be characterized by a set of physical response types, including

- Thermal loading
- Steady-state voltage
- Transient voltage
- Voltage stability
- Transient stability
- Damping of electromechanical oscillations
- Transient frequency
- Frequency stability
- Other (such as stability limits, reactive power reserve, relay margin, etc.)

Second, for each of the aforementioned response types, the system must operate within a specified performance range in order to maintain security. These are explained briefly in the following.

For thermal loading, the flow (measured by either current or power) on each transmission circuit (line or transformer) must be within its rating. For different system conditions, different ratings may be used. For example, at the normal (pre-contingency) condition, the standard rating is applied, while at the post-contingency condition, an emergency rating may be used. For online DSA applications, dynamic ratings adjusted at real time with measured temperature and wind speed may be used to achieve better accuracy.

For steady-state voltage, the magnitudes of voltages in a system must be within specified ranges. Values of these ranges may depend on the contingency condition, region of the system, and voltage levels. Commonly used ranges are between 0.95 and 1.05 pu for pre-contingency condition and between 0.9 and 1.1 pu for post-contingency condition.

For transient voltage, the main concerns are to avoid under-voltage and overvoltage protection relays to react following a contingency and to ensure the proper voltage recovery to restore loads. A transient voltage criterion is normally defined as the maximum time for which the system voltages are allowed to be continuously below or over specified thresholds.

For voltage stability, a system must be able to settle at a new equilibrium following a contingency. If this is not achievable, voltage collapse will occur, either in a transient time frame due to the inability to recover voltages, or in steady-state time frame due to the inability to supply the required load.

For transient stability, a system must be transiently stable following any contingency, that is, all synchronous generators must remain in synchronism when reaching the new equilibrium.

For damping of electromechanical oscillations, the criterion is usually set for the damping ratio of an oscillatory mode (Kundur, 1994). Although a positive damping ratio is sufficient for the oscillations to damp out, a minimum value (say between 3% and 5%) is usually required to provide reasonable margin (CIGRE, 1996).

For transient frequency, the main concern is to avoid under- and over-frequency protection relays to react following a contingency. Transient frequency criterion is normally defined as the maximum time for which the system frequency is allowed to be continuously below or over specified thresholds.

For frequency stability, a system must be able to settle at a new equilibrium following a contingency that involves loss of load or generation. This criterion is usually checked for small systems or for an islanding situation in a large interconnected system. The key to maintaining frequency stability is the proper coordination between load and generation shedding schemes after a large active power imbalance occurs in the system as a result of a contingency.

For DSA, different performance criteria may be required for different types of contingencies, and to a lesser degree for different types of transmission systems (EHV, HV, distribution, etc.). Such performance criteria are normally set by regulatory bodies and reliability coordination councils for large interconnected systems. For example, the North American Electric Reliability Corporation (NERC) issues a set of performance transmission planning (TPL) standards for its members to enforce (NERC, 2009).

16.4 Modeling

In order to perform analysis for the phenomena of interest in DSA, it is necessary to formulate mathematical models that capture the fundamental characteristics to be analyzed. Such models must cover all system components that are essential in considering the required security criteria and the modeling methods must also be consistent with the types of performance to be examined. For example, for SSA, static algebraic models are used for loads. In DSA, however, dynamic load components often need to be added to capture the required phenomena (such as slow voltage recovery). The entire modeling work for DSA consists of three steps:

1. Determine appropriate models to be included for the specific study.
2. Build the models to be used.
3. Validate the models.

The following sections provide comments on general modeling requirements for major power system components for DSA. The emphasis is on the provision of a modeling guideline, rather than the detailed derivation and presentation of the mathematical models, which can be found in a number of good references (Fouad and Vittal, 1992; Kundur, 1994; Taylor, 1994; Sauer and Pai, 1998; Anderson and Fouad, 2002).

16.4.1 Power System Network

A power system network described here refers to transmission lines, cables, transformers, reactive compensation devices (either series or shunt reactors/capacitors) connected to form the power grid. For the moment let us exclude more complex devices such as HVDC transmission (and other advanced forms of transmission technologies), generation, and loads.

As a good approximation, for purpose of DSA (Kundur, 1994), a power system network with the aforementioned components can be represented mathematically by a set of algebraic equations, establishing the relationship of bus voltage and current injection phasors evaluated at the nominal frequency. Such a model is adequate for both static and dynamic analysis. In special situations, additional modeling considerations are required, for example,

- Transformer under-load tap movements: This is important when studying voltage stability. Static and dynamic models can be included to account for such controls. Similar modeling practice is required for phase shifter transformers.
- Switched shunts: Some shunt compensation devices are installed in a set of reactor/capacitor blocks, which can be switched in or out of the system automatically, depending on the set control strategy. This type of devices may be critical in evaluating system voltage performance, and when necessary, their control strategy and reactor/capacitor switching should be included in the system model.
- Network frequency variation: As mentioned earlier, the network algebraic equation is usually established at the nominal system frequency. This is acceptable except for some islanding situations in which system frequencies can be significantly higher or lower than the nominal value during the transients. In such situations, it might be necessary to compensate the network admittances to account for the frequency fluctuations.

16.4.2 Generators

Generators are the most important component in a power system. In static analysis, modeling of generators (all types) is quite simple, usually supplying a specified active power while holding the voltage at a specified bus. In addition to this, the capabilities of a generator may be specified simply as the minimum and maximum active and reactive power it can supply, and for some studies details of such capabilities can be provided by lookup tables describing the relationship between its active and reactive power outputs.

For dynamic analysis, however, the generator modeling can be complex. The modeling of a synchronous machine includes the following main features (Kundur, 1994):

- Two differential equations to describe rotor mechanical dynamics
- Depending on modeling details required, as many as four differential equations to describe field and damper winding dynamics on the rotor
- Algebraic equations to relate machine terminal voltages, currents, and fluxes
- Magnetic saturation effects

Therefore, a synchronous machine may be represented by a set of differential equations of the second to sixth order, along with a few algebraic equations. In addition to these, the following models are usually included with the synchronous machine models to form the entire generator model:

- Exciter/AVR model
- The turbine-governor model
- Power system stabilizer (PSS) model
- Minimum and maximum excitation limiter models (for dynamic voltage stability studies)

Some of the recently developed renewable generation technologies do not use synchronous machines. For example, one of the main wind generation technologies is based on the doubly fed asynchronous generator, while a solar PV array can be represented as a controlled power source behind a voltage sourced converter. Models of these devices have been developed and used in DSA (GE Energy, 2005; Xue et al., 2009).

16.4.3 Loads

Loads are another important component in a power system, and its modeling presents different types of challenges, due largely to the difficulties in deriving an accurate yet simple aggregated model for all associated load components connected at a feeder in a substation. In addition, the load profile and composition is constantly changing both diurnally and seasonally. Furthermore, the load composition (i.e., types of load, industrial vs. residential, etc.) varies from region to region. In practice, reasonable approximations are made to capture the main characteristics required for the system performance to be studied. Two types of load models are often used in DSA:

1. Static models: These models are described by algebraic equations. The most commonly used is the so-called ZIP model in which the model is a weighted linear combination of constant impedance, constant current, and constant power load components. More complicated static models may include components that are general functions of voltage and frequency.
2. Dynamic models: Since the majority of the dynamic components in loads are various types of induction motors, most dynamic load models take the format of a composite load model including aggregated induction motor models. Inclusion of dynamic load models in DSA is critical when assessing voltage stability phenomena, particularly transient voltage performance.

It is quite common that a mixture of static and dynamic models is used for loads in DSA, with the weight factors set based on seasons and regions where the loads are located (Kosterev and Meklin, 2006).

16.4.4 Advanced Transmission Technologies

Advanced transmission technologies include HVDC transmission, flexible AC transmission system (FACTS), etc. These devices usually include complex controls and thus dynamic models described by differential equations are required to study their performances. In SSA, equivalent static models are available for these devices (Gyugyi and Hingorani, 1999). Dynamic models for FACTS devices are also available in the literature (Pourbeik et al., 2006).

16.4.5 Protective Devices

Protective devices are widely used in power systems to protect power equipment and systems. In DSA, performed for system planning purposes (i.e., off-line DSA), protective devices are usually not explicitly included in the system models. It is assumed in this case that either they would operate (e.g., when clearing faults), or they would not operate (when prohibited by planning standards). In DSA performed for system operation purposes (or online DSA), however, it is usually required to explicitly model protective devices, particularly those designed to prevent the system from losing stability (Wang et al., 2008). Such protective devices include generation rejection schemes, load shedding schemes, cross-tripping of transmission lines, out-of-step relays, special protection schemes (SPS), etc. In most cases, inclusion of these devices in online DSA has two objectives: (1) ensuring that the system maintains stability when the system conditions designed for the protective devices do occur and (2) determining the appropriate arming and tripping parameters (also referred to as lookup tables) for the specified system conditions so that the effectiveness of corrective actions can be guaranteed if the designed for events do occur.

16.4.6 Model Validation

As seen from previous sections, a complete power system model for DSA includes models for many components in the system. It is very important to ensure that these components are represented by appropriate models and the parameters in these models are suitable for the intended studies. Model validation is the process by which the mathematical models used for power system analysis are verified against the measured response of the actual equipment. There are various approaches to model validation (Pourbeik, 2010). There are two aspects to verifying the appropriateness of the model:

- Checking the consistency of models through simulation-based methods
- Validating that equipment models do emulate the expected dynamic response of the device through measurement-based methods

Simulation-based method uses various computer analysis programs to test the models (and in most situations the steady-state system conditions) for which DSA is performed. These may be simple tests to check basic model and data consistency. For example,

- Unusual transmission line reactance to resistance (X/R) ratios and transformer tap ratios
- Flows on transmission lines exceeding their surge impedance loading
- Very small time constants in dynamic models
- Inconsistent ratings of generator and its controls causing limit violations at the steady-state condition

More elaborated tests can be performed to help identify hidden problems in models. The following are some of these tests:

- No-fault simulation test: In this test, a time-domain simulation is performed without applying a disturbance. It is expected that the system responses stay unchanged from the starting steady-state values. If not, the models or the initial system conditions may be incorrect, or they may not be consistent.
- Device eigenvalue test: In this test, a component is decoupled from the system (e.g., by connecting to an infinite bus) and the eigenvalues of its linearized dynamic model are computed. An unstable or critically stable mode from such a model may indicate incorrect parameters and operating conditions.

Measurement-based model validation methods, on the other hand, validate models by measuring the responses of physical equipment from the designated field tests. From the measured responses, the model and its parameters for the equipment can be derived or validated. There are well-established procedures in conducting tests and model validation for commonly used equipment such as generators (WECC, 2006). More recently, methods have been developed and demonstrated for using online disturbance monitoring for the validation of power equipment (Pourbeik, 2009, 2010).

16.5 Analysis Methods

Depending on the types of performance concerned, or the criteria to be applied, different analysis methods can be used for DSA. An overview of these methods is provided in the following sections. Details of most of these methods can be found in a number of good references (Fouad and Vittal, 1992; Kundur, 1994; Taylor, 1994; Sauer and Pai, 1998; Wehenkel, 1998; Pavella et al., 2000; Anderson and Fouad, 2002).

16.5.1 Power Flow Analysis

Power flow analysis refers to the determination of the steady-state operation condition of a power system for a given set of network configuration, controls, and known inputs. It is used for most of the SSA tasks, for example, thermal loading and voltage analysis. Mathematically, the power flow analysis problem is formulated as a set of nonlinear algebraic equations, and well-developed solution methods are available, including the fast decoupled method and the Newton–Raphson method. Power flow analysis is also the foundation of many of the more advanced analyses for DSA. For example, it provides the initial condition for the time-domain simulations in transient stability analysis.

16.5.2 P–V Analysis and Continuation Power Flow Methods

In voltage stability analysis, it is often required to determine the voltage stability margin, for example, the maximum power that can be transferred from generation (source) to meet demand (sink).

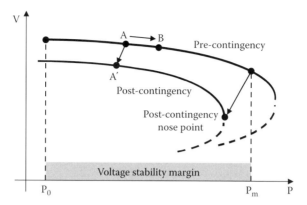

FIGURE 16.3 P–V analysis.

This can be done by power–voltage (P–V) analysis as shown in Figure 16.3. This analysis consists of the following steps:

1. Start from a solved power flow (A) at the pre-contingency condition for a specified transfer level (a pair of matching source/sink conditions).
2. Apply the contingencies to be considered and solve the power flow (A′).
3. If all post-contingency power flows can be solved without security violations, increase the transfer level by a preset step and solve the power flow (B). Go to step 2 to repeat the contingency analysis. If any of post-contingency power flows cannot be solved, or can be solved but with security violations (such as low voltages), the voltage stability limit is found.

The aforementioned process results in a series of P–V curves, corresponding to the pre-contingency and all post-contingency conditions. The voltage stability margin is then defined as the distance from the current operation condition (P_0) and the condition on the pre-contingency P–V curve corresponding to the node point of the most critical post-contingency P–V curve:

$$\text{Voltage stability margin (\%)} = \frac{P_m - P_0}{P_0} \times 100$$

One of the challenges in performing P–V analysis is to ensure the convergence of the power flow solution close to the nose point as the equations become ill-conditioned. A special power flow solution method known as continuation power flow (CPF) was developed to overcome this issue (Ajjarapu and Christy, 1992).

16.5.3 Time-Domain Simulations

The problems for transient stability and other security criteria related to transient performance are commonly solved by the so-called time-domain simulations. This refers to the application of numerical integration algorithms to solve the set of nonlinear differential equations that describe the dynamics of a system model (Dommel and Sato, 1972). Different from the modeling assumptions and techniques used in the electromagnetic transient simulations, the simulations for DSA give the following system responses:

- Positive sequence RMS phasor values (where applicable).
- The valid frequency range is approximately 0.1–5 Hz.
- The typical time frame for the analysis is 10–20 s for transient stability analysis, which may be extended to minutes for dynamic voltage stability analysis when appropriate models are included.

Time-domain simulations are performed with three basic sets of input data:

- A solved power flow case, which provides the network topology, parameters, and initial operating conditions of the system
- A set of dynamic models that match the components in the power flow case
- A set of contingencies to be applied as disturbances during simulations

The basic outputs from simulations are the time-domain responses of various physical quantities in the system, such as generator rotor angles, speeds, bus voltages, flows in transmission lines, load powers, etc. More advanced applications of time-domain simulations include use of special techniques to post-process simulation results in order to obtain more insightful results to determine transient performance of the system. Typical advanced applications include

- Calculation of critical clearance time (CCT) of a fault. This is useful to rank contingencies and thus to find weak regions in a system that are prone to transient instability.
- Identification of critical oscillatory modes captured in time-domain responses using the Prony method (Hauer, 1991). This can be used as a complementary method with eigenvalue analysis for small-signal stability analysis.
- Determination of stability limits (IEEE, 1999).

16.5.4 Eigenvalue Analysis

For studies of low-frequency oscillations, the frequency-domain method using eigenvalue analysis is a common choice in addition to time-domain simulations. In this method, the nonlinear differential equations describing the system dynamics are linearized around an operating condition and eigenvalues corresponding to electromechanical oscillations are computed from the linearized model. These eigenvalues, together with other relevant information such as eigenvectors and transfer function zeros and residues, give valuable information on the characteristics of the oscillations in the system and provide the directions on how to improve damping for these oscillations (Rogers, 2000).

Another application of eigenvalue analysis is for voltage stability analysis (Gao et al., 1992). The smallest eigenvalue of the power flow Jacobian matrix gives good indication of how close the system is to its nose point (or stability limit). In addition, the eigenvector associated with this eigenvalue contains information on the mode of voltage instability, that is, the region in the system that is likely to experience voltage collapse. This is very useful in practice for the understanding of voltage stability problems and for deriving remedial controls.

16.5.5 Direct Methods

Direct methods refer to a class of methods that assess the transient stability of a power system, and also give a measure of degree of stability ("stability margin"), based on partial responses of the system obtained from time-domain simulations. Two main types of such methods have been developed. The first is often called the transient energy function (TEF) method (Pai, 1989; Fouad and Vittal, 1992). The idea is to replace the numerical integration by stability criteria. The value of a suitably designed Lyapunov function V is calculated at the instant of the last switching in the system and compared to a previously determined critical value V_{cr}. If V is smaller than V_{cr}, the post-fault transient process is stable. The second type of the direct methods is called EEAC (or SIME), in which a parametric single-machine-infinite-bus (SMIB) system is constructed from the time-domain simulation results and the system stability is determined from this SMIB system (Xue et al., 1989; Pavella et al., 2000). Both types of direct methods have found applications in transient stability analysis, particularly in online DSA (Fang and Xue, 2000; Chiang et al., 2010).

16.5.6 Other Methods

In addition to the methods previously described, many others have been proposed and applied to DSA problem, including

- V–Q analysis for voltage stability assessment (Kundur, 1994)
- Probabilistic methods (Anderson and Bose, 1983)
- Expert system method (El-Kady et al., 1990)
- Neural network methods (El-Keib and Ma, 1995; Mansour et al., 1997; Chen et al., 2000)
- Pattern recognition (Hakim, 1992)
- Decision tree methods (Wehenkel, 1998)

16.6 Control and Enhancements

The adoption of security concepts for electric power systems clearly separates the two functions of assessment and control. Assessment is the analysis necessary to determine the outcome of a credible contingency. Control is the operator intervention or automatic actions that are designed for use to avoid the contingency entirely, or to remedy unacceptable post-contingency conditions. When the controls are implemented, they may then become a part of the assessment analysis through a modification of the contingency descriptions.

Controls may be classified into three types. Preventative control is the action taken to maneuver the system from the alert state back to the normal state. This type of control may be slow, and may be guided by extensive analysis. Emergency control is the action taken when the system has already entered the emergency state. This type of control must be fast and guided by predefined automatic remedial schemes. Restorative control is the action taken to return the system from the restorative state to the normal state. This type of control may be slow, and may be guided by analysis and predefined remedial schemes.

16.7 Off-Line DSA

In off-line DSA analysis, detailed analysis is performed for a vast number of credible contingencies and a variety of operating conditions, subject to a set of specified security criteria. In the basic form, this off-line analysis is used to determine the security status of one system condition and this is referred to as the *basecase analysis*. The results of a basecase analysis indicate whether or not the system is secure at the studied system condition, and if not secure, which security criteria are violated and the contingencies under which the security violations occur. Typically, the basecase condition used is the most onerous one, for most systems this is the forecasted peak load system condition.

Extending the basecase analysis leads to *transfer analysis*, in which power transfer limits across important system interfaces are determined in a way similar to the P–V analysis shown in Section 16.5.2. Such transfers are defined by a set of sources and sinks, and are measured by the flows on the key interfaces. The power transfer limits are defined as the maximum transfer levels at which no security criteria are violated. The limits so computed may be used in system planning or operation. Since the analysis is performed off-line, there is no severe restriction on computation time and therefore detailed analysis can be done for a wide range of conditions and contingencies.

A power transfer can be one dimensional, in which one source-and-sink pair of variables are defined and changed to determine the transfer limit (a single number), or two dimensional, in which two independent sources (or sinks) and one dependent sink (or source) variables are defined and changed to determine the transfer limit (a two-dimensional monogram). Higher-dimensional transfers are possible but they are normally analyzed by reducing to a series of one- or two-dimensional transfers. Figure 16.4 illustrates a two-dimensional transfer limit monogram.

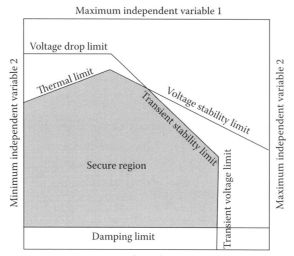

FIGURE 16.4 Two-dimensional transfer limit monogram.

16.8 Online DSA

Online DSA is essentially a technology that takes a real-time snapshot of a power system condition from the energy management system (EMS), performs desired security assessment in near real time (within a specified time cycle), and provides the operators with warnings of abnormal situations, secure operation regions (stability limits), as well as remedial measure recommendations, if applicable.

Although the online DSA concept started as an extension of the off-line DSA, it is significantly more challenging to implement, yet offers many additional benefits. After continued research and development, during the past couple of decades, the online DSA technology is becoming mature. More and more practical applications have been reported (Morison et al., 2004; Vittal et al., 2005; Wang and Morison, 2006; CIGRE, 2007; Savulescu, 2009). This section summarizes the general architecture of an online DSA system, its potential applications areas, and the performance expectation.

An online DSA system may consist of up to six main functional modules shown in Figure 16.5:

1. Measurements: This is to obtain the real-time system condition. This function is also part of the EMS. While the measurements from the traditional SCADA can generally meet the online DSA input data requirements, the latest data collection technology (e.g., PMU-based WAMS) can provide much better and accurate system conditions. This can greatly improve the online DSA application quality.
2. Modeling: This is to assemble a set of models suitable for DSA. This is a critical module in online DSA. Some functions in this module (such as the State Estimator) are also part of an EMS; others may be functions specific for the online DSA, including
   a. External network and dynamics equivalencing
   b. Matching real-time system data with non-real-time data (such as dynamic models)
   c. Contingency definition based on real-time bus/breaker status
   d. Extended modeling capabilities, for example, inclusion of relays and special protection systems (SPS), automatic detection and handling of unusual or problematic system conditions, consideration of real-time in- and out-of-service status of control devices (AVR, PSS, SPS, etc.)
   e. Creation of modified system conditions for forecast or study mode analysis
3. Computation: This is the computation engine of the online DSA. Three main analysis options are usually included
   a. Security assessment of the basecase conditions (real time, forecast, study mode, etc.)

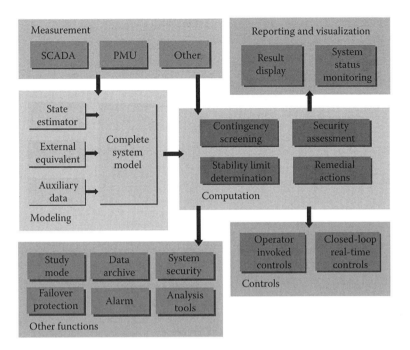

FIGURE 16.5 Function modules in an online DSA system.

   b. Determination of the stability limits
   c. Identification of applicable remedial actions to handle insecure contingencies and to increase stability limits, if necessary

   In addition to the aforementioned analysis options, advanced computational techniques are often used to meet performance requirements. For example, use of contingency screening techniques (Demaree et al., 1994; Chadalavada et al., 1997; Vaahedi et al., 1999; Chiang et al., 2010) and distributed computation technique (Moshref et al., 1999).

4. Reporting and visualization: This includes display and visualization of DSA results as well as reports for the operational status of the DSA system. Careful attention has been paid to the visualization of DSA results. This does not only focus on the information to be presented, but also on how it is presented. Web-based and geographical display methods are becoming more and more popular (Alstom, 2010).
5. Control: Various controls (such as generator rejection) are used as remedial actions to ensure system security. Online DSA can be integrated with such controls to provide settings calculated in real time and even to send the arming signals when the system condition requires such controls to react (Pai and Sun, 2008). For this application, only a limited number of examples are available; however, this is considered as one of the main attractions of online DSA and it is expected to become more mature as the DSA technology and power grid infrastructure is improved.
6. Other functions: These are a collection of functions to improve the reliability, usability, and applicability of an online DSA system. Some of these play important roles in the deployment of online DSA. For example,
   a. System security to be compliant with NERC cyber security standards (NERC, 2010)
   b. Data archiving to provide history cases ready for use in study mode
   c. Integration with other analysis functions (such as oscillation monitoring based on PMU measurements) to extend online DSA functionality

Online DSA can be used in a number of application fields. Clear identification of the requirements to have an online DSA system, or in other words, the benefits that an online DSA system can offer, provides the motivation and ultimately the objectives for implementing such a system. Depending on different ways in which an online DSA system is applied, the following functions and the associated benefits can be achieved.

16.8.1 Monitor System Security

This is the basic function with which the online DSA system checks the security status of a real-time system condition for possible and credible contingencies. All or part of the security criteria described in Section 16.3 can be applied. The operator is alerted for appropriate actions should any defined security criteria be violated. This function is particularly useful when the system is operating in a region not studied off-line. The reference (Savulescu, 2009) includes operational experiences indicating potential unstable system conditions that were not known to the operator.

16.8.2 Determine Stability Limits

Traditionally, stability limits are calculated off-line using planning system models and augmented with significant margins to ensure system security. It is well known that such an approach is often overly conservative, resulting in lost transfer capacity. Limits calculated by online DSA using real-time data can be much more accurate and reliable to use and thus eliminate the need for overly conservative stability limit margins.

16.8.3 Recommend Preventative and Corrective Control Actions

When a system is insecure, or does not have sufficient security margin, or is not able to transfer the required power due to a stability limit, preventative and/or corrective control actions (wherever available) should be dispatched to mitigate the problem. Online DSA is ideal for such tasks and its results are usually much better than those obtained from the lookup tables prepared off-line. As an example, such a function is implemented in PJM's online DSA system (Tong and Wang, 2006).

16.8.4 Handling Distributed and Variable Generation

Generation technologies such as wind and solar generation have become increasingly popular as a form of clean and renewable energy. These generation technologies tend to exhibit significant variability in their output and thus systems with high penetration of such generation may be more challenging to operate due to the characteristics of such generation as compared to traditional synchronous-generation-based fossil-fuel plants, for example,

- Renewables are often located in unpopulated remote region requiring long transmission.
- The output of wind and solar PV generation plants are highly variable due to fluctuations in wind and solar light, which presents a significant challenge for maintaining load/generation balance for systems with high penetration of these energy resources.
- New generation technologies are adopted, which are typically nonsynchronous, even nonrotary in the case of solar PV.
- Hitherto, these generation resources typically do not provide primary frequency regulation, and some do not provide inertial response either. Also, depending on the plant design they may not provide significant voltage regulation.*

---

* Certainly, many modern wind turbine generator technologies do offer functionality such as low-voltage ride-through and voltage regulation at the point of interconnection of the wind power plant. Also, many vendors have demonstrated and developed functionality for their wind turbine generators to provide inertial response and frequency regulation, particularly for over frequency events; however, whether such functions are deployed in the field is a function of many factors including regional power system market design.

Assessment of Power System Stability and Dynamic Security Performance

These factors can raise concerns from system operators on the amount of such generation that can be allowed at any given time without compromising system security. The solution to these concerns is often to perform quick online assessment of system conditions in order to determine the optimal dispatch of this type of generation (e.g., under certain conditions, a percentage of variable generation may need to be curtailed). Online DSA is able to perform such analysis with a turnaround speed in the order of minutes. Dudurych (2010) shows an example of such applications implemented at the Irish national grid company (EirGrid).

16.8.5 Verify Special Protection Systems

SPS are designed to protect a system from losing stability for severe contingencies. The design and commissioning of an SPS can only be performed for limited system conditions, and as such SPS malfunctions may occur in special system conditions. Inclusion of SPS models in online DSA (Wang et al., 2008) can put these systems under constant testing, thus verifying their functions and ensuring their correct action during various system conditions.

16.8.6 Settle Transactions in Power Market

One of the problems in operating a power market is to settle power transaction requests and this must be done very quickly with all security criteria respected. Therefore, settlement often needs to take consideration of stability so that required security margin is met. One of the successful applications in this area is the online DSA system at ERCOT (Rosales et al., 2003), which computes voltage stability limits and post them for use by the ERCOT power market.

16.8.7 Determine Active and Reactive Power Reserves

Determination of active and reactive power reserve is important for system operation. For example, for small systems whose frequency fluctuation can be sufficiently large when a large unit is tripped, maintaining proper active power reserve is important to ensure frequency stability of the system. In this case, online DSA can play a critical role in determining the optimal active power reserve while respecting the frequency criteria of the system. The application described in (Dudurych, 2010) actually has this function, which determines the appropriate level of active power reserve required to maintain the specified frequency criteria.

16.8.8 Help in Scheduling Equipment Maintenance

Equipment maintenance represents an unusual system operation condition and if such maintenance is to occur in stressed system condition, care must be taken to ensure security. Outage scheduling is traditionally done using planning models and, similarly to stability limit determination; this often results in overly conservative conclusions. Online DSA provides a much better alternative, in which the real-time system condition is modified to reflect the required maintenance. This not only gives more accurate results, but also makes the studies more efficient.

16.8.9 Calibrate and Validate Power System Models

One of the continued pursuits of power system engineers is to develop system models that can capture reasonably well the required system characteristics. Unfortunately, this is not always achieved easily. The typical examples are the postmortem analysis following the 1996 and 2003 North America blackouts, in which initial system models assembled after the incidents were found to be unable to replicate the system responses recorded. This prompted programs to improve the quality of system models, such as the WECC generator model testing and validation initiative. Online DSA provides a great way to calibrate and validate such models by comparing the computer simulation results with field measurements (such as from PMU). In fact, model validation has been one of the strong motivations behind the implementation of some of the recent online DSA projects in North America.

16.8.10 Prepare Models for System Studies

Traditionally, power system studies performed off-line use data prepared from system planning. It is increasingly recognized that such models do not give good representation of the practical system condition, particularly when used to perform short-term operational planning studies. One way to overcome this problem is to start such studies with the real-time models assembled by online DSA. This approach is used recently by a major ISO in the United States, which is in the process of switching from the planning cases to the real-time cases prepared by online DSA for short-term operational planning studies.

16.8.11 Perform System Restoration

When a system is in an emergency state with loss of many elements, restoration tasks must be performed. The system in this form of state is usually not studied using off-line models and, therefore, little guidance is available with regard to security during system restoration. Online DSA can provide a vital role in this process, as shown in the experience documented in (Viikinsalo et al., 2006).

16.8.12 Perform Postmortem Analysis of Incidents

When an incident does strike on a power system with widespread impact, it is inevitable that a postmortem analysis is required. An online DSA system is usually capable of archiving system conditions periodically, together with all necessary auxiliary data to perform stability analysis. Such archived data can provide the quick and good starting point for a postmortem analysis.

Since online DSA is usually required to be done within a specified time cycle, the speed of performance is a severe constraint in addition to the other technical challenges. Typical performance goals are to complete a computation cycle within 5–20 min after a real-time snapshot is available. For example, the PJM online Transient Stability Analysis and Control system (Tong and Wang, 2006) has a computation cycle of 15 min, within which 3000 contingencies are processed in the basecase analysis and 40 stability limits are determined in the transfer analysis for real-time cases of 13,500 buses and 2,500 generators. Such performance is made possible by using two computation techniques: (1) early termination of simulations if the stability of the system can be decided and (2) distribution of computations in multiple simultaneous sessions in an array of computer servers.

16.9 Status and Summary

A few recent publications provide a detailed description of the status of DSA tools (Vittal et al., 2005; CIGRE, 2007), particularly for online DSA. With the increase in transactions on the bulk power system there is a critical need to determine system security in an online setting and also perform remedial controls if the analysis indicates that the system is insecure. In recent years, the industry has seen the development and deployment of large renewable generation projects (particularly wind generation). Thus, the stability properties of the system are changing since these new generation technologies (particularly wind and solar PV) have different characteristics to conventional synchronous generators. Thus, the analysis of power system steady-state and dynamic security continues to be of vital importance and a field of study that is continually evolving in order to understand the nature and behavior of a continually changing power system.

Stability problems may not happen frequently, but their impact, when they do happen, can be enormous. Most of the time, off-line studies are performed to determine conservative limits. In the new environment, the responsibility of monitoring system stability, preferably through online applications, may be vested with the independent system operator (ISO) and regional transmission organization (RTO).

A special panel session was held at the 2010 IEEE PES general meeting, at which six papers (Chiang et al., 2010; Dudurych, 2010; Loud et al., 2010; Neto et al., 2010; Wu et al., 2010; Yao and Atanackovic, 2010) were presented on the utility application experience for online DSA. These papers represent a snapshot of the current state-of-the-art on this technology. It is clear that, as an important component in the modernization of power system control centers, with the current trend of smart grid development worldwide (Zhang et al., 2010), more and more applications of online DSA technology will appear in the near future.

References

Ajjarapu, V. and Christy, C., The continuation power flow: A tool for steady state voltage stability analysis, *IEEE Transactions on Power Systems*, 7(1), 416–423, February 1992.

Alstom e-terra*vision* brochure, available from http://www.alstom.com/grid/eterravision/(accessed on November 16, 2010).

Anderson, P.M. and Bose, A., A probabilistic approach to power system stability analysis, *IEEE Transactions on Power Apparatus and Systems*, PAS-102(8), 2430–2439, August 1983.

Anderson, P.M. and Fouad, A.A., *Power System Control and Stability*, 2nd edn., Wiley-IEEE Press, Piscataway, NJ, October 2002.

Chadalavada, V., Vittal, V., Ejebe, G.C., Irisarri, G.D., Tong, J., Pieper, G., and McMullen, M., An on-line contingency filtering scheme for dynamic security assessment, *IEEE Transactions on Power Systems*, 12(1), 153–159, February 1997.

Chen, L., Tomsovic, K., Bose, A., and Stuart, R., Estimating reactive margin for determining transfer limits, *Proceedings, 2000 IEEE Power Engineering Society Summer Meeting, 2000*, Seattle, WA. IEEE, Vol. 1, July 2000.

Chiang, H.D., Tong, J., and Tada, Y., On-line transient stability screening of 14,000-bus models using TEPCO-BCU: Evaluations and methods, a paper presented at a panel session at the *IEEE PES General Meeting*, Minneapolis, MN, July 2010, pp. 1–8.

CIGRE Technical Brochure No. 111, Analysis and Control of Power System Oscillations, CIGRE Task Force 38.01.07, Paris, France, 1996 (www.e-cigre.org).

CIGRE Technical Brochure No. 325, Review of On-Line Dynamic Security Assessment Tools and Techniques, CIGRE Working Group C4.601, Paris, France, 2007 (www.e-cigre.org).

Demaree, K., Athay, T., Chung, K., Mansour, Y., Vaahedi, E., Chang, A.Y., Corns, B.R., and Garett, B.W., An on-line dynamic security analysis system implementation, *IEEE Transactions on Power Systems*, 9(4), 1716–1722, November 1994.

Dommel, H.W. and Sato, N., Fast transient stability solutions, *IEEE Transactions on Power Apparatus and Systems*, 91, 1643–1650, July/August 1972.

Dudurych, I.M., On-line assessment of secure level of wind on the Irish power system, a paper presented at a panel session at the *IEEE PES General Meeting*, Minneapolis, MN, July 2010, pp. 1–7.

Dy Liacco, T.E., The adaptive reliability control system, *IEEE Transactions on Power Apparatus and Systems*, PAS-86(5), 517–531, May 1967.

El-Kady, M.A., Fouad, A.A., Liu, C.C., and Venkataraman, S., Use of expert systems in dynamic security assessment of power systems, *Proceedings 10th PSCC*, Graz, Austria, pp. 913–920, 1990.

El-Keib, A.A. and Ma, X., Application of artificial neural networks in voltage stability assessment, *IEEE Transactions on Power System*, 10(4), 1890–1896, November 1995.

Fang, Y.J. and Xue, Y.S., An on-line pre-decision based transient stability control system for the Ertan power system, *Powercon 2000*, Perth, Australia, October 22–26, 2006, December 4–7, 2000, Vol. 1, pp. 287–292.

Fouad, A.A. (Chairman. IEEE PES Working Group on DSA), Dynamic security assessment practices in North America, *IEEE Transactions on Power Systems*, 3(3), 1310–1321, August 1988.

Fouad, A.A. and Vittal, V., *Power System Transient Stability Analysis Using the Transient Energy Function Method*, Prentice-Hall, Englewood Cliffs, NJ, 1992.

Gao, B., Morison, G.K., and Kundur, P., Voltage stability evaluation using modal analysis, *IEEE Transactions on Power Systems*, 7(4): 1529–1542, November 1992.

GE Energy, Modeling of GE wind turbine-generators for grid studies, Version 3.4b, March 4, 2005.

Gyugyi, L. and Hingorani, N., *Understanding FACTS: Concepts and Technology of Flexible AC Transmission Systems*, IEEE, New York, 1999.

Hakim, H., Application of pattern recognition in transient security assessment, *Journal of Electrical Machines and Power Systems*, 20, 1–15, 1992.

Hauer, J.F., Application of Prony analysis to the determination of model content and equivalent models for measured power systems response, *IEEE Transactions on Power Systems*, 6, 1062–1068, August 1991.

IEEE Special Publication, Techniques for Power System Stability Limit Search, IEEE Catalog Number 99TP138, 1999.

IEEE/CIGRE Joint Task Force on Stability Terms and Definitions, Definition and classification of power system stability, *IEEE Transactions on Power Systems*, 19(2), 1387–1401, August 2004.

Kosterev, D. and Meklin, A., Load modeling in WECC, *PSCE*, Atlanta, GA, October 2006.

Kundur, P., *Power System Stability and Control*, McGraw Hill, New York, 1994.

Loud, L., Guillon, S., Vanier, G., Huang, J.A., Riverin, L., Lefebvre, D., and Rizzi, J.-C., Hydro-Québec's challenges and experiences in on-line DSA applications, a paper presented at a panel session at the *IEEE PES General Meeting*, Minneapolis, MN, July 2010.

Mansour, Y., Chang, A.Y., Tamby, J., Vaahedi, E., Corns, B.R., and El-Sharkawi, M.A., Large scale dynamic security screening and ranking using neuron networks, *IEEE Transactions on Power Systems*, 12(2), 954–960, May 1997.

Morison, K., Wang, L., and Kundur, P., Power system security assessment, *IEEE Power and Energy Magazine*, 2(5), 30–39, September/October 2004.

Moshref, A., Howell, R., Morison, G.K., Hamadanizadeh, H., and Kundur, P., On-line voltage security assessment using distributed computing architecture, *Proceedings of the International Power Engineering Conference*, Singapore, May 24–26, 1999.

NERC, Reliability standards for transmission planning TPL-001 to TPL-006, available from www.nerc.com (accessed on May 18, 2009).

NERC, Critical infrastructure protection standards CIP-001 to CIP-009, available from www.nerc.com (accessed on October 12, 2010).

Neto, C.A.S., Quadros, M.A., Santos, M.G., and Jardim, J., Brazilian system operator online security assessment system, a paper presented at a panel session at the *IEEE PES General Meeting*, Minneapolis, MN, July 2010.

Pai, M.A., *Energy Function Analysis for Power System Stability*, Kluwer Academic publishers, Boston, MA, 1989.

Pai, A.C. and Sun, J., BCTC's experience towards a smarter grid—Increasing limits and reliability with centralized intelligence remedial action schemes, *Proceedings of IEEE Canada Electrical Power and Energy Conference*, Vancouver, British Columbia, Canada, October 6–7, 2008.

Pavella, M., Ernst, D., and Ruiz-Vega, D., *Transient Stability of Power Systems—A Unified Approach to Assessment and Control*, Kluwer Academic Publishers, Boston, MA, 2000.

Pourbeik, P., Automated parameter derivation for power plant models from system disturbance data, *Proceedings of the IEEE PES General Meeting*, Calgary, Alberta, Canada, July 2009.

Pourbeik, P., Approaches to validation of power system models for system planning studies, *Proceedings of the IEEE PES General Meeting*, Minneapolis, MN, July 2010, pp. 1–10.

Pourbeik, P., Boström, A., and Ray, B., Modeling and application studies for a modern static var system installation, *IEEE Transactions on Power Delivery*, 21(1), 368–377, January 2006.

Rogers, G., *Power System Oscillations*, Kluwer Academic publishers, Boston, MA, 2000.

Rosales, R.A., Sadjadpour, A., Gibescu, M., Morison, K., Hamadani, H., and Wang, L., ERCOT's implementation of on-line dynamic security assessment, a paper presented at a panel session at the *IEEE PES meeting*, Toronto, Ontario, Canada, 2003.

Sauer, P.W. and Pai, M.A., *Power System Dynamics and Stability*, Prentice-Hall, Upper Saddle River, NJ, 1998.

Savulescu, S., ed., *Real-Time Stability Assessment in Modern Power System Control Centers*, IEEE Press, Piscataway, NJ, 2009.

Taylor, C.W., *Power System Voltage Stability*, McGraw-Hill, New York, 1994.

Tong, J. and Wang, L., Design of a DSA tool for real time system operations, *PowerCon 2006*, Chongqing, China, October 22–26, 2006, pp. 1–5.

Vaahedi, E., Fuches, C., Xu, W., Mansour, Y., Hamadanizadeh, H., and Morison, K., Voltage stability contingency screening and ranking, *IEEE Transaction on Power Systems*, 14(1), 256–265, February 1999.

Viikinsalo, J., Martin, A., Morison, K., Wang, L., and Howell, F., Transient security assessment in real-time at Southern Company, a paper presented at a panel session at the *IEEE PSCE Conference*, Atlanta, GA, October 2006, pp. 13–17.

Vittal, V., Sauer, P., and Meliopoulos, S., On-line transient stability assessment scoping study, Final Project Report, PSERC Publication 05-04, Power Systems Engineering Research Center, February 2005.

Wang, L., Howell, F., and Morison, K., A framework for special protection system modeling for dynamic security assessment of power systems, *Powercon Conference*, New Delhi, India, October 12–15, 2008.

Wang, L. and Morison, K., Implementation of online security assessment, *IEEE Power & Energy Magazine*, 4, 24–59, September/October 2006.

WECC, Generating unit model validation policy, available from www.wecc.biz (accessed on October 27, 2006).

Wehenkel, L., *Automatic Learning Techniques in Power Systems*, Kluwer Academic Publishers, Boston, MA, 1998.

Wu, W., Zhang, B., Sun, H., and Zhang, Y., Development and application of on-line dynamic security early warning and preventive control system in China, a paper presented at a panel session at the *IEEE PES General Meeting*, Minneapolis, MN, July 2010, pp. 1–7.

Xue, Y., Van Custem, T., and Ribbens-Pavella, M., Extended equal area criterion justifications, generalizations, applications, *IEEE Transactions on Power Systems*, 4(1), 44–52, February 1989.

Xue, J., Yin, Z., Wu, B., and Peng, J., Design of PV array model based on EMTDC/PSCAD, *Proceedings of Power and Energy Engineering Conference, 2009*, Wuhan, China, pp. 1–5, 2009.

Yao, Z. and Atanackovic, D., Issues on security region search by online DSA, a paper presented at a panel session at the *IEEE PES General Meeting*, Minneapolis, MN, July 2010, pp. 1–4.

Zhang, P., Li, F., and Bhatt, N., Next-generation monitoring, analysis, and control for the future smart control center, *IEEE Transaction on Smart Grid*, 1(2), 186–192, September 2010.

17
Power System Dynamic Interaction with Turbine Generators

17.1	Introduction 17-1
17.2	Subsynchronous Resonance... 17-2
	Known SSR Events • SSR Terms and Definitions • SSR Physical Principles • SSR Mitigation • SSR Analysis • SSR Countermeasures • Fatigue Damage and Monitoring • SSR Testing • Summary
17.3	Device-Dependent Subsynchronous Oscillations 17-17
	HVDC Converter Controls • Variable Speed Motor Controllers • Power System Stabilizers • Renewable Energy Projects and Other Interactions
17.4	Supersynchronous Resonance 17-18
	Known SPSR Events • SPSR Physical Principles • SPSR Countermeasures
17.5	Device-Dependent Supersynchronous Oscillations 17-20
	Known DDSPSO Events • DDSPSO Physical Principles • DDSPSO Countermeasure
17.6	Transient Shaft Torque Oscillations 17-21
References 17-22

Bajarang L. Agrawal
Arizona Public Service Company

Donald G. Ramey (retired)
Siemens Corporation

Richard G. Farmer
Arizona State University

17.1 Introduction

Turbine generators for power production are critical parts of electric power systems, which provide power and energy to the user. The power system can range from a single generator and load to a complex system. A complex system may contain hundreds of power lines at various voltage levels and hundreds of transformers, turbine generators, and loads. When the power system and its components are in the normal state, the synchronous generators produce sinusoidal voltages at synchronous frequency (60 Hz in the United States) and desired magnitude. The voltages cause currents to flow at synchronous frequency through the power system to the loads. The only current flowing in the generator rotor is the direct current in the generator field. Mechanical torque on the turbine-generator rotor produced by the turbine is constant and unidirectional. There is a reaction torque produced by the magnetic field in the generator, which balances the mechanical torque and maintains constant speed. The system is said to be in synchronism and there is no dynamic interaction between the power system and the turbine generators.

At other times, the system and its components are disturbed, thereby causing a periodic exchange of energy between the components of the power system. If there is a periodic exchange of energy between a turbine generator and the power system, we will refer to this energy exchange as power system dynamic interaction with a turbine generator. When this occurs, the magnetic interaction in the generator, together with motion of the generator rotor, results in oscillating torques on the shafts of the turbine generator. If the frequency of these torques is equal to, or near, one of the natural mechanical frequencies of the turbine generator, excessive mechanical stress may occur along the turbine-generator rotor at critical locations. In addition, excessive voltage and current may occur in the generator and power system. Turbine-generator components known to be affected by such interaction are shafts, turbine blades, and generator retaining rings.

There have been several dramatic events resulting from power system dynamic interaction with turbine generators, including significant turbine-generator damage. Analysis of these events has made the power engineering community aware of the potential for even more extensive turbine-generator damage from power system dynamic interaction. For these reasons, methods have been developed to identify and analyze the potential for power system dynamic interaction and countermeasures have been developed to control such interaction.

This chapter addresses the types of power system dynamic interaction with turbine generators that have been identified as potentially hazardous. For each type of interaction there is a discussion of known events, physical principles, analytic methods, possible countermeasures, and references. The types of interaction to be addressed are

- Subsynchronous resonance
- Induction generator effect
- Device-dependent subsynchronous oscillations
- Supersynchronous resonance
- Device-dependent supersynchronous oscillations
- Transient shaft torque oscillations

For all of these interactions, except for the induction generator effect, the natural frequencies and mode shapes for turbine-generator rotor systems are critical factors. As generating plants age, modifications may be made that modernize or allow uprating of the units. Typical changes that can have significant effects on the rotor dynamics are replacement of shaft driven exciters with static excitation systems and replacement of turbine rotors. In a few instances electric generators or generator rotors have been replaced. All of these changes have the potential for either reducing or increasing the dynamic interaction for the specific turbine generator. It is important that system engineers, new equipment design engineers, and service engineers all be aware of the interactions that are addressed in this chapter and of the potential for their occurrence at a specific plant.

17.2 Subsynchronous Resonance

Series capacitors have been used extensively since 1950 as a very effective means of increasing the power transfer capability of a power system that has long (150 miles or more) transmission lines. Series capacitors provide a capacitive reactance in series with the inherent inductive reactance of a transmission line, thereby reducing the effective inductive reactance. Series capacitors significantly increase transient and steady-state stability limits, in addition to being a near-perfect means of var and voltage control. One transmission project, consisting of 1000 miles of 500 kV transmission lines, estimates that the application of series capacitors reduced the project cost by 25%. Until about 1971, it was generally believed that up to 70% series compensation could be used in any transmission line with little or no concern. However, in 1971 it was learned that series capacitors can create an adverse interaction between the series compensated electrical system and the spring-mass mechanical system of the turbine generators. This effect is called *subsynchronous resonance* (SSR) since it is the result of a resonant condition, which has a natural frequency below the fundamental frequency of the power system [1].

17.2.1 Known SSR Events

In 1970, and again in 1971, a 750 MW cross-compound Mohave turbine generator in southern Nevada experienced shaft damage. The damage occurred when the system was switched so that the generator was radial to the Los Angeles area on a 176 mile, series-compensated 500 kV transmission line. The shaft damage occurred in the slip ring area of the high-pressure turbine generator. Metallurgical analysis showed that the shaft had experienced cyclic fatigue, leading to plasticity. Fortunately, the plant operators were able to shut the unit down before there was a shaft fracture. In each case, the turbine generator had to be taken out of service for several months for repairs [2]. Intensive investigation in the electric power industry led to the conclusion that the Mohave events were caused by an SSR condition referred to as *torsional interaction*. Torsional interaction created sustained torsional oscillations in the second torsional mode, which has a stress concentration point in the slip ring area of the affected turbine generator.

17.2.2 SSR Terms and Definitions

A set of terms and definitions has been developed so engineers can communicate clearly using consistent terminology. Following are definitions for the most commonly used terms. These are consistent with the terms and definitions presented in Ref. [3].

Subsynchronous: Electrical or mechanical quantities associated with frequencies below the synchronous frequency of a power system.

Supersynchronous: Electrical or mechanical quantities associated with frequencies above the synchronous frequency of a power system.

Subsynchronous resonance: The resonance between a series-capacitor-compensated electric system and the mechanical spring-mass system of the turbine generator at subsynchronous frequencies.

Self-excitation: The sustainment or growth of response of a dynamic system without externally applied excitation.

Induction generator effect: The effect of having subsynchronous positive sequence currents in the armature of a synchronously rotating generator.

Torsional interaction: Self-excitation of the combined mechanical spring-mass system of a turbine generator and a series-capacitor-compensated electric network when the subsynchronous rotor motion developed torque is of opposite polarity and greater in magnitude than the mechanical damping torque of the rotor.

Torque amplification: The amplification of turbine-generator shaft torque at one or more of the natural frequencies of the rotor system caused by transient oscillations at subsynchronous natural frequencies of series-capacitor-compensated transmission systems or unfavorable timing of switching events in the electric network.

Subsynchronous oscillation: The exchange of energy between the electric network and the mechanical spring-mass system of the turbine generator at subsynchronous frequencies.

Torsional mode frequency: A natural frequency of the mechanical spring-mass system of the turbine generator in torsion.

Torsional damping: A measure of the decay rate of torsional oscillations.

Modal model: The mathematical spring-mass representation of the turbine-generator rotor corresponding to one of its mechanical natural torsional frequencies.

Torsional mode shape: The relative angular position or velocity at any instant of time of the individual rotor masses of a turbine-generator unit during torsional oscillation at a natural frequency.

17.2.3 SSR Physical Principles

For this discussion the simplest possible system will be considered with a single turbine generator connected to a single series-compensated transmission line as shown in Figure 17.1. The turbine generator has only two masses connected by a shaft acting as a torsional spring. There are damping elements

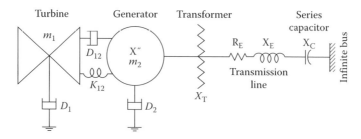

FIGURE 17.1 Turbine generator with series-compensated transmission line. (From IEEE Committee Report, *IEEE Transactions on Power Apparatus and Systems*, PAS-104, 1326, 1985. © 1984 IEEE. With permission.)

between the two masses and each mass has a damping element. The electrical system of Figure 17.1 has a single resonant frequency, f_{er}, and the mechanical spring-mass system has a single natural frequency, f_n. It must be recognized that the electrical system may be a complex grid with many series-compensated lines resulting in numerous resonance frequencies $f_{er1}, f_{er2}, f_{er3}$, etc. Likewise, the turbine generator may have several masses connected by shafts (springs), resulting in several natural torsional frequencies (torsional modes) f_{n1}, f_{n2}, f_{n3}, etc. Even so, the system of Figure 17.1 is adequate to present the physical principles of SSR.

SSR is a phenomenon that results in significant energy exchange between the electric system and a turbine generator at one of the natural frequencies of the turbine generator below the synchronous frequency, f_o. When the electric system of Figure 17.1 is series compensated, there will be one subsynchronous natural frequency, f_{er}. For any electric system disturbance, there will be armature current flow in the three phases of the generator at frequency f_{er}. The positive sequence component of these currents will produce a rotating magnetic field at an angular electrical speed of $2\pi f_{er}$. Currents are induced in the rotor winding due to the relative speed of the aforementioned rotating field and the speed of the rotor. The resulting rotor current will have a frequency of $f_r = f_o - f_{er}$. A subsynchronous rotor current creates induction generator effect as will be discussed further in Section 17.2.3.1. The armature magnetic field, rotating at an angular frequency of f_{er}, interacts with the rotor's DC field, rotating at an angular frequency of f_o, to develop an electromagnetic torque component on the generator rotor at an angular frequency of $f_o - f_{er}$. This torque component contributes to torsional interaction, which will be discussed further in Section 17.2.3.2, and to torque amplification, which will be discussed further in Section 17.2.3.3 [3].

17.2.3.1 Induction Generator Effect

Induction generator effect involves only the electric system and the generator (does not involve turbines). For an induction machine the effective rotor resistance as seen from the armature and external power system is given by the following equations:

$$R'_r = \frac{R_r}{s} \qquad (17.1)$$

$$s = \frac{f_{er} - f_o}{f_{er}} \qquad (17.2)$$

where
 R'_r is the apparent rotor resistance viewed from the armature
 R_r is the rotor resistance
 s is the slip
 f_{er} is the frequency of the subsynchronous component of current in the armature
 f_o is the synchronous frequency

Combining Equations 17.1 and 17.2 yields

$$R'_r = \frac{R_r f_{er}}{f_{er} - f_o} \qquad (17.3)$$

Since f_{er} is subsynchronous it will always be less than f_o. Therefore, the effective generator resistance as viewed from the armature circuit will always be negative. If this equivalent resistance exceeds the sum of the positive armature resistance and system resistance at the resonant frequency f_{er}, the armature currents can be sustained or growing. This is known as induction generator effect [1,12].

17.2.3.2 Torsional Interaction

Torsional interaction involves both the electrical and the mechanical systems. Both systems have one or more natural frequency. The electrical system natural frequency is designated f_{er} and the mechanical spring-mass system natural frequency is designated f_n. Generator rotor oscillations at a natural torsional frequency, f_n, induce armature voltage components of subsynchronous frequency, $f_{en}^- = f_o - f_n$, and supersynchronous frequency, $f_{en}^+ = f_o + f_n$. When the frequency of the subsynchronous component of armature voltage, f_{en}^-, is near the electric system natural frequency, f_{er}, the resulting subsynchronous current flowing in the armature is phased to produce a rotor torque that reinforces the initial rotor torque at frequency f_n. If the resultant torque exceeds the inherent damping torque of the turbine generator for mode n, sustained or growing oscillations can occur. This is known as torsional interaction. For a more detailed mathematical discussion of torsional interaction, see Refs. [4,5].

17.2.3.3 Torque Amplification

When there is a major disturbance in the electrical system, such as a short circuit, there are relatively large amounts of electrical energy stored in the transmission line inductance and series capacitors. When the disturbance is removed from the system, the stored energy will be released in the form of current flowing at the electrical system resonant frequency, f_{er}. If all, or a portion of the current, flows through a generator armature, the generator rotor will experience a subsynchronous torque at a frequency $f_o - f_{er}$. If the frequency of this torque corresponds to one of the torsional modes of the turbine-generator spring-mass system, the spring-mass system will be excited at that natural torsional frequency and cyclic shaft torque can grow to the endurance limit in a few cycles. This is referred to as torque amplification. For more in-depth treatments of torque amplification, see Refs. [6,7].

17.2.4 SSR Mitigation

If series capacitors are to be applied, or seriously considered, it is essential that SSR control be thoroughly investigated. The potential for SSR must be evaluated and the need for countermeasures determined. When a steam-driven turbine generator is connected directly to a series-compensated line, or a grid containing series-compensated lines, a potential for SSR problems exists. There are three types of series-capacitor applications for which SSR would not be expected. The first type occurs when the turbine generator includes a hydraulic turbine. In this case, the ratio of generator mass to turbine mass is relatively high, resulting in larger modal damping and modal inertia than exists for steam turbine generators [8]. The second type of series-capacitor application that is generally free from SSR concerns has turbine generators connected to an uncompensated transmission system, which is overlaid by a series-compensated transmission system. The California–Oregon transmission system is of this type with a 500 kV system that has 70% series compensation overlaying an uncompensated 230 kV transmission system. Turbine generators are connected to the 230 kV system. Extensive study of this system has failed to identify any potential SSR problems. The third type involves series-capacitor-compensation levels below 20%. There have been no potential SSR problems identified for compensation levels below 20%.

For those series-capacitor applications that are identified as having potential SSR problems, an SSR countermeasure will be required. Such countermeasures can range from a simple operating procedure to equipment costing millions of dollars. Numerous SSR countermeasures have been proposed and several have been applied [9]. Fortunately, for every series-capacitor installation investigated, an effective SSR countermeasure has been identified.

An orderly approach to planning and providing SSR mitigation has been proposed [1]. This includes the following five steps.

17.2.4.1 Screening Studies

Screening studies need to be made to determine the potential SSR problems for every turbine generator near a series-capacitor installation. These studies will probably need to be conducted using estimated data for torsional damping and modal frequencies for the turbine generator unless the turbine generator is in place and available for testing. Accurate modal frequencies and damping can only be obtained from tests, although manufacturers will usually provide their best estimate. The most popular analytic tool for screening studies is the frequency-scanning technique. This technique can provide an approximate assessment of the potential and severity for the three types of SSR: induction generator effect, torsional interaction, and torque amplification [10]. To conduct the frequency-scan studies, the positive sequence model for the power system is required. Generator impedance as a function of frequency is needed and may be estimated. The best estimate for turbine-generator torsional damping and a spring-mass model are required. The manufacturer will usually provide an estimated spring-mass-damping model. If the screening study is conducted using estimated data for the turbine generator, data sensitivity should be examined.

17.2.4.2 Accurate Studies

If screening studies indicate any potential SSR problem, additional studies are required using the most accurate data as they become available from the manufacturer and from tests. The frequency-scan program may be adequate for assessment of induction generator effect and torsional interaction, but an eigenvalue study is desirable if large capital expenditures are being considered for self-excitation countermeasures. If the screening studies show any potential for torque amplification, detailed studies should be conducted to calculate the shaft torque levels to be expected and the probability of occurrence. The torque amplification studies should be made using the most current spring-mass-damping models from the manufacturer. The studies can be updated, as more accurate data become available from tests. The well-known electromagnet transient program (EMTP) is usually used for these studies.

17.2.4.3 SSR Interim Protection

If series capacitors are to be energized prior to acquiring accurate data from turbine-generator tests and the aforementioned studies indicate a potential SSR problem, interim protection must be provided. Such protection might consist of reduced levels of series compensation, operating procedures to avoid specific levels of series compensation and/or transmission line configurations, and/or relays to take the unit off-line in the event an SSR condition is detected. These precautions should also be taken when a new turbine generator is added to an existing series-compensated transmission system if studies show potential for SSR concerns.

17.2.4.4 SSR Tests

Some SSR testing will be required unless the studies discussed previously show no or very low probability for the hazards of SSR. The torsional natural frequencies of the spring-mass system can probably be measured through monitoring during normal turbine generator and system operation. To measure modal damping it is necessary to operate the turbine generator at varying load levels while stimulating the spring-mass system. Testing will be discussed in more detail in Section 17.2.8.

17.2.4.5 Countermeasure Requirements

The countermeasure selection must assure that sustained or growing oscillations do not occur and it may involve an analysis of the acceptable fatigue life expenditure (FLE) for damped oscillations. See Section 17.2.5.3.5 for a discussion of FLE. Implementation of the selected countermeasures requires careful coordination. If the countermeasures involve hardware, the effectiveness of the hardware should be determined by testing. Countermeasures will be presented in more detail in Section 17.2.6.

17.2.5 SSR Analysis

SSR analysis involves the identification of all system and generator operating conditions that result in SSR conditions and the determination of the severity by calculating the negative damping and shaft torque amplification. The primary computer programs used in the industry for SSR analysis are frequency scanning, eigenvalue, and transient torque (EMTP). Some program validation has been made in the industry by comparing the results of these analytic methods with test results [11].

17.2.5.1 Frequency Scanning

The frequency-scanning technique involves the determination of the driving point impedance over the frequency range of interest as viewed from the neutral of the generator being studied [10]. For frequency scanning the following modeling is required:

A positive sequence model of the power system, including series compensation, as viewed from the generator terminals.

The generator being studied is represented by its induction generator equivalent impedance as a function of slip. These data can generally be obtained from the generator manufacturer. If not, an approximation is presented in Ref. [12]. Other generators in the system are generally modeled by their short-circuit equivalent. Load is generally represented by the short-circuit equivalent impedance viewed from the transmission system side of the transformer connecting the transmission and distribution networks.

Figure 17.2 is a typical output from a frequency-scanning program. The plots consist of the reactance and resistance as a function of frequency as viewed from the generator neutral. In addition, the 60 Hz

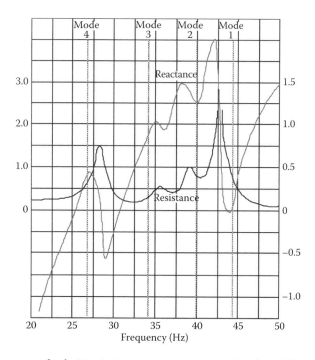

FIGURE 17.2 Frequency scan for the Navajo Project generator connected to the 500 kV system. (From Anderson, P.M. and Farmer, R.G., Subsynchronous resonance, *Series Compensation of Power Systems*, PBLSH!, San Diego, CA, 1996. With permission.)

complements of the modal frequencies have been superimposed and labeled by mode number. The use of frequency scanning to evaluate the three types of SSR will be presented in the following.

17.2.5.1.1 Induction Generator Effect

Frequency scanning is an excellent tool for the analysis of induction generator effect. Induction generator effect is indicated when the frequency scan shows that the reactance crosses zero at frequencies corresponding to negative resistance. Such points can be identified by inspection from frequency-scan plots. This is an indication of growing currents and voltage oscillation in the electrical system. The generator will be subjected to oscillating torque, but these oscillations will not be amplified in the rotor system unless the frequency nearly coincides with a torsional natural frequency. Such problems typically occur for radial operating conditions with series compensation in the radial line.

17.2.5.1.2 Torsional Interaction

When a resonant frequency of the electrical system, as viewed from the generator neutral, corresponds to the 60 Hz complement of one of the turbine-generator modal frequencies, negative damping of the turbine generator exists. If this negative damping exceeds the positive modal damping of the turbine generator, sustained or growing shaft torque would be experienced. Such negative damping can be approximated from frequency-scanning results according to Ref. [4].

Using the method of Ref. [4], the amount of negative damping for torsional mode n is directly related to the conductance, G_n, for that mode and can be calculated by the following approximate formula:

$$\Delta\sigma_n = \frac{60 - f_n}{8 f_n H_n} G_n \tag{17.4}$$

where
$\Delta\sigma_n$ is the negative damping for mode n in rad/s
H_n is the equivalent pu stored energy for a pure modal oscillation (see Ref. [10])
G_n is the pu conductance of the electrical system including the generator on the generator MVA base at $(60 - f_n)$ Hz:

$$G_n = \frac{R_n}{R_n^2 + X_n^2}$$

where
R_n is the resistance from frequency scan at $(60 - f_n)$ Hz
X_n is the reactance from frequency scan at $(60 - f_n)$ Hz

Equation 17.4 neglects the damping due to the supersynchronous components of current. This is generally negligible. Equation 6.4 in Ref. [1] includes the supersynchronous effect. Reference [10] includes a sample calculation for H_n.

The existence and severity of torsional interaction can now be determined by comparing the negative damping, $\Delta\sigma_n$, determined from frequency scanning for mode n, with the natural mechanical damping of the turbine generator for mode n. In equation form, this is

$$\sigma_{net} = \sigma_n - \Delta\sigma_n \tag{17.5}$$

where
σ_{net} is the net torsional damping for mode n
σ_n is the turbine-generator damping for mode n
$\Delta\sigma_n$ is the negative damping for mode n due to torsional interaction

If the net damping, σ_{net}, is negative, torsional interaction instability for mode n is indicated at the operating condition being studied. From the same frequency-scan case $\Delta\sigma_n$ can be calculated for all other active modes and then compared with the natural damping, σ_n, for the corresponding mode. This provides an indication of the severity of torsional interaction for the operating condition (case) being studied. This process should be repeated for all credible operating conditions that are envisioned.

The natural torsional frequencies and modal damping for the turbine generator will only be known accurately if the machine has been tested. If estimated data are being used, the possible variations should be accounted for. The simplest way to account for variations in modal frequency is to apply margin. One way is to calculate the maximum conductance for Equation 17.4 within a frequency range. Reference [10] suggests a frequency range of ±1 Hz of the predicted modal frequency. Experience has shown that estimated modal damping can significantly vary from the measured damping. Hence, unless the estimated damping values are based upon measurements from other similar units, a very conservative value of damping should be used in the studies.

The frequency-scanning technique, as used to calculate negative damping, has been validated through comparison with test results. There has been reasonable correlation, as shown in Refs. [10,11], when the turbine-generator model parameters are accurate. Frequency scanning is a cost-effective means to study induction generator effect and torsional interaction. The results must be used with care. If the study results indicate that very small positive damping exists, but there are large reactance dips indicating significant resonance conditions [10], tests should be conducted to validate the study results prior to making a final decision not to implement any countermeasures. Also, if frequency-scanning studies indicate only a small negative damping, it is prudent to validate the studies by tests prior to committing to costly countermeasures or series-compensation reduction [1].

17.2.5.1.3 Torque Amplification

Frequency scanning cannot be used to quantify the torque to be expected for a specific disturbance but it is a very good tool for determining the potential for torque amplification problems and the system configurations that need to be investigated in detail. Reference [10] suggests that, if a frequency-scan case shows a significant reactance dip within ±3 Hz of the 60 Hz complement of a modal frequency of the turbine generator, torque amplification should be investigated. This provides an excellent screening tool for developing a list of cases to be studied for torque amplification. The frequency-scan results in Figure 17.2 suggest potential torque amplification for Modes 1 and 2. The largest reactance dip is near Mode 1, but is slightly detuned. The reactance dip for Mode 2 is smaller but is nearly perfectly tuned. The system configuration represented by Figure 17.2 was studied using EMTP (electromagnetic transients program) and found to have serious torque amplification problems (see Ref. [22]).

17.2.5.2 Eigenvalue Analysis

Eigenvalue analysis for SSR is straightforward for torsional interaction and induction generator effect since they can be analyzed by linear methods [1]. The approach is as follows:

1. Model the power system by its positive sequence model.
2. Model the generator electrical circuits.
3. Model the turbine-generator spring-mass system with zero damping.
4. Calculate the eigenvalues of the interconnected systems.
5. The real component of eigenvalues that correspond to the subsynchronous modes of the turbine-generator spring-mass system shows the severity of torsional interaction.
6. The real component of eigenvalues that correspond to only electric system resonant frequencies shows the severity of the induction generator effect problem.

The eigenvalues to be analyzed for torsional interaction can be identified by comparing the imaginary part of each eigenvalue with the modal frequencies of the spring-mass system. The corresponding real part of the eigenvalue is a quantitative indication of the damping for that mode. If the eigenvalue has a negative real part, positive damping is indicated. If it has a positive real part, negative damping is indicated. The real part of the eigenvalue is a direct measure of the positive or negative damping for each mode. Adding the calculated damping algebraically to the inherent modal damping results in the net modal damping for the system. For a mathematical treatment of modeling for eigenvalue analysis, see Ref. [5].

17.2.5.3 Transient Analysis

Transient analysis is required to determine the potential for SSR torque amplification. The well-known EMTP is very well suited for such analysis [13]. There are various versions of the program. Bonneville Power Administration (BPA) developed the program and has added contributions from other engineers and upgraded it through the years. A version referred to as ATP is in the public domain. Several other versions of the EMTP are commercially available. EMTP provides for detailed modeling of those elements required for assessing the severity of SSR torque amplification. This includes the power system, the generators in the system, and the mechanical model of the turbine generator being studied.

17.2.5.3.1 EMTP Power System Model

Three-phase circuits, a neutral circuit, and a ground connection model the electrical elements of the power system. The data for the model can generally be provided in the form of phase components or symmetrical components. Special features of series capacitors can be modeled, including capacitor protection by gap flashing or nonlinear resistors. Load is usually included in a short-circuit equivalent circuit at the point where it connects to the portion of the network being modeled in detail.

17.2.5.3.2 EMTP Generator Model

The electrical model for a synchronous generator being studied in EMTP is a two-axis Park's equivalent with several rotor circuits on the direct and quadrature axes. The input data can be in the form of either winding data or conventional stability data. The generator data can be obtained from the manufacturer in the form of conventional stability data. All generators in the system, other than the study generator, can generally be represented by a voltage source and impedance without affecting the study accuracy. For a detailed treatment of generator modeling for SSR analysis, see Ref. [5].

17.2.5.3.3 EMTP Turbine-Generator Mechanical Model

The turbine-generator mechanical model in EMTP consists of lumped masses, spring constants, and dampers. For torque amplification studies mechanical damping is not a critical factor. The peak shaft torque would be expected to only vary by about 10% over a range of damping from zero to maximum [1]. Hence, the turbine-generator mechanical damping is generally neglected in EMTP studies.

17.2.5.3.4 Critical Factors for Torque Amplification

The most important use of EMTP for SSR analysis is to find the peak transient shaft torque that is to be expected when series capacitors are applied. It is necessary to understand that the major torque amplification events due to SSR will occur either during a power system fault or after the clearing of a power system fault. The energy stored in series capacitors during a fault will be discharged as subsynchronous frequency current that can flow in a generator armature, creating amplified subsynchronous torque. The peak shaft torque to be expected depends on many factors. Experience has shown that the dominant factors that should be varied during a torque amplification study are electric system tuning, fault location, fault clearing time, and capacitor control parameters and the largest transient torques occur when the unit is fully loaded. For a detailed discussion on system tuning and faults, see Ref. [1]. For information on capacitor controls, see Ref. [7].

17.2.5.3.5 Computing Fatigue Life Expenditure

When the torque of a turbine-generator shaft exceeds a certain minimum level (endurance limit), fatigue life is expended from the shaft during each torsion cycle. The machine manufacturer can generally furnish an estimate of FLE per cycle corresponding to shaft torque magnitude for each shaft. When plotted this is referred to as an *S–N* curve. EMTP can then be used to predict the FLE for a specific system disturbance. One method requires the complete simulation and FLE calculation of an event over approximately 30 s, which may be time consuming and cumbersome, if numerous scenarios are to be investigated. An alternate simplified method requires some approximation. For this method EMTP studies are conducted to find the peak shaft torque that will occur for a given scenario. Since the peak shaft torques generally occur within 0.5 s, EMTP simulation and FLE calculation time are minimized. It is assumed that after the shaft torque has peaked, it will decay at a rate corresponding to the mechanical damping of the excited modes. The FLE for the simulated event can then be calculated from knowledge of the peak torque, the decay rate, and the *S–N* curve. This gives conservative estimates of FLE. It is important to recognize that FLE for each incident is accumulative. When the accumulated FLE reaches 100%, the shaft is expected to experience cracks at its surface but not gross failure. For more detail on computing FLE, see Ref. [1].

17.2.5.4 Data for SSR Analysis

Data requirements for SSR analysis consist of system data and turbine-generator data.

17.2.5.4.1 System Data

System data for eigenvalue and frequency-scanning studies are generally of the same form as the positive sequence data used for power flow, short-circuit, and power system stability studies. The data may require refinement to account for the resistance variations with frequency and for system equivalents. The classical short-circuit equivalent may not be adequate when the equivalent system includes series capacitors. In such cases an RLC equivalent might be developed. It should be checked with the frequency-scanning program to determine if the equivalent reasonably approximates the driving point impedance of the system it is to represent over the frequency range of interest (10–50 Hz). Large load centers near the machine being analyzed may need to be represented by a special equivalent. For one outstanding case, where the apparent impedance as viewed from the study generator terminal was actually measured over the frequency range of 15–45 Hz, it was found that the Phoenix, Arizona load must be modeled to provide a good equivalent [14]. In that case, it was found that the following load model could form an accurate equivalent:

Sixty percent of the total load consists of induction motor load with x_d'' of 0.135 pu.
Forty percent of the load is purely resistive.

The validity of such a model for other locations has not been determined.

For torque amplification studies using EMTP, the system data requirements are much more extensive since all three phases and ground are represented. In EMTP, the series capacitors can be modeled in detail, including the capacitor protective equipment. For more detail on system data for SSR analysis, see Ref. [1].

17.2.5.4.2 Turbine-Generator Data

The IEEE SSR working group has developed a set of recommended SSR data items that should be furnished by the turbine-generator manufacturer. These are generally the minimum data required for SSR studies. Following is a description of the three types of data:

Generator electrical model

1. Resistance and reactance as a function of frequency for the generator as viewed from the generator terminals. This should include armature and rotor circuits.
2. Typical stability format data for the "Park's equivalent" generator model.

Turbine-generator mechanical model

1. The inertia constant for each turbine element, generator, and exciter
2. The spring constants for each shaft connecting turbine elements, generator, and exciter

3. The natural torsional frequencies and mode shapes as determined for the mechanical model defined by items 1 and 2
4. The modal damping as a function of load corresponding to the mechanical model defined by items 1 and 2

Life expenditure curves: For each shaft connecting the turbine elements, generator, and exciter a plot of the life expended per transient incident as a function of the peak oscillating torque, or an S–N curve showing torque versus number of cycles to crack initiation or crack propagation. The manufacturers should provide all assumptions made in the preparation of these curves.

For more detail on turbine-generator modeling, see Refs. [1,5].

17.2.6 SSR Countermeasures

If series capacitors are to be used and SSR analysis shows that damaging interactions may exist for one or more system configurations, countermeasures must be provided, even if the probability of an SSR event is low. Such countermeasures may not completely eliminate turbine-generator shaft FLE. Even so, prudent countermeasure selections can probably limit the FLE of any shaft to less than 100% over the expected life of the turbine generator. A strategy for SSR countermeasure selection should be formulated during the SSR analysis stage so that it can be used as a guide for the studies to be conducted. Reference [15] presents one utility's guidelines that were developed to guide countermeasure selection, including the required SSR studies.

Numerous SSR countermeasures have been studied [9,16], and twelve or so, have been applied. Following is a description of the countermeasures known to have been applied with references for each. These are separated into unit-tripping and nonunit-tripping types.

17.2.6.1 Unit-Tripping SSR Countermeasures

The following countermeasures will cause the generator to be electrically separated from the power system when a hazardous condition is detected.

Torsional motion relay [17,18]: Such relays typically derive their input from rotor motion at one or two places on the turbine generator. Rotor motion signals are typically obtained from toothed wheels mounted on the shaft. The signal is first conditioned and then analyzed for presence of modal components. The trip logic is based upon the level of signal and rate of growth. One needs to have the turbine-generator stress versus cycles to failure information to properly set the relays.

The torsional-motion-based relays are usually very effective in protecting against torsional interaction type of SSR problems. However, these relays may not be fast enough to protect against the worst case of torque amplification problem. The newer torsional-motion-based relays are microprocessor based compared to the older relays that were analog-type relays.

Armature current relay [19,20]: The armature-based relays use generator current as the input signal and condition the input signal to filter out the normal 50/60 Hz component. The signal is then filtered to derive the modal component of the current. The tripping logic is based upon the level of SSR current and rate of growth. Since these relays use armature current as the input, they are capable of protecting against torsional interaction, induction generator effect, and torque amplification types of SSR problems.

Since the SSR current is a function of system impedance, it is usually necessary to set the relays very sensitive to be able to protect against all possible SSR conditions while protecting against the torsion interaction problems. One disadvantage of setting them very sensitive is the possibility of false trips.

Unit-tripping logic schemes [21]: The unit-tripping logic scheme is usually a hardwired logic scheme, which will take the unit off-line if predetermined system conditions exist. Such schemes can be used only if there are low-probability conditions for which SSR conditions exist and one is reasonably sure that there are no other unknown system conditions for which an SSR condition can occur. Since it is difficult to assess all possible conditions for which an SSR condition may exist, this countermeasure should be applied very carefully.

17.2.6.2 Nonunit-Tripping SSR Countermeasures

The following SSR countermeasures will provide varying levels of SSR protection without electrically separating the generator from the power system. Each countermeasure is designed to offer protection for specific SSR concerns and the choice of which one to employ is based on the nature and severity of the concern. The static blocking filter provides the broadest range of protection, but it has both the highest price and most demanding maintenance requirement.

- Static blocking filter [22,23]
- Dynamic stabilizer [24–26]
- Excitation system damper [27,28]
- Turbine-generator modifications [1]
- Pole face Amortisseur windings [9,22]
- Series-capacitor bypassing [7]
- Coordinated series-capacitor control with loading [9]
- Operating procedures [1]

17.2.6.3 Thyristor-Controlled Series Capacitor

The thyristor-controlled series capacitor (TCSC) is a capacitor in series with the transmission line, with a thyristor pair and small reactor in parallel with the capacitor. It can function as a series capacitor if the thyristors are blocked, as a series reactor if the thyristors fully conduct, or as a variable impedance when the duty cycle of the thyristors is varied. The device has been applied to improve stability in weak AC networks and to protect the series capacitor from transient overvoltage. It is expected that TCSCs will be used to control SSR interactions in the future.

Two installations in the United States have demonstrated control algorithms for SSR concerns [29,30]. These projects were in locations where there was a low probability of sustained or growing oscillations, but they provided both demonstrations of control algorithms and equipment installation and operation. They also provide information about required ratings for the components of the TCSC and reliability of the power electronic components, the cooling systems, and the control systems.

There have been a large number of technical studies and papers describing control algorithms, equipment sizes, and the most effective location in the network for TCSC installations. A sample of this information is contained in Refs. [31–33]. These circuits are considered to be the most effective means to directly control SSR.

17.2.7 Fatigue Damage and Monitoring

Fatigue damage of turbine-generator shafts is certainly undesirable, but it may not be practical to completely avoid it. Therefore, it is important to understand the consequences of fatigue damage, and to know how to quantify any fatigue damage experienced so that gross shaft failure is avoided [3,18].

The consequences of high cycle fatigue and low cycle fatigue differ. In the case of high cycle fatigue, characterized by a large number of small amplitude stress cycles where purely elastic deformation occurs, there is no permanent deformation and no irreparable damage. It is said that 100% FLE occurs when cracks are initiated at the stress concentration points on the shaft surfaces. When this point is reached, cracks will be propagated as additional torsional stresses above the endurance limit (at the stress concentration point at the end of the crack) are experienced. This does not mean that shaft failure will occur when 100% FLE is reached. On the contrary, the ultimate strength of the shaft in torsion is not significantly reduced. It does mean that cracks will be expected to increase in number and size if appropriate action is not taken. Fortunately, machining the shaft surface to remove the cracks can effectively restore the total shaft integrity. Cracks can be identified by visual inspection at stress concentration points on the shaft. Even so, it may be very costly to shut a unit down for a visual inspection following an incident suspected to result in significant FLE. For this reason, torsional monitoring techniques have been developed to provide a permanent history of torque experienced by each turbine-generator shaft. The most likely phenomenon leading to high cycle fatigue is sustained torsional interaction where the torsional amplitude is limited by nonlinear damping.

In the case of low cycle fatigue, characterized by a small number of very large amplitude stress cycles for which plastic deformation occurs, the consequences may be quite different from those in high cycle fatigue. When plastic deformation occurs, there is irreversible shaft deformation in torsion (kink). In the most severe cases this can result in a bending moment being applied to the shaft each revolution. If the unit continues to operate in this condition, shaft failure in the bending mode may occur. If a monitor detects low cycle fatigue and there is a corresponding increase in lateral vibration, the shaft should be inspected as soon as practical. The most likely phenomenon leading to low cycle fatigue is torque amplification.

Shaft torque monitoring techniques have been developed, which will provide a permanent history of the approximate torque experienced by each turbine-generator shaft. This information is extremely useful in making a decision following a unit trip by SSR relay action or an unusual event such as out-of-phase synchronization. The options are as follows:

1. Take the unit off-line and inspect the shaft.
2. Inspect the shaft at the next scheduled outage.
3. Synchronize, load the unit, and continue to operate without interruption.

A wrong decision could cause significant shaft damage or an unnecessary unit outage. Several methods have been developed for monitoring shaft torque as reported in the literature [18,34–38].

17.2.8 SSR Testing

The analytic methods and corresponding software for SSR analysis can be very detailed, but they have limited value unless the required data for the electric system, generators, and turbine-generator spring-mass-damping system are available and are reasonably accurate. It has been found from tests that the torsional frequencies are usually within 1 Hz of that predicted by the manufacturer. This implies that the spring-mass model data are reasonably accurate. The turbine-generator manufacturers estimate torsional damping, but testing has shown that damping predictions, when compared with measured data, may have large variations. Therefore, little confidence can be placed in predicted damping unless it is based upon measured data from similar units. Accurate torsional damping values can only be obtained from tests.

SSR tests can vary in their complexity, depending on their purpose, availability of turbine-generator rotor motion monitoring points, type of generator excitation system, power system configuration, and other factors. The minimum and simplest tests are those used to identify the natural torsional frequencies of a turbine generator. Tests to measure torsional damping are more difficult, particularly at high loading. Various types of tests may be devised to test the effectiveness of countermeasures.

17.2.8.1 Torsional Mode Frequency Tests

These tests generally involve a spectrum analysis of rotor motion or shaft strain in torsion at points that respond to all active modes of interest. Rotor motion signals can be obtained by demodulating the output of a proximity probe mounted adjacent to a toothed wheel on the rotor. Shaft strain is obtainable from strain gauges fixed to the shaft [39]. With the use of digital spectrum analyzers, natural torsional frequencies can be measured by merely recording the appropriate signals during normal operation of the turbine-generator unit without any special switching [1].

17.2.8.2 Modal Damping Tests

The most successful method for measuring damping is to excite the spring-mass system by some means and then measure the natural decay rate following removal of the stimulus. Two methods have been used to excite the torsional modes. These methods are referred to as the "impact method" and the "steady-state method."

The impact method requires the application of an electrical torque transient to the turbine generator being tested. The transient must be large enough to allow the decay rate of each modal response to be measured during ring down. Rotor motion is generally the preferred signal, but shaft stress has also been used successfully. Since the transient excites all modes, a series of narrowband and band-reject filters are applied to the signal to separate the response into the modal components of interest. Switching series-capacitor banks or line switching can create the required transient. Synchronizing the generator to the power system may also provide adequate stimulus. Such tests are described in Refs. [39,40].

The steady-state method uses a sinusoidal input signal to the voltage regulator of the excitation system, which produces a sinusoidal component of generator field voltage. Some types of excitation systems can create a large enough sinusoidal response of the generator rotor to provide meaningful analysis. The frequency of the signal is varied to obtain a pure modal response of rotor motion or shaft strain. When a steady-state condition with pure mode stimulus has been obtained, the stimulus is removed and the decaying modal oscillation is recorded and plotted. The decay rate is a measure of the modal damping. This process is repeated for each torsional mode of interest. The steady-state method is the preferred method since pure modes can be excited. This method is only applicable to generators whose excitation system has sufficient gain and speed of response to produce a significant torque from the voltage regulator input signal. Such tests are reported in detail in Refs. [39,40].

The damping measured from either of the two test methods is the net damping of the coupled mechanical and electrical systems. Depending on the system configuration during the damping tests, the measured damping may include positive or negative damping due to interaction of the mechanical and electrical systems. This effect can be calculated from eigenvalue studies or from frequency-scanning studies in conjunction with the interaction Equation 17.4. To obtain the true mechanical damping, the measured damping must be corrected to account for the interaction in accordance with the following equation:

$$\sigma_n = \sigma_{meas} \pm \Delta\sigma_n \tag{17.6}$$

where
σ_n is the mechanical modal damping for mode n
σ_{meas} is the measured damping from tests
$\Delta\sigma_n$ is the positive or negative damping due to interaction

It is usually important to have measured torsional damping of all active subsynchronous modes as a function of load, ranging from no load to full load. It is often more difficult to obtain full load damping because modal response decreases as damping increases and damping generally increases with load. It may be impossible to obtain adequate torsional excitation at full load. See Figure 17.3 for results of such tests reported in Ref. [39]. Fortunately, it is the damping values at low loads that are of most interest because they represent the most severe interaction conditions.

17.2.8.3 Countermeasure Tests

Testing the effectiveness of any countermeasure to be applied is important, but may not be feasible. For example, if a countermeasure is to limit loss of shaft life for the most severe transient, it is not reasonable to conduct such a test. Tests for effectiveness of torsional interaction countermeasures are practical and should be made whenever possible. One method is to conduct damping tests, as described in Section 17.2.8.2, with the countermeasure in service and the system configured to yield significant negative damping due to torsional interaction. Such tests are described in Refs. [23,41].

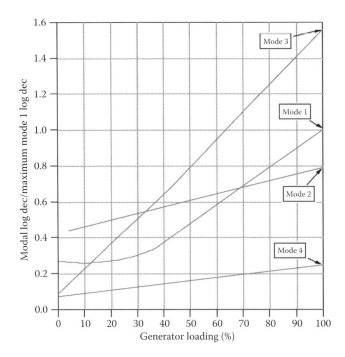

FIGURE 17.3 Variations of modal damping as a function of generator load for Navajo Generators. (From Anderson, P.M. and Farmer, R.G., Subsynchronous resonance, *Series Compensation of Power Systems*, PBLSH!, San Diego, CA, 1996. With permission.)

If SSR relays are to be applied, it may be possible to initiate a unit trip by SSR relay action under controlled conditions to verify proper operation. This has been accomplished, at least at one plant, by reducing the relay settings to a very sensitive level, and then causing rotor oscillations by the steady-state method previously described. The stimulus can be increased to a level of sustained modal oscillations that will cause the relay to pick up. For the reduced setting, the shaft torques are kept below the endurance limit. Such a test provides confidence in both the relay capabilities to initiate a unit trip and the correct wiring of the circuits from the relay output to the circuit breaker trip coils.

17.2.9 Summary

Consideration must be given to the potential for SSR whenever series capacitors are to be applied. The ability to analyze and control SSR for the extreme problems encountered has been clearly demonstrated over the last 30 years. Various countermeasures for SSR control have been developed and successfully applied. In many cases, the sole SSR protection can be provided by relays. Monitoring is valuable for the units exposed to SSR. It provides a permanent history of the torques experienced by the shafts and the accumulative shaft life expenditure. Such information can be used to schedule shaft inspection and maintenance, as required, to maintain shaft integrity. Continuous monitoring of SSR countermeasure performance by modern digital equipment can also be cost-effective.

If potential SSR problems are identified when series-capacitor applications are considered, there is a clear course established by the utility industry. Analytical methods are available for either cursory or detailed analysis. Countermeasure selection guidelines used by others are available. Testing methods have been developed that vary from simple monitoring to sophisticated signal processing and system switching. SSR can be controlled, thus making it possible to benefit from the distinct advantages of series capacitors.

17.3 Device-Dependent Subsynchronous Oscillations

Device-dependent subsynchronous oscillations have been defined as interaction between turbine-generator torsional systems and power system components. Such interaction with turbine generators has been observed with DC converter controls, variable speed motor controllers, and power system stabilizers (PSS). There is potential for such interaction for any wide bandwidth power controller located near a turbine generator.

17.3.1 HVDC Converter Controls

In 1977, tests were conducted to determine the interaction of the Square Butte HVDC converter in North Dakota with the Milton Young #2 turbine generator. It was found that both the high-gain power modulation control and the HVDC firing angle control destabilized the first torsional mode of the turbine generator at 11.5 Hz. Fortunately, it was also found that the basic HVDC controls created growing torsional oscillations of the turbine generator in the first torsional mode for a specific system configuration that nearly isolated the turbine generator and HVDC converter from the rest of the AC network. Careful analysis of this phenomenon shows that any HVDC converter has the potential for creating subsynchronous torsional oscillations in turbine generators that are connected to the same bus as the HVDC converter. The potential reduces as the impedance between the two increases or as additional AC circuits are connected. The HVDC system appears as a load to the turbine generator. The load would be positively damped for crude firing angle control. Successful converter operation requires sophisticated firing angle control. This sophisticated control may make the converter appear as a negatively damped load in the range of 2–20 Hz. The potential problem of HVDC converter control interaction with turbine generators can be investigated by eigenvalue analysis. If negative damping is expected, the problem may be solved by retuning converter controls. Also a subsynchronous damping controller has been conceptually designed as reported in Ref. [42]. Reference [43] describes a field test and analysis of interaction between a turbine generator and a HVDC system.

17.3.2 Variable Speed Motor Controllers

In 1979 and 1980, a European fossil fired power plant experienced subsynchronous oscillations of a 775 MW, 3000 RPM turbine generator. The plant was equipped with variable speed drives for the boiler feedwater pumps. The pump drives are equipped with six-pulse subsynchronous converter cascades. For such a converter, the load power to the motors has a component at six times the motor slip frequency. At specific load levels, the feedwater pump speed is such that the load has a component whose frequency corresponds to the 50 Hz complement of one of the natural torsional frequencies of the turbine generator. Under these conditions, the pump load acts as a continuous torsional stimulus of the turbine generator. FLE could occur under such conditions, depending on the magnitude of the torsional oscillations. A torsional stress monitor detected the aforementioned event. Modeling and analysis in EMTP or similar programs could probably predict such an event, but there is no known record of such an analysis. The countermeasure applied to the problem described controls feed water pump speed to avoid speeds that would excite the natural torsional modes of the turbine generator [44].

17.3.3 Power System Stabilizers

In 1969, a 500 MW unit was commissioned at the Lambton Generating Station. A PSS was added some time later to provide positive damping for the local mode of about 1.67 Hz. The PSS derived its input signal from rotor motion at a point adjacent to the generator mass. When the PSS was initially tested, sustained 16.0 Hz torsional oscillations of the generator were observed; 16.0 Hz corresponds to the first torsional mode of the turbine-generator mechanical system [45]. From analysis and simulation it was determined that if the torsional oscillations were allowed to continue, severe turbine-generator shaft damage would occur. It was also learned that any generator rotor motion at the first torsional mode (16.0 Hz) creates a 16.0 Hz signal input to the PSS. The gain and phase of the PSS and the excitation system created an oscillating torque on the generator at 16.0 Hz, which reinforced the initiating 16.0 Hz oscillation. This type of problem can be analyzed using either eigenvalue or EMTP-type computer programs, which have provisions for modeling the turbine-generator mechanical system.

There are various countermeasures that can be applied to deal with the PSS problem. The countermeasures used at Lambton consisted of moving the rotor motion sensing location to a point of the spring-mass system, which has no torsional motion, or provides positive damping, at the active torsional modes. In addition, a 16.35 Hz notch filter was included in the PSS to drastically reduce the gain for the first torsional mode. Others use a high-order low-pass filter or a wideband band-reject filter in the PSS loop to insure that torsional oscillations are not generated by the PSS.

17.3.4 Renewable Energy Projects and Other Interactions

As renewable energy projects become larger, the potential for dynamic interactions with the transmission network may become an issue. An October 2009 incident in West Texas illustrates the concern. During this incident the clearing of a single line to ground fault in the 345 kV network left two wind farms connected through a single series-compensated transmission line. The resulting subsynchronous frequency current oscillations initiated bypass of the series capacitors in 1.5 s after fault clearing. The incident resulted in damage to the power electronic converters that supply current to the rotors of several double-fed induction generators at the wind farms.

Relay records of transient current in the 345 kV line showed oscillations at multiple subsynchronous frequencies with the principle frequency being approximately 25 Hz. These oscillations began when the fault was cleared and grew to a magnitude greater than the pre-fault line current in 0.2 s. They were sustained until the series capacitors were bypassed. This is characteristic of the SSR induction generator effect previously described. The wind generators would have shared the current and it would have been reflected into their rotor circuits and the connected converters.

A complete simulation of this event and analysis of the potential for SSR in other systems is complicated by the presence of the power electronic converters that are widely used in renewable energy projects. Large wind power projects mostly use either double-fed induction generators or induction generators and full electronic converters to connect to the transmission network. The dynamic models need to consider both the rotor and stator circuits for the electric machine and the control circuits for the converter. When protective measures such as TCSCs are considered, the dynamics of these controllers must also be considered for the analysis. If the electric generator has rotor windings, the rotor circuit will be similar to that of a synchronous generator with field and damper circuits on both the direct and quadrature axes.

Although there was no reported mechanical damage to the wind turbines during the reported SSR incident, the torque pulsations would have been as large as the rated torque for the units. The frequencies of these pulsations were approximately 60 Hz minus the oscillation frequencies of the line current. Had there been resonant frequencies of the mechanical system near these frequencies, there is a high likelihood of mechanical damage. Evaluation of this potential requires mechanical models similar to those needed for large steam turbines.

In general, any device that controls or responds rapidly to power or speed variations in the subsynchronous frequency range is a potential source for excitation of subsynchronous oscillations. The technical literature includes the effect of governor characteristics on turbine-generator shaft torsionals [46] and subsynchronous torsional interactions with static var compensators (SVCs) [47].

17.4 Supersynchronous Resonance

The term supersynchronous resonance (SPSR) is used here to refer to a torsional resonant condition of a turbine-generator mechanical system at a frequency greater than the frequency corresponding to rated turbine speed and power system rated frequency. Such a resonant condition can be excited from the power system. There have been at least three incidents of turbine blade failure contributed to the excitation of turbine-generator torsional modes that are very near to twice the AC operating frequency (120 Hz for 60 Hz AC systems).

The excitation for these events is the double frequency torque that results from unbalanced phase currents in the AC system. In per unit the magnitude of this torque is very nearly equal to the magnitude of the negative sequence AC current. This value is dependent on transmission line design and balance in system loads. For most systems it is less that 2% of rated torque, but it may increase for some contingencies. The excitation frequency will also vary due to variations in synchronous frequency. This variation is most pronounced in very weak systems and in isolated systems.

17.4.1 Known SPSR Events

In 1985, a turbine generator outside the United States experienced the failure of eight blades in the last stage of an 1800 RPM low-pressure turbine with 43 in. last-stage blades. The blades failed at the root attachments to the rotor disk due to high cycle fatigue. A 1 year outage was required to repair the unit. In 1993, a turbine generator in the United States experienced the failure of two blades in the next to last row of an 1800 RPM low-pressure turbine with 38 in. last-stage blades. The blades failed at the dovetails on the rotor disk. A 49-day outage was required to repair the unit. The turbine-generator units for both incidents were from the same manufacturer and both have relatively long turbine blades on 1800 RPM low-pressure turbines. Similar events occurred in the 1970s to an 1800 RPM turbine generator from a different manufacturer [48].

17.4.2 SPSR Physical Principles

Long turbine blades, such as the 38 and 43 in. blades on 1800 RPM low-pressure turbines, often have a natural vibration frequency near 120 Hz when coupled to the rotor disk. A blade disk with a natural frequency near 120 Hz may be excited by torsional oscillations near 120 Hz [49]. Although individual turbines are designed to avoid 120 Hz natural torsional frequencies (torsional modes) with at least 0.5 Hz margin, the complex modes of coupled shaft systems at these frequencies are difficult to calculate with sufficient accuracy. The following scenario can contribute to turbine blade failure due to high cycle fatigue.

Negative sequence current flows in the generator armature due to unbalanced loads, untransposed lines, or unbalanced faults. The resulting magnetic flux interacting with the field flux results in a double AC system frequency electromagnetic torque applied to the generator rotor. This will excite torsional oscillations if there is a torsional mode of the shaft system at this frequency with sufficient net torque along the generator rotor. Torsional oscillations, at points along the shaft where long turbine blades are attached, can excite blade vibration if the blade-disk natural frequency is approximately the same as the rotor mode frequency (120 Hz for 60 Hz systems) and the coupled mode can be excited by torque applied to the generator rotor. Continuous blade vibration, or numerous transient events, will initiate cracks at the stress concentration points and finally blades will fail.

For the aforementioned scenario, generally there is a torsional mode within 0.5 Hz of 120 Hz. Turbine-generator designers have made efforts to avoid torsional frequencies near 120 Hz but have not always had the technology to accurately calculate the frequencies for the higher torsional modes near 120 Hz. The 1993 blade failure has been contributed to an undetected torsional mode within 0.5 Hz of 120 Hz. For the 1985 blade failure, there were no natural modes within 0.5 Hz of 120 Hz but the turbine generator was operating in a relatively small power system whose frequency varied significantly. These frequency variations, in conjunction with negative sequence generator current, excited the torsional modes that were 1–2 Hz away from 120 Hz.

Tests and experience have shown that generators experience continuous negative sequence current ranging from 1% to 3%. Of course, much higher negative sequence currents occur during unbalanced fault conditions. Therefore, if the blade-disk natural frequencies are near 120 Hz, it is essential that there are no natural torsional frequencies between 119.5 and 120.5 Hz that can be excited by torque applied to the generator rotor. The turbine-generator manufacturer calculates the blade-disk natural frequencies and the torsional natural frequencies. Unfortunately, the calculated frequencies may not be sufficiently accurate to determine if blade failure is to be expected. The turbine-generator natural frequencies can be accurately determined from tests. An off-line test has been devised, which will accurately show the natural torsional frequencies at no load. This is called a ramp test and consists of monitoring torsional strain at critical points while negative-sequence current flows in the generator armature circuit and the turbine-generator speed is accelerated. The negative sequence current is induced by shorting two generator terminals and controlling field voltage with a separate power supply. The ramp test and other tests are described in Refs. [48,50]. Using accurate test data, an analytic model can be developed by an iterative process. The resulting analytic model can be used to find appropriate countermeasures.

17.4.3 SPSR Countermeasures

The countermeasures that have been applied to avoid turbine blade failure, caused by SPSR, involve either moving natural torsional frequencies away from 120 Hz or changing the mode shapes, of modes near 120 Hz, so that they are not excited by electrical torque applied to the generator rotor. This has been successfully accomplished by several methods. One is to braze the tie wires on all last-stage blades. This modification may increase torsional frequencies and it alters the participation of individual blades in the oscillation. A second countermeasure involves adding a mass ring at an appropriate location along the torsional spring-mass system. This modification may reduce the critical torsional frequencies and it will change the mode shapes. Other methods include machining critical sections along the shaft system and changing the generator pole face slotting to move frequencies and modify mode shapes. Tests to determine the natural torsional frequencies following modifications should be made to verify the analytic model [48,50]. A relay has been proposed to alarm or trip the turbine-generator for combinations of negative sequence current and off-nominal frequency operation deemed to be excessive.

17.5 Device-Dependent Supersynchronous Oscillations

There have been a series of events that resulted in turbine-generator damage due to SPSR stimulated by a power system device. This type of interaction is referred to as device-dependent supersynchronous oscillations (DDSPSO).

17.5.1 Known DDSPSO Events

The Comanche Unit 2 near Pueblo, Colorado went into service in 1975 and during the period of 1987–1994, the unit suffered generator damage. In 1987, there was a crack in the generator shaft. In 1993, there were two failures of the rotating exciter. In 1994, there was a retaining ring failure resulting in serious rotor and stator damage [51]. All have been contributed to the same phenomena.

17.5.2 DDSPSO Physical Principles

Comanche Unit 2 is about 3 miles from 2 to 60 MVA steel mill arc furnaces. The arc furnaces have an SVC for flicker control. It has been found that the SVC had a control loop instability that caused negative sequence current to flow in the armature of Comanche 2 at a frequency near 55 Hz. The instability resulted in a 5 Hz amplitude modulation of the 60 Hz SVC current. This modulation created upper and lower sidebands of 55 and 65 Hz in all three phases, but in reverse rotation. The 65 Hz component did not appear outside the SVC delta winding but a 55 Hz negative sequence component flowed in the generator armature. The frequency of this component varied between 54 and 58 Hz, depending on the steel mill operating conditions. This produced a component of electromagnetic torque in the frequency range of 114–118 Hz. The natural torsional frequency for Mode 6 of Comanche 2 was about 118 Hz prior to the retaining ring failure. The mode shape for Mode 6 shows large displacement at the two ends of the generator. Therefore, stimulus from the SVC created torsional oscillations was sufficiently large and sustained to result in high cycle fatigue of the generator shaft, rotating exciter, and retaining ring before the root cause of the problem was found.

17.5.3 DDSPSO Countermeasure

Extensive testing was performed to determine natural modal frequencies for the turbine generator, the components of armature current, and the arc furnace and SVC stimulus. Once the root cause of the problem was determined, it was a simple matter to retune the control circuit of the SVC [51].

17.6 Transient Shaft Torque Oscillations

Turbine-generator design has been guided for many years by a simple requirement for the strength of the shaft system. This requirement in the American National Standards Institute (IEEE/ANSI) Standard C50.13 states that

> A generator shall be designed so that it can be fit for service after experiencing a sudden short circuit of any kind at its terminals while operating at rated load and 1.05 per unit rated voltage, provided that the fault is limited by the following conditions:
> - The maximum phase current does not exceed that obtained from a three-phase sudden short circuit.
> - The stator winding short time thermal requirements are not exceeded [52].

Although this requirement does not refer to the turbine-generator shaft, this transient has been used for many years to verify the shaft design. It has generally been assumed that the more frequent shocks resulting from more remote short circuits, out-of-phase synchronizing, and transmission line switching would have low enough magnitudes and be infrequent enough that fatigue issues did not have to be considered. Experience has verified this assumption. While there have been reports of damage to coupling faces and coupling bolts, there have not been many reports of more severe damage.

When shaft damage due to dynamic interactions between turbine generators and transmission systems became a concern, more detailed analysis of the effects of short-circuit and line switching transients was performed [53]. This analysis generally confirmed that system disturbances did not result in larger shaft torques than terminal short circuits unless these disturbances involved dynamic interactions, series capacitors, or multiple switching events. The most common scenario for multiple switching events is fault clearing in the transmission system. When a fault occurs in the network, turbine generators experience a step change in torque. The fault clearing 50–150 ms later produces a second step change usually in the opposite direction. This second shock can reinforce oscillations initiated by the fault if the timing coincides with critical timing for one of the shaft natural frequencies. Unsuccessful high-speed reclosing events coupled with low damping for shaft oscillations would provide more opportunities for further amplification. For this reason turbine-generator manufacturers requested that the practice of high-speed reclosing be discontinued for multiphase faults on transmission lines connected to generating stations.

In 2004 damage was reported at one of the older operating nuclear stations in the United States [54]. Both turbine generators at this station had to be removed from service when changes in shaft vibration became unacceptable. Relatively long cracks were discovered emanating from coupling keyways in the generator shaft. The analysis determined that the likely cause of these cracks was multiple torsional events during the lifetime of the machines. The geometry of the keyway between the generator coupling and shaft was also considered to be a factor that resulted in high stress concentration. There is no detailed history of the specific transients, but repairs including a redesign of the coupling and keyways make the units less susceptible to further damage. This experience renewed industry interest in transient shaft torque oscillations and suggests that further analysis and monitoring are warranted.

References

1. Anderson, P.M. and Farmer, R.G., Subsynchronous resonance, *Series Compensation of Power Systems*, PBLSH!, San Diego, CA, 1996, Chapter 6.
2. Hall, M.C. and Hodges, D.A., Experience with 500-kV subsynchronous resonance and resulting turbine generator shaft damage at Mohave Generating Station, *Analysis and Control of Subsynchronous Resonance*, IEEE PES Special Publication 76 CH 1066-0-PWR, 1976, pp. 22–29.
3. IEEE Committee Report, Terms, definitions and symbols for subsynchronous resonance, *IEEE Transactions on Power Apparatus and Systems*, PAS-104, 1326–1334, June 1985.
4. Kilgore, L.A., Ramey, D.G., and Hall, M.C., Simplified transmission and generation system analysis procedures for subsynchronous resonance, *IEEE Transactions on Power Apparatus and Systems*, PAS-96, 1840–1846, November/December 1977.
5. Anderson, P.M., Agrawal, B.L., and Van Ness, J.E., *Subsynchronous Resonance in Power Systems*, IEEE Press, New York, 1990.
6. Joyce, J.S., Kulig, T., and Lambrecht, D., Torsional fatigue of turbine-generator shafts caused by different electrical system faults and switching operations, *IEEE Transactions on Power Apparatus and Systems*, PAS-97, 965–977, September/October 1978.
7. IEEE Committee Report, Series capacitor controls and settings as a countermeasure to subsynchronous resonance, *IEEE Transactions on Power Apparatus and Systems*, PAS-101, 1281–1287, June 1982.
8. Anderson, G., Atmuri, R., Rosenqvist, R., and Torseng, S., Influence of hydro units generator-to-turbine ratio on damping of subsynchronous oscillations, *IEEE Transactions on Power Apparatus and Systems*, PAS-103, 2352–2361, August 4, 1984.
9. IEEE Committee Report, Countermeasures to subsynchronous resonance problems, *IEEE Transactions on Power Apparatus and Systems*, PAS-99, 1810–1818, September/October 1980.
10. Agrawal, B.L. and Farmer, R.G., Use of frequency scanning technique for subsynchronous resonance analysis, *IEEE Transactions on Power Apparatus and Systems*, PAS-98, 341–349, March/April 1979.
11. IEEE Committee Report, Comparison of SSR calculations and test results, *IEEE Transactions on Power Systems*, PWRS-4, 336–344, February 1, 1989.
12. Kilgore, L.A., Elliott, L.C., and Taylor, E.T., The prediction and control of self-excited oscillations due to series capacitors in power systems, *IEEE Transactions on Power Apparatus and Systems*, PAS-96, 1840–1846, November/December 1977.
13. Gross, G. and Hall, M.C., Synchronous machine and torsional dynamics simulation in the computation of electro-magnetic transients, *IEEE Transactions on Power Apparatus and Systems*, PAS-97, 1074–1086, July/August 1978.
14. Agrawal, B.L., Demcko, J.A., Farmer, R.G., and Selin, D.A., Apparent impedance measuring system (AIMS), *IEEE Transactions on Power Systems*, PWRS-4(2), 575–582, May 1989.
15. Farmer, R.G. and Agrawal, B.L., Guidelines for the selection of subsynchronous resonance countermeasures, *Symposium on Countermeasures for Subsynchronous Resonance*, IEEE Special Publication 81TH0086-9-PWR, pp. 81–85, July 1981.
16. IEEE Committee Report, Reader's guide to subsynchronous resonance, *IEEE Transactions on Power Systems*, PWRS-7(1), 150–157, February 1992.
17. Bowler, C.E.J., Demcko, J.A., Menkoft, L., Kotheimer, W.C., and Cordray, D., The Navajo SMF type subsynchronous resonance relay, *IEEE Transactions on Power Apparatus and Systems*, PAS-97(5), 1489–1495, September/October 1978.

18. Ahlgren, L., Walve, K., Fahlen, N., and Karlsson, S., *Countermeasures against Oscillatory Torque Stresses in Large Turbogenerators*, CIGRÉ Session, Paris, France, 1982.
19. Sun, S.C., Salowe, S., Taylor, E.R., and Mummert, C.R., A subsynchronous oscillation relay—Type SSO, *IEEE Transactions on Power Apparatus and Systems*, PAS-100, 3580–3589, July 1981.
20. Farmer, R.G. and Agrawal, B.L., Application of subsynchronous oscillation relay—Type SSO, *IEEE Transactions on Power Apparatus and Systems*, PAS-100(5), 2442–2451, May 1981.
21. Perez, A.J., Mohave Project subsynchronous resonance unit tripping scheme, *Symposium on Countermeasures for Subsynchronous Resonance*, IEEE Special Publication 81TH0086-9-PWR, pp. 20–22, 1981.
22. Farmer, R.G., Schwalb, A.L., and Katz, E., Navajo Project report on subsynchronous resonance analysis and solution, *IEEE Transactions on Power Apparatus and Systems*, PAS-96(4), 1226–1232, July/August 1977.
23. Bowler, C.E.J., Baker, D.H., Mincer, N.A., and Vandiveer, P.R., Operation and test of the Navajo SSR protective equipment, *IEEE Transactions on Power Apparatus and Systems*, PAS-95, 1030–1035, July/August 1978.
24. Ramey, D.G., Kimmel, D.S., Dorney, J.W., and Kroening, F.H., Dynamic stabilizer verification tests at the San Juan Station, *IEEE Transactions on Power Apparatus and Systems*, PAS-100, 5011–5019, December 1981.
25. Ramey, D.G., White, I.A., Dorney, J.H., and Kroening, F.H., Application of dynamic stabilizer to solve an SSR problem, *Proceedings of American Power Conference*, 43, 605–609, 1981.
26. Kimmel, D.S., Carter, M.P., Bednarek, J.N., and Jones, W.H., Dynamic stabilizer on-line experience, *IEEE Transactions on Power Apparatus and Systems*, PAS-103(1), 198–212, January 1984.
27. Bowler, C.E.J. and Baker, D.H., Concepts of countermeasures for subsynchronous supplementary torsional damping by excitation modulation, *Symposium on Countermeasures for Subsynchronous Resonance*, IEEE Special Publication 81TH0086-9-PWR, pp. 64–69, 1981.
28. Bowler, C.E.J. and Lawson, R.A., Operating experience with supplemental excitation damping controls, *Symposium on Countermeasures for Subsynchronous Resonance*, IEEE Special Publication 81TH0086-9-PWR, pp. 27–33, 1981.
29. Christl, N., Hedin, R., Sadek, K., Lützelberger, P., Krause, P.E., Mckenna, S.M., Montoya, A.H., and Torgerson, D., *Advanced Series Compensation (ASC) with Thyristor Controlled Impedance*, CIGRÉ Session 34, Paper 14/37/38-05, Paris, France, 1992.
30. Piwko, R.J., Wagner, C.A., Kinney, S.J., and Eden, J.D., Subsynchronous resonance performance tests of the slatt thyristor-controlled series capacitor, *IEEE Transactions on Power Delivery*, 11, 1112–1119, April 1996.
31. Hedin, R.A., Weiss, S., Torgerson, D., and Eilts, L.E., SSR characteristics of alternative types of series compensation schemes, *IEEE Transactions on Power Systems*, 10(2), 845–851, May 1995.
32. Pilotto, L.A.S., Bianco, A., Long, W.F., and Edris, A.-A., Impact of TCSC control methodologies on subsynchronous oscillations, *IEEE Transactions on Power Delivery*, 18(1), 243–252, January 2003.
33. Kakimoto, N. and Phongphanphanee, A., Subsynchronous resonance damping control of thyristor-controlled series capacitor, *IEEE Transactions on Power Delivery*, 18(3), 1051–1059, July 2003.
34. Walker, D.N., Placek, R.J., Bowler, C.E.J., White, J.C., and Edmonds, J.S., Turbine-generator shaft torsional fatigue and monitoring, *CIGRÉ Meeting*, Paper 11-07, 1984.
35. Joyce, J.S. and Lambrecht, D., Monitoring the fatigue effects of electrical disturbances on steam turbine-generators, *Proceedings of American Power Conference*, 41, 1153–1162, 1979.
36. Stein, J. and Fick, H., The torsional stress analyzer for continuously monitoring turbine-generators, *IEEE Transactions on Power Apparatus and Systems*, PAS-99(2), 703–710, March/April 1980.
37. Ramey, D.G., Demcko, J.A., Farmer, R.G., and Agrawal, B.L., Subsynchronous resonance tests and torsional monitoring system verification at the Cholla station, *IEEE Transactions on Power Apparatus and Systems*, PAS-99(5), 1900–1907, September 1980.

38. Agrawal, B.L., Demcko, J.A., Farmer, R.G., and Selin, D.A., Shaft torque monitoring using conventional digital fault recorders, *IEEE Transactions on Power Systems*, 7(3), 1211–1217, August 1992.
39. Walker, D.N. and Schwalb, A.L., Results of subsynchronous resonance test at Navajo, *Analysis and Control of Subsynchronous Resonance*, IEEE Special Publication 76 CH 1066-0-PWR, pp. 37–45, 1976.
40. Walker, D.N., Bowler, C.E.J., Jackson, R.L., and Hodges, D.A., Results of the subsynchronous resonance tests at Mohave, *IEEE Transactions on Power Apparatus and Systems*, PAS-94(5), 1878–1889, September/October 1975.
41. Tang, J.F. and Young, J.A., Operating experience of Navajo static blocking filter, *Symposium on Countermeasures for Subsynchronous Resonance*, IEEE Special Publication 81TH0086-9-PWR, pp. 23–26, 1981.
42. Bahrman, M.P., Larsen, E.V., Piwko, R.J., and Patel, H.S., Experience with HVDC—Turbine-generator interaction at SquareButte, *IEEE Transactions on Power Apparatus and Systems*, PAS-99, 966–975, 1980.
43. Mortensen, K., Larsen, E.V., and Piwko, R.J., Field test and analysis of torsional interaction between the Coal Creek turbine-generator and the CU HVDC system, *IEEE Transactions on Power Apparatus and Systems*, PAS-100, 336–344, January 1981.
44. Lambrecht, D. and Kulig, T., Torsional performance of turbine generator shafts especially under resonant excitation, *IEEE Transactions on Power Apparatus and Systems*, PAS-101(10), 3689–3702, October 1982.
45. Watson, W. and Coultes, M.E., Static exciter stabilizing signals on large generators—Mechanical problems, *IEEE Transactions on Power Apparatus and Systems*, PAS-92, 204–211, January/February 1973.
46. Lee, D.C., Beaulieu, R.E., and Rogers, G.J., Effect of governor characteristics on turbo-generator shaft torsionals, *IEEE Transactions on Power Apparatus and Systems*, PAS-104, 1255–1259, June 1985.
47. Piwko, R.J., Rostamkolai, N., Larsen, E.V., Fisher, D.A., Mobarak, M.A., and Poitras, A.E., Subsynchronous torsional interactions with static var compensators—Concepts and practical implications, *IEEE Transactions on Power Systems*, 5(4), 1324–1332, November 1990.
48. Raczkowski, C. and Kung, G.C., Turbine-generator torsional frequencies—Field reliability and testing, *Proceedings of the American Power Conference*, Chicago, IL, Vol. 40, 1978.
49. Kung, G.C. and LaRosa, J.A., Response of turbine-generators to electrical disturbances, Presented at the *Steam Turbine-Generator Technology Symposium*, Charlotte, NC, October 4–5, 1978.
50. Evans, D.G., Giesecke, H.D., Willman, E.C., and Moffitt, S.P., Resolution of torsional vibration issues for large turbine generators, *Proceedings of the American Power Conference*, Chicago, IL, Vol. 57, 1985.
51. Andorka, M. and Yohn, T., Vibration induced retaining ring failure due to steel mill—Power plant electromechanical interaction, *IEEE 1996 Summer Power Meeting Panel Session on Steel-Making, Inter-Harmonics and Generator Torsional Impacts*, Denver, CO, July 1996.
52. IEEE Power and Energy Society (IEEE)/ANSI Standard C50.13, *Standard for Cylindrical-Rotor 50 and 60 Hz, Synchronous Generators Rated 10 MVA and Above*, 2005.
53. Ramey, D.G., Sismour A.C., and Kung, G.C., *Important Parameters in Considering Transient Torques on Turbine-Generator Shaft Systems*, IEEE Power & Energy Society (PES) Special Publication 79TH0059-6-PWR, pp. 25–31, 1979.
54. Exelon Corporation Internal Report, Root cause report, Dresden unit 2 & 3 generator cracked rotors caused by intermittent oscillating torsional loads requiring unit shutdowns.

18
Wind Power Integration in Power Systems

Reza Iravani
University of Toronto

18.1	Introduction	**18**-1
18.2	Background..	**18**-1
18.3	Structure of Wind Turbine Generator Units............................	**18**-2
	Fixed-Speed WTG • Variable-Speed WTG • Control of Type-3 and Type-4 WTG Units	
18.4	Wind Power Plant Systems...	**18**-7
	Onshore WPPs • Offshore WPPs	
18.5	Models and Control for WPPs..	**18**-11
	WPP Models • WPP Control	
	References..	**18**-16

18.1 Introduction

This chapter provides a brief and general overview of the following subjects:

- The status of wind energy for power system integration.
- The main four types of wind power generating units, namely Type-1 to Type-4 units and their control structures.
- Onshore and offshore wind power plants (WPPs) including various options for power collection and transmission system for offshore WPPs.
- Models for system studies of wind turbine generator (WTG) units and WPPs, with emphasis on Type-3 and Type-4 wind generation technologies.
- Control of WPPs. A set of references is also included at the end of the chapter to cover the background materials and the state-of-the-art research and developments in the field.

18.2 Background

Wind energy is globally one of the fastest growing energy resource types, although it still accounts for a relatively small portion of the total world electricity supply. The global average cumulative wind power capacities for 2005 and 2010 were about 60 and 130 GW, respectively, and the prediction for 2030 is about 1120 GW. The average wind-based electricity generations for 2005 and 2010 were about 124 and 299 TWh, respectively, and for 2030 is estimated to be about 2768 TWh [1,2]. The driving forces behind this rapid growth are competitive and declining cost of wind energy, increase and volatility of energy cost from most other mainstream resources, favorable energy policies mainly to combat adverse environmental effects of conventional electric power plants, competition in the electric power industry to offer "clean electricity," resource diversity/insurance against electricity shortage, and rapid improvements and new developments in technologies for efficient harvesting of wind power and more flexible integration of large WPPs in electric power systems. Since early 1970s and particularly after mid-1990s, the technology to harvest wind power has continuously and rapidly evolved such that currently wind power is widely considered as an integral part of the electricity supply system [1].

The building block to harvest wind power is a WTG unit that includes four main components: a wind turbine, an electric machine, a power-electronic converter/conditioner, and a WTG-level controller. A wind farm is a cluster of often identical WTG units that are collectively interfaced to the host power system at a point of interconnection (POI), through a collector system. Thus, the wind farm is viewed as a single entity at the POI. In the early installations, for example, 1980s and early 1990s, a wind farm was expected to only extract maximum power from the wind regime and inject it into the host system. The system was expected to address the wind power intermittency, power/voltage fluctuations, flicker, harmonics, and the operational issues associated with the wind farm(s). Furthermore, the wind farm could have been disconnected during transients and reconnected afterward. The system was also expected to deal with the impacts of sudden power fluctuations due to the wind farm disconnection and reconnection.

Subsequently, grid codes [3] evolved with the gradual increase in the sizes and power ratings of WTG units, increase in the number of WTG units in the wind farms and, consequently, the increase in the total capacity of each farm, and the presence of multiple wind farms in the power system; it then became apparent and necessary for the wind farms, similarly to the conventional power plants, to have active participation in the control and operation of the host power system. This initiated significant R&D, from short-term wind forecast to the development of elaborate control/protection strategies, to enable active involvement of wind farms in the control and operation of the grid during both steady-state conditions and dynamic regimes. Hereinafter, we use the more appropriate term wind power plant (WPP), instead of wind farm, to refer to a cluster of WTG units that are designed to collectively interact with, and to assist the operation of, the host power system in the steady-state, dynamic, and transient conditions.

The dominant WTG technology, particularly for applications in WPPs, is based on the horizontal axis, three-bladed, upwind turbine structure [4]. The technology has significantly evolved during the last two decades. For example, the state-of-the art commercially available WTG unit in 1995 was rated at about 750 kW and the rotor diameter of about 50 m. In 2010, however, WTG units of 7.5 MW with rotor diameter of 150 m were in operation.

Based on the size and applications, WTG units can be divided into four categories: (1) small WTG units, with capacities of up to about 15 kW, for homes, small farms, and battery charging applications; (2) medium-size WTG units with capacities ranging from 15 kW to about 750 kW, mainly as distributed generators for large farms, small communities, and hybrid wind-diesel applications; (3) large-size WTG units, with capacities ranging from 750 kW to about 2.5 MW, as distributed generators and for small- to medium-size WPPs; and (4) extra-large-size WTG units, with capacities in the range of 2.5 MW up to about 7.5 MW, for medium- to large-size onshore and offshore large-size WPP applications. The focus of this manuscript is mainly on the extra-large-size and to some extent on the large-size categories.

18.3 Structure of Wind Turbine Generator Units

WTG units are classified into fixed-speed and variable-speed WTG units.

18.3.1 Fixed-Speed WTG

In the fixed-speed WTG technology, the rotor speed is determined by the grid frequency, regardless of the wind speed, and the generator is an induction generator that is directly interfaced to the host electrical system through a transformer. The induction generator is often equipped with an electronic soft starter and shunt capacitor banks for the generator reactive power compensation. To increase efficiency, the generator rotor can have two sets of windings corresponding to two distinct speeds associated with the wind high speed (e.g., four-pole operation) and low speed (e.g., eight-pole operation).

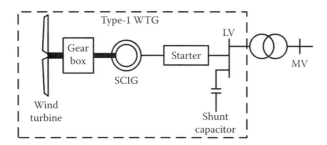

FIGURE 18.1 A schematic diagram of a Type-1 (fixed-speed) WTG unit.

The main features of the fixed-speed WTG unit are simplicity, robustness of the components, and relatively low cost; its drawbacks are the need for reactive power, excessive mechanical stresses, and significant fluctuations in the output quantities due to the fixed-speed nature of operation. The fixed-speed WTG technology is also referred to as the Type-1 WTG technology in the technical literature [5].

Figure 18.1 shows a schematic diagram of the Type-1 WTG unit. The unit is composed of a wind turbine, a gear box, a squirrel-cage induction generator (SCIG), a shunt capacitor, and an interface transformer. The Type-1 technology also includes a starter for smooth grid connection. The Type-1 WTG unit requires a fairly stiff voltage at its point of connection (PC) to the system to provide acceptable operation. Moreover, its mechanical structure should be designed for high mechanical stresses. The Type-1 WTG technology with stall control, pitch control, and active stall control was widely adopted in early 1990s. The fixed-speed WTG was the conventional system in the early 1990s; however, it is not the structure of choice for large-size WTG energy conversion system units and WPPs.

18.3.2 Variable-Speed WTG

Due to the major limitations of Type-1 (fixed-speed) WTG, the variable-speed WTG units were subsequently developed. As compared with the fixed-speed WTG structure and over its range of operation, the variable-speed WTG technology is aerodynamically more efficient, enables the generator to provide a fairly constant torque, has lower drive-train mechanical stresses, increases the energy capture, and reduces power/voltage fluctuations at the PC and consequently introduces less detrimental power quality effects. Depending on the range of speed variations, the variable-speed WTG units are classified under Type 2, Type 3, and Type 4 [5].

18.3.2.1 Type 2: Limited Variable-Speed WTG

Figure 18.2 depicts a schematic diagram of the Type-2 WTG unit. In this type, a fairly limited range for variable-speed operation is achieved through the rotor variable resistance. As shown in Figure 18.2, the Type-2 WTG includes a wind turbine, a gearbox, a wound-rotor induction generator (WRIG) equipped with a rotor resistor adjustment device, a shunt capacitor system for var control, and an interface transformer.

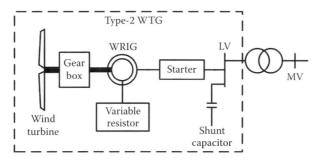

FIGURE 18.2 A schematic diagram of a Type-2 (limited variable-speed) WTG unit.

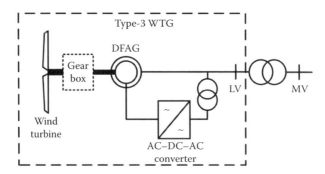

FIGURE 18.3 A schematic diagram of a Type-3 (variable-speed) WTG unit.

The variable resistor device changes the effective rotor resistance and thus enables slip control, typically up to 10%. The turbine of the Type-2 WTG unit is often equipped with a pitch-angle control. Similar to the Type-1 WTG, the Type-2 WTG configuration is not the preferred choice for large-size WTG units and WPPs.

18.3.2.2 Type 3: Variable-Speed WTG Unit (Partially Rated Converter System)

Figure 18.3 presents a schematic diagram of the Type-3 WTG unit; the system is composed of a pitch-controlled wind turbine, a gearbox, a doubly fed asynchronous generator (DFAG) [6] of which the stator is directly connected to the host power system, and an AC–AC power-electronic energy conversion system that connects the three-phase rotor circuit to the power system. Practically, the AC–AC converter is composed of two cascaded AC–DC power-electronic voltage-sourced converters (VSCs), which form a bidirectional power-flow AC–DC–AC conversion system. Since the converter system only handles the rotor power, its rating can be significantly smaller (about 0.3 times) than the machine MVA rating. This is one of the economical advantages of the Type-3 unit as compared with the Type-4 WTG unit. The Type-3 WTG unit typically provides variable-speed operation from about −40% to about +30% of the nominal power system frequency. The main drawbacks of the Type-3 configuration are its requirements for slip ring and protection against system faults. The Type-3 WTG unit is commercially available for multi-MW, for example, 3.6 MW, and widely used for large-size WPPs.

18.3.2.3 Type 4: Variable-Speed WTG Unit (Fully Rated Converter System)

Figure 18.4 shows a schematic diagram of a Type-4 WTG unit. In the Type-4 WTG technology, a pitch-controlled wind turbine is mechanically interfaced to the generator, either through a gearbox (conventional scheme) or directly (direct-drive scheme). The generator structures corresponding to the conventional and the direct-drive schemes are substantially different. For the conventional structure, the generator is a high-speed, for example, a four-pole, machine, and thus requires a gearbox. By contrast, in the direct-driven structure, the generator is a low-speed, for example, an 84-pole machine, and thus directly interfaced to the rotor shaft. The generator of Type-4 unit can be a wound-rotor (conventional

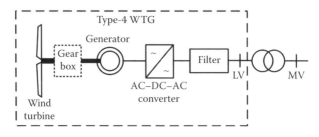

FIGURE 18.4 A schematic diagram of a Type-4 (variable-speed) WTG unit.

field-controlled) synchronous machine (WRSM), a permanent magnet synchronous machine (PMSM), or a wound-rotor asynchronous machine (WRAM) [7].

In the Type-4 WTG technology, an AC–AC conversion system interfaces the Type-4 generator electrical output terminals to the system. The converter system transfers the whole power of the WTG unit from the generator to the power system and, in contrast to that of Type-3 WTG, is rated at the full WTG power. The AC–AC converter system is most often composed of two cascaded, power-electronic, AC–DC converters that have a common DC link. The converter enables the full range of variable-speed operation for the turbine generator and reactive power control at the PC. The Type-4 and Type-3 WTG technologies are the preferred structures for large-size WTG units and WPPs.

18.3.3 Control of Type-3 and Type-4 WTG Units

This section provides a general overview of the most widely used control systems for Type-3 and Type-4 WTG units for interface to a power system.

18.3.3.1 Control of Type-3 WTG System

As explained in Section 18.3.2.2, the DFAG rotor circuit of the Type-3 WTG unit (Figure 18.3) is connected to the power system through a converter system. The converter system is composed of two cascaded AC–DC VSCs that share the same DC-link capacitor and can permit bidirectional power flow between the DFAG rotor circuit and the power system. Conceptually, each VSC is capable of controlling its real/reactive power and frequency and thus enables various modes of operation/control of the Type-3 WTG system through the variable frequency operation of the rotor. The converter system of the Type-3 WTG also provides a parallel path to the stator circuit for exchange of power between the DFAG and the power system [2,8–10]. The Type-3 WTG unit is practically designed to operate over a relatively limited range of frequency, that is, 30% of the rated speed, and thus it is operated almost as a constant speed WTG unit at low and high wind speeds.

One strategy to control the Type-3 WTG unit is to decompose the rotor current into orthogonal direct-quadrature components to control the generator torque and the terminal voltage [2]. One current component is used to regulate the torque, based on rotor speed measurements, for extraction of maximum power from wind. The other current component can be used for voltage control or power factor control through the rotor-side converter. Alternatively, the grid-side converter (Figure 18.3) can also be used to control the voltage.

Another control strategy for the Type-3 WTG is based on the rotor flux magnitude and angle control (FMAC) to adjust the terminal voltage and the output power [2,9]. Figure 18.5 shows a block diagram of the control system. In Figure 18.5, the reference value for the generator internal voltage magnitude is specified based on the terminal voltage, and the error is used to generate the magnitude of the rotor voltage. Similarly, rotor speed measurements are exploited to generate the power reference from

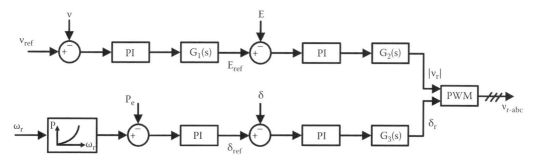

FIGURE 18.5 A block diagram of the "flux magnitude and angle control" of a Type-3 WTG unit.

the turbine power-speed characteristic, and then the error is used to generate the reference angle. The angle error is used to generate the angle for the rotor voltage.

18.3.3.2 Control of Type-4 WTG System

The main functions of the control system of a Type-4 WTG unit (Figure 18.4) are to enable the turbine to extract the maximum power from wind, to deliver the extracted power to the system at the nominal frequency, and to provide reactive-power/voltage support at the PC (depending on the system requirements). The details of the control functions are determined based on

- The generator type, that is, the asynchronous machine, the field-controlled (conventional) synchronous machine, and the permanent magnet synchronous machine [11–13]
- The power-electronic system structure, that is, either the cascaded conventional AC–DC rectifier, DC–DC converter, and VSC or the two cascaded VSCs with a common DC link

For a given rating, as compared with the conventional field-controlled synchronous machine Type-4 WTG, the PMSM Type-4 WTG has the following features. It is smaller in size and lighter in weight, has lower rotor losses, does not require slip rings, and is structurally simpler since the need for field excitation mechanism and the corresponding control system is eliminated. However, the permanent magnet synchronous generator is not able to control its terminal voltage/reactive power.

Figure 18.6 shows a high-level control diagram of a Type-4 WTG that uses a PMSM and an AC–DC–AC converter system. The converter system is composed of a diode rectifier, a chopper for boosting the rectifier DC voltage, and a VSC. The DC-link voltage controller provides control over the DC–DC chopper to regulate the DC-capacitor voltage at the desired level, based on the measured capacitor voltage. The VSC is controlled by using the rotor speed for maximum wind power extraction based on the turbine characteristics. The VSC control can be achieved based on either the concept of load-angle control method [14] or the dq-frame current-control method [15]. The dq-frame control approach is based on the machine dynamic model and inherently can provide a more refined and superior dynamic performance.

Figure 18.7 shows a high-level control diagram of the Type-4 WTG unit that uses a PMSM and a VSC-based converter system. Based on the rotor speed, the generator-side VSC controls the generator for maximum power extraction from the wind regime [2,15]. The grid-side VSC uses

- The DC-link voltage measurement to regulate the DC-link voltage at a desired, prespecified value and thus transfer the extracted wind power to the AC side
- The voltage measurement at the AC side (or the desired reactive power injection) to control the AC side voltage (or inject reactive power into the AC system)

Figures 18.8 and 18.9 present the dq-frame based controls of the generator-side and the grid-side converters of Type-4 WTG unit that uses a conventional field-controlled synchronous machine [2,14]. At the generator side, the d-axis can be oriented with the rotor magnetic axis and the q-axis can lead the

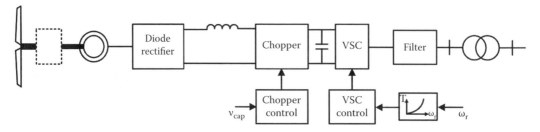

FIGURE 18.6 A schematic diagram of the Type-4 WTG control system using a diode rectifier, a DC–DC chopper, and DC–AC VSC as the conversion system.

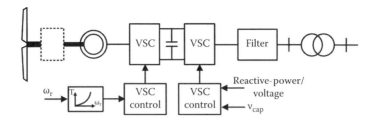

FIGURE 18.7 A schematic diagram of the Type-4 WTG control using two cascaded VSC units as the converter system.

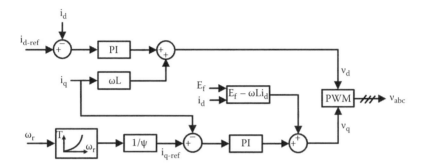

FIGURE 18.8 A dq-frame-based control block diagram of the generator-side VSC of Type-4 WTG unit.

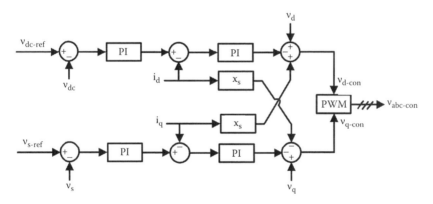

FIGURE 18.9 A dq-frame-based control block diagram of the power system side VSC of the Type-4 WTG unit.

d-axis by 90° [6]. Details of the dq-frame based control of a Type-4 WTG that exploits a conventional asynchronous (induction) generator are given in Ref. [16].

18.4 Wind Power Plant Systems

A WPP is either an onshore or an offshore installation. There is a continuously growing interest in the offshore WPPs due to the generally higher offshore wind speed, which translates into higher wind power production, lack of vast spaces in many jurisdictions for installation of large onshore WPPs, and lack/absence of the WTG visual impacts. However, the initial investment cost and the maintenance costs for offshore installations are significantly higher as compared with those of the onshore WPPs. The future offshore WPPs are anticipated to be in the range of 250–1000 MW and using large-size (more than 3 MW) WTG units [3].

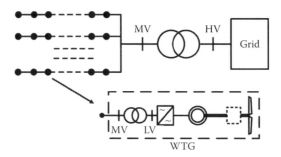

FIGURE 18.10 A schematic diagram of an onshore WPP.

18.4.1 Onshore WPPs

A typical onshore WPP (Figure 18.10) consists of tens of identical WTG units, based often on either Type-3 or Type-4 WTG units. The units are connected in parallel through the AC collector system of the plant. The collector system is typically a three-phase, medium-voltage (up to 36 kV), underground cable and overhead line that connects all the WTG units within the WPP to the plant POI. Each WTG is typically larger than 1.5 MW, and its output voltage is within the range of 400 V to 5 kV. Each WTG unit uses a step-up transformer for connection to the collector system at its PC. Additional apparatus may be necessary for the operation of the WPP, for example, the var compensator and/or the energy storage system. These apparatus are often connected at the POI of the plant. Well-established standards/guidelines and significant experience for the design and installation of onshore WPPs do exist. The high-voltage side of the POI is typically either at the transmission or the sub-transmission voltage level and may also directly supply loads.

18.4.2 Offshore WPPs

An AC collector system, conceptually similar to that for the onshore WPP of Figure 18.10, is also often considered for a relatively small-size offshore WPP (60 MW) that is located fairly close to the shore. For such a WPP, the substation will be onshore and constitutes the POI of the offshore WPP. The collector system is based on AC submarine cables with the voltage of up to 36 kV [3]. Depending on the distance between the WPP and the onshore substation, a higher voltage, for example, 45 kV, may be considered.

As the installed capacity and distance of the offshore WPP increase, it becomes viable to have an offshore substation that collects the power from the WTG units. The total power is then transmitted onshore and delivered to the power system (Figure 18.11). Since the offshore substation does not supply any customer load, the POI for the offshore WPP can be considered at the onshore substation where the WPP delivers power to the host grid (Figure 18.11). Thus, the two substations and the connecting transmission system can be considered as part of the WPP. This arrangement provides further flexibility to control the WPP and meet the WPP operational requirements at POI. In the system of Figure 18.11, high-voltage AC (HVAC), line-commutated converter (LCC), high-voltage DC (HVDC), and voltage-sourced converter HVDC (VSC-HVDC) systems are viable options for offshore to onshore transmission of the electric power [17].

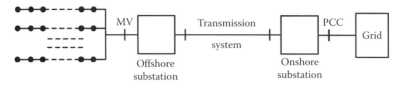

FIGURE 18.11 A schematic representation of an offshore WPP and its dedicated power transmission system.

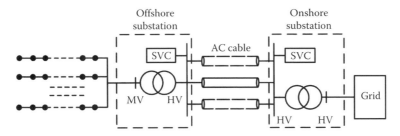

FIGURE 18.12 A schematic representation of an offshore WPP and its AC transmission system.

18.4.2.1 AC Collector System with HVAC Transmission

Figure 18.12 shows a schematic diagram of an offshore WPP and its dedicated HVAC transmission system for interface to the grid [18]. The offshore medium voltages (MVs) and HVs can be in the range of 30 and 150 kV, respectively, and each AC line is a three-phase submarine cable. Depending on the length of the AC line, reactive-power compensators may be required at both substations.

18.4.2.2 AC Collector System with LCC-HVDC Transmission

An alternative to the AC transmission scheme of Figure 18.12 is the point-to-point LCC-HVDC system of Figure 18.13 [19,20]. Since the LCC requires reactive power for operation, the offshore substation requires diesel generator and/or reactive power supply, for example, by means of a static compensator (STATCOM), to maintain the operation under all conditions, including light wind conditions.

18.4.2.3 AC Collector System with VSC-HVDC Transmission

Figure 18.14 shows a schematic diagram of an offshore WPP that utilizes an AC collector system and transmits the collected power through a dedicated VSC-HVDC to an onshore substation [3,18,20]. In contrast to the LCC-HVDC, the VSC-HVDC does not require a "stiff" AC system at the POI and thus can reduce the structural complexity, weight, volume, and facilitate the WPP operation. Although the losses of a VSC is comparatively higher than that of a LCC, the gap is rapidly closing and the technical advantages of the VSC over LCC, for example, capability for four-quadrant operation, independent control of active and reactive power capability, reduced filtering requirements, and the black-start capability, have rendered the VSC-HVDC technology a viable option for transferring the power from large-size offshore WPPs.

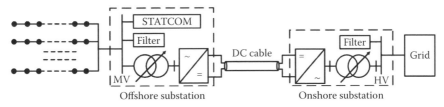

FIGURE 18.13 A schematic diagram of an offshore WPP and its LCC-HVDC transmission system.

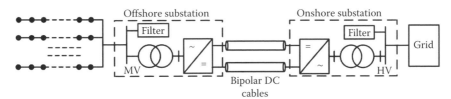

FIGURE 18.14 A schematic diagram of an offshore WPP and its dedicated VSC-HVDC power transmission system.

18.4.2.4 DC Collector System and DC Transmission

Ongoing development in power semiconductor switches and the availability of efficient DC–DC converters for high power applications in conjunction with the VSC technology have potentially enabled the use of DC collector system for offshore WPPs [3,21,22]. The DC collector system provides flexibility to alter and simplify the DC transmission system of the WPP. Figure 18.15 shows a schematic diagram of an offshore WPP with medium- or low-voltage DC collector. This configuration replaces the DC–AC converter of each Type-4 WTG unit with a step-up DC–DC converter and also eliminates the need for the WTG step-up transformer. The DC–DC converter at the offshore substation steps up the collector DC voltage for the DC transmission system.

The number of DC–DC converters in the system of Figure 18.15 can be reduced based on the arrangement shown in Figure 18.16 in which a set of physically close WTG units are aggregated by a single DC–DC converter [3]. The system of Figure 18.16 reduces the conversion system for each WTG unit to a single AC–DC converter and can result in higher efficiency and reduced control and operational complexity. Another possible configuration for the DC collector and DC transmission system is shown in Figure 18.17 where each WTG unit embeds a DC–DC conversion system that provides the desired voltage for HVDC transmission and eliminates the need for the offshore DC–DC converter.

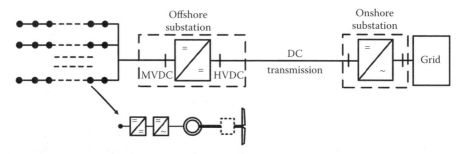

FIGURE 18.15 A schematic diagram of an offshore WPP with a DC collector system.

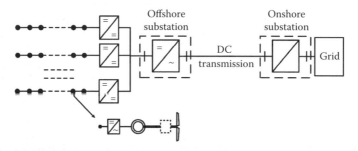

FIGURE 18.16 A schematic diagram of an offshore WPP with aggregated WTG units through DC–DC converters.

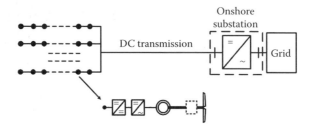

FIGURE 18.17 A schematic diagram of an offshore WPP with a DC collector system and "distributed" DC–DC converters.

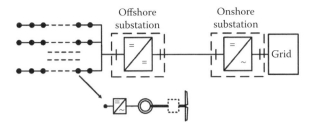

FIGURE 18.18 A schematic diagram of an offshore WPP with DC collector and a centralized DC–DC conversion system.

Figure 18.18 shows a centralized DC–DC converter for the offshore WPP in which the DC collector system is directly connected to the DC side of each WTG unit. Similar to the configuration of Figure 18.16, this configuration also simplifies the conversion system of each WTG unit and its control system.

Other options for the collector and transmission of wind power from offshore WPPs are based on

- Medium-voltage, low-frequency AC collector, that is, from 16 2/3 to 25 Hz, then HV transmission at the same frequency, and finally frequency conversion to the system nominal frequency at the onshore substation [23]
- Multiterminal VSC-HVDC system [24,25]

18.5 Models and Control for WPPs

18.5.1 WPP Models

Impact assessment studies, planning, operation, and performance evaluation/verification of the WPP require a host of analytical and simulation tools necessitating the development of appropriate WPP models for each type of system studies. Traditionally, power system studies are divided into two categories, that is, steady-state and dynamic analyses. The steady-state analysis is based on the solution of the system nonlinear, algebraic equation to determine power flow in the system. The dynamic analysis is based on the solution of the system algebraic-differential equations. The dynamic analysis, in turn, is divided into the following two subgroups:

1. Small-signal dynamic analysis, which is based on linearization of the system nonlinear equations about a steady-state operating point. Based on the linearized equations (model), the system eigenstructure provides an analytical platform for power system control design.
2. Large-signal dynamic analysis based on step-by-step numerical integration of the power system nonlinear equations (model). The large-signal dynamic analysis approach has been widely used to investigate transient behavior and performance of power system apparatus, and the overall system, within a wide range of frequencies.

The power system dynamic phenomena cover a wide frequency range, from a fraction of a hertz, for example, intra-area oscillations of aggregates of generating units, up to hundreds of kilo-hertz, for example, lightning phenomenon and electromagnetic transients in gas-insulated substations (GIS). Low-frequency dynamic phenomena propagate in the electrical system, over wide geographical areas, and involve a large number of apparatus. However, due to the low-frequency nature of the phenomena, the model required to represent each apparatus is of a relatively low order. On the contrary, high-frequency dynamic phenomena of the power system are confined to a relatively small portion of the system and do not affect a vast geographical area. However, the component models for the analysis of high-frequency phenomena must provide accurate representation over a wide frequency range, and thus are of a significantly higher order as compared with those required for the analysis of low-frequency phenomena.

Mathematical models of the conventional power system apparatus, for example, turbine-generator units, transformers, transmission lines, HVDC converters, and flexible AC transmission systems (FACTS) controllers, for analyses of various dynamic phenomena, in different frequency ranges, have been extensively developed and evaluated during the past few decades. Correspondingly, a wide range of analysis tools, predominantly based on digital-computer time-domain simulation (numerical integration) approach, also have been developed and commercialized. These tools are used for short-, mid- and long-term angle (transient) stability studies, voltage stability studies, short-circuit studies, relay/protection coordination, and analysis of electromagnetic transients.

Presence of multiple WPPs and their anticipated depth of wind power penetration into power systems during the last decade have presented a major challenge in terms of development of WPP models for different types of system studies and integration of such models in the production grade software tools. The WPP models are necessary to conduct impact assessment studies, planning, and reliability evaluation of power systems that embed WPPs. The challenges in the development of models of WPPs are

- Internal electrical layout of a WPP that, in contrast to the conventional power system apparatus, for example, thermal turbine-generator units and transformers, is itself a large-size electrical entity with multiple overhead line and underground cables, large number of multi-mass turbine-generator units, and power-electronic converters
- Options for AC or DC collector systems
- Integration of the HVDC transmission between an offshore WPP and the host system as part of the offshore WPP system and its implication on the control/protection and the operational strategies of the WPP-HVDC
- Various types and technologies for the WTG units, for example, Type 3 and Type 4, and even multiplicity of converter configurations and control protection strategies for each WTG type
- Ongoing evolution and development of the WTG technologies, performance characteristics, and their control functions and strategies
- Multiplicity and lack of uniformity of grid codes in terms of control/protection and operational requirements/constraints during steady state and dynamics
- Multiplicity of WPP control options depending on the type and structure of its WTG units for real power, reactive power, voltage, and frequency control, and damping system oscillations

The WPP-related power system studies are divided into two categories, that is, (1) internal to the WPP in which the external host system, with respect to the WPP POI, is represented by an appropriate equivalent for the envisioned study, and (2) external to the WPP in which the WPP must be represented by an equivalent at the POI with respect to the external host power system. The former category of studies is currently carried out based on detailed time-domain simulation studies [26] and requires detailed information of the WPP electrical system structure and intimate knowledge of the constituent WTGs that can only be provided by the WTG manufacturer. The latter category of studies is carried out by the utility system and often requires a generic equivalent model of the WPP at the POI to adequately represent the WPP in terms of the study objectives.

The state-of-the-art model developments associated with the latter category provide a set of generic models of WPPs based on Type-1–Type-4 WTG technologies for power system stability studies [5,27–31] and is considered as the work in progress. The WPP model is a scaled version of the simplified model of the identical WTG units in the plant [28,29,32], and represents the positive-sequence behavior of the plant for system stability studies.

Figure 18.19 shows a block representation of the Type-3 WTG unit [28,33] for power system stability studies and includes the models of the generator and the converter, the wind turbine, and the WTG-level controls. To represent the WPP model, the model of Figure 18.19 is scaled and also equipped with the WPP-level controls that provide command signals for the WTG units. Details of each block including pitch control model, turbine aerodynamic model, typical power-speed characteristic, real-power control model, reactive-power control model, and the generator and converter model are described in [28].

Wind Power Integration in Power Systems

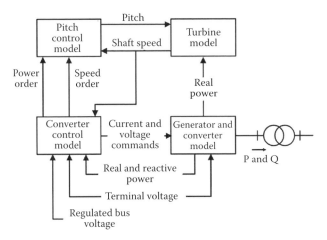

FIGURE 18.19 A block diagram representation of the Type-3 WTG model for power system stability studies.

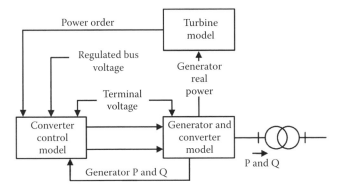

FIGURE 18.20 A block diagram representation of the Type-4 WTG model for power system stability studies.

Figure 18.20 depicts a block-diagram representation of the Type-4 WTG [28,33] and includes the generator/converter model, converter control model, and the wind-turbine model. Similar to that of Type 3, the WPP model associated with the Type-4 WTG model of Figure 18.20 must be properly scaled and equipped with the WPP-level controls. The details of the block of Figure 18.20, that is, the generator and converter model, the converter current limit model, and the wind-turbine model are also available in [28].

Scaling the Type-3/Type-4 WTG models to deduce the corresponding WPP models inherently assumes the WTG units in the plant are under the same wind regime and operate identically. This assumption is not necessary valid, particularly in the case of large WPPs where the WTGs are spread over a fairly large geographical area and consequently are subject to different wind speeds. In such cases, the WPP can be modeled by multiple equivalents where each equivalent model represents a set of electrically and geographically close WTG units. Furthermore, as the frequency of interest increases, the degree of accuracy of the scaled WTG model to represent a WPP decreases. For example, a scaled Type-3 WTG model may provide acceptable results for low-frequency inertial oscillatory mode of a WPP with respect to the power system, whereas it may not be accurate enough to model torsional dynamics of the mechanical drive train of a Type-3-based WPP.

Depending on the difference of the wind regimes of the WTG units, the WPP-level control may impose different control commands on the WTG units and result in further deviation of the operational conditions of the WTG units with respect to each other. This, in turn, reduces the accuracy of the equivalent WPP model based on scaling the model of a single WTG unit. The generic models of Figures 18.19 and 18.20 only represent the conceptual operational control and concepts and do not include the vendor-specific protection, control limits, and control functions that may be activated under specific operating scenarios. This may introduce further inaccuracy in the model results. Thus, the user must be cognizant of the applications of the adopted models. There are vendor-specific, detail WTG models that are made available to technology users for detailed system studies. However, such models are not available in the public domain and the focus of this document is on public-domain models.

18.5.2 WPP Control

The WPP control system consist of two control levels [34,35], that is, the WPP-level control and the WTG-level control of individual WTG units (Figure 18.21). The WPP-level control receives control commands from the power system operator, local measurements, and the status of available wind power from individual WTG units within the WPP. The output of the WPP-level controller comprises the set points for the WTG-level controls. Based on the set points from the WPP-level control, each WTG-level control determines the corresponding real-/reactive-power output components. If the WPP includes energy storage and reactive-power compensation units, for example, battery energy storage and var compensating units, the WPP-level control should also coordinate their operation with the WTG units.

The WPP-level control is to enable a WPP, with respect to the host power system at the PCC, to operate analogous to the conventional thermal and hydro power plants, and actively enhance steady-state and dynamic operation of the power system according to the requirements specified by grid codes [36]. The grid codes define operational and connection requirements for power plants, that is, conventional plants and WPPs, large loads, and ancillary service providers. The requirements can significantly vary depending on the power system characteristics, depth of wind power penetration, and the type and technology of WTG units within the WPP. Such requirements are fairly new and as such evolving and include

- Real-power/frequency control
- Reactive-power/voltage control
- Fault ride-through capability

18.5.2.1 WPP-Level Real-Power Control

The purpose of WPP-level real-power controller [37], Figure 18.22, is to adjust the injected real power at the POI. The inputs to the real-power controller are real-power control demand of the system, measured

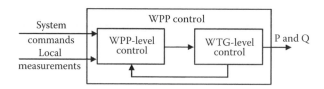

FIGURE 18.21 A typical structure of the WPP control system.

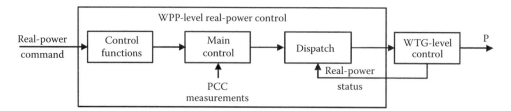

FIGURE 18.22 A block diagram representation of the WPP real-power controller.

real-power injection at the POI, and available real power of each WTG unit. If the POI is equipped with an energy storage system as part of the WPP, the POI measurement also must consider the storage system and the WPP-level real-power control must include provision for its coordinated control. The WPP-level control scheme of Figure 18.22 comprises three main blocks [34,37]:

- Control function block [38]: This block, based on the system real-power demand, determines to activate the delta control, the balance control, or the rate-limiter control. The output of the control function is the reference real power of the WPP at POI and supplied to the main real-power control block (Figure 18.22).
- Main control block [34]: This block compares the reference signal obtained from the control function block with the corresponding POI measured signal and provides the error signal to a PI controller. The output signal is the WPP real-power output reference signal that is supplied to the dispatch block (Figure 18.22).
- Dispatch block [37,39,40]: This block distributes the received WPP real-power reference signal to the WTG units within the WPP either directly without checking the overall available WPP real power [39] or based on an optimization process to meet the system demanded real power [40].

18.5.2.2 WPP-Level Reactive-Power Control

Figure 18.23 shows a block diagram of the WPP reactive-power controller, which is functionally similar to that of real-power controller of Figure 18.22 and includes the WPP-level and the WTG-level controls [34]. The input signal from the system is the reactive-power demand, the voltage magnitude, or the power factor at the PCC. The input signal and the associated reactive-power control strategy is selected based on the power system characteristics, for example, X/R ratio and the short-circuit ratio at POI. The POI measurement may include voltage and reactive-power signals. The WPP-level controller includes two main blocks (Figure 18.23):

1. Main control block: This block receives the system command and the corresponding local measurement signals, determines the error, and processes the error signal through a PI controller. Depending on the input signal, the output of the block is either the desired WPP reactive power or the voltage magnitude reference signal [37] (Figure 18.24). The main control block of Figure 18.24

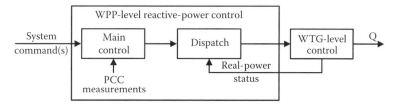

FIGURE 18.23 A block diagram representation of the WPP reactive-power controller.

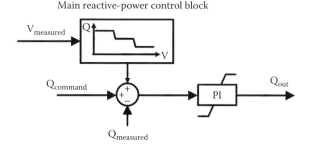

FIGURE 18.24 The main control block of the system of Figure 18.23.

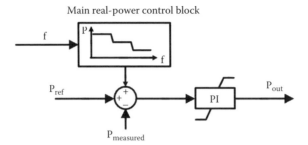

FIGURE 18.25 The WPP-level main control block of the system of Figure 18.22, including an auxiliary frequency deviation signal for frequency control.

can be augmented by a voltage-based reactive-power correction to impose voltage constraints at POI. A similar augmentation, based on a reactive-power auxiliary signal, can be envisioned when voltage control strategy is adopted. An alternative structure to the main block of Figure 18.24 has been described in [39].

2. Dispatch block: This block receives the output of the main control block and the status of available power at each WTG unit and determines the reactive-power set point or voltage set point for each WTG unit within the WPP. It should be noted that voltage set point also should be converted to an equivalent reactive-power set point. The reactive-power set point can be determined based on the WTG available reactive power [38], directly based on the output signal of the main control block [40], or based on an optimized strategy [40,41].

Coordinated reactive-power control of the WPP is to satisfy the grid code voltage and reactive-power requirements at the POI and the internal performance of the WPP. One strategy to achieve this objective is to include a slow control loop at the WPP-level control and a fast voltage control loop at each WTG-level control [42].

18.5.2.3 WPP Frequency Control

Subsequent to a disturbance that subjects the power system to frequency deviations from the nominal power frequency, the system power plants including the WPPs are expected to respond to the frequency excursions and settle the frequency to a viable value. If the frequency is not the rated frequency, the automatic generation control (AGC) takes over and brings it back to the nominal value. The main control block of Figure 18.22 can be augmented by an auxiliary real-power command signal, based on a power/frequency droop characteristic, to provide frequency control through the WPP-level controller (Figure 18.25). The frequency control can be directly implemented in the WTG real-power control and by utilizing the inertial response or the real-power reserve of the unit or pitch-angle control. The real-power command of each WTG unit, based on the system low-frequency components, also can be modulated to provide damping for power system oscillations, analogous to the action of power system stabilizer (PSS).

References

1. IEA Wind Energy, Annual Report 2008, July 2009, ISBN 0-9786383-3-6.
2. O. Anaya-Lara, N. Jenkins, J. Ekanayake, P. Cartwright, and M. Hughes, *Wind Energy Generation—Modeling and Control*, Wiley, Chichester, U.K., 2009, ISBN 978-0470-7433-1.
3. T. Ackermann, *Wind Power in Power Systems*, Wiley, Chichester, U.K., 2005, ISBN 0-470-85508-8.
4. S. Heier, *Grid Integration of Wind Energy Conversion System*, Wiley, Chichester, U.K., 1988, ISBN 0-471-97143-x.
5. R. Piwko, E. Camm, A. Ellis, E. Muljadi, R. Zavadil, R. Walling, M. O'Mally, G. Irwin, and S. Saylors, A whirl of activity, *IEEE Power and Energy Magazine*, 7(6): 26–35, November/December 2009.

6. P.C. Krause, O. Wasynczuk, and S.D. Sudhoff, *Analysis of Electrical Machinery and Drive Systems*, Wiley, New York, 2002, ISBN 0-471-14326-X.
7. L.H. Hansen, L. Helle, F. Blaabjerg, E. Ritchie, S. Munk-Nielson, H. Bindner, P. Sorensen, and B. Bak-Jensen, Conceptual survey of generators and power electronics for wind turbines, Riso-R-1-1205 (EN), Riso National Laboratory, Roskilde, Denmark, December 2001.
8. J.B. Ekanayake, L. Holdsworth, X. Wu, and N. Jenkins, Dynamic modeling of doubly-fed induction generator wind turbine, *IEEE Transactions on Power System*, PWRD-18(2): 803–809, April 2003.
9. F.M. Hughes, O. Anaya-Lara, N. Jenkins, and G. Strbac, Control of DFIG-based wind generation for power network support, *IEEE Transactions on Power Systems*, PWRS-20(24): 1958–1966, November 2005.
10. O. Anaya-Lara, F.M. Hughes, N. Jenkins, and G. Strbac, Rotor flux magnitude and control strategy for doubly fed induction generators, *Wind Energy*, 9(5): 479–495, 2006.
11. V. Akhmatov, V. Nielson, and A.H. Pedersen, Variable-speed wind turbines with multi-pole synchronous permanent generators—Part 1: Modelling and in dynamic simulation tools, *Wind Engineering*, 27: 531–548, 2003.
12. A. Grauers, Design of direct driven permanent magnet generators for wind turbines, PhD thesis, Chalmers University of Technology, Gothenburg, Sweden, No. 292L, 1996.
13. Z. Chen and E. Spooner, Grid interface options for variable-speed permanent magnet generator, *IEE Proceedings Electric Power Application*, 145(4): 273–283, 1998.
14. P. Kundur, *Power System Stability and Control*, McGraw Hill, New York, 1994, ISBN 0-07-03598-x.
15. A. Yazdani and R. Iravani, *Voltage-Sourced Converters in Power Systems—Modeling, Control, and Applications*, Wiley-IEEE, Hoboken, NJ, 2010, ISBN 978-0-470-52156-4.
16. A. Tabesh, Dynamic modelling and analysis of multi-machine power systems including wind farms, PhD thesis, University of Toronto, Toronto, Ontario, Canada, 2005.
17. M. Hausler and F. Owman, AC or DC for connecting offshore wind farms to the transmission grid? *Proceeding of 3rd International Workshop on Large-Scale Integration of Wind Power and Transmission Network for Offshore Wind Farms*, Royal Institute of Technology, Stockholm, Sweden, 2002.
18. E. Ericksoon, P. Halvarsson, D. Wensky, and M. Hausler, System approach on designing an offshore windpower grid connection, *Proceeding of 4th International Workshop on Large-Scale Integration of Wind Power and Transmission Network for Offshore Wind Farms*, Royal Institute of Technology, Stockholm, Sweden, 2003.
19. P. Cartwright, L. Xu, and C. Saase, Grid integration of large offshore wind farms using hybrid HVDC transmission, *Proceedings of the Nordic Wind Power Conference*, Chalmers University of Technology, Gothenburg, Sweden, 2004.
20. S. Weigel, B. Weise, and M. Poller, Control of offshore wind farms with HVDC grid connection, *Proceedings of the 9th International Workshop on Large-Scale Integration of Wind Power into Power Systems and Transmission Networks for Offshore Wind Power Plants*, pp. 419–426, Quebec, Canada, October 2010.
21. J. Pan, L. Qi, J. Li, M. Reza, and K. Srivastava, DC connection of large-scale wind farms, *Proceedings of the 9th International Workshop on Large-Scale Integration of Wind Power into Power Systems and Transmission Networks for Offshore Wind Power Plants*, pp. 435–441, Quebec, Canada, October 2010.
22. C. Meyer, M. Hing, A. Peterson, and R.W. Doncker, Control and design of DC grids for offshore wind farms, *IEEE Transactions on Industry Applications*, IAS-43(6): 1474–1482, November/December 2007.
23. T. Schutte, B. Gustavsson, and M. Strom, The use of low-frequency AC for offshore wind farms, *Proceedings of the 2nd International Workshop on Transmission Networks for Offshore Wind Farms*, Royal Institute of Technology, Stockholm, Sweden, 2001.
24. K. Rudin, H. Abildgaard, A.G. Orths, and Z.A. Styczynski, Analysis of operational strategies for multi-terminal VSC-HVDC system, *Proceedings of the 9th International Workshop on Large-Scale Integration of Wind Power into Power Systems and Transmission Networks for Offshore Wind Power Plants*, pp. 411–418, Quebec, Canada, October 2010.

25. C. Ismunandar, A.A. van der Meer, M. Gibescu, R.L. Hendrik, and W.L. Kling, Control of multi-terminal VSC-HVDC for wind power integration using voltage-margin method, *Proceedings of the 9th International Workshop on Large-Scale Integration of Wind Power into Power Systems and Transmission Networks for Offshore Wind Power Plants*, pp. 427–434, Quebec, Canada, October 2010.
26. R. Gagnon, G. Turmel, C. Larose, J. Brochu, G. Sybille, and M. Fecteau, Large-scale real-time simulation of wind power plants into Hydro-Quebec power system, *Proceedings of the 9th International Workshop on Large-Scale Integration of Wind Power into Power Systems and Transmission Networks for Offshore Wind Power Plants*, pp. 73–80, Quebec, Canada, October 2010.
27. CIGRE Technical Brochure, Modeling and dynamic performance of wind generations as it relates to power system control and dynamic performance, CIGRE Study Committee C4, August 2007.
28. Joint Report—WECC WG on dynamic Performance of Wind Power Generation and IEEE WG on Dynamic Performance of Wind Power Generation, Description and technical specifications for generic WTD models—A status report, to be published in the *IEEE PES Transactions on Power Systems* (in press).
29. E. Muljadi, C.P. Butterfield, A. Ellis, J. Mechenbier, J. Hocheimer, R. Young, N. Miller, R. Zavadil, and J.C. Smith, Equivalencing the collector system of a large wind power plant, *IEEE PES General Meeting*, Montreal, Quebec, Canada, 2006.
30. E. Muljadi and E. Ellis, Validation of the wind power plant models, *IEEE PES General Meeting*, Pittsburgh, PA, 2008.
31. IEEE Task Force Report—Ah hoc TF on Wind Generation Model Validation of IEEE PES WG on Dynamic Performance of Wind Power Generation, Model validation for wind turbine generator models, to be published in the *IEEE Transactions on Power Systems* (in press).
32. Y. Kazachkov and S. Stapleton, Does the generic dynamic simulation wind turbine model exist? *Wind Power Conference*, Denver, CO, May 2005.
33. K. Clark, N.W. Miller, and J.J. Sanchez-Gasca, Modeling of GE wind turbine-generator for grid studies, General Electric International, Schenectady, NY, Version 4.5, April 2010.
34. M. Altin, R. Teodorescu, B.B. Jenson, P. Rodriguez, F. Iov, and P.C. Kjaer, Wind power plant control—An overview, *Proceedings of the 9th International Workshop on Large-Scale Integration of Wind Power into Power Systems and Transmission Networks for Offshore Wind Power Plants*, pp. 581–588, Quebec, Canada, October 2010.
35. O. Goksu, R. Teodorescu, P. Rodriguez, and L. Helle, A review of the state of the art in control of variable-speed wind turbines, *Proceedings of the 9th International Workshop on Large-Scale Integration of Wind Power into Power Systems and Transmission Networks for Offshore Wind Power Plants*, pp. 589–596, Quebec, Canada, October 2010.
36. M. Altin, G. Guksu, R. Teodorescu, P. Rodriguez, B.B. Jenson, and L. Helle, Overview of recent grid codes for wind power integration, *Proceeding of Optimization of Electrical and Electronic Equipment Conference*, Brasov, Romania, 2010.
37. A. Hansen, P. Sorensen, F. Iov, and F. Blaabjerg, Centralized power control of wind farm with doubly fed induction generators, *Renewable Energy*, 27: 351–359, 2006.
38. J.R. Kristoffersen and P. Christiansen, Horns Rev offshore windfarm: Its main controller and remote control system, *Wind Energy*, 27: 351–359, 2003.
39. J. Rodriguez-Amenedo, S. Arnalte, and J. Burgos, Automatic generation control of wind farm with variable speed wind turbines, *IEEE Transactions on Energy Conversion*, 17(2): 279–284, May 2002.
40. R.G. de Almeida, E. Castronuovo, and J. Pecas Lopez, Optimum generation control in wind parks when carrying out system operator request, *IEEE Transactions on Power Systems*, 21(2): 718–725, 2006.
41. General Electric Patent, Windfarm collector system loss optimization, patent #EP2108828A2, 2007.
42. J. Footmann, M. Wilch, F. Koch, and I. Erlrich, A novel centralized wind farm collector utilizing voltage control capability of wind turbines, *Proceeding of Power System Computational Conference*, Glasgow, Scotland, 2008.

19
Flexible AC Transmission Systems (FACTS)

19.1	Introduction	19-1
19.2	Concepts of FACTS	19-2
19.3	Reactive Power Compensation in Transmission Lines	19-3
19.4	Static var Compensator	19-4
	Operating Principle • Voltage Control by SVC • SVC Applications	
19.5	Thyristor-Controlled Series Compensation	19-12
	Operating Principle • TCSC Applications	
19.6	Static Synchronous Compensator	19-18
	Operating Principle • STATCOM Applications	
19.7	Static Series Synchronous Compensator	19-20
	Operating Principle • SSSC Applications	
19.8	Unified Power Flow Controller	19-22
	Operating Principle • UPFC Applications	
19.9	FACTS Controllers with Energy Storage	19-25
	Superconducting Magnetic Energy Storage • Battery Energy Storage System	
19.10	Coordinated Control of FACTS Controllers	19-26
	Coordination between Multiple FACTS Controllers • Coordination with Conventional Equipment for Long-Term Voltage-var Management	
19.11	FACTS Installations to Improve Power System Dynamic Performance	19-31
19.12	Conclusions	19-31
	References	19-32

Rajiv K. Varma
University of Western Ontario

John Paserba
Mitsubishi Electric Power Products, Inc.

19.1 Introduction

Flexible AC transmission systems (FACTS) are defined by IEEE as "alternating current transmission systems incorporating power-electronic based and other static controllers to enhance controllability and increase power transfer capability" [1]. FACTS Controllers allow flexibility in the operation of a transmission network to accomplish increased stability margins [2–11].

The typical power transfer limits in a transmission line based on thermal, steady-state stability, transient/angular stability, and damping considerations are illustrated in Figure 19.1 [5]. In certain scenarios the limits imposed from damping viewpoint may, for example, be more constraining than transient/angular stability. Evidently, the transmission line is utilized to a much lower extent than that dictated by its thermal limit. An increase in power transmission along the same corridor can only be achieved by constructing new lines at a very high cost. FACTS provide a potential capability of enhancing the

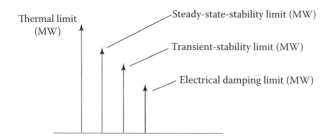

FIGURE 19.1 Illustration of different limits of power transfer along a line. (From Mathur, R.M. and Varma, R.K., *Thyristor-Based FACTS Controllers for Electrical Transmission Systems*, Wiley-IEEE Press, New York, February 2002.)

stability limits of the existing transmission line up to its thermal limit. The construction of a new line can therefore be avoided or delayed.

FACTS are utilized in transmission systems for accomplishing one or more of the following objectives:

- Increase/control of power transmission capacity in a line and for preventing loop flows
- Improvement of system transient stability limit
- Enhancement of system damping
- Mitigation of subsynchronous resonance (SSR)
- Alleviation of voltage instability
- Limiting short-circuit currents
- Improvement of HVDC converter terminal performance
- Grid integration of wind power generation systems

FACTS Controllers are broadly classified into the following major categories:

1. Thyristor-based FACTS:
   a. Static var compensator (SVC)
   b. Thyristor-controlled series compensation (TCSC)
   Some other FACTS Controllers in this category are thyristor-controlled phase angle regulator (TCPAR), thyristor-controlled voltage limiter (TCVL), and thyristor-controlled braking resistor (TCBR).
2. Voltage-sourced converter (VSC)-based FACTS:
   a. Static synchronous compensator (STATCOM)
   b. Static synchronous series compensator (SSSC)
   c. Unified power flow controller (UPFC)

The interline power flow controller (IPFC) is another Controller in this category. Power electronic converter technologies have facilitated integration of energy storage resulting in two important FACTS Controllers—battery energy storage systems (BESS) and superconducting magnetic energy storage systems (SMES).

This chapter presents a brief overview of operating principles and the applications of some major FACTS Controllers.

19.2 Concepts of FACTS

The concept of FACTS technology [4,5] can be explained through a simple transmission system having a line of reactance X and voltages $V_1 \angle \delta$ and $V_2 \angle 0$ across its two ends, as depicted in Figure 19.2.

The power transfer across the transmission line is given by

$$P = \left[\frac{(V_1 V_2)}{X}\right] \sin \delta \qquad (19.1)$$

Flexible AC Transmission Systems (FACTS)

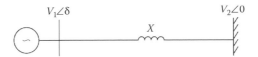

FIGURE 19.2 Concept of FACTS. (From Mathur, R.M. and Varma, R.K., *Thyristor-Based FACTS Controllers for Electrical Transmission Systems*, Wiley-IEEE Press, New York, February 2002.)

The concepts of different FACTS Controllers evolve from an understanding of the aforementioned relationship. The various options for enhancing power transfer capacity are

1. Increasing V_1 and V_2
   This choice is limited as the voltages can typically be varied within ±5%–6% of a nominal transmission voltage level.
2. Decreasing X
   This can be accomplished by
   a. Installing a new parallel line, which can be extremely expensive and requires right-of-way permissions to be obtained, which may be quite time consuming
   b. Providing midline shunt reactive compensation for voltage control—this leads to the concept of shunt FACTS—SVC and STATCOM
   c. Inserting series capacitors in the line to compensate the line inductive voltage drop—this is the principle of TCSC
   d. Injecting a voltage in series with the line in phase opposition to the inductive voltage drop—this leads to the concept of SSSC
3. Controlling the angular difference δ across transmission line. This action can be performed by the TCPAR.

19.3 Reactive Power Compensation in Transmission Lines

The voltage profiles of a lossless line, which is voltage regulated at both ends, are depicted in Figure 19.3. Maximum voltage rise is experienced at the line midpoint for lightly loaded condition, whereas maximum voltage dip is noticed at line midpoint for heavily loaded scenario.

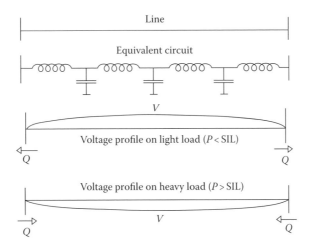

FIGURE 19.3 Voltage profile along a long transmission line. (From IEEE Substations Committee, Tutorial on static var compensators, Module 1, *IEEE PES T&D Conference & Exposition*, New Orleans, LA, April 20–22, 2010.)

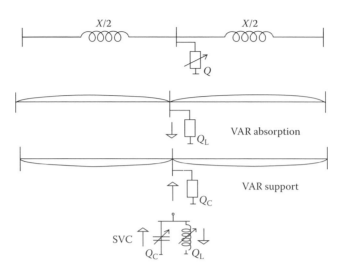

FIGURE 19.4 Voltage profile along a long transmission line with midpoint reactive power compensation. (From IEEE Substations Committee, Tutorial on static var compensators, Module 1, *IEEE PES T&D Conference & Exposition*, New Orleans, LA, April 20–22, 2010.)

The midline voltage can be regulated by reactive power absorption during light loads and by reactive power injection during heavy loads as demonstrated in Figure 19.4. In both cases, the reactive power exchange has to be variable and controlled to obtain a desired bus voltage. This is the essential function of a SVC [12–16].

19.4 Static var Compensator

The SVC is the most widely employed FACTS Controller [5,12–14]. It exchanges continuously controlled reactive power (ranging from inductive to capacitive) with the transmission system to regulate specific parameters of the system—typically bus voltage. The essential elements of an SVC are a thyristor-controlled reactor (TCR). To understand the concept a single phase TCR is shown in Figure 19.5. The TCR comprises two antiparallel connected thyristors in series with a fixed inductor. A symmetrical variation in the firing angle of both thyristors T_1 and T_2 from 90° to 180° results in a non-sinusoidal thyristor current I_{TCR}, which has symmetrical pulses on both sides. Different variables relating to the TCR operation are depicted in Figure 19.6 [5,12,13]. For a constant sinusoidal supply voltage V_S, a variation in the fundamental frequency

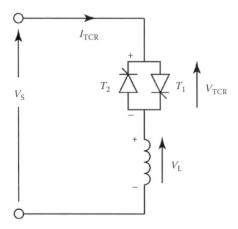

FIGURE 19.5 SVC: A single-phase TCR. (From Mathur, R.M. and Varma, R.K., *Thyristor-Based FACTS Controllers for Electrical Transmission Systems*, Wiley-IEEE Press, New York, February 2002.)

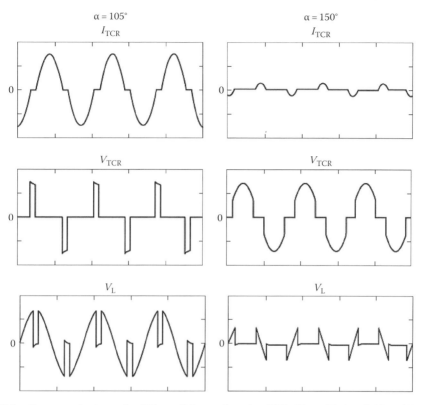

FIGURE 19.6 Current and voltages for different firing angles α in a TCR. (From Mathur, R.M. and Varma, R.K., *Thyristor-Based FACTS Controllers for Electrical Transmission Systems*, Wiley-IEEE Press, New York, February 2002.)

component of TCR current is achieved with varying firing angles, thereby resulting in a variable inductive susceptance. The relationship between the TCR susceptance and firing angle is nonlinear [12,13]. Three single-phase TCRs are connected in delta configuration to prevent the flow of triplen harmonics into the transmission network. Also, the inductor is generally split into two—one on each side of the thyristor pair to protect the thyristor valves from any short circuits across the inductor [5].

While a TCR provides continuously controlled lagging reactive power, capacitors are connected in parallel with the TCR to provide a continuously controlled leading reactive power capability for the SVC. The shunt capacitor can be connected in fixed mode resulting in fixed capacitor TCR (FC-TCR) or may be mechanically switched resulting in mechanically switched capacitor TCR (MSC-TCR) configuration. Generally, the FC portion of a TCR-based SVC acts as a filter for the harmonics generated by the nonlinear operation of the TCR.

The most versatile SVC involves a thyristor switched capacitor (TSC) in shunt with a TCR, known as TSC-TCR. This configuration results in both lower losses and a smaller size inductor as compared to the FC-TCR [13], albeit at a higher cost. Figure 19.7 illustrates the general configuration of a TCR-TSC-type SVC [5,15]. Small series inductors are provided in the TSC branches to provide filtering of the characteristic harmonics generated by the TCR [13,14]. In some cases an additional high pass filter may also be installed. The SVC typically operates at a medium voltage due to the voltage ratings of the thyristors and is usually always connected to the high-voltage transmission system through a coupling transformer.

The SVC controls the high-side bus voltage to a specified level V_{ref}. The three-phase bus voltage is measured and a DC voltage V_{meas} typically equivalent to the RMS voltage is obtained for comparison with the reference voltage. Adequate filters are provided on both sides of the voltage magnitude transducer to preclude any network resonant modes and other harmonics/noise in the bus voltage. Filtering of the

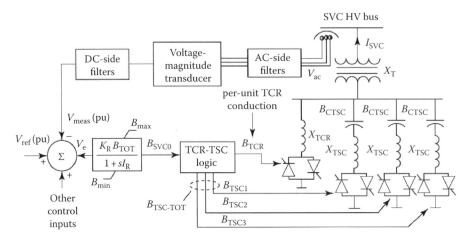

FIGURE 19.7 Basic configuration of a TSC-TCR SVC. *Note:* $B_{min} = B_{SVC}$ at the TCR only. $B_{max} = B_{SVC}$ at all TSCs only. $B_{TOT} = B_{max} - B_{min}$. K_R is the static gain (full range for the voltage change of $1/K_R$). T_R is the regulator time constant. B_{SVC0} is the net susceptance at the SVC HV bus. (From Mathur, R.M. and Varma, R.K., *Thyristor-Based FACTS Controllers for Electrical Transmission Systems*, Wiley-IEEE Press, New York, February 2002.)

measured signal is important as it can help avoid controller instabilities [15–17]. The error voltage is passed through a regulator (typically proportional integral or a gain with time constant) and a desired SVC susceptance B_{SVC} is computed. The TCR-TSC logic utilizes this input to determine the number of TSCs to be activated and the firing angle of the TCR.

Several other control inputs are provided at the summing junction. The most important one is control signal proportional to the SVC current, which implements a droop or slope in the operating characteristics of the SVC [5,14,15]. Other auxiliary control signals are provided after processing to their own controllers to impart various other functionalities to the SVC, such as damping control, SSR control, etc. [5,15].

19.4.1 Operating Principle

The steady-state and dynamic voltage–current characteristics of the SVC are shown in Figure 19.8 [5,14,15]. The SVC maintains a constant voltage, although with a slight slope (typically 1%–3%), by varying its current continuously over its entire linear controllable range. An intersection of the system load line with the *V–I* characteristic determines the SVC operating point. Outside of the controllable range, the SVC behaves as a fixed capacitor on the low-voltage side and performs like a fixed inductor on the high-voltage side. If the bus voltage continues to rise, the thyristor valve current is limited by appropriate firing control to protect the valves.

The steady-state *V–I* characteristic of the SVC is similar to the dynamic *V–I* characteristic, except for a deadband in voltage. The real function of the SVC is to provide dynamic reactive power support during disturbances. If the steady-state bus voltages tend to change (in absence of disturbances) the SVC will have a tendency to traverse far into its controllable range to provide steady-state voltage support. This will leave little reactive power capability for SVC to respond during faults. The deadband with its associated susceptance control prevents such an occurrence [18]. This highlights two important aspects of SVC control operation:

- SVCs are meant to provide dynamic voltage support not steady-state voltage support.
- SVCs are floating in steady state (i.e., do not exchange reactive power with the system).

This is discussed more in Section 19.10.2.

Flexible AC Transmission Systems (FACTS)

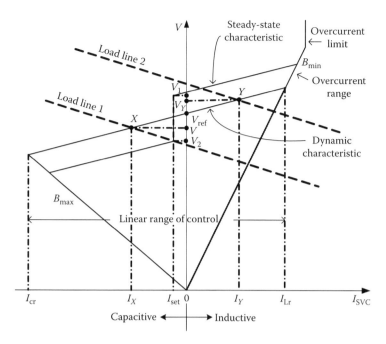

FIGURE 19.8 *V–I* characteristics of SVC.

19.4.2 Voltage Control by SVC

Figure 19.9 demonstrates the Thevenin's equivalent of a power system viewed from the SVC bus. The compensated bus voltage V_C is given by

$$V_C = V_S - jX_S\, I_{SVC} \tag{19.2}$$

The phasor diagrams (a) and (b) depict the voltage regulation provided by inductive and capacitive operations of the SVC, respectively. The contribution of the SVC to voltage control is given by the product $X_S\, I_{SVC}$.

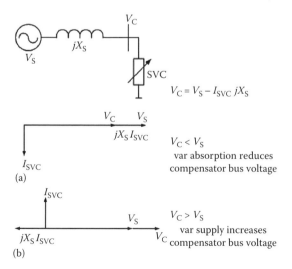

FIGURE 19.9 Voltage regulation by SVC: (a) inductive operation of SVC and (b) capacitive operation of SVC. (From IEEE Substations Committee, Tutorial on static var compensators, Module 1, *IEEE PES T&D Conference & Exposition*, New Orleans, LA, April 20–22, 2010.)

This implies that the SVC is more effective for voltage control when

1. I_{SVC} is large, that is, the reactive power capability of the SVC is high.
2. X_S is large, that is, system is weak and has a high short-circuit impedance. The SVC is, therefore, more effective when it is really needed, that is, in weak systems, which tend to undergo large voltage fluctuations.

The other important features of SVC control are as follows:

1. SVC control systems are typically optimized to provide fastest response during weak system configurations [12,15].
2. SVC controller response slows down as the system strength increases [5,14].

Adaptive features, such as gain optimizer, may need to be installed to provide a fast SVC response for a wide range of operating conditions [19].

The aforementioned principles of voltage control also apply to STATCOM.

19.4.3 SVC Applications

SVC is the most widely employed FACTS Controller and has several applications [5]. These are described in the following.

19.4.3.1 Increasing Power Transmission Capacity in a Line

The power transfer capacity of a long line shown in Figure 19.2 is given by Equation 19.1. If V_1 and V_2 are considered to be 1 pu and δ to be 90°, the maximum power transfer $P_{12\,max}$ is expressed by

$$P_{12\,max} = \frac{1}{X} \tag{19.3}$$

Figure 19.10a shows the same line compensated at midpoint by an SVC that regulates the voltage at V_m.
The corresponding power transfer over the line is given by

$$P = \left[\frac{V_1 V_m}{0.5X}\right] \sin\left(\frac{\delta}{2}\right) \tag{19.4}$$

If V_1, V_2, and V_m are considered to be 1 pu and δ to be 180°, the maximum power transfer $P'_{12\,max}$ is expressed by

$$P'_{12\,max} = \frac{2}{X} \tag{19.5}$$

The voltage regulation provided by the midline SVC thus doubles the maximum power transfer capacity. This is a great advantage of an SVC. However, this doubling requires an ideal SVC, that is, an SVC of a very large size [12]. The variation of power flow P in the uncompensated line, power flow P_C in the compensated line, and the reactive power Q_{SVC} of the ideal SVC are depicted in Figure 19.10b. A realistic size SVC although may not double the power transfer, it still increases the power transfer capacity substantially [5].

19.4.3.2 Improvement of System Transient Stability Limit

Equal area criteria for transient stability when applied to the power angle curve of a transmission system compensated by SVC as shown in Figure 19.10, demonstrates that the transient stability margin increases substantially even with a realistic size SVC [14].

Flexible AC Transmission Systems (FACTS)

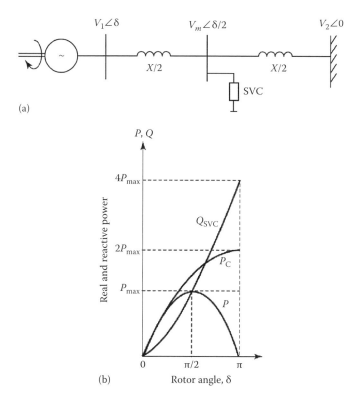

FIGURE 19.10 (a) A transmission line compensated at midpoint by an SVC and (b) Variation of real power flow without and with an ideal SVC, and reactive power of the ideal SVC. (From Mathur, R.M. and Varma, R.K., *Thyristor-Based FACTS Controllers for Electrical Transmission Systems*, Wiley-IEEE Press, New York, February 2002.)

19.4.3.3 Enhancement of System Damping

SVCs are primarily utilized for voltage control and do not inherently contribute to system damping [4,5,12]. However, by incorporating auxiliary control SVCs can significantly improve the electrical damping of power systems [14,15,20]. This control concept is presented through a single machine infinite bus system with a midline located SVC, as illustrated in Figure 19.11.

- If $d(\Delta\delta)/dt$ is positive, that is, generator rotor is accelerating due to built-up kinetic energy, the SVC terminal voltage V_m is controlled to increase generator electrical power output (refer Equation 19.4).
- If $d(\Delta\delta)/dt$ is negative, that is, rotor is decelerating due to loss of kinetic energy, the SVC terminal voltage V_m is controlled to decrease generator electrical power output (refer Equation 19.4).

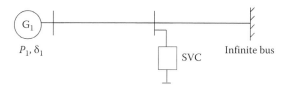

FIGURE 19.11 Single machine infinite bus system compensated at midpoint by an SVC. (From IEEE Substations Committee, Tutorial on static var compensators, *IEEE PES T&D Conference & Exposition*, New Orleans, LA, April 20–22, 2010.)

The distinguishing feature of this damping control is that the SVC bus voltage is not maintained rigidly constant but is modulated in response to auxiliary signals that contain information on rotor oscillations.

19.4.3.3.1 Choice of Auxiliary Signals for Damping Control

The different auxiliary signals [5,15,18] utilized in SVCs for damping control can be classified into two categories:

1. *Local signals*
   a. Line current
   b. Real power flow
   c. Bus frequency
   d. Synthesized remote signals based on local signals and knowledge of system parameters [20]
2. *Remote signals* (telecommunicated through dedicated fiber-optic cables or by phasor measurement units (PMUs) [21])
   a. Rotor angle/speed deviation of a remote generator
   b. Angle/frequency difference between remote voltages at the two ends of the transmission line

The selected auxiliary control signal should be effective for power flow in either direction [15]. The features of an appropriate damping signal in terms of observability, controllability, and effect of SVC operation on control signal itself are presented in Chapter 9.

19.4.3.3.2 Case Study

A two-area study system with four generators and a midline located SVC is shown in Figure 19.12 [22,23]. A five-cycle three-phase fault at bus 8 causes the system to become unstable without SVC, as depicted in Figure 19.13a. The effectiveness of four different signals for damping the interarea mode of oscillation is compared in Figure 19.12. These are (a) no damping signal (SVC on pure voltage control), (b) line current magnitude signal, (c) remote speed signal of generator 3, and (d) remote speed signal of generator 2. It is determined a priori from participation factor analysis that the generator 3 contributes much more to the interarea mode oscillation than generator 2. The most effective signal is the remote signal corresponding to speed of generator 3, followed by local line current magnitude signal, and then by the remote signal corresponding to speed of generator 2. Pure SVC voltage control provides least damping to the oscillatory mode. The effectiveness of these signals is in correspondence with the criteria for selection of control signals stated earlier.

Controllers need to be designed to damp specific oscillatory modes [5, Chapter 9]. This requires the filtering of the required oscillatory mode from the obtained power system raw signal. The damping control remains idle during steady state and gets activated when oscillatory modes get excited.

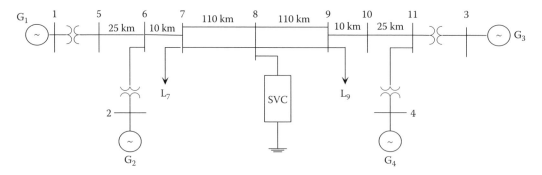

FIGURE 19.12 Two-area study system. (From IEEE Substations Committee, Tutorial on static var compensators, Module 1, *IEEE PES T&D Conference & Exposition*, New Orleans, LA, April 20–22, 2010; Kundur, P., *Power System Stability and Control*, McGraw-Hill, Inc., New York, 1994.)

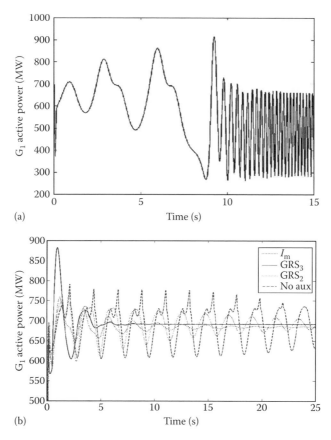

FIGURE 19.13 (a) System response without SVC. (b) System response with different SVC auxiliary control signals. (From IEEE Substations Committee, Tutorial on static var compensators, Module 1, *IEEE PES T&D Conference & Exposition*, New Orleans, LA, April 20–22, 2010.)

19.4.3.4 Mitigation of Subsynchronous Resonance

A multiple steam turbine driven synchronous generator connected to a series compensated line can potentially be subject to SSR [24]. An SVC located at the terminals of the generator with a subsynchronous damping controller can effectively suppress torsional oscillations in the frequency range 10–40 Hz [25,26]. In these cases the SVC is essentially a shunt-connected inductance, the current through which is modulated in response to the rotor oscillations or generator speed.

19.4.3.5 Alleviation of Voltage Instability

SVCs with voltage control can provide dynamic reactive power support to transmission systems that are either weak and/or are typically connected to induction motor loads [27]. Such systems can potentially undergo voltage instability due to the import of substantially large amounts of reactive power over the lines especially during transient events. Studies show that simple reactive power support by fixed capacitors in unable to prevent voltage instability in such situations whereas SVCs with dynamic voltage control can effectively enhance voltage stability [27,28].

19.4.3.6 Improvement of HVDC Converter Terminal Performance

HVDC transmission systems connected to weak AC systems present unique problems of stability, temporary overvoltages, and recovery after faults. These problems are related to the reactive power needs of the conventional HVDC converters, which can be as high as 60% of their nominal active power rating [8].

Dynamic reactive power support from SVC at the HVDC terminals can alleviate some of these problems by providing [8,29]

1. Voltage regulation especially during peak power transfers
2. Voltage support during HVDC link recovery subsequent to large disturbances such as faults
3. Suppression of temporary overvoltages typically resulting from load rejections

SVCs, however, do not help increasing the system short-circuit level unlike synchronous condensers, and hence in some cases are not preferred despite their fast response. Also, SVCs have an inherent dead time as they can only be line commutated and must wait for the current to go to zero before a new firing angle can be implemented. Due to this characteristic feature, SVCs are unable to control the first voltage peak that may occur within their dead time of typically a quarter cycle [14,29].

19.4.3.7 Grid Integration of Wind Power Generation Systems

Wind farms utilizing self-excited induction generators need external reactive power support for satisfactory operation specially during starting operations and recovery from faults. If the AC system to which they are connected is weak, the wind farm may not recover from line faults and may adversely impact neighboring generators [30,31]. SVCs are installed close to several such wind farms for dynamic reactive power support and voltage regulation [32].

19.5 Thyristor-Controlled Series Compensation

A thyristor-controlled series compensation (TCSC) is a series-controlled capacitive reactance controller that can provide continuous control of power on a transmission line [33–36]. The basic TCSC module consists of a series capacitor C in parallel with a TCR L_S as shown in Figure 19.14a. The TCR is controlled such that its current increases the effective voltage across the fixed series capacitor. This enhanced voltage changes the effective value of the series capacitive reactance with respect to the same line current. An actual TCSC module also includes protective equipment normally installed with series capacitors, such as metal oxide varistor (MOV), circuit breaker (CB), and an ultrahigh speed contact (UHSC) across the valve as depicted in Figure 19.14b [5].

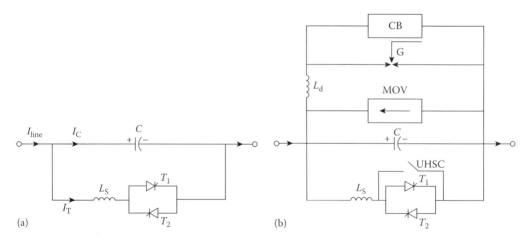

FIGURE 19.14 TCSC module: (a) basic module and (b) practical module. (From Mathur, R.M. and Varma, R.K., *Thyristor-Based FACTS Controllers for Electrical Transmission Systems*, Wiley-IEEE Press, New York, February 2002.)

19.5.1 Operating Principle

There are basically three modes of TCSC operation [5,36] described in the following and illustrated in Figure 19.15.

1. Bypassed-thyristor mode
   The thyristors are made fully conducting with a conduction angle of 180°, and the TCSC module behaves like a parallel combination of capacitor and inductor. However, the net current through the module is inductive since the susceptance of reactor is chosen to be greater than that of the capacitor. The bypassed-thyristor mode is employed for control purposes and also for limiting short-circuit currents.
2. Blocked-thyristor mode
   In this mode, the firing pulses to the thyristor valves are blocked and the thyristors turn off as soon as the current through them goes through a zero. The TCSC module is thus reduced to a fixed series capacitor and the net TCSC reactance is capacitive. This mode is also termed as "waiting" mode. In this mode, the DC-offset voltages of the capacitors are monitored and quickly discharged utilizing a DC offset control [5].
3. Partially conducting thyristor mode or Vernier mode
   This mode of operation allows the TCSC to behave either as a continuously controllable capacitive reactance or as a continuously controllable inductive reactance. It is achieved by varying the thyristor pair firing angle in an appropriate range. A smooth transition from capacitive to inductive mode is, however, not permitted owing to the resonant region between the two modes as depicted in Figure 19.16. The firing angle α of the forward-looking thyristor is limited within the range $\alpha_{min} \leq \alpha \leq 180°$. This provides a continuous vernier control of TCSC module reactance. The loop current increases as α is decreased from 180° to α_{min}. The maximum TCSC reactance permissible with $\alpha = \alpha_{min}$ is typically 2.5–3 times the capacitor reactance at fundamental frequency.

The TCSC can also be operated in an "inductive vernier mode" by having a high level of thyristor conduction. In this mode, the circulating current reverses its direction and the TCSC presents an overall inductive impedance.

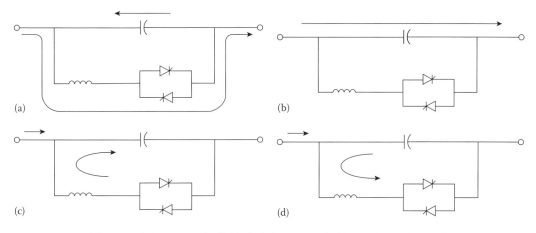

FIGURE 19.15 (a) Bypassed-thyristor mode, (b) blocked-thyristor mode, (c) partially conducting thyristor (capacitive vernier mode), and (d) partially conducting thyristor (inductive vernier mode). (From Mathur, R.M. and Varma, R.K., *Thyristor-Based FACTS Controllers for Electrical Transmission Systems*, Wiley-IEEE Press, New York, February 2002.)

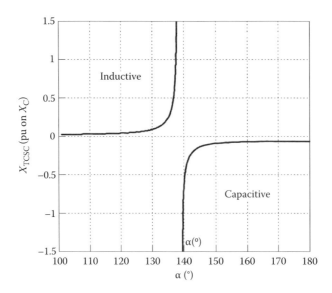

FIGURE 19.16 Reactance characteristic of a TCSC. (From Mathur, R.M. and Varma, R.K., *Thyristor-Based FACTS Controllers for Electrical Transmission Systems*, Wiley-IEEE Press, New York, February 2002.)

The TCSC voltage, line current, capacitor current, TCR current, and valve voltage for the capacitive vernier mode and inductive vernier mode are depicted in Figure 19.17a and b, respectively. The TCSC voltage in inductive mode operation has more harmonics, yet the line current is sinusoidal [5].

Although both SVC and TCSC have a capacitor connected in parallel with a TCR, the fundamental difference between them is that the SVC voltage is sinusoidal but the SVC current is not, whereas the TCSC line current is sinusoidal and TCSC voltage is not. In TCSC, the harmonics tend to circulate within the FC-TCR loop and do not propagate into the network.

The TCSC provides both continuously controllable capacitive and inductive operation but is unable to transition smoothly from one mode to the other due to the resonance region. This can be obviated by splitting a single TCSC into more than one module and independently operating them in capacitive and inductive modes [35,36]. The reactance–current capability curves for a TCSC with different number of modules are illustrated in Figure 19.18. Although the continuously controllable range increases for the same total MVA rating of TCSC, the cost also increases.

The TCSC operates on either open loop control or closed loop control [37,38]. In the open loop control, the TCSC reactance can be varied to provide a desired level of series compensation to achieve power flow control. In the closed loop constant current control, the TCSC reactance is varied to maintain the line current at a desired magnitude. The TCSC can also be controlled to maintain a constant angular difference across a transmission line [11].

19.5.2 TCSC Applications

19.5.2.1 Improvement of System Power Transfer Capacity

The TCSC is usually employed in conjunction with fixed series capacitors. This presents a comparatively lower cost alternative than a totally TCSC implementation to achieve a similar level of series compensation. Unlike SVC, there is a greater flexibility in locating a TCSC in a line for providing series compensation. The TCSC can rapidly change the series compensation level of a transmission line, thus resulting in a desired increased magnitude of power flow. For a system similar to that shown in Figure 19.2 and having a TCSC compensated line, the power flow P_{12} is given by

Flexible AC Transmission Systems (FACTS)

$$P_{12} = \left[\frac{(V_1 V_2)}{(X - X_C)}\right] \sin \delta \qquad (19.6)$$

where X_C is the controlled TCSC reactance together with fixed series capacitor reactance.

19.5.2.2 Enhancement of System Damping

The reactance of the TCSC can be controlled in response to system oscillations such that the net damping of those oscillatory modes gets increased. The principle of damping is similar to that described for SVC. The supplementary signals that could be employed for modulating the impedance of TCSC are

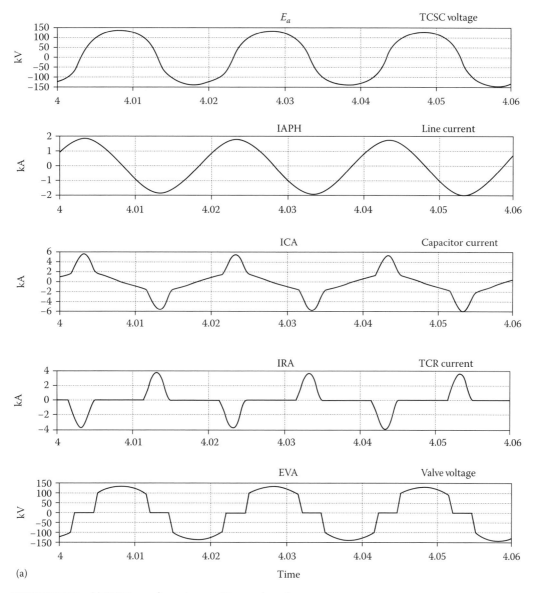

FIGURE 19.17 (a) TCSC waveforms in capacitive mode and

(*continued*)

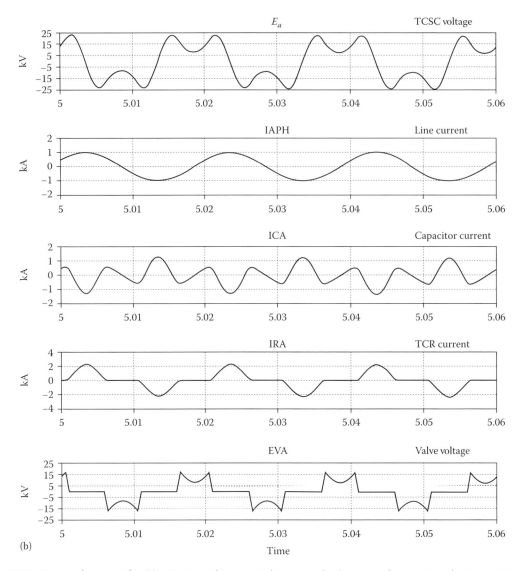

FIGURE 19.17 (continued) (b) TCSC waveforms in inductive mode. (From Mathur, R.M. and Varma, R.K., *Thyristor-Based FACTS Controllers for Electrical Transmission Systems*, Wiley-IEEE Press, New York, February 2002.)

also similar as for SVC, except that bus voltage can also be used a modulating signal for TCSC [5]. The TCSCs are very effective in damping interarea oscillations [39]. The damping enhancement by a TCSC is illustrated in Figure 19.19 [5].

19.5.2.3 Mitigation of Subsynchronous Resonance

Although TCSC provides capacitive compensation at fundamental system frequency, it presents an inherent resistance and inductive reactance at subsynchronous frequencies, which not only detunes the SSR condition but also helps in damping the subsynchronous oscillations [40]. The SSR mitigating capability of the TCSC can be further increased if reactance modulation schemes based on line power or line current are employed [5].

Flexible AC Transmission Systems (FACTS)

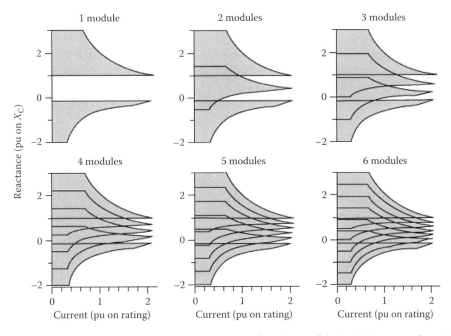

FIGURE 19.18 Reactance current capability characteristics of a multi-module TCSC. (From Mathur, R.M. and Varma, R.K., *Thyristor-Based FACTS Controllers for Electrical Transmission Systems*, Wiley-IEEE Press, New York, February 2002.)

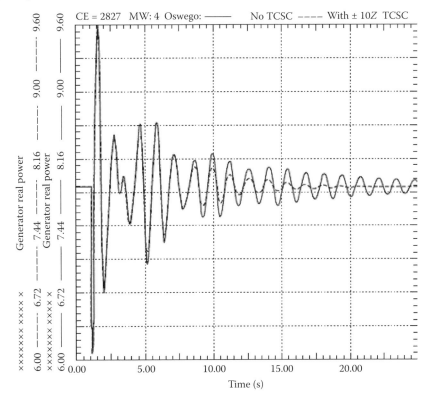

FIGURE 19.19 Damping enhancement by a TCSC. (From Mathur, R.M. and Varma, R.K., *Thyristor-Based FACTS Controllers for Electrical Transmission Systems*, Wiley-IEEE Press, New York, February 2002.)

19.5.2.4 Prevention of Voltage Instability

TCSCs can substantially enhance the loadability of transmission networks, thus avoiding voltage instabilities at existing power transfer levels [28]. While reducing line reactance, the TCSC also generates increased amount of reactive power with line current, which influences bus voltages positively.

The other FACTS controllers in the thyristor-based FACTS category [4,11] are TCPAR [4,41], TCVL, and TCBR, although these have not been employed in practical applications as SVC and TCSC.

19.6 Static Synchronous Compensator

A STATCOM provides continuously controlled reactive power generation as well as absorption purely by means of electronic processing of voltage and current waveforms in a VSC [42,43]. Physical capacitor banks and shunt reactors are not needed for generation and absorption of reactive power, giving STATCOM a compact design, a small footprint, as well as low harmonic noise and low magnetic impacts.

19.6.1 Operating Principle

A single line STATCOM power circuit is shown in Figure 19.20a. A capacitor is typically connected at the DC side to provide a constant DC voltage. The VSC is connected to utility bus through a transformer. As shown in Figure 19.20b the STATCOM can be visualized as an adjustable voltage source behind a reactance, much like a synchronous condenser [43].

Reactive power exchange between the converter and the AC system can be controlled by varying the amplitude of the three-phase output voltage, E_s, of the converter as illustrated in Figure 19.20c. If the amplitude of the output voltage is increased above that of the utility bus voltage, E_t, then a current flows through the reactance from the converter to the AC system, and the converter generates capacitive reactive power for the AC system. If the amplitude of the output voltage is decreased below the utility bus voltage, then the current flows from the AC system to the converter, and the converter absorbs inductive reactive power from the AC system. The corresponding voltage waveforms are depicted in Figure 19.21.

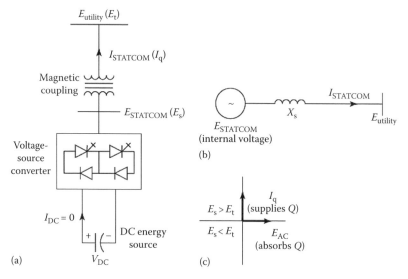

FIGURE 19.20 (a) STATCOM power circuit, (b) STATCOM equivalent circuit, and (c) principle of reactive power exchange. (From Mathur, R.M. and Varma, R.K., *Thyristor-Based FACTS Controllers for Electrical Transmission Systems*, Wiley-IEEE Press, New York, February 2002.)

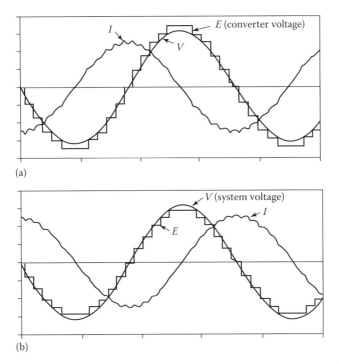

FIGURE 19.21 (a) STATCOM capacitive operation and (b) STATCOM inductive operation. (From Erinmez, I.A. and Foss, A.M., (eds.), Static synchronous compensator (STATCOM), Working Group 14.19, *CIGRE Study Committee 14*, Document No. 144, August 1999.)

If the output voltage equals the AC system voltage, the reactive power exchange becomes zero, and the STATCOM is said to be in a floating state.

A STATCOM provides the desired reactive power by exchanging the instantaneous reactive power among the phases of the AC system. The mechanism by which the converter internally generates and/or absorbs the reactive power can be understood by considering the relationship between the output and input powers of the converter. The converter switches connect the DC input circuit directly to the AC output circuit. Therefore, the net instantaneous power at the AC output terminals must always be equal to the net instantaneous power at the DC input terminals (neglecting losses). Assume that the converter is operated to supply only reactive output power. In this case, the real power provided by the DC source as input to the converter has to be zero. The DC source supplies no reactive power as input to the converter, that is, plays no part in the generation of reactive output power by the converter. Thus, the converter simply interconnects the three output terminals in such a manner that the reactive output currents can flow freely among them and that a circulating reactive power exchange is established among the phases [5].

The real power that the converter exchanges at its AC terminals with the AC system must be supplied to or absorbed from its DC terminals by the DC capacitor. Adjusting the phase shift between the converter output voltage and the AC system voltage can similarly control real power exchange between the converter and the AC system. If a DC energy storage system such as battery is connected at the DC side instead of DC capacitor, the converter can supply real power to the AC system from its DC energy storage if the converter output voltage angle is made to lead that of the AC system voltage. It can absorb real power from the AC system for the DC system if its voltage angle lags the AC system voltage angle. It is necessary to have a DC capacitor connected across the input terminals of the converter to primarily provide a circulating current path as well as act as a voltage source. The magnitude of the capacitor is chosen such that the DC voltage across its terminals remains fairly constant during the entire operation.

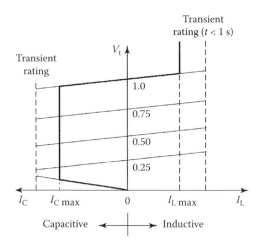

FIGURE 19.22 *V–I* characteristic of a STATCOM. (From Mathur, R.M. and Varma, R.K., *Thyristor-Based FACTS Controllers for Electrical Transmission Systems*, Wiley-IEEE Press, New York, February 2002.)

The VSC has the same rated current capability when operating with capacitive or inductive reactive current. Therefore, a VSC having a certain MVA rating gives the STATCOM twice the dynamic range in Mvar [44–46].

A typical *V–I* characteristic of a STATCOM is depicted in Figure 19.22 [5]. As can be seen, the STATCOM can supply both the capacitive and the inductive compensation and is able to independently control its output current over the rated maximum capacitive or inductive range irrespective of the amount of the AC system voltage. The STATCOM can provide full capacitive reactive power at any practical system voltage. This characteristic of a STATCOM reveals another strength of this technology: it is capable of yielding full output of capacitive generation almost independently of the system voltage (constant current output at lower voltages). This is particularly useful in situations where the STATCOM is needed to support the system voltage during recovery after faults.

19.6.2 STATCOM Applications

The STATCOM is used for all the applications for which SVC is employed as described in Section 19.4.3 [43]. The performance of STATCOM can be comparatively faster and better than SVC due to the turn-on and turn-off capability of its switching devices (IGBT or GTO), as compared to the only turn-on capability of thyristors.

19.7 Static Series Synchronous Compensator

Static synchronous series compensator, commonly known as SSSC or S³C, is a series-connected synchronous voltage source, which can vary the effective impedance of a transmission line by injecting a voltage with an appropriate phase angle in relation to the line current [47,48]. The SSSC comprises a multi-pulse VSC with a DC capacitor or a DC energy storage system as depicted in Figure 19.23a. The SSSC is connected in series with the transmission line. The valve-side voltage rating is higher than the line-side voltage rating of the coupling transformer to reduce the required current rating of the self-commutating valves. Also, the valve-side winding is delta connected to provide a path for third harmonics to flow. Solid-state switches are provided on the valve side to bypass the VSC during situations of very large current flow in the transmission line or when the VSC is inoperative. The basic DC voltage for conversion to AC is provided by the capacitor. The DC–AC conversion is achieved by pulse width

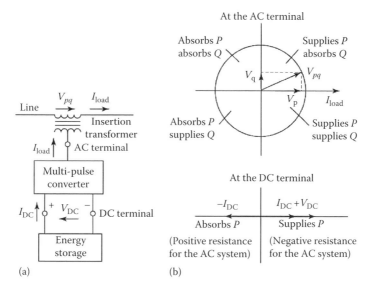

FIGURE 19.23 (a) SSSC power circuit and (b) SSSC operating modes. (From Mathur, R.M. and Varma, R.K., *Thyristor-Based FACTS Controllers for Electrical Transmission Systems*, Wiley-IEEE Press, New York, February 2002.)

modulation techniques [11]. The DC capacitor rating is chosen to minimize the ripple in DC voltage. An MOV is installed across the DC capacitor to limit its voltage and provide protection to the valves.

19.7.1 Operating Principle

The operating modes of SSSC are graphically illustrated in Figure 19.23b. The SSSC has the capability of exchanging both real and reactive power with the transmission system [48,49]. For instance, if the injected voltage is in phase with the line current, then it exchanges real power. However, if a voltage is injected in quadrature with the line current, then reactive power is exchanged—either absorbed or generated.

19.7.2 SSSC Applications

19.7.2.1 Power Flow Control

SSSC can very effectively control power flow in a transmission line due to its ability to vary the voltage drop in a transmission line by injecting suitable quadrature voltages [10,49].

19.7.2.2 Damping of Power Oscillations

SSSC emerges as a potentially more beneficial controller in comparison to a TCSC. This is because of its ability to not only modulate the line reactance but also the line resistance in consonance with the power swings, thereby imparting an enhanced damping to the generators participating in the power oscillations.

This concept of damping control is described in the following [47]:

- If $d\delta/dt > 0$, that is, the generators are accelerating, the SSSC output is controlled to increase the power drawn from the generator, thereby decreasing its kinetic energy. This is achieved by enhancing the series capacitive compensation provided by SSSC.
- If $d\delta/dt < 0$, that is, the generators are decelerating, the SSSC output is controlled to provide an inductive compensation in the line to decrease the power transmitted from the generator.

If the SSSC is equipped with a DC energy storage device, it can exchange real power with the system. An additional mode of damping control becomes possible.

- If $d\delta/dt > 0$, that is, the generators accelerate, SSSC is controlled to absorb real power from the system. This effectively introduces a positive (apparent) resistance in the transmission network.
- If $d\delta/dt < 0$, that is, the generators decelerate, the SSSC is controlled to inject real power into the system. This is equivalent to the introduction of a negative (virtual) resistance in the transmission circuit.

19.7.2.3 Mitigation of SSR

The SSSC is a unique device in that it provides capacitive series compensation to reduce the line series reactance and thereby assists in improving power transfer capacity. However, since its reactive compensation is introduced by an appropriately injected quadrature voltage and not by a physical capacitor it does not introduce SSR.

19.7.2.4 Alleviation of Voltage Instability

The SSSC can rapidly reduce the line series impedance by appropriate quadrature voltage injection and thereby obviate the possibility of voltage instability.

The convertible static compensator at NYPA is an existing installation that operates as an SSSC much of the time [4,50,51].

19.8 Unified Power Flow Controller

The UPFC is the most versatile FACTS Controller developed so far with all encompassing capabilities of voltage regulation, series compensation and phase shifting. It can independently control both the real and reactive power flows in a transmission line at an extremely rapid rate [52–55].

The UPFC is configured as shown in Figure 19.24 [5]. It comprises two VSCs coupled through a common DC terminal. One VSC, converter 1, is connected in shunt with the line through a coupling transformer and the other VSC, converter 2, is inserted in series with the transmission line through an interface transformer. The DC voltage for both converters is provided by a common DC capacitor. The series converter is controlled to inject a voltage V_{pq} in series with the line, which can be varied between 0 and V_{pqmax}. Moreover, the phase angle of the phasor V_{pq} can be independently varied between 0° and 360°. In this process the series converter exchanges both real and reactive power with the

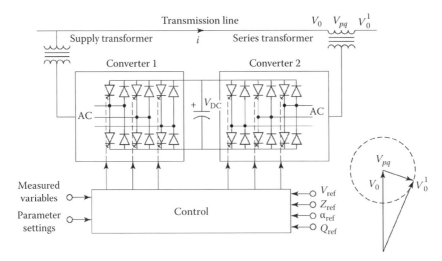

FIGURE 19.24 Implementation of a UPFC with two VSCs. (From Mathur, R.M. and Varma, R.K., *Thyristor-Based FACTS Controllers for Electrical Transmission Systems*, Wiley-IEEE Press, New York, February 2002.)

transmission line. While the reactive power is internally generated/absorbed by the series converter, the real power generation/absorption is made feasible by the DC energy storage device, that is, the capacitor.

The shunt-connected converter 1 is mainly used to supply the real power demand of converter 2, which it derives from the transmission line itself. The shunt converter maintains the voltage of the DC bus constant. Thus, the net real power drawn from the AC system is equal to the losses of the two converters and their coupling transformers. In addition, the shunt converter functions like a STATCOM and regulates the terminal voltage of the interconnected bus by generating/absorbing requisite amount of reactive power.

19.8.1 Operating Principle

The realization of various power flow control functions by UPFC is illustrated in Figure 19.25 [4]. Figure 19.25a depicts the addition of a general voltage phasor V_{pq} to the existing bus voltage V_0 at an angle varying between 0° and 360°. Voltage regulation is effected if V_{pq} (=ΔV_0) is generated in phase with V_0, as shown in Figure 19.25b. A combination of voltage regulation and series compensation is implemented in Figure 19.25c where V_{pq} is the sum of a voltage-regulating component ΔV_0 and a series compensation providing voltage component V_c that lags the line current by 90°. The phase shifting process is illustrated in Figure 19.25d where the UPFC generated voltage V_{pq} is a combination of voltage regulating component ΔV_0 and a phase shifting voltage component V_α. The function of V_α is to change the phase angle of the regulated voltage phasor $V_0 + \Delta V$, by an angle α. A simultaneous attainment of all the aforementioned three power flow control functions is depicted in Figure 19.26. The controller of UPFC can select either one or a combination of the aforementioned three functions as its control objective, depending upon system requirements.

19.8.2 UPFC Applications

The UPFC being the most versatile FACTS Controller is capable of performing all the functions of different FACTS Controllers described earlier, very efficiently [56–59]. However, in this section a case study of its superior power flow control and damping enhancement capability in a system shown in Figure 19.27 is presented [5]. Two areas exchange power through two transmission lines of unequal power transfer capacity, one of which operates at 345 kV and the other at 138 kV. The power transmission capability is determined by the transient stability considerations of the 345 kV line. The UPFC is installed in the 138 kV network. A three-phase-to-ground fault is applied on the 345 kV line for four cycles and the line is disconnected after the fault. The maximum stable power flow possible in the 138 kV line without UPFC is shown in Figure 19.28 to be 176 MW. However, with UPFC the power transfer can be increased to 357 MW.

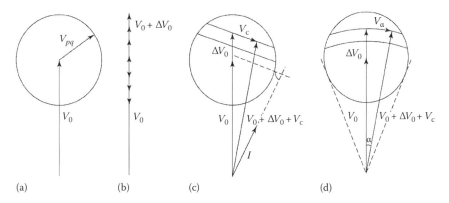

FIGURE 19.25 Phasor diagram showing the general concepts of series voltage injection. (a) Series-voltage injection, (b) terminal-voltage regulation, (c) terminal-voltage and line-impedance regulation, and (d) terminal-voltage and phase-angle regulation. (From Hingorani, N.G. and Gyugyi, L., *Understanding FACTS*, IEEE Press, New York, 1999; Mathur, R.M. and Varma, R.K., *Thyristor-Based FACTS Controllers for Electrical Transmission Systems*, Wiley-IEEE Press, New York, February 2002.)

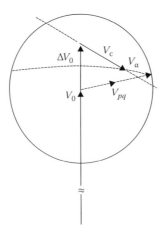

FIGURE 19.26 Simultaneous regulation of terminal voltage, line impedance, and phase angle by appropriate series voltage injection. (From Hingorani, N.G. and Gyugyi, L., *Understanding FACTS*, IEEE Press, New York, 1999; Mathur, R.M. and Varma, R.K., *Thyristor-Based FACTS Controllers for Electrical Transmission Systems*, Wiley-IEEE Press, New York, February 2002.)

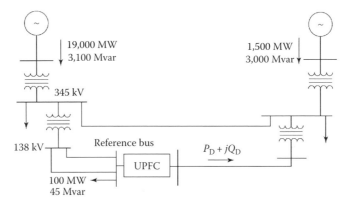

FIGURE 19.27 Study system with UPFC. (From Mathur, R.M. and Varma, R.K., *Thyristor-Based FACTS Controllers for Electrical Transmission Systems*, Wiley-IEEE Press, New York, February 2002.)

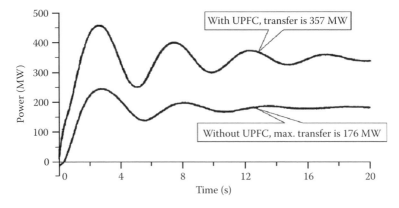

FIGURE 19.28 Increase of power transfer and improvement in power oscillation damping with UPFC damping control. (From Mathur, R.M. and Varma, R.K., *Thyristor-Based FACTS Controllers for Electrical Transmission Systems*, Wiley-IEEE Press, New York, February 2002.)

19.9 FACTS Controllers with Energy Storage

There are two main FACTS Controllers in this category: SMES [60,61] and BESS [60,62]. An energy storage device is interfaced to the DC bus of a VSC, possibly through a secondary power conversion stage.

19.9.1 Superconducting Magnetic Energy Storage

SMES involves a current circulating in a superconducting coil, which is converted to a DC voltage and applied to the DC side of the VSC. This must be achieved through additional power electronics. In SMES, a high amount of energy can be made available to the power system over a very short period of time.

Following are the major applications for SMES:

1. Power quality improvement
    a. The power system performance is substantially improved if the power loss from transmission outages can immediately be replenished by energy obtained from the SMES.
2. Power system damping enhancement
    a. The damping of electromechanical oscillations and transient disturbances can be effectively attained with SMES through rapid extraction/injection of real power from/to the system.

19.9.2 Battery Energy Storage System

BESS is similar to the SMES except that it may not require the additional power electronics on the DC side. The battery provides an energy source for the STATCOM that can also charge the battery when energy from it is not required.

Applications for BESS are similar to SMES. Other applications include

1. Black start
    a. If sufficient stored energy is available, BESS can be applied to generate voltage to which some load can be connected and some nearby generators can be synchronized.
2. Peak shaving
    a. For loads having a high peak value and a poor load factor, energy storage can be an effective means to supply power at peak loads. This is especially true with generation systems whose energy source is intermittent or dependent on uncontrollable factors such as wind farms assuming that adequate energy storage capacity is available for this purpose.
3. Wind farm storage
    a. With intermittent sources such as wind farms, BESS can be applied to store electricity when it is cheap and release it when electricity prices are more profitable.
4. Delaying new transmission facilities
    a. In situations where radial feeders supply growing loads, a BESS located near the load can store energy overnight from the transmission feeder and release it to the load during the day when the transmission capacity would likely be exceeded. Also, the VSC interfacing the battery energy storage can provide voltage support. In this manner, new transmission facilities can be delayed or avoided.
5. Rapid changes in power schedule
    a. Some power systems do not have any substantial fast responding generators such as hydroelectric plant. The increasing demands of electricity markets may need fast schedule changes at the beginning and end of a transaction period. Battery energy storage can provide the fast schedule allowing enough time for the relatively slowly responding online generators to modify their power outputs.

19.10 Coordinated Control of FACTS Controllers

19.10.1 Coordination between Multiple FACTS Controllers

FACTS Controllers may tend to adversely interact due to their fast controls. Usually controls are tuned optimally assuming the remaining power system to be passive. However, these control parameters may not be optimal when dynamics of other controllers are existent such as other FACTS Controllers, HVDC systems, or power system stabilizers (PSS) [63–68]. For instance, independently optimized control systems of two FACTS Controllers may result in an unstable operation when the two FACTS Controllers are made to operate together in a system [65]. Control coordination, therefore, implies simultaneous tuning of different controllers (operating with similar speed of response) to result in an overall positive improvement in their control performance. Control interactions may occur between the following devices:

- Control systems of individual FACTS
- FACTS controls and network
- FACTS controls and high voltage direct current (HVDC) links

Further, these interactions are a function of network strength and choice of controller parameters. Some of these interactions can only be detected by electric magnetic transients program (EMTP) and not by conventional eigenvalue programs, which model network by algebraic equations. It is, therefore, important to study the potential of adverse interactions between these fast responding FACTS Controllers and optimize their control systems in a simultaneous and coordinated manner.

19.10.2 Coordination with Conventional Equipment for Long-Term Voltage-var Management

19.10.2.1 Coordination Concepts

The primary control objective of an SVC or STATCOM is to support the bus voltage to which it is connected by injecting or absorbing reactive power. This is accomplished by a regulator using bus-measurement feedback, typically bus terminal voltage. The typical step-response time of such a FACTS Controller for this primary function of voltage control is on the order of 50 ms (time from step input to steady state).

As an example, Figure 19.29 is a primary control of a STATCOM [69]. The figure illustrates a primary control that has two main portions, namely, an automatic voltage regulator (AVR) with bus-voltage feedback, and an automatic reactive power regulator (AQR) with a reactive-power-output feedback, along with associated limiters. Also shown in this figure is an AVR with an available input for an auxiliary voltage signal, such as for a power swing damping control and an auxiliary input for the AQR,

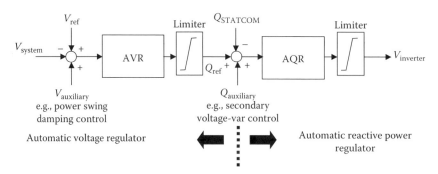

FIGURE 19.29 Functional block diagram of the primary control of a STATCOM. (From Paserba, J.J., Secondary voltage-var controls applied to static compensators (STATCOMs) for fast voltage control and long term var management, *Proceedings of the IEEE PES Summer Power Meeting*, Chicago, IL, July 2002.)

Flexible AC Transmission Systems (FACTS)

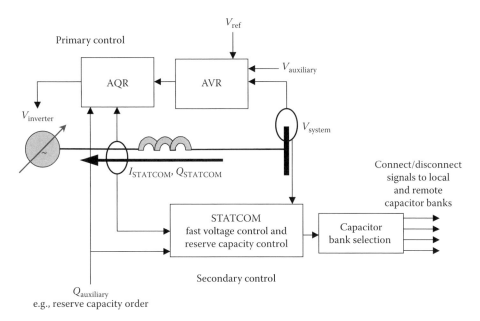

FIGURE 19.30 Functional diagram of the primary and secondary control of a STATCOM (similar concepts can be applied to an SVC). (From Paserba, J.J., Secondary voltage-var controls applied to static compensators (STATCOMs) for fast voltage control and long term var management, *Proceedings of the IEEE PES Summer Power Meeting*, Chicago, IL, July 2002.)

which can be used for a coordination function for local and remote capacitor banks for fast voltage control and long-term var management, as discussed in the following.

The main purpose of secondary controls applied to an SVC or STATCOM is to ensure that the compensator maintains an adequate range of dynamic capability for major system disturbances. The output of the secondary controls calls for the switching of capacitor banks to "reset" the reactive power output of the shunt reactive compensator to a prespecified level after a system event (long term), or during the course of a daily load cycle (long term), or during an event for voltage control (fast). The concept of the primary and secondary control is illustrated in Figure 19.30.

Reference [70] discusses the concept of coordinating a shunt compensator with local voltage-var control devices such as load-tap changers (LTCs) and capacitor banks, for long-term voltage-var management. Reference [70] introduced the concepts of long-term voltage-var management for any one of the following three objectives:

- Resetting a shunt compensator by a simple reactive power runback function so that it would be available for the "next" dynamic event on the system
- Improving the overall system voltage profile by coordinating the shunt compensator with local LTCs and/or capacitor banks
- Reducing LTC tap movements by coordinating the shunt compensator with local LTCs and/or capacitor banks

Reference [70] discusses the advantages and disadvantages of applying secondary controls to STATCOMS for each of the aforementioned objectives.

19.10.2.2 Example Installation

19.10.2.2.1 Description of a STATCOM System

A STATCOM was installed to provide compensation for heavy increases in summertime electric usage, which have rendered the existing system increasingly vulnerable to disturbances. The requirements

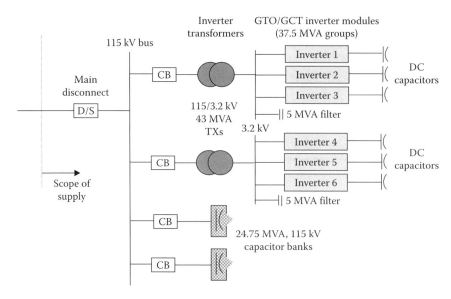

FIGURE 19.31 STATCOM system one-line diagram (CB, circuit breaker; D/S, disconnect switch). (From Paserba, J.J., Secondary voltage-var controls applied to static compensators (STATCOMs) for fast voltage control and long term var management, *Proceedings of the IEEE PES Summer Power Meeting*, Chicago, IL, July 2002.)

(i.e., the purpose of the STATCOM) can be categorized as dynamic reactive compensation needed for fast voltage support during critical contingencies.

As shown in Figure 19.31, the STATCOM system installation consists of two groups of VSCs (37.5 MVA each) and two sets of shunt capacitors (24.75 Mvar each). Each 37.5 MVA converter group consists of three sets of 12.5 MVA modules plus a 5 Mvar harmonic filter, with a nominal phase-to-phase AC voltage of 3.2 kV and a DC link voltage of 6000 V. The two STATCOM groups are connected to the 115 kV system via two three-phase inverter transformers rated at 43 MVA, 3.2 kV/115 kV.

In addition to the primary control requirements described earlier, there were secondary power system control issues associated with this STATCOM installation. The secondary control issues concerned both reserve capacity control and fast voltage control. Therefore, the STATCOM control is coordinated with several local and remote capacitor banks to perform these secondary control functions. The STATCOM control monitors and switches (in or out) seven other capacitor banks: four local 24.75 Mvar banks at STATCOM substation and three remote 24.75 Mvar banks. There are also provisions built into the controller for two future banks at the STATCOM substation.

19.10.2.2.2 Fast Voltage Control

As illustrated in Figure 19.32, the secondary control function for fast voltage control monitors the voltage error of the STATCOM from the primary control (AVR), and if the error exceeds a threshold for a specified time, then a connect (for low voltage conditions) or disconnect (for high voltage conditions) signal is given. The panel of the STATCOM controller for the fast voltage control is shown in Figure 19.33. This figure shows that the available settings are for the voltage error (typically ±2%), a time for how long the voltage error must be exceeded (typically seconds), and a time interval before a subsequent switch signals can be given (typically tens of seconds or a few minutes). There are separate timer settings for connect and disconnect control actions.

Since the monitored voltage error is based on the STATCOM substation, this fast voltage control is primarily for severe system conditions when the STATCOM is pushed into its limits. Thus, an action of capacitor bank switching can move the STATCOM back into its controllable range.

Flexible AC Transmission Systems (FACTS)

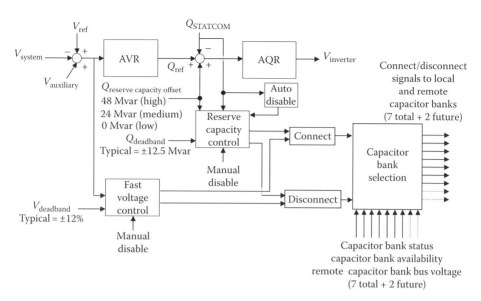

FIGURE 19.32 Functional block diagram of the overall voltage-var control for the STATCOM. (From Paserba, J.J., Secondary voltage-var controls applied to static compensators (STATCOMs) for fast voltage control and long term var management, *Proceedings of the IEEE PES Summer Power Meeting*, Chicago, IL, July 2002.)

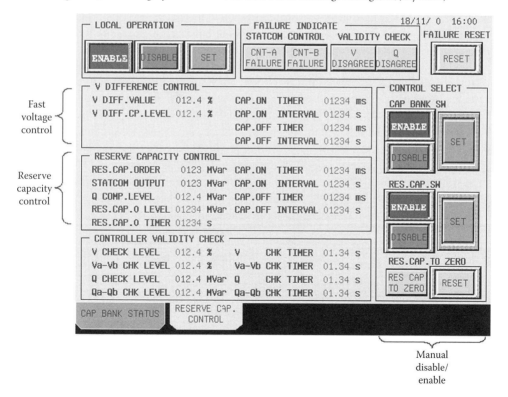

FIGURE 19.33 Secondary control panel (fast voltage control and reserve capacity control) of the STATCOM controller (note all input values are shown here as "01234..." before factory settings were in place. Values shown in this panel for the STATCOM secondary control are settable by the utility). (From Paserba, J.J., Secondary voltage-var controls applied to static compensators (STATCOMs) for fast voltage control and long term var management, *Proceedings of the IEEE PES Summer Power Meeting*, Chicago, IL, July 2002.)

19.10.2.2.3 Reserve Capacity Control

The reserve capacity control is designed to enable the operating point of the STATCOM inverters to be offset into the inductive region so that a desired "net capacitive range" or "reserve capacity" can be achieved. Reserve capacity is defined as the available net change in STATCOM inverter output toward the capacitive region from a given operating point. For example, if the STATCOM inverters are operating with zero net output, the reserve capacity will be equal to the maximum output rating of the inverters (75 Mvar). If the operating point is biased into the inductive region, for example, to 24 or 48 Mvar inductive, then the reserve capacity will be 99 or 123 Mvar, respectively. The reserve capacity of the STATCOM can be selected by the operator to one of three positions: high, medium, and low, which add inductive offsets of 48, 24, and 0 Mvar, respectively, to the operating setpoint of the STATCOM. This is illustrated in Figure 19.32.

The desired reserve capacity is a function of the system loading conditions with generally higher reserve capacity (i.e., more biasing into the inductive region) being required under heavy load conditions. Under light load conditions the system requirements for reserve capacity are lower and it is advantageous to operate the STATCOM at the low or medium reserve capacity settings to reduce the losses. The reserve capacity requirement is achieved by automatically connecting or disconnecting shunt capacitors at the local STATCOM and remote substations.

The panel for the STATCOM controller for the reserve capacity control is shown in Figure 19.33. The capacitor banks selection logic is discussed in the next subsection.

19.10.2.2.4 Capacitor Bank Selection

The STATCOM secondary controls (fast voltage control or the reserve capacity control) sends a signal when a capacitor bank switching event (connect or disconnect) is being requested. The algorithm adopted for the STATCOM first switches all capacitor banks at the local substation with the "first-on/last-off" logic. For the remote capacitor banks at the remote substations, they are switched on or off based on their bus voltage (e.g., lowest voltage on first, highest voltage off first). If a selected capacitor bank is already online at the specified substation or is disabled, the selection controller searches for the next one in the hierarchy. The capacitor banks status panel of the STATCOM control is shown in Figure 19.34.

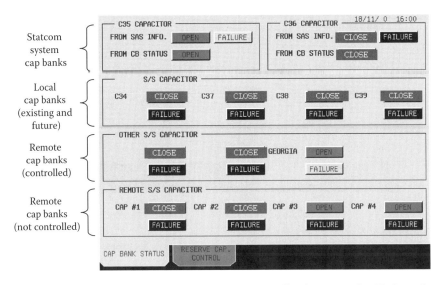

FIGURE 19.34 Capacitor bank status panel of the STATCOM controller. (From Paserba, J.J., Secondary voltage-var controls applied to static compensators (STATCOMs) for fast voltage control and long term var management, *Proceedings of the IEEE PES Summer Power Meeting*, Chicago, IL, July 2002.)

Flexible AC Transmission Systems (FACTS)

There are provisions for control of four capacitor banks currently at the STATCOM substation (two of which are associated with the STATCOM installation) plus two future banks at the local substation, plus for the three remote banks. The information transmitted from the local and remote capacitor bank substations into the selection logic of the STATCOM secondary control, illustrated in Figure 19.32, is as follows:

- Capacitor bank status
- Capacitor bank availability
- Remote capacitor bank bus voltage

To avoid frequent switching of the capacitor banks for the fast voltage control, the capacitor bank selection logic has voltage deadbands, settable by the utility on the STATCOM control panels, as illustrated in Figure 19.33. The initial settings of the controller are as follows:

- Q reserve capacity level: medium (24 Mvar)
- Q control deadband: ±12.5 Mvar
- Q control cap On/Off timer: 10 s
- Q control cap On/Off interval: 55 s
- Q control auto-disable: over +100 Mvar total output (STATCOM plus caps) for 20 s
- V control deadband: ±2%
- V control cap On/Off timer: 10 s
- V control cap On/Off interval: 30 s

19.10.2.3 Other Installations

The concept of coordination of FACTS Controllers with conventional equipment for long-term voltage-var management has been applied at numerous installations of SVC and STATCOM and by multiple manufacturers [71–78].

19.11 FACTS Installations to Improve Power System Dynamic Performance

In today's utility environment, financial and market forces are, and will continue to, demand a more optimal and profitable operation of the power system. To achieve both operational reliability and financial profitability, it has become clear that more efficient utilization and control of the existing transmission system infrastructure is required. Power-electronic-based FACTS Controllers can provide technical solutions to address these new operating challenges being presented today. FACTS Controllers allow for improved transmission system operation with minimal infrastructure investment, environmental impact, and implementation time, compared to the construction of new transmission lines.

References [79–96] provide a decade's worth of utility installations and technology advancement, both large and small, of FACTS Controllers to improve power system dynamic performance.

19.12 Conclusions

FACTS Controllers are very effective in improvement of power system performance. In real systems, FACTS Controllers are mostly utilized in conjunction with conventional passive devices or traditional solutions to make the best use of financial resources while not compromising with the technical merit.

A complete set of bibliographies published on FACTS is provided in Refs. [97–130].

References

1. A. Edris, R. Adapa, M.H. Baker, L. Bohmann, K. Clark, K. Habashi, L. Gyugyi, J. Lemay, A.S. Mehraban, A.K. Myers, J. Reeve, F. Sener, D.R. Torgerson, and R.R. Wood, Proposed terms and definitions for flexible AC transmission system (FACTS), *IEEE Transactions on Power Delivery*, 12(4), 1848–1853, October 1997.
2. IEEE Power Engineering Society/CIGRE, *FACTS Overview*, Publication 95-TP-108, IEEE Press, New York, 1995.
3. IEEE Power Engineering Society, *FACTS Applications*, Publication 96-TP-116-0, IEEE Press, New York, 1996.
4. N.G. Hingorani and L. Gyugyi, *Understanding FACTS*, IEEE Press, New York, 1999.
5. R.M. Mathur and R.K. Varma, *Thyristor-Based FACTS Controllers for Electrical Transmission Systems*, Wiley-IEEE Press, New York, February 2002.
6. Y.H. Song and A.T. Johns, *Flexible AC Transmission Systems (FACTS)*, IEEE Press, London, U.K., 1999.
7. E. Acha, C.R. Fuerte-Esquivel, H. Ambriz-Perez, and C. Angeles-Camacho, *FACTS Modelling and Simulation in Power Networks*, John Wiley & Sons, Ltd., London, U.K., 2004.
8. V.K. Sood, *HVDC and FACTS Controllers: Applications of Static Converters in Power Systems*, Springer, London, U.K., 2004.
9. X.-P. Zhang, C. Rehtanz, and B. Pal, *Flexible AC Transmission Systems: Modelling and Control*, Springer, Berlin, Germany, 2006.
10. K.K. Sen and M.L. Sen, *Introduction to FACTS Controllers: Theory, Modeling, and Applications*, Wiley-IEEE Press, Hoboken, NJ, 2009.
11. K.R. Padiyar, *FACTS Controllers in Power Transmission and Distribution*, New Age International Publishers, New Delhi, India, 2007.
12. IEEE Power Engineering Society, *Application of Static var Systems for System Dynamic Performance*, IEEE Special Publication 87TH0187-5-PWR, IEEE PES, New York, 1987.
13. T.J.E. Miller, *Reactive Power Control in Electric Systems*, John Wiley & Sons, Inc., New York, 1982.
14. I.A. Erinmez, (ed.), *Static var Compensators*, Working Group 38-01, Task Force No. 2 on SVC, CIGRE, Paris, France, 1986.
15. EPRI Report TR 100696, Improved static VAR compensator control, Final Report Project 2707-01, General Electric Company, Schenectady, NY, June 1992.
16. IEEE Substations Committee, Tutorial on static var compensators, Module 1, *IEEE PES T&D Conference & Exposition*, New Orleans, LA, April 20–22, 2010.
17. L. Gerin-Lajoie, G. Scott, S. Breault, E.V. Larsen, D.H. Baker, and A.F. Imece, Hydro-Quebec multiple SVC application control stability study, *IEEE Transactions on Power Delivery*, 5(3), 1543–1551, July 1990.
18. IEEE Special Stability Controls Working Group, Static var compensator models for power flow and dynamic performance simulation, *IEEE Transactions on Power Systems*, 9(1), 229–239, February 1994.
19. J. Belanger, G. Scott, T. Anderson, and S. Torseng, Gain supervisor for thyristor controlled shunt compensators, CIGRE Paper 38-01, August 1984.
20. K.R. Padiyar and R.K. Varma, Damping torque analysis of static var system controllers, *IEEE Transactions on Power Systems*, 6(2), 458–465, May 1991.
21. Electric Power Research Institute (EPRI), Real time phasor measurement for monitoring and control, EPRI Report TR-103640S, December 1993.
22. P. Kundur, *Power System Stability and Control*, McGraw-Hill, Inc., New York, 1994.
23. R. Peng, R.K. Varma, and J. Jiang, New static var compensator (SVC) based damping control using remote generator speed signal, *International Journal of Energy Technology and Policy*, 4(3/4), 255–273, 2006.
24. K.R. Padiyar, *Analysis of Subsynchronous Resonance in Power Systems*, Kluwer Academic Publishers, Boston, MA, 1999.

25. N.C.A. Samra, R.F. Smith, T.E. McDermott, and M.B. Chidester, Analysis of thyristor controlled shunt SSR counter measures, *IEEE Transactions on Power Apparatus and Systems*, 104(3), 584–597, March 1985.
26. D.G. Ramey, D.S. Kimmel, J.W. Dorney, and F.H. Kroening, Dynamic stabilizer verification tests at the San Juan station, *IEEE Transactions on Power Apparatus and Systems*, 100, 5011–5019, December 1981.
27. A.E. Hammad and M.Z. El-Sadek, Prevention of transient voltage instabilities due to induction motor loads by static var compensators, *IEEE Transactions on Power Systems*, 4(3), 1182–1190, August 1989.
28. C.A. Canizares and Z.T. Faur, Analysis of SVC and TCSC controllers in voltage collapse, *IEEE Transactions on Power Systems*, 14(1), 158–165, February 1999.
29. O.B. Nayak, A.M. Gole, D.G. Chapman, and J.B. Davis, Dynamic performance of static and synchronous compensators at an HVDC inverter bus in a very weak AC system, Paper 93 SM 447-3 PWRD, Presented at the *IEEE PES 1993 Summer Meeting*, Vancouver, British Columbia, Canada, July 1993.
30. P. Pourbeik, R.J. Koessler, D.L. Dickmander, and W. Wong, Integration of large wind farms into utility grids (Part 2—Performance issues), *Power Engineering Society General Meeting 2003*, Toronto, Ontario, Canada, Vol. 3, pp. 1520–1525, July 13–17, 2003.
31. R.K. Varma, S. Auddy, and Y. Semsedini, Mitigation of subsynchronous resonance in a series-compensated wind farm using FACTS Controllers, *IEEE Transactions on Power Delivery*, 23(3), 1645–1654, July 2008.
32. S. Irokawa, L. Andersen, D. Pritchard, and N. Buckley, A coordination control between SVC and shunt capacitor for wind farm, *CIGRE 2008*, Paris, France, 2008.
33. E. Larsen, C. Bowler, B. Damsky, and S. Nilsson, Benefits of thyristor controlled series compensation, *International Conference on Large High Voltage Electric Systems*, Paper 14/37/38-04, CIGRE, Paris, France, 1992.
34. CIGRE Working Group 14.18, Thyristor controlled series compensation, Technical Brochure, CIGRE, Paris, France, 1996.
35. E.V. Larsen, K. Clark, S.A. Miske Jr., and J. Urbanek, Characteristics and rating considerations of thyristor controlled series compensation, *IEEE Transactions on Power Delivery*, 9(2), 992–1000, April 1994.
36. J.J. Paserba, N.W. Miller, E.V. Larsen, and R.J. Piwko, A thyristor controlled series compensation model for power system stability analysis, *IEEE Transactions on Power Delivery*, 10(3), 1471–1478, July 95.
37. N. Martins, H.J.C.P. Pinto, and J.J. Paserba, TCSC controls for line power scheduling and system oscillation damping—Results for a small example system, *Proceedings of the 14th Power System Control Conference*, Trondheim, Norway, June 28–July 2, 1999.
38. E.V. Larsen, J.J. Sanchez-Gasca, and J.H. Chow, Concepts for design of FACTS Controllers to damp power swings, *IEEE Transactions on Power Systems*, 10(2), 948–955, May 1995.
39. C. Gama, R.L. Leoni, J. Gribel, R. Fraga, M.J. Eiras, W. Ping, A. Ricardo, J. Cavalcanti, and R. Tenorio, Brazilian north south interconnection—Application of thyristor controlled series compensation (TCSC) to damp inter-area oscillation mode, CIGRE Paper 14-101, Paris, France, 1998.
40. W. Zhu, R. Spee, R.R. Mohler, G.C. Alexander, W.A. Mittelstadt, and D. Maratukulam, An EMTP study of SSR mitigation using the thyristor controlled series capacitor, *IEEE Transactions on Power Delivery*, 10(3), 1479–1485, July 1995.
41. M.R. Iravani and D. Maratukulam, Review of semiconductor-controlled (static) phase shifters for power system applications, *IEEE Transactions on Power Systems*, 9(4), 1833–1839, 1994.
42. L. Gyugyi et al., Advanced static var compensator using gate turn-off thyristors for utility applications, CIGRE Paper 23–203, 1990.
43. I.A. Erinmez and A.M. Foss (eds.), Static synchronous compensator (STATCOM), Working Group 14.19, *CIGRE Study Committee 14*, Document No. 144, August 1999.

44. C. Schauder, M. Gernhardt, E. Stacey, T. Lemak, L. Gyugyi, T. Cease, and A. Edris. TVA STATCON project: Design, installation, and commissioning, CIGRE Paper 14–106, 1996.
45. C.D. Schauder and H. Mehta, Vector analysis and control of advanced static var compensators, *IEE Proceedings C*, 140(4), 299–306, 1993.
46. C.D. Schauder et al., Development of a ±100 MVAR static condenser for voltage control of transmission lines, *IEEE Transactions on Power Delivery*, 10(3), 1486–1493, July 1995.
47. L. Gyugyi, C.D. Schauder, and K.K. Sen, Static synchronous series compensator: A solid-state approach to the series compensation of transmission lines, *IEEE Transactions on Power Delivery*, 12, 406–417, January 1997.
48. K.K. Sen, SSSC-static synchronous series compensator: Theory, modelling and applications, *IEEE Transactions on Power Delivery*, 13(1), 241–246, January 1998.
49. C.J. Hatziadoniu and A.T. Funk, Development of a control scheme for series connected solid state synchronous voltage source, *IEEE Transactions on Power Delivery*, 11(2), 1138–1144, April 1996.
50. L. Gyugyi, K.K. Sen, and C.D. Schauder, The interline power flow controller Concept: A new approach to power flow management in transmission systems, *IEEE Transactions on Power Delivery*, 14(3), 1115–1123, July 1999.
51. B. Faradanesh and A. Schuff, Dynamic studies of the NYS transmission system with the Marcy CSC in the UPFC and IPFC configurations, *IEEE Conference on Transmission and Distribution*, New York, Vol. 3, pp. 1175–1179, September 2003.
52. L. Gyugyi, A unified power flow control concept for flexible AC transmission systems, *IEE Proceedings C*, 139(4), 323–331, July 1992.
53. CIGRE Task Force 14-27, Unified Power Flow Controller, CIGRE Technical Brochure, Paris, France, 1998.
54. L. Gyugyi, C.D. Schauder, S.L. Williams, T.R. Reitman, D.R. Torgerson, and A. Edris, The unified power flow controller: A new approach to power transmission control, *IEEE Transactions on Power Delivery*, 10(2), 1085–1093, April 1995.
55. C.D. Schauder, L. Gyugyi, M.R. Lund, D.M. Hamai, T.R. Reitman, D.R. Torgerson, and A. Edris, Operation of the unified power flow controller (UPFC) under practical constraints, *IEEE Transactions on Power Delivery*, 13(2), 630–639, April 1998.
56. M. Rahman et al., UPFC application on the AEP system: Planning considerations, *IEEE Transactions on Power Systems*, 12(4), 1695–1071, November 1997.
57. B.A. Renz et al., AEP unified power flow controller performance, *IEEE Transactions on Power Delivery*, Paper PE-042—PWRD-0-12-1998, 14(4), 1374–1381, 1999.
58. B.A. Renz et al., World's first unified power flow controller on the AEP system, CIGRE Paper No. 14–107, 1998.
59. K.K. Sen and E.J. Stacey, UPFC–Unified power flow Controller: Theory, modelling and applications, Paper Presented in *IEEE PES 1998 Winter Meeting*, Tampa, FL, 1998. IEEE, New York.
60. P.F. Ribeiro, B.K. Johnson, M.L. Crow, A. Arsoy, and Y. Liu, Energy storage systems for advanced power applications, *Proceedings of the IEEE*, 89(12), 1744–1756, 2001.
61. L. Chen, Y. Liu, A.B. Arsoy, P.F. Ribeiro, M. Steurer, and M.R. Iravani, Detailed modeling of superconducting magnetic energy storage (SMES) system, *IEEE Transactions on Power Delivery*, 21(2), 699–710, 2006.
62. Z. Yang, C. Shen, L. Zhang, M.L. Crow, and S. Atcitty, Integration of a StatCom and battery energy storage, *IEEE Transactions on Power Systems*, 16(2), 254–260, 2001.
63. CIGRE Working Group 14.29, Coordination of controls of multiple FACTS/HVDC links in the same system, CIGRE Technical Brochure 149, Paris, France, December 1999.
64. CIGRE Task Force 38.02.16, Impact of interactions among power system controls, CIGRE Technical Brochure 166, Paris, France, August 2000.
65. EPRI Report TR-109969, Analysis of control interactions on FACTS assisted power systems, Final Report, Electric Power Research Institute (EPRI), Palo Alto, CA, January 1998.

66. E.V. Larsen, D.H. Baker, A.F. Imece, L. Gerin-Lajoie, and G. Scott, Basic aspects of applying SVCs to series compensated AC transmission lines, *IEEE Transactions on Power Delivery*, 5(3), 1466–1473, July 1990.
67. P. Pourbeik and M.J. Gibbard, Simultaneous coordination of power system stabilizers and FACTS device stabilizers in a multimachine power system for enhancing dynamic performance, *IEEE Transactions on Power Systems*, 13(2), 473–479, May 1998.
68. J.J. Sanchez-Gasca, Coordinated control of two FACTS devices for damping interarea oscillations, *IEEE Transactions on Power Systems*, 13(2), 428–434, May 1998.
69. J.J. Paserba, Secondary voltage-var controls applied to static compensators (STATCOMs) for fast voltage control and long term var management, *Proceedings of the IEEE PES Summer Power Meeting*, Chicago, IL, July 2002.
70. J.J. Paserba, D.J. Leonard, N.W. Miller, S.T. Naumann, M.G. Lauby, and F.P. Sener, Coordination of a distribution level continuously controlled compensation device with existing substation equipment for long term var management, *IEEE Transactions on Power Delivery*, 9(2), 1034–1040, April 1994.
71. G. Reed, J. Paserba, T. Croasdaile, M. Takeda, N. Morishima, Y. Hamasaki, L. Thomas, and W. Allard, STATCOM application at VELCO Essex substation, Panel session on FACTS applications to improve power system dynamic performance, *Proceedings of the IEEE PES T&D Conference and Exposition*, Atlanta, GA, October/November 2001.
72. G. Reed, J. Paserba, T. Croasdaile, R. Westover, S. Jochi, N. Morishima, M. Takeda, T. Sugiyama, Y. Hamazaki, T. Snow, and A. Abed, SDG&E Talega STATCOM Project-system analysis, design, and configuration, Panel session on FACTS technologies: Experiences of the past decade and developments for the 21st century in Asia and the world, *Proceedings of the IEEE PES T&D-Asia Conference and Exposition*, Yokahama, Japan, October 2002.
73. D. Sullivan, J. Paserba, G. Reed, T. Croasdaile, R. Pape, D. Shoup, M. Takeda, Y. Tamura, J. Arai, R. Beck, B. Milošević, S.-M. Hsu, and F. Graciaa, Design and application of a static VAR compensator for voltage support in the Dublin, Georgia area, *Proceedings of the IEEE PES T&D Conference and Exposition*, Dallas, TX, May 2006.
74. D.J. Sullivan, J. Paserba, J.J. Reed, G.F. Croasdaile, T. Westover, R. Pape, R. M. Takeda, S. Yasuda, H. Teramoto, Y. Kono, K. Kuroda, K. Temma, W. Hall, D. Mahoney, D. Miller, and P. Henry, Voltage control in southwest Utah with the St. George static var system, *Proceedings of the IEEE PES Power System and Exposition*, Atlanta, GA, October 2006.
75. D. Sullivan, J. Paserba, T. Croasdaile, R. Pape, M. Takeda, S. Yasuda, H. Teramoto, Y. Kono, K. Temma, A. Johnson, R. Tucker, and T. Tran, Dynamic voltage support with the rector SVC in California's San Joaquin Valley, *Proceedings of the IEEE Transmission and Distribution Conference*, Chicago, IL, April 2008.
76. D. Sullivan, R. Pape, J. Birsa, M. Riggle, M. Takeda, H. Teramoto, Y. Kono, K. Temma, S. Yasuda, K. Wofford, P. Attaway, and J. Lawson, Managing fault-induced delayed voltage recovery in metro Atlanta with the barrow county SVC, *Proceedings of the IEEE PES Power Systems Conference and Exposition*, Seattle, WA, March 2009.
77. A. Scarfone, B. Oberlin, J. Di Luca Jr., D. Hanson, C. Horwill, and M. Allen, Dynamic performance studies for a ±150 Mvar STATCOM for northeast utilities, *Proceedings of the IEEE PES Transmission and Distribution Conference and Exposition*, Dallas, TX, September 2003.
78. A. Oskoui, B. Mathew, J.-P. Hasler, M. Oliveira, T. Larsson, Å. Petersson, and E. John, Holly STATCOM–FACTS to replace critical generation, operational experience, *Proceedings of the IEEE PES Transmission and Distribution Conference and Exposition*, Dallas, TX, April 2006.
79. IEEE PES, Power System Dynamic Performance Committee, Panel Session on modeling, simulation, and applications of FACTS Controllers in angle and voltage stability studies, *Proceedings of the IEEE PES 2000 Winter Meeting*, Singapore, January 2000.
80. IEEE PES, Power System Dynamic Performance Committee, Panel session on FACTS applications to improve power system dynamic performance, *Proceedings of the T&D Conference and Exposition*, Atlanta, GA, September 2001.

81. IEEE PES, Power System Dynamic Performance Committee, Panel session on power system stability controls using power electronic devices, *Proceedings of the IEEE PES Summer Power Meeting*, Chicago, IL, July 2002.
82. IEEE PES, Power System Dynamic Performance Committee, Panel session on FACTS technologies: Experiences of the past decade and developments for the 21st century in Asia and the world, *Proceedings of the IEEE PES T&D-Asia Conference and Exposition*, Yokahama, Japan, October 2002.
83. IEEE PES, Power System Dynamic Performance Committee, Panel session on FACTS VSC applications for improving power system performance, *Proceedings of the IEEE PES General Meeting*, Toronto, Ontario, Canada, July 2003.
84. IEEE PES, Power System Dynamic Performance Committee, Panel Session on SVC refurbishment & life extension, *Proceedings of the IEEE PES General Meeting*, Toronto, Ontario, Canada, July 2003.
85. IEEE PES, Power System Dynamic Performance Committee, Panel session on FACTS applications to improve power system dynamic performance, *Proceedings of the T&D Conference and Exposition Meeting*, Dallas, TX, September 2003.
86. IEEE PES, Power System Dynamic Performance Committee, Special technical session on FACTS fundamentals, *Proceedings of the IEEE PES T&D Conference and Exposition*, Dallas, TX, September 2003.
87. IEEE PES, Power System Dynamic Performance Committee, Special technical session on FACTS fundamentals, *Proceedings of the IEEE PES General Meeting*, Denver, CO, June 2004.
88. IEEE PES, Power System Dynamic Performance Committee, Panel session on FACTS/power electronic applications to improve power system dynamic performance, *Proceedings of the T&D Conference and Exposition*, Dallas, TX, April 2006.
89. IEEE PES, Power System Dynamic Performance Committee, Panel session on FACTS/power electronic applications to improve power system dynamic performance, *Proceedings of the Power Systems Conference and Exposition* (*PSCE*), Atlanta, GA, October/November 2006.
90. IEEE PES, Power System Dynamic Performance Committee, Panel session on Planning and implementing FACTS Controllers for improving power system performance, *Proceedings of the IEEE PES General Meeting*, Tampa, FL, June 2007.
91. IEEE PES, Power System Dynamic Performance Committee, Panel session on FACTS/power electronic applications to improve power system dynamic performance, *Proceedings of the T&D Conference and Exposition*, Chicago, IL, April 2008.
92. IEEE PES, Power System Dynamic Performance Committee, Panel session on Network solutions using FACTS, *Proceedings of the Power Systems Conference and Exposition*, Seattle, WA, March 2009. Sponsored by the T&D Committee.
93. IEEE PES, Power System Dynamic Performance Committee, Panel session on FACTS/power electronic applications to improve power system dynamic performance, *Proceedings of the Power Systems Conference and Exposition*, Seattle, WA, March 2009.
94. IEEE PES, Power System Dynamic Performance Committee, Panel session on network solutions using FACTS, *Proceedings of the Transmission and Distribution Conference and Exposition*, New Orleans, LA, April 2010.
95. IEEE PES, Power System Dynamic Performance Committee, Panel session on FACTS/power electronic applications to improve power system dynamic performance, *Proceedings of the Transmission and Distribution Conference and Exposition*, New Orleans, LA, April 2010.
96. IEEE PES, Power System Dynamic Performance Committee, Panel session on FACTS/power electronic installations, *Proceedings of the Power Systems Conference and Exposition*, Phoenix, AZ, March 2011.
97. S.A. Rahman, R.K. Varma, and W.H. Litzenberger, Bibliography of FACTS applications for grid integration of wind and PV solar power systems: 1995–2010, IEEE working group report, *Proceedings of the IEEE PES General Meeting*, Detroit, MI, 2011.

98. J. Berge, R.K. Varma, and W.H. Litzenberger, Bibliography of FACTS 2009–2010—Part I, IEEE working group report, *Proceedings of the IEEE PES General Meeting*, Detroit, MI, 2011.
99. J. Berge, R.K. Varma, and W.H. Litzenberger, Bibliography of FACTS 2009–2010—Part II, IEEE working group report, *Proceedings of the IEEE PES General Meeting*, Detroit, MI, 2011.
100. J. Berge, S.S. Rangarajan, R.K. Varma, and W.H. Litzenberger, Bibliography of FACTS 2009–2010—Part III, IEEE working group report, *Proceedings of the IEEE PES General Meeting*, Detroit, MI, 2011.
101. J. Berge, S.S. Rangarajan, R.K. Varma, and W.H. Litzenberger, Bibliography of FACTS 2009–2010—Part IV, IEEE working group report, *Proceedings of the IEEE PES General Meeting*, Detroit, MI, 2011.
102. I. Axente, R.K. Varma, and W.H. Litzenberger, Bibliography of FACTS 1998: IEEE working group report, *Proceedings of the IEEE PES General Meeting*, Detroit, MI, 2011.
103. I. Axente, R.K. Varma, and W.H. Litzenberger, Bibliography of FACTS 1999: IEEE working group report, *Proceedings of the IEEE PES General Meeting*, Detroit, MI, 2011.
104. I. Axente, R.K. Varma, and W.H. Litzenberger, Bibliography of FACTS 2000—Part 1, IEEE working group report, *Proceedings of the IEEE PES General Meeting*, Detroit, MI, 2011.
105. I. Axente, R.K. Varma, and W.H. Litzenberger, Bibliography of FACTS 2000—Part 2, IEEE working group report, *Proceedings of the IEEE PES General Meeting*, Detroit, MI, 2011.
106. R. Varma, W. Litzenberger, and I. Axente, Bibliography of FACTS: 2006–2007—Part I IEEE working group report, *Proceedings of the IEEE PES General Meeting*, Minneapolis, MN, July 2010.
107. R. Varma, W. Litzenberger, and I. Axente, Bibliography of FACTS: 2006–2007—Part II IEEE working group report, *Proceedings of the IEEE PES General Meeting*, Minneapolis, MN, July 2010.
108. R. Varma, W. Litzenberger, and I. Axente, Bibliography of FACTS: 2006–2007—Part III IEEE working group report, *Proceedings of the IEEE PES General Meeting*, Minneapolis, MN, July 2010.
109. R. Varma, W. Litzenberger, and S.A. Rahman, Bibliography of FACTS: 2008—Part I IEEE working group report, *Proceedings of the IEEE PES General Meeting*, Minneapolis, MN, July 2010.
110. R. Varma, W. Litzenberger, and S.A. Rahman, Bibliography of FACTS: 2008—Part II IEEE working group report, *Proceedings of the IEEE PES General Meeting*, Minneapolis, MN, July 2010.
111. R. Varma, W. Litzenberger, and S.A. Rahman, Bibliography of FACTS: 2008—Part III IEEE working group report, *Proceedings of the IEEE PES General Meeting*, Minneapolis, MN, July 2010.
112. R. Varma, J. Berge, and W. Litzenberger, Bibliography of FACTS 2009—Part 1, IEEE working group report, *Proceedings of the IEEE PES General Meeting*, Minneapolis, MN, July 2010.
113. R. Varma, J. Berge, and W. Litzenberger, Bibliography of FACTS 2009—Part 2, IEEE working group report, *Proceedings of the IEEE PES General Meeting*, Minneapolis, MN, July 2010.
114. R. Varma, J. Berge, and W. Litzenberger, Bibliography of FACTS 2009—Part 3, IEEE working group report, *Proceedings of the IEEE PES General Meeting*, Minneapolis, MN, July 2010.
115. R.K. Varma, Elements of FACTS Controllers, Panel paper presented in FACTS panel session on FACTS fundamentals, *Proceedings of the 2010 IEEE PES T&D Conference & Exposition*, New Orleans, LA, April 20–22, 2010.
116. R.K. Varma, W. Litzenberger, A. Ostadi, and S. Auddy, Bibliography of FACTS 2005–2006 part I: IEEE working group report, *Proceedings of the IEEE PES General Meeting*, Tampa, FL, June 2007.
117. R.K. Varma, W. Litzenberger, A. Ostadi, and S. Auddy, Bibliography of FACTS 2005–2006 part II: IEEE working group report, *Proceedings of the IEEE PES General Meeting*, Tampa, FL, June 2007.
118. R.K. Varma, W. Litzenberger, and J. Berge, Bibliography of FACTS 2001—Part I: IEEE working group report, *Proceedings of the IEEE PES General Meeting*, Tampa, FL, June 2007.
119. R.K. Varma, W. Litzenberger, and J. Berge, Bibliography of FACTS 2001—Part II: IEEE working group report, *Proceedings of the IEEE PES General Meeting*, Tampa, FL, June 2007.
120. R.K. Varma, W. Litzenberger, and J. Berge, Bibliography of FACTS 2002—Part I: IEEE working group report, *Proceedings of the IEEE PES General Meeting*, Tampa, FL, June 2007.
121. R.K. Varma, W. Litzenberger, and J. Berge, Bibliography of FACTS 2002—Part II: IEEE working group report, *Proceedings of the IEEE PES General Meeting*, Tampa, FL, June 2007.

122. R.K. Varma, W. Litzenberger, and J. Berge, Bibliography of FACTS 2003–Part I: IEEE working group report, *Proceedings of the IEEE PES General Meeting*, Tampa, FL, June 2007.
123. R.K. Varma, W. Litzenberger, and J. Berge, Bibliography of FACTS 2003—Part II: IEEE working group report, *Proceedings of the IEEE PES General Meeting*, Tampa, FL, June 2007.
124. R.K. Varma, W. Litzenberger, S. Auddy, J. Berge, A.C. Cojocaru, and T. Sidhu, Bibliography of FACTS: 2004–2005—Part I: IEEE working group report, *Proceedings of the IEEE PES General Meeting*, Montreal, Quebec, Canada, 2006.
125. R.K. Varma, W. Litzenberger, S. Auddy, J. Berge, A.C. Cojocaru, and T. Sidhu, Bibliography of FACTS: 2004–2005—Part II: IEEE working group report, *Proceedings of the IEEE PES General Meeting*, Montreal, Quebec, Canada, 2006.
126. R.K. Varma, W. Litzenberger, S. Auddy, J. Berge, A.C. Cojocaru, and T. Sidhu, Bibliography of FACTS: 2004–2005—Part III: IEEE working group report, *Proceedings of the IEEE PES General Meeting*, Montreal, Quebec, Canada, 2006.
127. W. Litzenberger, R.K. Varma, and J.D. Flanagan (eds.), An annotated bibliography of high voltage direct-current (HVDC) transmission and flexible AC transmission systems (FACTS) 1996–1997, on behalf of the IEEE Working Group on HVDC and FACTS Bibliography and Records, Presented at the *IEEE Power Engineering Society Summer Meeting*, San Diego, CA, July 1998, 521pp. E.P.R.I. and Bonneville Power Administration (B.P.A.), Portland, OR.
128. W. Litzenberger and R.K. Varma (eds.), An annotated bibliography of high voltage direct-current (HVDC) transmission and flexible AC transmission systems (FACTS) 1994–1995, on behalf of the IEEE Working Group on HVDC and FACTS Bibliography and Records, Presented at the *IEEE Power Engineering Society Summer Meeting*, Denver, CO, July 1996, 346pp. Department of Energy, Washington, DC, and Bonneville Power Administration (B.P.A.), Portland, OR.
129. W.H. Litzenberger (ed.), An annotated bibliography of high-voltage direct-current transmission and flexible AC transmission (FACTS) devices, 1991–1993, Bonneville Power Administration and Western Area Power Administration, Portland, OR, 1994.
130. W.H. Litzenberger (ed.), An annotated bibliography of high-voltage direct-current transmission, 1989–1991. Bonneville Power Administration and Western Area Power Administration, Portland, OR, 1992.

III

Power System Operation and Control

Bruce F. Wollenberg

20 **Energy Management** *Neil K. Stanton, Jay C. Giri, and Anjan Bose* 20-1
   Power System Data Acquisition and Control • Automatic Generation Control •
   Load Management • Energy Management • Security Control • Operator Training
   Simulator • Trends in Energy Management • References • Further Information

21 **Generation Control: Economic Dispatch and Unit Commitment**
   Charles W. Richter Jr. 21-1
   Economic Dispatch • The Unit Commitment Problem • Summary of Economical
   Generation Operation • References

22 **State Estimation** *Jason G. Lindquist and Danny Julian* 22-1
   State Estimation Problem • State Estimation Operation • Example State Estimation
   Problem • Defining Terms • References

23 **Optimal Power Flow** *Mohamed E. El-Hawary* 23-1
   Conventional Optimal Economic Scheduling • Conventional OPF Formulation •
   OPF Incorporating Load Models • SCOPF Including Load Modeling • Operational
   Requirements for Online Implementation • Conclusions • References

24 **Security Analysis** *Nouredine Hadjsaid* 24-1
   Definition • Time Frames for Security-Related Decision • Models •
   Determinist vs. Probabilistic • Appendix A • Appendix B • References

Bruce F. Wollenberg received his BSc in electrical engineering and MSc in electric power engineering from Rensselaer Polytechnic Institute, Troy, New York in 1964 and 1966, respectively. He subsequently received his PhD in systems engineering from the University of Pennsylvania, Philadelphia, Pennsylvania in 1974.

From 1966 to 1974, Dr. Wollenberg worked at Leeds and Northrup Company in Philadelphia, Pennsylvania, where he was responsible for the development of an operator power flow, constrained economic dispatch, and contingency analysis software. From 1974 to 1984, he worked at Power Technologies Inc. (PTI) in Schenectady, New York, where he was responsible for the development of PTI's power system simulator for operations. He conducted research on optimal power flow, contingency selection, power system planning software, cost benefit studies for energy management systems, and advanced state estimation algorithms. He also served as an adjunct professor in the Department of Electric Power Engineering at Rensselaer Polytechnic Institute from 1979 to 1984.

From 1984 to 1989, Dr. Wollenberg worked at Control Data Corporation's Energy Management Systems Division in Plymouth, Minnesota, where he was responsible for the development of artificial intelligence applications. He consulted on advanced research projects in energy management system application software and he participated in the power system operator training simulator project for the Electric Power Research Institute. Dr. Wollenberg was also active in teaching and served as an adjunct professor in the Department of Electrical Engineering at the University of Minnesota from 1984 to 1989.

In 1989, he was appointed to a professorship at the University of Minnesota in Minneapolis, Minnesota, where his interests involved the development of large-scale network solution algorithms using vector processing supercomputers, the extension of traditional power system control techniques to incorporate spot pricing algorithms and distributed computing technologies, and the application of expert systems to enhance the information presented to power system operators using real-time computers.

He has coauthored, with Allen Wood, the Wiley textbook *Power Generation, Operation, and Control*.

Dr. Wollenberg is a member of the National Academy of Engineering and a life fellow of the IEEE. He is a member of Tau Beta Pi, Eta Kappa Nu, and Sigma Xi honorary societies; and he is the former chair of the Power Systems Engineering Committee of the IEEE Power Engineering Society.

20
Energy Management

20.1	Power System Data Acquisition and Control 20-3
20.2	Automatic Generation Control 20-4
	Load Frequency Control • Economic Dispatch • Reserve Monitoring • Interchange Transaction Scheduling
20.3	Load Management 20-6
20.4	Energy Management 20-6
20.5	Security Control 20-7
20.6	Operator Training Simulator 20-8
	Energy Control System • Power System Dynamic Simulation • Instructional System
20.7	Trends in Energy Management 20-9
	References 20-10
	Further Information 20-10

Neil K. Stanton
Stanton Associates

Jay C. Giri
ALSTOM Grid, Inc.

Anjan Bose
Washington State University

Energy management is the process of monitoring, coordinating, and controlling the generation, transmission, and distribution of electrical energy. The physical plant to be managed includes generating plants that produce energy fed through transformers to the high-voltage transmission network (grid), interconnecting generating plants, and load centers. Transmission lines terminate at substations that perform switching, voltage transformation, measurement, and control. Substations at load centers transform to subtransmission and distribution levels. These lower-voltage circuits typically operate radially, that is, no normally closed paths between substations through subtransmission or distribution circuits. (Underground cable networks in large cities are an exception.)

Since transmission systems provide negligible energy storage, supply and demand must be balanced by either generation or load. Production is controlled by turbine governors at generating plants, and automatic generation control (AGC) is performed by control center computers remote from generating plants. Load management, sometimes called demand-side management, extends remote supervision and control to subtransmission and distribution circuits, including control of residential, commercial, and industrial loads.

Events such as lightning strikes, short circuits, equipment failure, or accidents may cause a system fault. Protective relays actuate rapid, local control through operation of circuit breakers before operators can respond. The goal is to maximize safety, minimize damage, and continue to supply load with the least inconvenience to customers. Data acquisition provides operators and computer control systems with status and measurement information needed to supervise overall operations. Security control analyzes the consequences of faults to establish operating conditions that are both robust and economical.

Energy management is performed at control centers (see Figure 20.1), typically called system control centers, by computer systems called *energy management systems* (EMSs). Data acquisition and remote control is performed by computer systems called *supervisory control and data acquisition* (SCADA) systems.

FIGURE 20.1 Manitoba Hydro Control Center in Winnipeg, Manitoba, Canada. (Photo used with permission of ALSTOM ESCA Corporation, Bellevue, WA.)

These latter systems may be installed at a variety of sites including system control centers. An EMS typically includes a SCADA "front-end" through which it communicates with generating plants, substations, and other remote devices.

Figure 20.2 illustrates the applications layer of modern EMS as well as the underlying layers on which it is built: the operating system, a database manager, and a utilities/services layer.

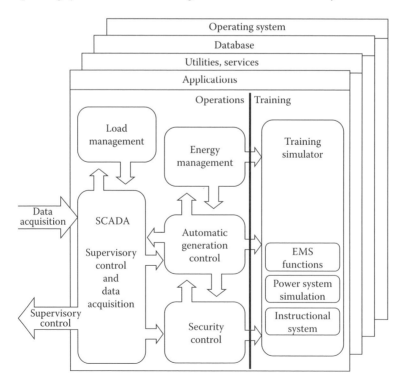

FIGURE 20.2 Layers of a modern EMS.

20.1 Power System Data Acquisition and Control

A SCADA system consists of a master station that communicates with remote terminal units (RTUs) for the purpose of allowing operators to observe and control physical plants. Generating plants and transmission substations certainly justify RTUs, and their installation is becoming more common in distribution substations as costs decrease. RTUs transmit device status and measurements to, and receive control commands and set point data from, the master station. Communication is generally via dedicated circuits operating in the range of 600–4800 bits/s with the RTU responding to periodic requests initiated from the master station (polling) every 2–10 s, depending on the criticality of the data.

The traditional functions of SCADA systems are summarized as follows:

- Data acquisition: provides telemetered measurements and status information to operator.
- Supervisory control: allows operator to remotely control devices, for example, open and close circuit breakers. A "select before operate" procedure is used for greater safety.
- Tagging: identifies a device as subject to specific operating restrictions and prevents unauthorized operation.
- Alarms: inform operator of unplanned events and undesirable operating conditions. Alarms are sorted by criticality, area of responsibility, and chronology. Acknowledgment may be required.
- Logging: logs all operator entry, all alarms, and selected information.
- Load shed: provides both automatic and operator-initiated tripping of load in response to system emergencies.
- Trending: plots measurements on selected timescales.

Since the master station is critical to power system operations, its functions are generally distributed among several computer systems depending on specific design. A dual computer system configured in primary and standby modes is most common. SCADA functions are listed without stating which computer has specific responsibility:

1. Manage communication circuit configuration.
2. Downline load RTU files.
3. Maintain scan tables and perform polling.
4. Check and correct message errors.
5. Convert to engineering units.
6. Detect status and measurement changes.
7. Monitor abnormal and out-of-limit conditions.
8. Log and time-tag sequence of events.
9. Detect and annunciate alarms.
10. Respond to operator requests to
    a. Display information
    b. Enter data
    c. Execute control action
    d. Acknowledge alarms
11. Transmit control action to RTUs.
12. Inhibit unauthorized actions.
13. Maintain historical files.
14. Log events and prepare reports.
15. Perform load shedding.

20.2 Automatic Generation Control

AGC consists of two major and several minor functions that operate online in real time to adjust the generation against load at minimum cost. The major functions are load frequency control (LFC) and economic dispatch (ED), each of which is described in the following. The minor functions are reserve monitoring, which assures enough reserve on the system; interchange scheduling, which initiates and completes scheduled interchanges; and other similar monitoring and recording functions.

20.2.1 Load Frequency Control

LFC has to achieve three primary objectives, which are stated in priority order as follows:

1. To maintain frequency at the scheduled value
2. To maintain net power interchanges with neighboring control areas at the scheduled values
3. To maintain power allocation among units at economically desired values

The first and second objectives are met by monitoring an error signal, called *area control error* (ACE), which is a combination of net interchange error and frequency error and represents the power imbalance between generation and load at any instant. This ACE must be filtered or smoothed such that excessive and random changes in ACE are not translated into control action. Since these excessive changes are different for different systems, the filter parameters have to be tuned specifically for each control area. The filtered ACE is then used to obtain the proportional plus integral control signal. This control signal is modified by limiters, deadbands, and gain constants that are tuned to the particular system. This control signal is then divided among the generating units under control by using participation factors to obtain *unit control errors* (UCEs).

These participation factors may be proportional to the inverse of the second derivative of the cost of unit generation so that the units would be loaded according to their costs, thus meeting the third objective. However, cost may not be the only consideration because the different units may have different response rates and it may be necessary to move the faster generators more to obtain an acceptable response. The UCEs are then sent to the various units under control and the generating units monitored to see that the corrections take place. This control action is repeated every 2–6 s.

In spite of the integral control, errors in frequency and net interchange do tend to accumulate over time. These time errors and accumulated interchange errors have to be corrected by adjusting the controller settings according to procedures agreed upon by the whole interconnection. These accumulated errors as well as ACE serve as performance measures for LFC.

The main philosophy in the design of LFC is that each system should follow its own load very closely during normal operation, while, during emergencies, each system should contribute according to its relative size in the interconnection without regard to the locality of the emergency. Thus, the most important factor in obtaining good control of a system is its inherent capability of following its own load. This is guaranteed if the system has adequate regulation margin as well as adequate response capability. Systems that have mainly thermal generation often have difficulty in keeping up with the load because of the slow response of the units.

The design of the controller itself is an important factor, and proper tuning of the controller parameters is needed to obtain "good" control without "excessive" movement of units. Tuning is system specific, and although system simulations are often used as aids, most of the parameter adjustments are made in the field using heuristic procedures.

20.2.2 Economic Dispatch

Since all the generating units that are online have different costs of generation, it is necessary to find the generation levels of each of these units that would meet the load at the minimum cost. This has to take into account the fact that the cost of generation in one generator is not proportional to its generation

Energy Management

level but is a nonlinear function of it. In addition, since the system is geographically spread out, the transmission losses are dependent on the generation pattern and must be considered in obtaining the optimum pattern.

Certain other factors have to be considered when obtaining the optimum generation pattern. One is that the generation pattern provides adequate reserve margins. This is often done by constraining the generation level to a lower boundary than the generating capability. A more difficult set of constraints to consider are the transmission limits. Under certain real-time conditions, it is possible that the most economic pattern may not be feasible because of unacceptable line flows or voltage conditions. The present-day ED algorithm cannot handle these security constraints. However, alternative methods based on optimal power flows have been suggested but have not yet been used for real-time dispatch.

The minimum cost dispatch occurs when the incremental cost of all the generators is equal. The cost functions of the generators are nonlinear and discontinuous. For the equal marginal cost algorithm to work, it is necessary for them to be convex. These incremental cost curves are often represented as monotonically increasing piecewise-linear functions. A binary search for the optimal marginal cost is conducted by summing all the generation at a certain marginal cost and comparing it with the total power demand. If the demand is higher, a higher marginal cost is needed, and vice versa. This algorithm produces the ideal set points for all the generators for that particular demand, and this calculation is done every few minutes as the demand changes.

The losses in the power system are a function of the generation pattern, and they are taken into account by multiplying the generator incremental costs by the appropriate penalty factors. The penalty factor for each generator is a reflection of the sensitivity of that generator to system losses, and these sensitivities can be obtained from the transmission loss factors.

This ED algorithm generally applies to only thermal generation units that have cost characteristics of the type discussed here. The hydro units have to be dispatched with different considerations. Although there is no cost for the water, the amount of water available is limited over a period, and the displacement of fossil fuel by this water determines its worth. Thus, if the water usage limitation over a period is known, say from a previously computed hydro optimization, the water worth can be used to dispatch the hydro units.

LFC and the ED functions both operate automatically in real time but with vastly different time periods. Both adjust generation levels, but LFC does it every few seconds to follow the load variation, while ED does it every few minutes to assure minimal cost. Conflicting control action is avoided by coordinating the control errors. If the UCEs from LFC and ED are in the same direction, there is no conflict. Otherwise, a logic is set to either follow load (permissive control) or follow economics (mandatory control).

20.2.3 Reserve Monitoring

Maintaining enough reserve capacity is required in case generation is lost. Explicit formulas are followed to determine the spinning (already synchronized) and ready (10 min) reserves required. The availability can be assured by the operator manually, or, as mentioned previously, the ED can also reduce the upper dispatchable limits of the generators to keep such generation available.

20.2.4 Interchange Transaction Scheduling

The contractual exchange of power between utilities has to be taken into account by the LFC and ED functions. This is done by calculating the net interchange (sum of all the buy and sale agreements) and adding this to the generation needed in both the LFC and ED. Since most interchanges begin and end on the hour, the net interchange is ramped from one level to the new over a 10 or 20 min period straddling the hour. The programs achieve this automatically from the list of scheduled transactions.

20.3 Load Management

SCADA, with its relatively expensive RTUs installed at distribution substations, can provide status and measurements for distribution feeders at the substation. Distribution automation equipment is now available to measure and control at locations dispersed along distribution circuits. This equipment can monitor sectionalizing devices (switches, interruptors, fuses), operate switches for circuit reconfiguration, control voltage, read customers' meters, implement time-dependent pricing (on-peak, off-peak rates), and switch customer equipment to manage load. This equipment requires significantly increased functionality at distribution control centers.

Distribution control center functionality varies widely from company to company, and the following list is evolving rapidly:

- Data acquisition: acquires data and gives the operator control over specific devices in the field. Includes data processing, quality checking, and storage.
- Feeder switch control: provides remote control of feeder switches.
- Tagging and alarms: provide features similar to SCADA.
- Diagrams and maps: retrieve and display distribution maps and drawings. Support device selection from these displays. Overlay telemetered and operator-entered data on displays.
- Preparation of switching orders: provides templates and information to facilitate preparation of instructions necessary to disconnect, isolate, reconnect, and reenergize equipment.
- Switching instructions: guides operator through execution of previously prepared switching orders.
- Trouble analysis: correlates data sources to assess scope of trouble reports and possible dispatch of work crews.
- Fault location: analyzes available information to determine scope and location of fault.
- Service restoration: determines the combination of remote control actions that will maximize restoration of service. Assists operator to dispatch work crews.
- Circuit continuity analysis: analyzes circuit topology and device status to show electrically connected circuit segments (either energized or de-energized).
- Power factor and voltage control: combines substation and feeder data with predetermined operating parameters to control distribution circuit power factor and voltage levels.
- Electrical circuit analysis: performs circuit analysis, single phase or three phase, balanced or unbalanced.
- Load management: controls customer loads directly through appliance switching (e.g., water heaters) and indirectly through voltage control.
- Meter reading: reads customers' meters for billing, peak demand studies, time of use tariffs. Provides remote connect/disconnect.

20.4 Energy Management

Generation control and ED minimize the current cost of energy production and transmission within the range of available controls. Energy management is a supervisory layer responsible for economically scheduling production and transmission on a global basis and over time intervals consistent with cost optimization. For example, water stored in reservoirs of hydro plants is a resource that may be more valuable in the future and should, therefore, not be used now even though the cost of hydro energy is currently lower than thermal generation. The global consideration arises from the ability to buy and sell energy through the interconnected power system; it may be more economical to buy than to produce from plants under direct control. Energy accounting processes transaction information and energy measurements recorded during actual operation as the basis of payment for energy sales and purchases.

Energy Management

Energy management includes the following functions:

- System load forecast: forecasts system energy demand each hour for a specified forecast period of 1–7 days.
- Unit commitment: determines start-up and shutdown times for most economical operation of thermal generating units for each hour of a specified period of 1–7 days.
- Fuel scheduling: determines the most economical choice of fuel consistent with plant requirements, fuel purchase contracts, and stockpiled fuel.
- Hydrothermal scheduling: determines the optimum schedule of thermal and hydro energy production for each hour of a study period up to 7 days while ensuring that hydro and thermal constraints are not violated.
- Transaction evaluation: determines the optimal incremental and production costs for exchange (purchase and sale) of additional blocks of energy with neighboring companies.
- Transmission loss minimization: recommends controller actions to be taken in order to minimize overall power system network losses.
- Security constrained dispatch: determines optimal outputs of generating units to minimize production cost while ensuring that a network security constraint is not violated.
- Production cost calculation: calculates actual and economical production costs for each generating unit on an hourly basis.

20.5 Security Control

Power systems are designed to survive all probable contingencies. A contingency is defined as an event that causes one or more important components such as transmission lines, generators, and transformers to be unexpectedly removed from service. Survival means the system stabilizes and continues to operate at acceptable voltage and frequency levels without loss of load. Operations must deal with a vast number of possible conditions experienced by the system, many of which are not anticipated in planning. Instead of dealing with the impossible task of analyzing all possible system states, security control starts with a specific state: the current state if executing the real-time network sequence; a postulated state if executing a study sequence. Sequence means sequential execution of programs that perform the following steps:

1. Determine the state of the system based on either current or postulated conditions.
2. Process a list of contingencies to determine the consequences of each contingency on the system in its specified state.
3. Determine preventive or corrective action for those contingencies that represent unacceptable risk.

Real-time and study network analysis sequences are diagramed in Figure 20.3.

Security control requires topological processing to build network models and uses large-scale AC network analysis to determine system conditions. The required applications are grouped as a network subsystem that typically includes the following functions:

- Topology processor: processes real-time status measurements to determine an electrical connectivity (bus) model of the power system network.
- State estimator: uses real-time status and analog measurements to determine the "best" estimate of the state of the power system. It uses a redundant set of measurements; calculates voltages, phase angles, and power flows for all components in the system; and reports overload conditions.
- Power flow: determines the steady-state conditions of the power system network for a specified generation and load pattern. Calculates voltages, phase angles, and flows across the entire system.
- Contingency analysis: assesses the impact of a set of contingencies on the state of the power system and identifies potentially harmful contingencies that cause operating limit violations.

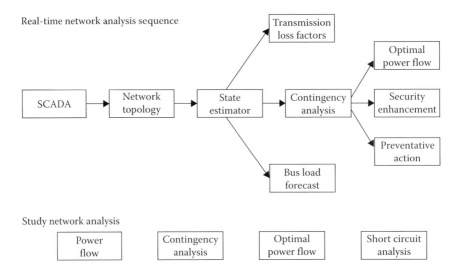

FIGURE 20.3 Real-time and study network analysis sequences.

- Optimal power flow: recommends controller actions to optimize a specified objective function (such as system operating cost or losses) subject to a set of power system operating constraints.
- Security enhancement: recommends corrective control actions to be taken to alleviate an existing or potential overload in the system while ensuring minimal operational cost.
- Preventive action: recommends control actions to be taken in a "preventive" mode before a contingency occurs to preclude an overload situation if the contingency were to occur.
- Bus load forecasting: uses real-time measurements to adaptively forecast loads for the electrical connectivity (bus) model of the power system network.
- Transmission loss factors: determines incremental loss sensitivities for generating units. Calculates the impact on losses if the output of a unit were to be increased by 1 MW.
- Short-circuit analysis: determines fault currents for single-phase and three-phase faults for fault locations across the entire power system network.

20.6 Operator Training Simulator

Training simulators were originally created as generic systems for introducing operators to the electrical and dynamic behavior of power systems. Today, they model actual power systems with reasonable fidelity and are integrated with EMS to provide a realistic environment for operators and dispatchers to practice normal, everyday operating tasks and procedures as well as experience emergency operating situations. The various training activities can be safely and conveniently practiced with the simulator responding in a manner similar to the actual power system.

An operator training simulator (OTS) can be used in an investigatory manner to recreate past actual operational scenarios and to formulate system restoration procedures. Scenarios can be created, saved, and reused. The OTS can be used to evaluate the functionality and performance of new real-time EMS functions and also for tuning AGC in an off-line, secure environment.

The OTS has three main subsystems (Figure 20.4).

20.6.1 Energy Control System

The energy control system (ECS) emulates normal EMS functions and is the only part of the OTS with which the trainee interacts. It consists of the SCADA system, generation control system, and all other EMS functions.

Energy Management

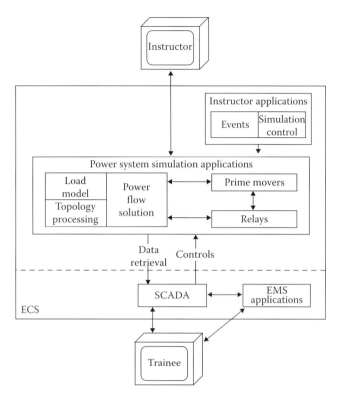

FIGURE 20.4 OTS block diagram.

20.6.2 Power System Dynamic Simulation

This subsystem simulates the dynamic behavior of the power system. System frequency is simulated using the "long-term dynamics" system model, where frequency of all units is assumed to be the same. The prime-mover dynamics are represented by models of the units, turbines, governors, boilers, and boiler auxiliaries. The network flows and states (bus voltages and angles, topology, transformer taps, etc.) are calculated at periodic intervals. Relays are modeled, and they emulate the behavior of the actual devices in the field.

20.6.3 Instructional System

This subsystem includes the capabilities to start, stop, restart, and control the simulation. It also includes making savecases, retrieving savecases, reinitializing to a new time, and initializing to a specific real-time situation.

It is also used to define event schedules. Events are associated with both the power system simulation and the ECS functions. Events may be deterministic (occur at a predefined time), conditional (based on a predefined set of power system conditions being met), or probabilistic (occur at random).

20.7 Trends in Energy Management

Recent advances in synchronized, sub-second measurement technologies and advanced visualization capabilities are dramatically improving the ability to manage grid operations more effectively. These advances result in greater automation of the grid, which in turn helps grid operators make better decisions to maintain grid integrity.

Automation of the grid is evolving toward more decentralized intelligent and localized control, which in turn moves toward a self-healing grid. A self-healing grid will work as the human body does when it quickly identifies an intrusion and deals with it locally to preserve the overall health of the rest of the body.

Already, globally synchronized measurements (in the sub-second range) are being used in control centers to facilitate early faster detection of problems and to make it easier to assess conditions across the expanse of the grid.

New control center applications are continually being developed to use this new type of synchronized measurement technology to further improve the ability to maintain the integrity of the power system. These applications will also be able to identify contingencies, unplanned events, and stability problems at a much faster rate, at the sub-second level.

The objective is to provide the operator "eyes" to always be aware of current system conditions and potential problems that might be lurking ahead. Speed is of the essence to be able to quickly navigate and drill down to the cause of a new problem. As the volume and frequency of measurement data grow—especially with the growth of sub-second synchronous measurement, such as phasor measurement units (PMUs)—it is of tantamount importance to convert this impending data tsunami into relevant useful information that can be concisely shown on an operator display screen and prompt decisions can be made with confidence.

Most control center operator decisions today are essentially *reactive*. Current information, as well as some recent history, is used to reactively make an assessment of the current state and its vulnerability. Operators then extrapolate from current conditions and postulate future conditions based on personal experience and planned forecast schedules.

The next step is to help operator's make decisions that are *preventive*. Once there is confidence in the ability to make reactive decisions, operators will need to rely on "what-if" analytical tools to be able to make decisions that will prevent adverse conditions if a specific contingency or disturbance were to occur. Thus, the focus shifts from "problem analysis" (reactive) to "decision making" (preventive).

The industry trend next foresees *predictive* decision making, and in the future, decisions will be *proactive*. This future system will use more accurate forecast information and more advanced analytical tools to be able to confidently predict system conditions and what-if scenarios, to be able to take action now, in order to preclude possibilities of problematic scenarios in the future.

References

Application of Optimization Methods for Economy/Security Functions in Power System Operations, IEEE tutorial course, IEEE Publication 90EH0328-5-PWR, 1990.
Distribution Automation, IEEE Power Engineering Society, IEEE Publication EH0280-8-PBM, 1988.
Energy Control Center Design, IEEE tutorial course, IEEE Publication 77 TU0010-9 PWR, 1977.
Erickson, C.J., *Handbook of Electrical Heating*, New York: IEEE Press, 1995.
Fundamentals of Load Management, IEEE Power Engineering Society, IEEE Publication EH0289-9-PBM, 1988.
Fundamentals of Supervisory Controls, IEEE tutorial course, IEEE Publication 91 EH0337-6 PWR, 1991.
Kleinpeter, M., *Energy Planning and Policy*, New York: Wiley, 1995.
Special issue on computers in power system operations, *Proc. IEEE*, 75, 12, 1987.
Turner, W.C., *Energy Management Handbook*, Lilburn, GA: Fairmont Press, 1997.

Further Information

Current innovations and applications of new technologies and algorithms are presented in the following publications:

- *IEEE Power Engineering Review* (monthly)
- *IEEE Transactions on Power Systems* (bimonthly)
- *Proceedings of the Power Industry Computer Application Conference* (biannual)

21 Generation Control: Economic Dispatch and Unit Commitment

Charles W. Richter Jr.
Charles Richter Associates, LLC

21.1 Economic Dispatch 21-1
   Economic Dispatch Defined • Factors to Consider in the EDC •
   EDC and System Limitations • The Objective of EDC •
   The Traditional EDC Mathematical Formulation • EDC
   Solution Techniques • An Example of Cost-Minimizing EDC •
   EDC and Auctions
21.2 The Unit Commitment Problem 21-7
   Unit Commitment Defined • Factors to Consider in Solving the UC
   Problem • Mathematical Formulation for UC • The Importance
   of EDC to the UC Solution • Solution Methods • A Genetic-Based
   UC Algorithm • Unit Commitment and Auctions
21.3 Summary of Economical Generation Operation 21-18
References 21-18

An area of power system control having a large impact on cost and profit is the optimal scheduling of generating units. A good schedule identifies which units to operate, and the amount to generate at each online unit in order to achieve a set of economic goals. These are the problems commonly referred to as the unit commitment (UC) problem, and the economic dispatch calculation, respectively. The goal is to choose a control strategy that minimizes losses (or maximizes profits), subject to meeting a certain demand and other system constraints. The following sections define EDC, the UC problem, and discuss methods that have been used to solve these problems. Realizing that electric power grids are complex interconnected systems that must be carefully controlled if they are to remain stable and secure, it should be mentioned that the tools described in this chapter are intended for steady-state operation. Short-term (less than a few seconds) changes to the system are handled by dynamic and transient system controls, which maintain secure and stable operation, and are beyond the scope of this discussion.

21.1 Economic Dispatch

21.1.1 Economic Dispatch Defined

An *economic dispatch calculation* (EDC) is performed to *dispatch*, or schedule, a set of online generating units to collectively produce electricity at a level that satisfies a specified demand in an economical manner. Each online generating unit may have many characteristics that make it unique, and which must be considered in the calculation. The amount of electricity demanded can vary quickly and the schedule produced by an EDC should leave units able to respond and adapt without major implications to cost or profit. The electric system may have limits (e.g., voltage, transmission, etc.) that impact the EDC and hence should be considered. Generating units may have prohibited generation levels at which resonant frequencies may cause damage or other problems to the system. The impact of transmission losses, congestion, and limits that may inhibit the ability to serve the load in a particular region from a particular generator (e.g., a low-cost generator) should be considered. The market structure within an operating region and its associated regulations must be considered in determining the specified demand, and in determining what constitutes economical operation. An independent system operator (ISO) tasked with maximizing social welfare would likely have a different definition of "economical" than does a generation company (GENCO) wishing to maximize its profit in a competitive environment. The EDC must consider all of these factors and develop a schedule that sets the generation levels in accordance with an economic objective function.

21.1.2 Factors to Consider in the EDC

21.1.2.1 The Cost of Generation

Cost is one of the primary characteristics of a generating unit that must be considered when dispatching units economically. The EDC is concerned with the short-term operating cost, which is primarily determined by fuel cost and usage. Fuel usage is closely related to generation level. Very often, the relationship between power level and fuel cost is approximated by a quadratic curve: $F = aP^2 + bP + c$. c is a constant term that represents the cost of operating the plant, b is a linear term that varies directly with the level of generation, and a is the term that accounts for efficiency changes over the range of the plant output. A quadratic relationship is often used in the research literature. However, due to varying conditions at certain levels of production (e.g., the opening or closing of large valves may affect the generation cost [Walters and Sheblé, 1992]), the actual relationship between power level and fuel cost may be more complex than a quadratic equation. Many of the long-term generating unit costs (e.g., costs attributed directly to starting and stopping the unit, capital costs associated with financing the construction) can be ignored for the EDC, since the decision to switch on, or *commit*, the units has already been made. Other characteristics of generating units that affect the EDC are the minimum and maximum generation levels at which they may operate. When binding, these constraints will directly impact the EDC schedule.

21.1.2.2 The Price

The price at which an electric supplier will be compensated is another important factor in determining an optimal economic dispatch. In many areas of the world, electric power systems have been, or still are, treated as a natural monopoly. Regulations allow the utilities to charge rates that guarantee them a nominal profit. In competitive markets, which come in a variety of flavors, price is determined through the forces of supply and demand. Economic theory and common sense tell us that if the total supply is high and the demand is low, the price is likely to be low, and vice versa. If the price is consistently below a GENCO's average total costs, the company may soon be bankrupt.

21.1.2.3 The Quantity Supplied

The amount of electric energy to be supplied is another fundamental input for the EDC. Regions of the world having regulations that limit competition often require electric utilities to serve all electric demand within a designated service territory. If a consumer switches on a motor, the electric supplier must provide the electric energy needed to operate the motor. In competitive markets, this *obligation to serve* is limited to those with whom the GENCO has a contract. Beyond its contractual obligations, the GENCO may be willing (if the opportunity arises) to supply additional consumer demand. Since the consumers have a choice of electric supplier, a GENCO determining the schedule of its own online generating units may choose to supply all, none, or only a portion of that additional consumer demand. The decision is dependent on the objective of the entity performing the EDC (e.g., profit maximization, improving reliability, etc.).

21.1.3 EDC and System Limitations

A complex network of transmission and distribution lines and equipment are required to move the electric energy from the generating units to the consumer loads. The secure operation of this network depends on bus voltage magnitudes and angles being within certain tolerances. Excessive transmission line loading can also affect the security of the power system network. Since superconductivity is a relatively new field, lossless transmission lines are expensive and are not commonly used. Therefore, some of the energy being transmitted over the system is converted into heat and is consequently lost. The schedule produced by the EDC directly affects losses and security; hence, constraints ensuring proper system operation must be considered when solving the EDC problem.

21.1.4 The Objective of EDC

In a regulated, vertically integrated, monopolistic environment, the obligated-to-serve electric utility performs the EDC for the entire service area by itself. In such an environment, providing electricity in an "economical manner" means minimizing the cost of generating electricity, subject to meeting all demand and other system operating constraints. In a competitive environment, the way an EDC is done can vary from one market structure to another. For instance, in a decentralized market, the EDC may be performed by a single GENCO wishing to maximize its expected profit given the prices, demands, costs, and other constraints described above. In a power pool, a central coordinating entity may perform an EDC to centrally dispatch generation for many GENCOs. Depending on the market rules, the generation owners may be able to mask the cost information of their generators. In this case, bids would be submitted for various price levels and used in the EDC.

21.1.5 The Traditional EDC Mathematical Formulation

Assuming operation under a vertically integrated, monopolistic environment, we must meet all demand, D. We must also consider minimum and maximum limits for each generating unit, P_i^{min} and P_i^{max}. We will assume that the fuel costs of the ith operating plant may be modeled by a quadratic equation as shown in Equation 21.1, and shown graphically in Figure 21.1. Note that the average fuel costs are also shown in Figure 21.1.

$$F_i = a_i P_i^2 + b_i P_i + c_i \quad \text{(fuel costs of ith generator)} \quad (21.1)$$

Thus, for N online generating units, we can write a Lagrangian equation, L, which describes the total cost and associated demand constraint, D.

$$L = F_T + \lambda \left(D - \sum_{i=1}^{N} P_i \right) = \sum_{i=1}^{N} \left(a_i P_i^2 + b_i P_i + c_i \right) + \lambda \cdot \left(D - \sum_{i=1}^{N} P_i \right)$$

$$F_T = \sum_{i=1}^{N} F_i \quad \text{(Total fuel cost is a summation of costs for all online plants)}$$

$$P_i^{min} \leq P_i \leq P_i^{max} \quad \text{(Generation must be set between the min and max amounts)} \quad (21.2)$$

Additionally, note that c_i is a constant term that represents the cost of operating the ith plant, b_i is a linear term that varies directly with the level of generation, P_i, and a_i are terms that account for efficiency changes over the range of the plant output.

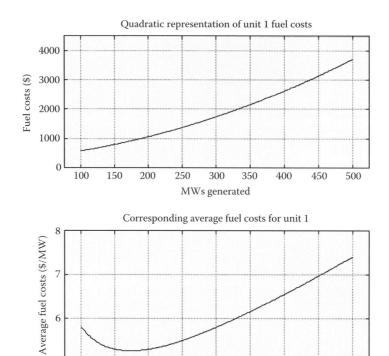

FIGURE 21.1 Relationship between fuel input and power output.

In this example, the objective will be to minimize the cost of supplying demand with the generating units that are online. From calculus, a minimum or a maximum can be found by taking the $N + 1$ derivatives of the Lagrangian with respect to its variables, and setting them equal to zero. The shape of the curves is often assumed well behaved—monotonically increasing and convex—so that determining the second derivative is unnecessary.

$$\frac{\partial L}{\partial P_i} = 2a_i P_i + b_i - \lambda = 0 \Rightarrow \lambda = 2a_i P_i + b_i \qquad (21.3)$$

$$\frac{\partial L}{\partial \lambda} = \left(D - \sum_{1}^{N} P_i\right) = 0 \qquad (21.4)$$

λ_i is the commonly used symbol for the "marginal cost" of the ith unit. At the margin of operation, the marginal cost tells us how many additional dollars the GENCO will have to spend to increase the generation by an additional MW. The marginal cost curve is an positively sloped line if a quadratic equation is being used to represent the fuel curve of the unit. The higher the quantity being produced, the greater the cost of adding an additional unit of the goods being produced. Economic theory says that if a GENCO has a set of plants and it wants to increase production by one unit, it should increase production at the plant that provides the most benefit for the least cost. The GENCO should do this until that plant is no longer providing the greatest benefit for a given cost. At that point it finds the plant now giving the highest benefit-to-cost ratio and increases its production. This is done until all plants are operating at the same marginal cost. When all unconstrained online plants have the same marginal cost, λ (i.e., $\lambda_1 = \lambda_2 = \cdots = \lambda_i = \cdots = \lambda_{SYSTEM}$), then the cost is at a minimum for that amount of generation. If there were binding constraints, it would prevent the GENCO from achieving that scenario.

If a constraint is binding on a particular unit (e.g., P_i becomes P_i^{max} when attempting to increase production), the marginal cost of that unit is considered to be infinite. No matter how much money is available to increase plant production by one unit, it cannot do so. (Of course, in the long term, things may be done that can reduce the effect of the constraint, but that is beyond the scope of this discussion.)

21.1.6 EDC Solution Techniques

There are many ways to obtain the optimum power levels that will achieve the objective for the EDC problem being considered. For very simple situations, one may solve the solution directly; but when the number of constraints that introduce nonlinearities to the problem grows, iterative search techniques become necessary. Wood and Wollenberg (1996) describe many such methods of calculating economic dispatch, including the graphical technique, the lambda-iteration method, and the first and second-order gradient methods. Another method that works well, even when fuel costs are not modeled by a simple quadratic equation, is the genetic algorithm.

In highly competitive scenarios, each inaccuracy in the model can result in losses to the GENCO. A very detailed model might include many nonlinearities (e.g., valve-point loading, prohibited regions of operation, etc.). Such nonlinearities may mean that it is not possible to calculate a derivative. If the relationship is not well-behaved, there may be no proof that the solution can ever be optimal. With greater detail in the model comes an increase in the amount of time to perform the EDC. Since the EDC is performed quite frequently (on the order of every few minutes), and because it is a real-time calculation, the solution technique should be quick. Since an inaccurate solution may produce a negative impact on the company profits, the solution should also be accurate.

21.1.7 An Example of Cost-Minimizing EDC

To illustrate how the EDC is solved via the graphical method, an example is presented here. Assume that a GENCO needs to supply 1000 MW of consumer demand, and that Table 21.1 describes the system online units that it is dispatching in a traditional, i.e., vertically integrated, monopolistic environment. Figure 21.2 shows the marginal costs of each of the units over their entire range. It also shows an aggregated marginal cost curve that could be called the system marginal cost curve. This aggregated system curve was created by a horizontal summation of the four individual graphs. Once the system curve is created, one simply finds the desired power level (i.e., 1000 MW) along the x-axis. Follow it up to the curve, and then look to the left. On the y-axis, the system marginal cost can be read. Since no limits were reached, each of the individual λ_is is the same as the system λ. The GENCO can find the λ_i on each of the unit curves and draw a line straight down from the point where the marginal cost, λ, crosses the curve to find its power level. The generation levels of each online unit are easily found and the solution is shown in the right-hand columns of Table 21.1. The procedure just described is the graphical method of EDC. If the system marginal cost had been above the diagonal portion of an individual unit curve, then we simply set that unit at its P_{max}.

TABLE 21.1 Generator Data and Solution for EDC Example

Unit Number	Unit Parameters					Solution		
	P_{min}	P_{max}	A	B	C	P_i (MW)	\$/MW (λ_i)	Cost \$/Hour
1	100	500	.01	1.8	300	233.2456	6.4649	1263.90
2	50	300	.012	2.24	210	176.0380	6.4649	976.20
3	100	400	.006	2.35	290	342.9094	6.4649	1801.40
4	100	500	.008	2.5	340	247.8070	6.4649	1450.80

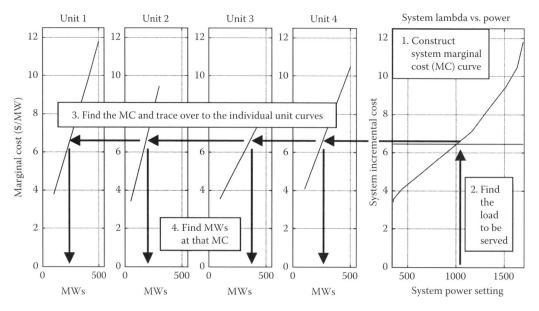

FIGURE 21.2 Unit and aggregated marginal cost curves for solving EDC with the graphical method.

21.1.8 EDC and Auctions

Competitive electricity markets vary in their operating rules, social objectives, and in the mechanism they use to allocate prices and quantities to the participants. Commonly, an auction is used to match buyers with sellers and to achieve a price that is considered fair. Auctions can be sealed bid, open out-cry, ascending ask English auctions, descending ask Dutch auctions, etc. Regardless of the solution technique used to find the optimal allocation, the economic dispatch is essentially performing the same allocation that an auction would. Suppose an auctioneer were to call out a price, and ask the participating/online generators how much power they would generate at that level. The reply amounts could be summed to determine the production level at that price. If all of the constraints, including demand, are met, then the most economical dispatch has been achieved. If not, the auctioneer adjusts the price and asks for

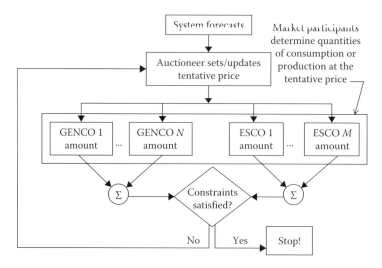

FIGURE 21.3 Economic dispatch and/or unit commitment as an auction.

the amounts at the new price. This procedure is repeated until the constraints are satisfied. Prices may ascend as in the English auction, or they may descend as in the Dutch auction. See Figure 21.3 for a graphical depiction of this process. For further discussion on this topic, the interested reader is referred to Sheblé (1999).

21.2 The Unit Commitment Problem

21.2.1 Unit Commitment Defined

The *unit commitment* (UC) problem is defined as the scheduling of a set of generating units to be on, off, or in stand-by/banking mode for a given period of time to meet a certain objective. For a power system operated by a vertically integrated monopoly, committing units is performed centrally by the utility, and the objective is to minimize costs subject to supplying all demand (and reserve margins). In a competitive environment, each GENCO must decide which units to commit, such that profit is maximized, based on the number of contracted MW; the additional MWh it forecasts that it can profitably wrest from its competitors in the spot market; and the prices at which it will be compensated.

A UC schedule is developed for N units and T periods. A typical UC schedule might look like the one shown in Figure 21.4. Since uncertainty in the inputs becomes large beyond 1 week into the future, the UC schedule is typically developed for the following week. It is common to consider schedules that allow unit-status change from hour to hour, so that a weekly schedule is made up of 168 periods. In finding an optimal schedule, one must consider fuel costs, which can vary with time, start-up and shut-down costs, maximum ramp rates, the minimum up-times and minimum down-times, crew constraints, transmission limits, voltage constraints, etc. Because the problem is discrete, the GENCO may have many generating units, a large number of periods may be considered, and because there are many constraints, finding an optimal UC is a complex problem.

21.2.2 Factors to Consider in Solving the UC Problem

21.2.2.1 The Objective of Unit Commitment

The objective of the unit commitment algorithm is to schedule units in the most economical manner. For the GENCO deciding which units to commit in the competitive environment, economical manner means one that maximizes its profits. For the monopolist operating in a vertically integrated electric system, economical means minimizing the costs.

21.2.2.2 The Quantity to Supply

In systems with vertically integrated monopolies, it is common for electric utilities to have an obligation to serve all demand within their territory. Forecasters provide power system operators an estimated amount of power demanded. The UC objective is to minimize the total operational costs subject to meeting all of this demand (and other constraints they may be considering).

```
UC schedule
Hour        1 2 3 4 5 6 ... T
Gen#1:      1 1 1 1 1 1 ... 0
Gen#2:      0 0 0 1 1 1 ... 1
Gen#3:      1 1 1 0 0 0 ... 1
    ...
Gen#N:      1 1 1 1 1 1 ... 0

0 = unit off-line    1 = unit online
```

FIGURE 21.4 A typical unit commitment schedule.

In competitive electric markets, the GENCO commits units to maximize its profit. It relies on spot and forward bilateral contracts to make part of the total demand known *a priori*. The remaining share of the demand that it may pick up in the spot market must be predicted. This market share may be difficult to predict since it depends on how its price compares to that of other suppliers.

The GENCO may decide to supply less demand than it is physically capable of. In the competitive environment, the obligation to serve is limited to those with whom the GENCO has a contract. The GENCO may consider a schedule that produces less than the forecasted demand. Rather than switching on an additional unit to produce one or two unsatisfied MW, it can allow its competitors to provide that 1 or 2 MW that might have substantially increased its average costs.

21.2.2.3 Compensating the Electricity Supplier

Maximizing profits in a competitive environment requires that the GENCO know what revenue is being generated by the sale of electricity. While a traditional utility might have been guaranteed a fixed rate of return based on cost, competitive electricity markets have varying pricing schemes that may price electricity at the level of the last accepted bid, the average of the buy, ask, and sell offer, etc. When submitting offers to an auctioneer, the GENCO's offer price should reflect its prediction market share, since that determines how many units they have switched on, or in banking mode. GENCOs recovering costs via prices set during the bidding process will note that the UC schedule directly affects the average cost, which indirectly affects the offering price, making it an essential input to any successful bidding strategy.

Demand forecasts and expected market prices are important inputs to the profit-based UC algorithm; they are used to determine the expected revenue, which in turn affects the expected profit. If a GENCO produces two UC schedules each having different expected costs and different expected profits, it should implement the one that provides for the largest profit, which will not necessarily be the one that costs the least. Since prices and demand are so important in determining the optimal UC schedule, price prediction and demand forecasts become crucial. An easy-to-read description of the cost-minimizing UC problem and a stochastic solution that considers spot markets has been presented in Takriti et al. (1997).

21.2.2.4 The Source of Electric Energy

A GENCO may be in the business of electricity generation, but it should also consider purchasing electricity from the market, if it is less expensive than its own generating unit(s). The existence of liquid markets gives energy trading companies an additional source from which to supply power that may not be as prevalent in monopolistic systems. See Figure 21.5. To the GENCO, the market supply curve can be thought of as a pseudo-unit to be dispatched. The supply curve for this pseudo-unit represents an aggregate supply of all of the units participating in the market at the time in question. The price forecast essentially sets the parameters of the unit. This pseudo-unit has no minimum uptime,

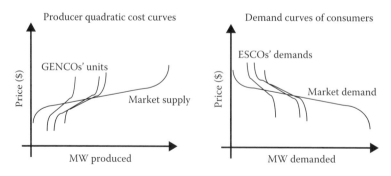

FIGURE 21.5 Treating the market as an additional generator and/or load.

minimum downtime, or ramp constraints; there are no direct start-up and shutdown costs associated with dispatching the unit.

The liquid markets that allow the GENCO to schedule an additional pseudo-unit, also act as a load to be supplied. The total energy supplied should consist of previously arranged bilateral or multilateral contracts arranged through the markets (and their associated reserves and losses). While the GENCO is determining the optimal unit commitment schedule, the energy demanded by the market (i.e., market demand) can be represented as another DISTCO or ESCO buying electricity. Each entity buying electricity should have its own demand curve. The market demand curve should reflect the aggregate of the demand of all the buying agents participating in the market.

21.2.3 Mathematical Formulation for UC

The mathematical formulation for UC depends upon the objective and the constraints that are considered important. Traditionally, the monopolist cost-minimization UC problem has been formulated (Sheblé, 1985):

$$\text{Minimize } F = \sum_{n}^{N} \sum_{t}^{T} \left[(C_{nt} + MAINT_{nt}) \cdot U_{nt} + SUP_{nt} \cdot U_{nt}(1 - U_{nt}) + SDOWN_{nt} \cdot (1 - U_{nt}) \cdot U_{nt-1} \right] \quad (21.5)$$

subject to the following constraints:

$$\sum_{n}^{N} (U_{nt} \cdot P_{nt}) = D_t \quad \text{(demand constraint)}$$

$$\sum_{n}^{N} (U_{nt} \cdot P\max_n) \geq D_t + R_i \quad \text{(capacity constraint)}$$

$$\sum_{n}^{N} (U_{nt} \cdot R\text{smax}_n) \geq R_t \quad \text{(system reserve constraint)}$$

When formulating the profit-maximizing UC problem for a competitive environment, the obligation-to-serve is gone. The demand constraint changes from an equality to an inequality ($\leq$). In the formulation presented here, we lump the reserves in with the demand. Essentially we are assuming that buyers are required to purchase a certain amount of reserves per contract. In addition to the above changes, formulating the UC problem for the competitive GENCO changes the objective function from cost-minimization to profit maximization as shown in Equation 21.6 below. The UC solution process is shown in block diagram form in Figure 21.6.

$$\text{Max } \Pi = \sum_{n}^{N} \sum_{t}^{T} (P_{nt} \cdot fp_t) \cdot U_{nt} - F \quad (21.6)$$

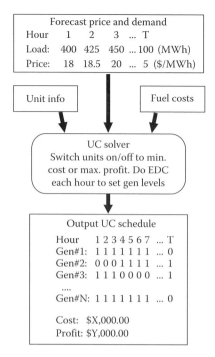

FIGURE 21.6 Block diagram of the UC solution process.

subject to:

$$D_t^{contracted} \leq \sum_{n}^{N} (U_{nt} \cdot P_{nt}) \leq D_t' \quad \text{(demand constraint w/out obligation-to-serve)}$$

$$Pmin_n \leq P_{nt} \leq Pmax_n \quad \text{(capacity limits)}$$

$$|P_{nt} - P_{n,t-1}| \leq Ramp_n \quad \text{(ramp rate limits)}$$

where individual terms are defined as follows:

U_{nt} is the up/down time status of unit n at time period t ($U_{nt}=1$ unit on, $U_{nt}=0$ unit off)
P_{nt} is the power generation of unit n during time period t
D_t is the load level in time period t
D_t' is the forecasted demand at period t (includes reserves)
$D_t^{contract}$ is the contracted demand at period t (includes reserves)
fp_t is the forecasted price/MWh for period t
R_t is the system reserve requirements in time period t
C_{nt} is the production cost of unit n in time period t
SUP_{nt} is the start-up cost for unit n, time period t
$SDOWN_{nt}$ is the shut-down cost for unit n, time period t
$MAINT_{nt}$ is the maintenance cost for unit n, time period t
N is the number of units
T is the number of time periods
$Pmin_n$ is the generation low limit of unit n
$Pmax_n$ is the generation high limit of unit n
$Rsmax_n$ is the maximum contribution to reserve for unit n

Although it may happen in certain cases, the schedule that minimizes cost is not necessarily the schedule that maximizes profit. Providing further distinction between the cost-minimizing UC for the monopolist and the profit maximizing competitive GENCO is the obligation-to-serve; the competitive GENCO may choose to generate less than the total consumer demand. This allows a little more flexibility in the UC schedules. In addition, our formulation assumes that prices fluctuate according to supply and demand. In cost-minimizing paradigms, it is assumed that leveling the load curve helps to minimize the cost. When maximizing profit, the GENCO may find that under certain conditions, it may profit more under a non-level load curve. The profit depends not only on cost, but also on revenue. If revenue increases more than the cost does, the profit will increase.

21.2.4 The Importance of EDC to the UC Solution

The economic dispatch calculation (EDC) is an important part of UC. It is used to assure that sufficient electricity will be available to meet the objective each hour of the UC schedule. For the monopolist in a vertically integrated environment, EDC will set generation so that costs are minimized subject to meeting the demand. For the price-based UC, the price-based EDC adjusts the power level of each online unit each has the same incremental cost (i.e., $\lambda_1 = \lambda_2 = \cdots = \lambda_i = \cdots = \lambda_T$). If a GENCO is operating in a competitive framework that requires its bids to cover fixed, start-up, shutdown, and other costs associated with transitioning from one state to another, then the incremental cost used by EDC must embed these costs. We shall refer to this modified marginal cost as a pseudo λ. The competitive generator will generate if the pseudo λ is less than or equal to the competitive price. A simple way to allocate the fixed and transitional costs that result in a \$/MWh figure is shown in Equation 21.7:

$$\lambda_t = fp_t - \frac{\sum_t \sum_n (\text{transition costs}) + \sum_t \sum_n (\text{fixed costs})}{\sum_t^T \sum_n^N P_{nt}} \qquad (21.7)$$

Other allocation schemes that adjust the marginal cost/price according to the time of day or price of power would be just as easy to implement and should be considered in building bidding strategies. Transition costs include start-up, shutdown, and banking costs, and fixed costs (present for each hour that the unit is on), which would be represented by the constant term in the typical quadratic cost curve approximation. For the results presented later in this chapter, we approximate the summation of the power generated by the forecasted demand.

The competitive price is assumed to be equal to the forecasted price. If the GENCO's supply curve is indicative of the system supply curve, then the competitive price will correspond to the point where the demand and supply curves cross. EDC sets the generation level corresponding to the point where the GENCO's supply curve crosses the demand curve, or to the point where the forecasted price is equal to the supply curve, whichever is lower.

21.2.5 Solution Methods

Solving the UC problem to find an optimal solution can be difficult. The problem has a large solution space that is discrete and nonlinear. As mentioned above, solving the UC problem requires that many economic dispatch calculations be performed. One possible way to determine the optimal schedule is to do an exhaustive search. Exhaustively considering all possible ways that units can be switched on or off for a small system can be done, but for a reasonably sized system this would take too long. Solving the UC problem for a realistic system generally involves using methods like Lagrangian relaxation, dynamic programming, genetic algorithms, or other heuristic search techniques. The interested reader may find many useful references regarding cost-minimizing UC for the monopolist in Sheblé and Fahd (1994) and Wood and Wollenberg (1996). Another heuristic technique that has shown much promise and that offers many advantages (e.g., time-to-solution for large systems and ability to simultaneously generate multiple solutions) is the genetic algorithm.

21.2.6 A Genetic-Based UC Algorithm

21.2.6.1 The Basics of Genetic Algorithms

A genetic algorithm (GA) is a search algorithm often used in nonlinear discrete optimization problems. The development of GAs was inspired by the biological notion of evolution. Initially described by John Holland, they were popularized by David Goldberg who described the basic genetic algorithm very well (Goldberg, 1989). In a GA, data, initialized randomly in a data structure appropriate for the solution to the problem, evolves over time and becomes a suitable answer to the problem. An entire population of candidate solutions (data structures with a form suitable for solving for the problem being studied) is "randomly" initialized and evolves according to GA rules. The data structures often consist of strings of binary numbers that are mapped onto the solution space for evaluation. Each solution (often termed a creature) is assigned a fitness—a heuristic measure of its quality. During the evolutionary process, those creatures having higher fitness are favored in the parent selection process and are allowed to procreate. The parent selection is essentially a random selection with a fitness bias. The type of fitness bias is determined by the parent selection method. Following the parent selection process, the processes of crossover and mutation are utilized and new creatures are developed that ideally explore a different area of the solution space. These new creatures replace less fit creatures from the existing population. Figure 21.7 shows a block diagram of the general GA.

21.2.6.2 GA for Price-Based UC

The algorithm presented here solves the UC problem for the profit maximizing GENCO operating in the competitive environment (Richter and Sheblé, 2000). Research reveals that various GAs have been used by many researchers in solving the UC problem (Kazarlis et al., 1995; Kondragunta, 1997). However, the algorithm presented here is a modification of a genetic-based UC algorithm for the cost-minimizing monopolist described in Maifeld and Sheblé (1996). Most of the modifications are to the fitness function, which no longer rewards schedules that minimize cost, but rather those that maximize profit. The intelligent mutation operators are preserved in their original form. The schedule format is the same. The algorithm is shown in block diagram format in Figure 21.8.

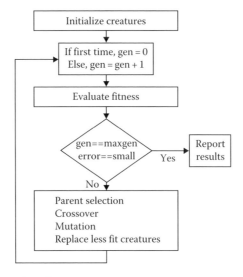

FIGURE 21.7 A simple genetic algorithm.

Generation Control: Economic Dispatch and Unit Commitment

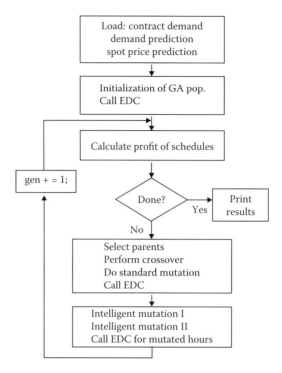

FIGURE 21.8 GA-UC block diagram.

The algorithm first reads in the contract demand and prices, the forecast of remaining demand, and forecasted spot prices (which are calculated for each hour by another routine not described here). During the initialization step, a population of UC schedules is randomly initialized. See Figure 21.9. For each member of the population, EDC is called to set the level of generation of each unit. The cost of each schedule is calculated from the generator and data read in at the beginning of the program. Next, the fitness (i.e., the profit) of each schedule in the population is calculated. "Done?" checks to see whether the algorithm as either cycled through for the maximum number of generations allowed, or whether other stopping criteria have been met. If done, then the results are written to a file; if not done, the algorithm goes to the reproduction process.

During reproduction, new schedules are created. The first step of reproduction is to select parents from the population. After selecting parents, candidate children are created using two-point crossover as shown in Figure 21.10. Following crossover, standard mutation is applied. Standard mutation involves turning a randomly selected unit on or off within a given schedule.

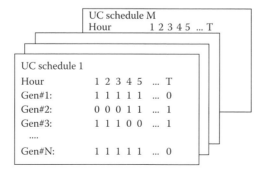

FIGURE 21.9 A population of UC schedules.

```
UC schedule parent 1                    UC schedule parent 2
Hour    1 2 3 4 5 ... T                 Hour    1 2 3 4 5 ... T
Gen#1:  1 1 1 1 1 ... 0                 Gen#1:  1 1 1 1 1 ... 0
Gen#2:  0 0 0 1 1 ... 1                 Gen#2:  1 1 1 1 1 ... 0
Gen#3:  1 1 1 0 0 ... 1                 Gen#3:  1 1 1 1 1 ... 0
Gen#4:  1 1 1 1 1 ... 0                 Gen#4:  1 1 1 1 1 ... 0
Gen#5:  0 0 0 1 1 ... 1                 Gen#5:  1 1 1 1 1 ... 0
Gen#6:  1 1 1 0 0 ... 1                 Gen#6:  1 1 1 1 1 ... 0

UC schedule child 1                     UC schedule child 2
Hour    1 2 3 4 5 ... T                 Hour    1 2 3 4 5 ... T
Gen#1:  1 1 1 1 1 ... 0                 Gen#1:  1 1 1 1 1 ... 0
Gen#2:  0 0 1 1 1 ... 1                 Gen#2:  1 1 0 1 1 ... 0
Gen#3:  1 1 1 1 1 ... 1                 Gen#3:  1 1 1 0 0 ... 0
Gen#4:  1 1 1 1 1 ... 0                 Gen#4:  1 1 1 1 1 ... 0
Gen#5:  0 0 1 1 1 ... 1                 Gen#5:  1 1 0 1 1 ... 0
Gen#6:  1 1 1 1 1 ... 1                 Gen#6:  1 1 1 0 0 ... 0
```

FIGURE 21.10 Two-point crossover on UC schedules.

An important feature of the previously developed UC-GA (Maifeld and Sheblé, 1996) is that it spends as little time as possible doing EDC. After standard mutation, EDC is called to update the profit only for the mutated hour(s). An hourly profit number is maintained and stored during the reproduction process, which dramatically reduces the amount of time required to calculate the profit over what it would be if EDC had to work from scratch at each fitness evaluation. In addition to the standard mutation, the algorithm uses two "intelligent" mutation operators that work by recognizing that, because of transition costs and minimum uptime and downtime constraints, 101 or 010 combinations are undesirable. The first of these operators would purge this undesirable combination by randomly changing 1s to 0s or vice versa. The second of these intelligent mutation operators purges the undesirable combination by changing 1 to 0 or 0 to 1 based on which of these is more helpful to the profit objective.

21.2.6.3 Price-Based UC-GA Results

The UC-GA is run on a small system so that its solution can be easily compared to a solution by exhaustive search. Before running the UC-GA, the GENCO needs to first get an accurate hourly demand and price forecast for the period in question. Developing the forecasted data is an important topic, but beyond the scope of our analysis. For the results presented in this section, the forecasted load and prices are taken to be those shown in Table 21.2. In addition to loading the forecasted hourly price and demand, the UC-GA program needs to load the parameters of each generator to be considered. We are modeling the generators with a quadratic cost curve (e.g., $A + B(P) + C(P)^2$), where P is the power level of the unit. The data for the two-generator case is shown in Table 21.3.

TABLE 21.2 Forecasted Demand and Prices for Two-Generator Case

Hour	Load Forecast (MWh)	Price Forecast ($/MWh)	Hour	Load Forecast (MWh)	Price Forecast ($/MWh)
1	285	25.87	8	328	8.88
2	293	23.06	9	326	9.12
3	267	19.47	10	298	8.88
4	247	18.66	11	267	25.23
5	295	21.38	12	293	26.45
6	292	12.46	13	350	25.00
7	299	9.12	14	350	24.00

TABLE 21.3 Unit Data for Two-Generator Case

	Generator 0	Generator 1
Pmin (MW)	40	40
Pmax (MW)	180	180
A (constant)	58.25	138.51
B (linear)	8.287	7.955
C (quadratic)	7.62e−06	3.05e−05
Bank cost ($)	192	223
Start-up cost ($)	443	441
Shut-down cost ($)	750	750
Min-uptime (h)	4	4
Min-downtime (h)	4	4

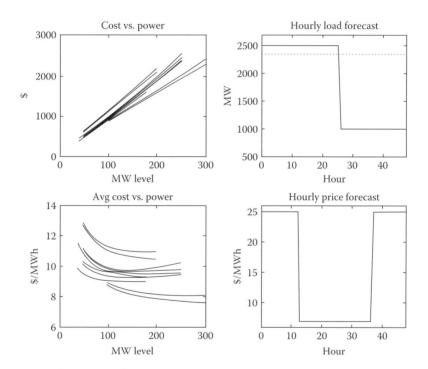

FIGURE 21.11 Data for 10-unit, 48 h case.

In addition to the 2-unit cases, a 10-unit, 48-h case is included in this chapter to show that the GA works well on larger problems. While dynamic programming quickly becomes too computationally expensive to solve, the GA scales up linearly with number of hours and units. Figure 21.11 shows the costs and average costs (without transition costs) of the 10 generators, as well as the hourly price and load forecasts for the 48 h. The data was chosen so that the optimal solution was known *a priori*. The dashed line in the load forecast represents the maximum output of the 10 units.

Before running the UC-GA, the user specifies the control parameters shown in Table 21.4, including the number of generating units and number of hours to be considered in the study. The "popsize" is the size of the GA population. The execution time varies approximately linearly with the popsize. The number of generations indicates how many times the GA will go through the reproduction phase. System reserve is the percentage of reserves that the buyer must maintain for each contract. Children per generation tells us how much of the population will be replaced each generation. Changing this can

TABLE 21.4 GA Control Parameters

Parameter	Setpoint	Parameter	Setpoint
No. of units	2	System reserve (%)	10
No. of hours	10	Children per generation	10
Popsize	20	UC schedules to keep	1
Generations	50	Random number seed	0.20

TABLE 21.5 Comparing UC-GA with Exhaustive Search

No. of Generators in Schedule	No. of Hours in Schedule	GA Finds Optimal Solution?	Solution Time for GA (s)	Solution Time Exhaustive Search (s)
2	10	Yes	0.5	674
2	12	Yes	2	6482
2	14	Yes	10	(Estimated) 62340
10	48	Yes	730	(Estimated) 2E138

affect the convergence rate. If there are multiple optima, faster convergence can trap the GA in a local suboptimal solution. "UC schedules to keep" indicates the number of schedules to write to file when finished. There is also a random number seed that is set between 0 and 1.

In the two-generator test cases, the UC-GA was run for the units listed in Table 21.3, and for the forecasted loads and prices listed in Table 21.2. The parameters listed in Table 21.4 were adjusted accordingly. To ensure that the UC-GA is finding optimal solutions, an exhaustive search was performed on some of the smaller cases. Table 21.5 shows the time to solution in seconds for the UC-GA and the exhaustive search methods. For small cases, the exhaustive search was performed and solution time

TABLE 21.6 The Best UC-GA Schedules of the Population

	Best Schedule for 2-Unit, 10-H Case
Unit 1	1111100000
Unit 2	0000000000
Cost	$17,068.20
Profit	$2,451.01
	Best schedule for 2-unit, 12-h case
Unit 1	111111000011
Unit 2	000000000000
Cost	$24,408.50
Profit	$4,911.50
	Best schedule found by UC-GA for 10-unit, 48-h case
Unit 1	111111111111000000000000000000000000111111111111
Unit 2	111111111111000000000000000000000000000000000000
Unit 3	111111111111000000000000000000000000000000000000
Unit 4	111111111111000000000000000000000000000000000000
Unit 5	111111111111000000000000000000000000000000000000
Unit 6	111111111111000000000000000000000000000000000000
Unit 7	111111111111000000000000000000000000111111111111
Unit 8	111111111111000000000000000000000000000000000000
Unit 9	111111111111000000000000000000000000111111111111
Unit 10	111111111111000000000000000000000000111111111111
Cost	$325,733.00
Profit	$676,267.00

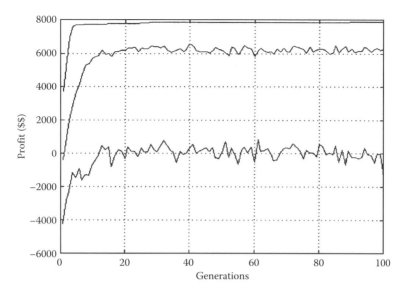

FIGURE 21.12 Max., min., and avg. fitness vs. GA generations for the two-generator, 14 h case.

compared to that of the UC-GA. Since the exhaustive search solution times were estimated to be prohibitively lengthy, the latter cases were not compared against exhaustive search solutions. Cases with known optimal solutions were used to verify that the UC-GA was, in fact, working for the large cases.

Table 21.6 shows the optimal UC schedules found by the UC-GA for selected cases. Figure 21.12 shows the maximum, minimum and average fitnesses (profit) during each generation of the UC-GA on the two-generator, 14-h/period case. The best individual of the population climbs quite rapidly to near the optimal solution. Half of the population is replaced each generation; often the child solutions are poor solutions, hence the minimum fitness tends to remain low over the generations, which is typical for GA optimization.

In the schedules shown in Table 21.6, it may appear as though minimum up- and downtime constraints are being violated. When calculating the cost of such a schedule, the algorithm ensures that the profit is based on a valid schedule by considering a zero surrounded by ones to be a banked unit, and so forth. In addition, note that only the best solution of the population for each of the cases is shown. The existence of additional valid solutions, which may have been only slightly suboptimal in terms of profit, is one of the main advantages of using the GA. It gives the system operator the flexibility to choose the best schedule from a group of schedules to accommodate things like forced maintenance.

21.2.7 Unit Commitment and Auctions

Regardless of the market framework, the solution method, and who is performing the UC, an auction can model and achieve the optimal solution. As mentioned previously in the section on EDC, auctions (which come in many forms, e.g., Dutch, English, sealed, double-sided, single-sided, etc.) are used to match buyers with sellers and to achieve a price that is considered fair. An auction can be used to find the optimal allocation, and the unit commitment algorithm essentially performs the same allocation that an auction would. Suppose an auctioneer was to call out a price, or a set of prices that is predicted for the schedule period. The auctioneer would then ask all generators how much power they would generate at that level. The generator must consider which units to switch on, and at what level to produce and sell. The reply amounts could be summed to determine the production level at that price. If all of the constraints, including demand, are met, then the most economical combination of units operating at the most economical settings has been found. If not, the auctioneer adjusts the price and asks for the amounts at the new price. This procedure is repeated until the constraints are satisfied. Prices may ascend as in the English auction, or they may descend as in the Dutch auction. See Figure 21.3 for a graphical depiction of this process. For further discussion on this topic, the interested reader is referred to Sheblé (1999).

21.3 Summary of Economical Generation Operation

Since the introduction of electricity supply to the public in the late 1800s, people in many parts of the world have grown to expect an inexpensive reliable source of electricity. Providing that electric energy economically and efficiently requires the generation company to carefully control their generating units, and to consider many factors that may affect the performance, cost, and profitability of their operation. The unit commitment and economic dispatch algorithms play an important part in deciding how to operate the electric generating units around the world. The introduction of competition has changed many of the factors considered in solving these problems. Furthermore, advancements in solution techniques offer a continuum of candidate algorithms, each having its own advantages and disadvantages. Research continues to push these algorithms further. This chapter has provided the reader with an introduction to the problems of determining optimal unit commitment schedules and economic dispatches. It is by no means exhaustive, and the interested reader is strongly encouraged to see the references at the end of the chapter for more details.

References

Goldberg, D., *Genetic Algorithms in Search, Optimization and Machine Learning*. Addison-Wesley Publishing Company, Inc., Reading, MA, 1989.

Kazarlis, S.A., Bakirtzis, A.G., and Petridis, V., A genetic algorithm solution to the unit commitment problem, *1995 IEEE/PES Winter Meeting*, 152-9 PWRS, New York, 1995.

Kondragunta, S., Genetic algorithm unit commitment program, MS thesis, Iowa State University, Ames, IA, 1997.

Maifeld, T. and Sheblé, G., Genetic-based unit commitment, *IEEE Trans. Power Syst.*, 11, 1359, August 1996.

Richter, C. and Sheblé, G., A Profit based unit commitment GA for the competitive environment, *IEEE Trans. Power Syst.*, 15(2), 715–721, 2000.

Sheblé, G., Unit commitment for operations, PhD dissertation, Virginia Polytechnic Institute and State University, Blacksburg, VA, March 1985.

Sheblé, G., *Computational Auction Mechanisms for Restructured Power Industry Operation*. Kluwer Academic Publishers, Boston, MA, 1999.

Sheblé, G. and Fahd, G., Unit commitment literature synopsis, *IEEE Trans. Power Syst.*, 9, 128–135, February 1994.

Takriti, S., Krasenbrink, B., and Wu, L.S.-Y., Incorporating fuel constraints and electricity spot prices into the stochastic unit commitment problem, IBM Research Report: RC 21066, Mathematical Sciences Department, T.J. Watson Research Center, Yorktown Heights, New York, December 29, 1997.

Walters, D.C. and Sheblé, G.B., Genetic algorithm solution of economic dispatch with valve point loading, *1992 IEEE/PES Summer Meeting*, 414-3, New York, 1992.

Wood, A. and Wollenberg, B., *Power Generation, Operation, and Control*. John Wiley & Sons, New York, 1996.

22
State Estimation

Jason G. Lindquist
Siemens Energy Automation

Danny Julian
ABB Inc.

22.1	State Estimation Problem...22-2
	Underlying Assumptions • Measurement Representations • Solution Methods
22.2	State Estimation Operation...22-7
	Network Topology Assessment • Error Identification • Unobservability
22.3	Example State Estimation Problem 22-9
	System Description • WLS State Estimation Process
22.4	Defining Terms 22-12
	References...22-12

An online AC power flow is a valuable application when determining the critical elements affecting power system operation and control such as overloaded lines, credible contingencies, and unsatisfactory voltages. It is the basis for any real-time security assessment and enhancement applications.

AC power flow algorithms calculate real and reactive line flows based on a multitude of inputs with generator bus voltages, real power bus injections, and reactive power bus injections being a partial list. This implies that in order to calculate the line flows using a power flow algorithm, all of the input information (voltages, real power injections, reactive power injections, etc.) must be known a priori to the algorithm being executed.

An obvious way to implement an online AC power flow is to telemeter the required input information at every location in the power system. This would require not only a large number of remote terminal units (RTUs), but also an extensive communication infrastructure to telemeter the data to the supervisory control and data acquisition (SCADA) system, both of which are costly. The generator bus voltages are usually readily available but the injection data are frequently lacking because it is much easier and cheaper to monitor the net injection at a bus than to measure separate injections directly. Also, this approach presents weaknesses for the online AC power flow that are due to meter accuracy and communication failure. An online power flow relying on a specific set of measurements could become unusable or give erroneous results if any of the predefined measurements became unavailable due to communication or measurement device failure. This is not a desirable outcome of an online application designed to alert system operators to unsecure conditions.

Given the obstacles of utilizing an online AC power flow, work was conducted in the late 1960s and early 1970s (Schweppe and Wildes, 1970) into developing a process of performing an online power flow using not just the limited data needed for the classical AC power flow algorithm, but using all available measurements. This work led to the *power system state estimator*, which uses not only the aforementioned voltages but other telemetered measurements such as real and reactive line flows, circuit breaker statuses, and transformer tap settings.

22.1 State Estimation Problem

State estimators perform a statistical analysis using a set of imperfect but redundant data telemetered from the power system to determine the state of the system. The state of the system is defined as a complete set of bus voltage magnitudes and phase angles. A state estimate solution is obtained by fitting the measurement set to a model of the network by minimizing the state estimator's objective function. A necessary requirement to obtain a complete representation of the system is to have the number of measurements, m, greater than the number of state variables, n. This is known as the observability criterion. Typically, m is two to three times the value of n, allowing for a considerable amount of redundancy in the measurement set.* Once a state estimate is obtained, all other system quantities can be calculated, including bus injections and branch flows.

22.1.1 Underlying Assumptions

A number of assumptions must be made to obtain a state estimate solution. A primary assumption is that the system is operating under normal conditions and balanced. A second assumption is made upon the system measurements. Telemetered measurements usually are corrupted since they are susceptible to noise. Statistical properties of the measurements allow certain assumptions to be made to estimate the true measured value. It is assumed that the measurement noise is normally distributed with an expected value of zero and that the correlation between measurements is zero (i.e., independent).† A variable, v, is said to be normal (or Gaussian) if its probability density function has the following form where σ is the standard deviation:

$$f(v) = \frac{1}{\sigma\sqrt{2\pi}} e^{-v^2/2\sigma^2} \tag{22.1}$$

The normal distribution (Figure 22.1) is used for the modeling of measurement errors since it is the distribution that results when many factors contribute to the overall error.

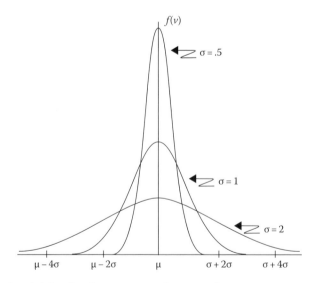

FIGURE 22.1   Normal probability distribution curve with a mean of μ.

---

* The number of state variables is equal to $(2n_b - 1)$, including n_b bus voltage magnitudes and $(n_b - 1)$ bus voltage angles. The phase angle of the reference bus is set to 0.0.
† In practice, measurements i and j are not necessarily independent since one measurement device may measure more than one value. Therefore, if the measurement device is bad, probably both measurements i and j are bad also.

State Estimation

Figure 22.1 also illustrates the effect of standard deviation on the normal density function. Standard deviation, σ, is a measure of the spread of the normal distribution about the mean (μ) and gives an indication of how many samples fall within a given interval around the mean. A large standard deviation implies there is a high probability that the measurement noise will take on large values. Conversely, a small standard deviation implies there is a high probability that the measurement noise will take on small values.

22.1.2 Measurement Representations

Since a measurement is not exact, it can be expressed with an error component:

$$z = z_T + v \qquad (22.2)$$

where
  z is the measured value
  z_T is the true value
  v is the measurement error that represents uncertainty in the measurement

The measured value, z, can be related to the states, x, by the measurement function, $h(x)$, in Equation 22.3:

$$z = h(x) + v \qquad (22.3)$$

The measurement function, $h(x)$, is a vector of nonlinear functions relating the measurements to the state variables, x. An example of the $h(x)$ vector can be shown using the transmission line in Figure 22.2.
The equations that relate the bus voltages to the branch flow between bus i and j are

$$P_{ij} = |\tilde{V}_i|^2 (g_{ij} + g_{ish}) - |\tilde{V}_i||\tilde{V}_j|\left[g_{ij}\cos(\delta_{ij}) + b_{ij}\sin(\delta_{ij})\right] \qquad (22.4)$$

$$Q_{ij} = -|\tilde{V}_i|^2 (b_{ij} + b_{ish}) - |\tilde{V}_i||\tilde{V}_j|\left[g_{ij}\sin(\delta_{ij}) - b_{ij}\cos(\delta_{ij})\right] \qquad (22.5)$$

In these equations, $|\tilde{V}_i|$ is the magnitude of the voltage at bus i, $|\tilde{V}_j|$ is the magnitude of the voltage at bus j, δ_{ij} is the phase angle difference between bus i and bus j, g_{ij} and b_{ij} are the conductance and susceptance of line i-j, respectively, and g_{ish} and b_{ish} are the shunt conductance and susceptance at bus i, respectively.

To define the bus injection equations, first define the bus-admittance matrix, $Y = G + jB$. The bus-admittance matrix, Y, defines the relationship between the net current injections vector, $\bar{I}$, and the complex bus voltages vector, $\bar{V}$:

$$\bar{I} = Y \cdot \bar{V} \qquad (22.6)$$

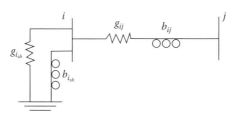

FIGURE 22.2 Transmission line representation.

The bus-admittance matrix is built by adding each branch admittance, $y_{ij} = g_{ij} + jb_{ij}$, values to the matrix Y:

$$Y'_{ii} = Y_{ii} + y_{ij}$$
$$Y'_{ij} = Y_{ij} - y_{ij}$$
$$Y'_{ji} = Y_{ji} - y_{ij}$$
$$Y'_{jj} = Y_{jj} + y_{ij}$$
(22.7)

The bus injection at bus i is then calculated by summing over all connected buses (i.e., $j \in N_i$):

$$P_i = |\tilde{V}_i| \sum_{j \in N_i} |\tilde{V}_j| \left[G_{ij} \cos(\delta_{ij}) + B_{ij} \sin(\delta_{ij}) \right]$$
$$Q_i = |\tilde{V}_i| \sum_{j \in N_i} |\tilde{V}_j| \left[G_{ij} \sin(\delta_{ij}) - B_{ij} \cos(\delta_{ij}) \right]$$
(22.8)

22.1.3 Solution Methods

The solution to the state estimation problem has been addressed by a broad class of techniques (Filho et al., 1990) and differs from power flow algorithms in two modes:

1. Certain input data are either missing or inexact.
2. The algorithm used for the calculation may entail approximations and approximate methods designed for high speed processing in the online environment.

In this section, two different solution methods to the state estimation problem will be introduced and described.

22.1.3.1 Weighted Least Squares

The most common approach to solving the state estimation problem is the method of weighted least squares (WLS) state estimation. This is accomplished by identifying the values of the state variables that minimize the performance index, J, which is the weighted sum of the square errors:

$$J = \bar{e}^T R^{-1} \bar{e}$$
(22.9)

The weighting factor is the inverse of the diagonal covariance matrix, R, of the measurements, Equation 22.10:

$$E\left[\overline{vv}^T\right] = R = \begin{bmatrix} \sigma_1^2 & 0 & 0 & 0 & 0 \\ 0 & \sigma_2^2 & 0 & 0 & 0 \\ 0 & 0 & \cdots & 0 & 0 \\ 0 & 0 & 0 & \cdots & 0 \\ 0 & 0 & 0 & 0 & \sigma_m^2 \end{bmatrix}$$
(22.10)

As a result, measurements of a higher quality have smaller variances resulting in their weights having higher values, while measurements with poor quality have smaller weights due to the correspondingly larger variance values.

State Estimation

By defining the error, e, in Equation 22.9 as the difference between the true measured value, z, and the estimated measured value, $\hat{z}$,

$$\bar{e} = \bar{z} - \hat{\bar{z}} \tag{22.11}$$

a new form for the performance index can be written as

$$J = \left(\bar{z} - \bar{h}(x)\right)^T R^{-1}\left(\bar{z} - \bar{h}(x)\right) \tag{22.12}$$

In order to minimize the performance index, J, a first-order necessary condition must hold, namely,

$$\left.\frac{\partial J}{\partial \bar{x}}\right|_{x^k} = 0 \tag{22.13}$$

Evaluating Equation 22.12 at the necessary condition gives the following:

$$H(x^k)^T R^{-1}\left(\bar{z} - \bar{h}(x)\right) = 0 \tag{22.14}$$

where $H(x)$ represents the $m \times n$ measurement Jacobian matrix evaluated at iteration k, Equation 22.13a*:

$$H(x) = \begin{bmatrix} \frac{\partial h_1}{\partial x_1} & \frac{\partial h_1}{\partial x_2} & \cdots & \frac{\partial h_1}{\partial x_n} \\ \frac{\partial h_2}{\partial x_1} & \frac{\partial h_2}{\partial x_2} & \cdots & \frac{\partial h_2}{\partial x_n} \\ \cdots & \cdots & \cdots & \cdots \\ \cdots & \cdots & \cdots & \cdots \\ \frac{\partial h_m}{\partial x_1} & \frac{\partial h_m}{\partial x_2} & \cdots & \frac{\partial h_m}{\partial x_n} \end{bmatrix}_{x^k} \tag{22.15}$$

An iterative solution is obtained by linearizing the relationship between the measurements and the state variables using the Taylor series expansion of the function $h(x)$ around a point x^k:

$$\bar{h}(x^k) = \bar{h}(x^k) + \Delta \bar{x}^k \frac{\partial \bar{h}(x^k)}{\partial \bar{x}} + \text{higher order terms} \tag{22.16}$$

This set of equations can be solved using an iterative approach such as Newton–Raphson's method. At each iteration, the state vector, x, is updated using Equation 22.17:

$$\bar{x}^{k+1} = \bar{x}^k + \left(H(x^k)^T R^{-1} H(x^k)\right)^{-1} H(x^k)^T R^{-1}\left(\bar{z} - \bar{h}(x^k)\right) \tag{22.17}$$

---

* m represents the number of measurements and n represents the number of states.

Convergence is obtained once Equation 22.18 is satisfied, where ε is some predetermined convergence factor:

$$\max(\bar{x}^{k+1} - \bar{x}^k) \leq \varepsilon \tag{22.18}$$

At convergence, the solution $\bar{x}^{k+1}$ corresponds to the WLS estimates of the state variables.

22.1.3.2 Linear Programming

Another solution method that addresses the state estimation problem is linear programming. Linear programming is an optimization technique that serves to minimize a linear objective function subject to a set of constraints:

$$\min\{\bar{c}^T \bar{x}\}$$
$$\text{s.t. } A\bar{x} = \bar{b} \tag{22.19}$$
$$\bar{x} \geq 0$$

There are many different techniques associated with solving linear programming problems including the simplex and interior point methods.

Since the objective function, as expressed in Equation 22.12, is quadratic in terms of the unknowns (states), it must be rewritten in a linear form. This is accomplished by first rewriting the measurement error, as expressed in Equation 22.3, in terms of a positive measurement error, v_p, and a negative measurement error, v_n:

$$\bar{z} = \bar{h}(x) + \bar{v}$$
$$= \bar{h}(x) + \bar{v}_p - \bar{v}_n \tag{22.20}$$

Restricting the positive and negative measurement errors to only nonnegative values insures that the problem is bounded. This was not a concern in the WLS approach since a quadratic function is convex and is guaranteed to contain a global minimum.

Using the new definition of a measurement described in Equation 22.20 and the inverse of the diagonal covariance matrix of the measurements for weights as described in the WLS approach, the objective function can now be written as

$$J = R^{-1}(\bar{v}_p + \bar{v}_n) \tag{22.21}$$

The constraints are the equations relating the state vector to the measurements as shown in Equation 22.20. Once again, since $h(x)$ is nonlinear, it must be linearized around a point x^k by expanding the Taylor series, as was performed previously in the WLS approach. The solution to the state estimation problem can then be determined by solving the following linear program:

$$\min\{R^{-1}(\bar{v}_p + \bar{v}_n)\}$$
$$\text{s.t. } \Delta \bar{z}^k - H(x^k)\Delta \bar{x}^k + \bar{v}_p - \bar{v}_n = 0 \tag{22.22}$$
$$\bar{v}_p \geq 0$$
$$\bar{v}_n \geq 0$$

where $H(x^k)$ represents the $m \times n$ measurement Jacobian matrix evaluated at iteration k as defined in Equation 22.15.

22.2 State Estimation Operation

State estimators are typically executed either periodically (i.e., every 5 min), on demand, or due to a status change such as a breaker operation isolating a line section. To illustrate the relationship of the state estimator with respect to other energy management system (EMS) applications, a simple depiction of an EMS is shown in Figure 22.3.

As shown, the state estimator receives inputs from the SCADA system and the network topology assessment applications and stores the state of the system in a central location (i.e., database). Power system applications, such as contingency analysis and optimal power flow, can then be executed based on the state of the system as computed by the state estimator.

22.2.1 Network Topology Assessment

Before the state estimator is executed in real time, the topology of the network is determined. This is accomplished by a system or **network processor** that establishes the configuration of the power system network based on telemetered breaker and switch statuses. The network processor normally addresses questions like

- Have breaker operations caused individual buses to either be split into two or more isolated buses, or combined into a single bus?
- Have lines been opened or restored to service?

The state estimator then uses the network determined by the network processor, which consists only of energized (online) lines and devices, as a basis for the calculations to determine the state of the system.

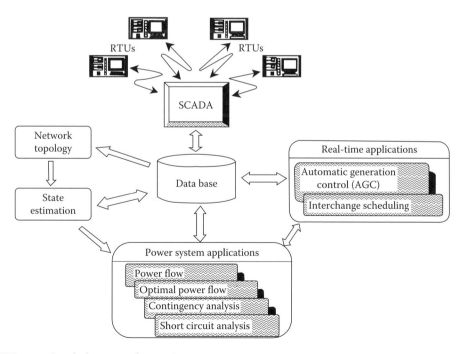

FIGURE 22.3 Simple depiction of an EMS.

22.2.2 Error Identification

Since state estimators utilize telemetered measurements and network parameters as a foundation for their calculations, the performance of the state estimator depends on the accuracy of the measured data as well as the parameters of the network model. Fortunately, the use of all available measurements introduces a favorable secondary effect caused by the redundancy of information. This redundancy provides the state estimator with more capabilities than just an online AC power flow; it introduces the ability to detect and identify "bad" data. Bad data can come from many sources, such as

- Approximations
- Simplified model assumptions
- Human data handling errors
- Measurement errors due to faulty devices (e.g., transducers, current transformers)

22.2.2.1 Telemetered Data

The ability to detect and identify bad measurements is an extremely useful feature of the state estimator. Without the state estimator, obviously wrong telemetered measurements would have little chance of being identified. With the state estimator, operation personnel can have a greater confidence that telemetered data are not grossly in error.

Data are tagged as "bad" when the estimated value is unreasonably different from the measured/telemetered value obtained from the RTU. As a simple example, suppose a bus voltage is measured to be 1.85 pu and is estimated to be 0.95 pu. In this case, the bus voltage measurement could be tagged as bad. Once data are tagged as bad, they should be removed from the measurement set before being utilized by the state estimator.

Most state estimators rely on a combination of pre-estimation and post-estimation schemes for detection and elimination of bad data. Pre-estimation involves gross bad data detection and consistency tests. Data are identified as bad in pre-estimation by the detection of gross measurement errors such as zero voltages or line flows that are outside reasonable limits using network topology assessment. Consistency tests classify data as valid, suspect, or raw for use in post-estimation analysis by using statistical properties of related measurements. Measurements are classified as valid if they pass a consistency test that separates measurements into subsets based on a consistency threshold. If the measurement fails the consistency test, it is classified as suspect. Measurements are classified as raw if a consistency test cannot be made and they cannot be grouped into any subset. Raw measurements typically belong to nonredundant portions of the complete measurement set.

Post-estimation involves performing a statistical analysis (e.g., hypothesis testing) on the normalized measurement residuals. A normalized residual is defined as

$$r_i = \frac{z_i - h_i(x)}{\sigma_i} \quad (22.23)$$

where σ_i is the ith diagonal term of the covariance matrix, R, as defined in Equation 22.10. Data are identified as bad in post-estimation typically when the normalized residuals of measurements classified as suspect lie outside a predefined confidence interval (i.e., fail the chi-square test).

22.2.2.2 Parameter Data

In parameter error identification, network parameters (i.e., admittances) that are suspicious are identified and need to be estimated. The use of faulty network parameters can severely impact the quality of state estimation solutions and cause considerable error. A requirement for parameter estimation is that all parameters be identifiable by measurements. This requirement implies the lines under consideration have associated measurements, thereby increasing the size of the measurement set by l, where l

is the number of parameters to be estimated. Therefore, if parameter estimation is to be performed, the observability criterion must be augmented to become $m \geq n + l$.

22.2.2.3 Topology Data

A fundamental assumption of power system state estimation is that the topology of the network is accurately known. When this is false, the solution obtained by the state estimator will not correspond to the true system and can result in divergence, false bad data detection, large measurement residuals, and false or undetected limit violations. There are two main categories of topology errors:

1. Branch status error—false inclusion or exclusion of a system branch from the network model
2. Substation configuration error—false merging or splitting of substation bus sections into a single or multiple buses

A topology error is the result of incorrect system switch and breaker statuses. Because these logical devices are not explicitly modeled in the network model used by the state estimator, detection and identification of topology errors are considerably more difficult than the error processing of measurements and parameter data.

22.2.3 Unobservability

By definition, a state variable is unobservable if it cannot be estimated. Unobservability occurs when the observability criterion is violated ($m < n$) and there are insufficient redundant measurements to determine the state of the system. Mathematically, the gain matrix $H(x^k)^T R^{-1} H(x^k)$ of Equation 22.17 becomes singular and cannot be inverted.

The obvious solution to the unobservability problem is to increase the number of measurements. The problem then becomes where and how many measurements need to be added to the measurement set. Adding additional measurements is costly since there are many supplementary factors that must be addressed in addition to the cost of the measuring device such as RTUs, communication infrastructure, and software data processing at the EMS. A number of approaches have been suggested that try to minimize the cost while satisfying the observability criterion (Baran et al., 1995; Park et al., 1998).

Another solution to address the problem of unobservability is to augment the measurement set with pseudo-measurements to reach an observability condition for the network. When adding pseudo-measurements to a network, the equation of the pseudo-measured quantity is substituted for actual measurements. In this case, the measurement covariance values in Equation 22.10 associated with these measurements should have large values that allow the state estimator to treat the pseudo-measurements as if they were measured from a very poor metering device. The most common type of pseudo-measurements used are bus injections. These pseudo-measurements can be created from historical information or generated from load forecasting and generation dispatch.

22.3 Example State Estimation Problem

This section provides a simple example to illustrate how the state estimation process is performed. The WLS method, as previously described, will be applied to a sample system.

22.3.1 System Description

A sample three-bus system is shown in Figure 22.4. Bus 1 is assumed to be the reference bus with a corresponding angle of zero. All other relevant system data are given in Table 22.1.

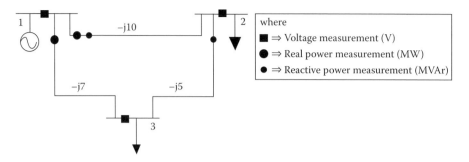

FIGURE 22.4 Sample three-bus power flow system.

TABLE 22.1 Sample System Data

Measurement Type	Measurement Location	Measurement Value (pu)	Measurement Covariance (σ)
$\|\tilde{V}\|$	Bus 1	1.02	0.05
$\|\tilde{V}\|$	Bus 2	1.0	0.05
$\|\tilde{V}\|$	Bus 3	0.99	0.05
P	Bus 1 – Bus 2	1.5	0.1
Q	Bus 1 – Bus 2	0.2	0.1
P	Bus 1 – Bus 3	1.0	0.1
Q	Bus 2 – Bus 3	0.1	0.1

22.3.2 WLS State Estimation Process

First, the states (x) are defined as the angles at bus 2 and bus 3 and the voltage magnitudes at all buses*:

$$\bar{x}^T = \begin{bmatrix} \delta_2 & \delta_3 & |\tilde{V}_1| & |\tilde{V}_2| & |\tilde{V}_3| \end{bmatrix}$$

The measurement vector is defined as

$$\bar{z}^T = \begin{bmatrix} |\tilde{V}_1| & |\tilde{V}_2| & |\tilde{V}_3| & P_{12} & Q_{12} & P_{13} & Q_{13} \end{bmatrix}$$

This gives a total of seven measurements and five states that satisfy the observability criterion requiring more measurements than states. Using the previously defined equations for the WLS state estimation procedure, the following can be determined:

$$R = \begin{bmatrix} (.05)^2 & 0 & 0 & 0 & 0 & 0 & 0 \\ 0 & (.05)^2 & 0 & 0 & 0 & 0 & 0 \\ 0 & 0 & (.05)^2 & 0 & 0 & 0 & 0 \\ 0 & 0 & 0 & (.10)^2 & 0 & 0 & 0 \\ 0 & 0 & 0 & 0 & (.10)^2 & 0 & 0 \\ 0 & 0 & 0 & 0 & 0 & (.10)^2 & 0 \\ 0 & 0 & 0 & 0 & 0 & 0 & (.10)^2 \end{bmatrix}$$

---

* The angle at bus 1 is not chosen as a state since it is designated as the reference bus.

$$H(x) = \begin{bmatrix} 0 & 0 & 1 & 0 & 0 \\ 0 & 0 & 0 & 1 & 0 \\ 0 & 0 & 0 & 0 & 1 \\ -10 & 0 & 0 & 0 & 0 \\ 0 & 0 & 10 & -10 & 0 \\ 0 & -7 & 0 & 0 & 0 \\ 0 & 0 & 0 & 5 & -5 \end{bmatrix}$$

A flat start is used as the initial guess for the state vector, with zero for the voltage angles and 1.0 for the voltage magnitudes. The state estimator converges to a solution after for iterations, Table 22.2.

The system values can be calculated from the estimated voltages and angles. Table 22.3 gives the bus voltage and injection values and Table 22.4 gives the branch power flow values. With the state of the system now known, other applications such as contingency analysis and optimal power flow may be performed. Notice the state estimation process results in the state of the system, just as when performing a power flow but without a priori knowledge of bus injections.

TABLE 22.2 State Estimation Solution

	Iteration			
State (pu)	0	1	2	3
δ_2	0.0	−0.150	−0.147	−0.147
δ_3	0.0	−0.143	−0.142	−0.143
$|\tilde{V}_1|$	1.0	1.023	1.015	1.016
$|\tilde{V}_2|$	1.0	1.003	1.007	1.007
$|\tilde{V}_3|$	1.0	0.984	0.988	0.987

TABLE 22.3 Bus Values

	Voltage		Generation		Load	
Bus Number	Magnitude (pu)	Angle (rad)	P (pu)	Q (pu)	P (pu)	Q (pu)
1	1.016	0.0	2.501	0.477	—	—
2	1.007	−0.147	—	—	−1.522	0.117
3	0.987	−0.143	—	—	−0.979	−0.221

TABLE 22.4 Branch Values

Branch Number	Bus		From Bus Injection		To Bus Injection	
	From	To	P (pu)	Q (pu)	P (pu)	Q (pu)
1	1	2	1.50	0.203	−1.50	0.019
2	1	3	1.0	0.274	−1.0	−0.125
3	2	3	−0.021	0.098	0.021	−0.096

22.4 Defining Terms

Remote terminal unit (RTU): Hardware that telemeters system-wide data from various field locations (i.e., substations, generating plants) to a central location.

State estimator: An application that uses a statistical process in order to estimate the state of the system.

State variable: The quantity to be estimated by the state estimator, typically bus voltage and angle.

Network processor: An application that determines the configuration of the power system based on telemetered breaker and switch statuses.

Supervisory control and data acquisition (SCADA): A computer system that performs data acquisition and remote control of a power system.

Energy management system (EMS): A computer system that monitors, controls, and optimizes the transmission and generation facilities with advanced applications. A SCADA system is a subset of an EMS.

References

Baran, M.E. et al., A meter placement method for state estimation, *IEEE Trans. Power Syst.*, 10(3), 1704–1710, August 1995.

Filho, M.B.D.C. et al., Bibliography on power system state estimation (1968–1989), *IEEE Trans. Power Syst.*, 5(3), 950–961, August 1990.

Park, Y.M. et al., Design of reliable measurement system for state estimation, *IEEE Trans. Power Syst.*, 3(3), 830–836, August 1998.

Schweppe, F.C. and Wildes, J., Power system static-state estimation I, II, III, *IEEE Trans. Power Appar. Syst.*, 89, 120–135, January 1970.

23
Optimal Power Flow

23.1	Conventional Optimal Economic Scheduling 23-2
23.2	Conventional OPF Formulation 23-3
	Application of Optimization Methods to OPF
23.3	OPF Incorporating Load Models 23-7
	Load Modeling • Static Load Models • Conventional OPF Studies Including Load Models • Security Constrained OPF Including Load Models • Inaccuracies of Standard OPF Solutions
23.4	SCOPF Including Load Modeling... 23-9
	Influence of Fixed Tap Transformer Fed Loads
23.5	Operational Requirements for Online Implementation......... 23-11
	Speed Requirements • Robustness of OPF Solutions with Respect to Initial Guess Point • Discrete Modeling • Detecting and Handling Infeasibility • Consistency of OPF Solutions with Other Online Functions • Ineffective "Optimal" Rescheduling • OPF-Based Transmission Service Pricing
23.6	Conclusions.. 23-14
	References... 23-14

Mohamed E.
El-Hawary
Dalhousie University

An optimal power flow (OPF) function schedules the power system controls to optimize an objective function while satisfying a set of nonlinear equality and inequality constraints. The equality constraints are the conventional power flow equations; the inequality constraints are the limits on the control and operating variables of the system. Mathematically, the OPF can be formulated as a constrained nonlinear optimization problem. This section reviews features of the problem and some of its variants as well as requirements for online implementation.

Optimal scheduling of the operations of electric power systems is a major activity, which turns out to be a large-scale problem when the constraints of the electric network are taken into account. This document deals with recent developments in the area emphasizing OPF formulation and deals with conventional OPF, accounting for the dependence of the power demand on voltages in the system, and requirements for online implementation.

The OPF problem was defined in the early 1960s (Burchett et al., 1982) as an extension of conventional economic dispatch (ED) to determine the optimal settings for control variables in a power network respecting various constraints. OPF is a static constrained nonlinear optimization problem, whose development has closely followed advances in numerical optimization techniques and computer technology. It has since been generalized to include many other problems. Optimization of the electric system with losses represented by the power flow equations was introduced in the 1960s (Carpentier, 1962; Dommel and Tinney, 1968). Since then, significant effort has been spent on achieving faster and robust solution methods that are suited for online implementation, operating practice, and security requirements.

OPF seeks to optimize a certain objective, subject to the network power flow constraints and system and equipment operating limits. Today, any problem that involves the determination of the instantaneous "optimal" steady state of an electric power system is referred to as an OPF problem. The optimal steady state is attained by adjusting the available controls to minimize an objective function subject to specified operating and security requirements. Different classes of OPF problems, designed for special-purpose applications, are created by selecting different functions to be minimized, different sets of controls, and different sets of constraints. All these classes of the OPF problem are subsets of the general problem. Historically, different solution approaches have been developed to solve these different classes of OPF. Commercially available OPF software can solve very large and complex formulations in a relatively short time, but may still be incapable of dealing with online implementation requirements.

There are many possible objectives for an OPF. Some commonly implemented objectives are

- Fuel or active power cost optimization
- Active power loss minimization
- Minimum control-shift
- Minimum voltage deviations from unity
- Minimum number of controls rescheduled

In fuel cost minimization, the outputs of all generators, their voltages, LTC transformer taps and LTC phase shifter angles, and switched capacitors and reactors are control variables. The active power losses can be minimized in at least two ways (Happ and Viearth, 1986). In both methods, all the above variables are adjusted except for the active power generation. In one method, the active power generation at the swing bus is minimized while keeping all other generation constant at prespecified values. This effectively minimizes the total active power losses. In another method, an actual expression for the losses is minimized, thus allowing the exclusion of lines in areas not optimized.

The behavior of the OPF solutions during contingencies was a major concern, and as a result, security constrained optimal power flow (SCOPF) was introduced in the early 1970s. Subsequently, online implementations became a new thrust in order to meet the challenges of new deregulated operating environments.

23.1 Conventional Optimal Economic Scheduling

Conventional optimal economic scheduling minimizes the total fuel cost of thermal generation, which may be approximated by a variety of expressions such as linear or quadratic functions of the active power generation of the unit. The total active power generation in the system must equal the load plus the active transmission losses, which can be expressed by the celebrated Kron's loss formula. Reserve constraints may be modeled depending on system requirements. Area and system spinning, supplemental, emergency, or other types of reserve requirements involve functional inequality constraints. The forms of the functions used depend on the type of reserve modeled. A linear form is evidently most attractive from a solution method point of view. However, for thermal units, the spinning reserve model is nonlinear due to the limit on a unit's maximum reserve contribution. Additional constraints may be modeled, such as area interchange constraints used to model network transmission capacity limitations. This is usually represented as a constraint on the net interchange of each area with the rest of the system (i.e., in terms of limits on the difference between area total generation and load).

The objective function is augmented by the constraints using a Lagrange-type multiplier lambda, λ. The optimality conditions are made up of two sets. The first is the problem constraints. The second set is based on variational arguments giving for each thermal unit:

$$\frac{\partial F_i}{\partial P_i} = \lambda \left[1 - \frac{\partial P_L}{\partial P_i} \right] \quad i = 1, \ldots, N \tag{23.1}$$

The optimality conditions along with the physical constraints are a set of nonlinear equations that requires iterative methods to solve. Newton's method has been widely accepted in the power industry as a powerful tool to solve problems such as the load flow and optimal load flow. This is due to its reliable and fast convergence, known to be quadratic.

A solution can usually be obtained within a few iterations, provided that a reasonably good initial estimate of the solution is available. It is therefore appropriate to employ this method to solve the present problem.

23.2 Conventional OPF Formulation

The OPF is a constrained optimization problem requiring the minimization of

$$f = (x, u) \tag{23.2}$$

subject to

$$g(x, u) = 0 \tag{23.3}$$

$$h(x, u) \leq 0 \tag{23.4}$$

$$u^{\min} \leq u \leq u^{\max} \tag{23.5}$$

$$x^{\min} \leq x \leq x^{\max} \tag{23.6}$$

Here $f(x, u)$ is the scalar objective function, $g(x, u)$ represents nonlinear equality constraints (power flow equations), and $h(x, u)$ is the nonlinear inequality constraint of vector arguments x and u. The vector **x** contains dependent variables consisting of bus voltage magnitudes and phase angles, as well as the MVAr output of generators designated for bus voltage control and fixed parameters such as the reference bus angle, noncontrolled generator MW and MVAr outputs, noncontrolled MW and MVAr loads, fixed bus voltages, line parameters, etc. The vector u consists of control variables including

- Real and reactive power generation
- Phase-shifter angles
- Net interchange
- Load MW and MVAr (load shedding)
- DC transmission line flows
- Control voltage settings
- LTC transformer tap settings

Examples of equality and inequality constraints are

- Limits on all control variables
- Power flow equations
- Generation/load balance
- Branch flow limits (MW, MVAr, MVA)
- Bus voltage limits
- Active/reactive reserve limits
- Generator MVAr limits
- Corridor (transmission interface) limits

The power system consists of a total of N buses, N_G of which are generator buses. M buses are voltage controlled, including both generator buses and buses at which the voltages are to be held constant. The voltages at the remaining (N − M) buses (load buses), must be found.

The network equality constraints are represented by the load flow equations:

$$P_i(V,\delta) - P_{gi} + P_{di} = 0 \tag{23.7}$$

$$Q_i(V,\delta) - Q_{gi} + Q_{di} = 0 \tag{23.8}$$

Two different formulation versions can be considered.

1. *Polar form*:

$$P_i(V,\delta) = |V_i| \sum_{1}^{N} |V_j||Y_{ij}|\cos(\delta_i - \delta_j - \phi_{ij}) \tag{23.9}$$

$$Q_i(V,\delta) = |V_i| \sum_{1}^{N} |V_j||Y_{ij}|\sin(\delta_i - \delta_j - \phi_{ij}) \tag{23.10}$$

$$Y_{ij} = |Y_{ij}|\underline{/\phi_{ij}} \tag{23.11}$$

where
P_i is the active power injection into bus i
Q_i is the reactive power injection into bus i
$|V_i|$ is the voltage magnitude of bus i
δ_i is the angle at bus i
$|\tilde{Y}_{ij}|, \varphi_{ij}$ is the magnitude and angle of the admittance matrix
P_{di}, Q_{di} is the active and reactive load on bus i

2. *Rectangular form*:

$$P_i(e,f) = e_i\left[\sum_{1}^{N}(G_{ij}e_j - B_{ij}f_j)\right] + f_i\left[\sum_{1}^{N}(G_{ij}f_j + B_{ij}e_j)\right] \tag{23.12}$$

$$Q_i(e,f) = f_i\left[\sum_{1}^{N}(G_{ij}e_j - B_{ij}f_j)\right] + e_i\left[\sum_{1}^{N}(G_{ij}f_j + B_{ij}e_j)\right] \tag{23.13}$$

where
e_i is the real part of complex voltage at bus i
f_i is the imaginary part of the complex voltage at bus i
G_{ij} is the real part of the complex admittance matrix
B_{ij} is the imaginary part of the complex admittance matrix

The control variables vary according to the objective being minimized. For fuel cost minimization, they are usually the generator voltage magnitudes, generator active powers, and transformer tap ratios. The dependent variables are the voltage magnitudes at load buses, phase angles, and reactive generations.

23.2.1 Application of Optimization Methods to OPF

Various optimization methods have been proposed to solve the OPF problem, some of which are refinements on earlier methods. These include

1. Generalized reduced gradient (GRG) method
2. Reduced gradient method
3. Conjugate gradient methods
4. Hessian-based method
5. Newton's method
6. Linear programming methods
7. Quadratic programming methods
8. Interior point methods

Some of these techniques have spawned production OPF programs that have achieved a fair level of maturity and have overcome some of the earlier limitations in terms of flexibility, reliability, and performance requirements.

23.2.1.1 Generalized Reduced Gradient Method

The GRG method, due to Abadie and Carpentier (1969), is an extension of the Wolfe's reduced gradient method (Wolfe, 1967) to the case of nonlinear constraints. Peschon et al. in 1971 and Carpentier in 1973 used this method for OPF. Others have used this method to solve the OPF problem since then (Lindqvist et al., 1984; Yu et al., 1986).

23.2.1.2 Reduced Gradient Method

A reduced gradient method was used by Dommel and Tinney (1968). An augmented Lagrangian function is formed. The negative of the gradient $\partial L/\partial u$ is the direction of steepest descent. The method of reduced gradient moves along this direction from one feasible point to another with a lower value of f, until the solution does not improve any further. At this point an optimum is found, if the Kuhn–Tucker conditions (1951) are satisfied. Dommel and Tinney used Newton's method to solve the power flow equations.

23.2.1.3 Conjugate Gradient Method

In 1982, Burchett et al. used a conjugate gradient method, which is an improvement on the reduced gradient method. Instead of using the negative gradient ∇f as the direction of steepest descent, the descent directions at adjacent points are linearly combined in a recursive manner.

$$\Gamma_k = -\nabla f + \beta_k \Gamma_{k-1} \quad \beta_0 = 0 \tag{23.14}$$

Here, β_k is the descent direction at iteration "k."

Two popular methods for defining the scalar value β_k are the Fletcher-Reeves method (Carpentier, 1973) and the Polak–Ribiere method (1969).

23.2.1.4 Hessian-Based Methods

Sasson (1969) discusses methods (Fiacco and McCormick, 1964; Lootsma, 1967; Zangwill, 1967) that transform the constrained optimization problem into a sequence of unconstrained problems. He uses a transformation introduced by Fletcher and Powell (1963). Here, the Hessian matrix is not evaluated directly. Instead, it is built indirectly starting initially with the identity matrix so that at the optimum point it becomes the Hessian itself.

Due to drawbacks of the Fletcher-Powell method, Sasson et al. (1973) developed a Hessian load flow with an extension to OPF. Here, the Hessian is evaluated and solved unlike in the previous method. The objective function is transformed as before to an unconstrained objective. An unconstrained objective is formed. All equality constraints and only the violating inequality constraints are included. The sparse nature of the Hessian is used to reduce storage and computation time.

23.2.1.5 Newton OPF

Newton OPF has been formulated by Sun et al. (1984), and later by Maria and Findlay (1987). An augmented Lagrangian is first formed. The set of first derivatives of the augmented objective with respect to the control variables gives a set of nonlinear equations as in the Dommel and Tinney method. Unlike in the Dommel and Tinney method where only a part of these are solved by the N-R method, here, all equations are solved simultaneously by the N-R method.

The method itself is quite straightforward. It is the method of identifying binding inequality constraints that challenged most researchers. Sun et al. use a multiply enforced, zigzagging guarded technique for some of the inequalities, together with penalty factors for some others. Maria and Findlay used an LP-based technique to identify the binding inequality set. Another approach is to use purely penalty factors. Once the binding inequality set is known, the N-R method converges in a very few iterations.

23.2.1.6 Linear Programming-Based Methods

LP methods use a linear or piecewise-linear cost function. The dual simplex method is used in some applications (Benthall, 1968; Shen and Laughton, 1970; Stott and Hobson, 1978; Wells, 1968). The network power flow constraints are linearized by neglecting the losses and the reactive powers, to obtain the DC load flow equations. Merlin (1972) uses a successive linearization technique and repeated application of the dual simplex method.

Due to linearization, these methods have a very high speed of solution, and high reliability in the sense that an optimal solution can be obtained for most situations. However, one drawback is the inaccuracies of the linearized problem. Another drawback for loss minimization is that the loss linearization is not accurate.

23.2.1.7 Quadratic Programming Methods

In these methods, instead of solving the original problem, a sequence of quadratic problems that converge to the optimal solution of the original problem are solved. Burchett et al. use a sparse implementation of this method. The original problem is redefined as simply, to minimize,

$$f(x) \tag{23.15}$$

subject to

$$g(x) = 0 \tag{23.16}$$

The problem is to minimize

$$g^T p + \frac{1}{2} p^T H p \tag{23.17}$$

subject to

$$Jp = 0 \tag{23.18}$$

where

$$p = x - x_k \tag{23.19}$$

Here, g is the gradient vector of the original objective function with respect to the set of variables "x." "J" is the Jacobian matrix that contains the derivatives of the original equality constraints with respect

Optimal Power Flow

to the variables, and "H" is the Hessian containing the second derivatives of the objective function and a linear combination of the constraints with respect to the variables. x_k is the current point of linearization. The method is capable of handling problems with infeasible starting points and can also handle ill-conditioning due to poor R/X ratios. This method was later extended by El-Kady et al. (1986) in a study for the Ontario Hydro System for online voltage/var control. A nonsparse implementation of the problem was made by Glavitsch and Spoerry (1983) and Contaxis et al. (1986).

23.2.1.8 Interior Point Methods

The projective scaling algorithm for linear programming proposed by N. Karmarkar is characterized by significant speed advantages for large problems reported to be as much as 50:1 when compared to the simplex method (Karmarkar, 1984). This method has a polynomial bound on worst-case running time that is better than the ellipsoid algorithms. Karmarkar's algorithm is significantly different from Dantzig's simplex method. Karmarkar's interior point rarely visits too many extreme points before an optimal point is found. The IP method stays in the interior of the polytope and tries to position a current solution as the "center of the universe" in finding a better direction for the next move. By properly choosing the step lengths, an optimal solution is achieved after a number of iterations. Although this IP approach requires more computational time in finding a moving direction than the traditional simplex method, better moving direction is achieved resulting in less iterations. Therefore, the IP approach has become a major rival of the simplex method and has attracted attention in the optimization community. Several variants of interior points have been proposed and successfully applied to OPF (Momoh, 1992; Vargas et al., 1993; Yan and Quintana, 1999).

23.3 OPF Incorporating Load Models

23.3.1 Load Modeling

The area of power systems load modeling has been well explored in the last two decades of the twentieth century. Most of the work done in this area has dealt with issues in stability of the power system. Load modeling for use in power flow studies has been treated in a few cases (Concordia and Ihara, 1982; IEEE Committee Report, 1973; IEEE Working Group Report, 1966; Iliceto et al., 1972; Vaahedi et al., 1987). In stability studies, frequency and time are variables of interest, unlike in power flow and some OPF studies. Hence, load models for use in stability studies should account for any load variations with frequency and time as well. These types of load models are normally referred to as dynamic load models. In power flow, OPF studies neglecting contingencies, and security-constrained OPF studies using preventive control, time, and frequency, are not considered as variables. Hence, load models for this type of study need not account for time and frequency. These load models are static load models.

In security-constrained OPF studies using corrective control, the time allowed for certain control actions is included in the formulation. However, this time merely establishes the maximum allowable correction, and any dynamic behavior of loads will usually end before any control actions even begin to function. Hence, static load models can be used even in this type of formulation.

23.3.2 Static Load Models

Several forms of static load models have been proposed in the literature, from which the exponential and quadratic models are most commonly used. The exponential form is expressed as

$$P_m = a_p V^{b_p} \tag{23.20}$$

$$Q_m = a_q V^{b_q} \tag{23.21}$$

The values of the coefficients a_p and a_q can be taken as the specified active and reactive powers at that bus, provided the specified power demand values are known to occur at a voltage of 1.0 per unit, measured at the network side of the distribution transformer. A typical measured value of the demand and the network side voltage is sufficient to determine approximately the values of the coefficients, provided the exponents are known. The range of values reported for the exponents vary in the literature, but typical values are 1.5 and 2.0 for b_p and b_q, respectively.

23.3.3 Conventional OPF Studies Including Load Models

Incorporation of load models in OPF studies has been considered in a couple of cases (El-Din et al., 1989; Vaahedi and El-Din, 1989) for the Ontario Hydro energy management system. In both cases, loss minimization was considered to be the objective. It is concluded by Vaahedi and El-Din (1989) that the modeling of ULTC operation and load characteristics is important in OPF calculations.

The effects of load modeling in OPF studies have been considered for the case where the generator bus voltages are held at prespecified values (Dias and El-Hawary, 1989). Since the swing bus voltage is held fixed at all times (and also the generator bus voltages in the absence of reactive power limit violations), the average system voltage is maintained in most cases. Thus, an increase in fuel cost due to load modeling was noticed for many systems that had a few (or zero) reactive limit violations, and a decrease for those with a noticeable number of reactive limit violations. Holding the generator bus voltages at specified values restricts the available degrees of freedom for OPF and makes the solution less optimal.

Incorporation of load models in OPF studies minimizing fuel cost (with all voltages free to vary within bounds) can give significantly different results when compared with standard OPF results. The reason for this is that the fuel cost can now be reduced by lowering the voltage at the modeled buses along with all other voltages wherever possible. The reduction of the voltages at the modeled buses lowers the power demand of the modeled loads and will thus give the lower fuel cost. When a large number of loads are modeled, the total fuel cost may be lower than the standard OPF. However, a lowering of the fuel cost via a lowering of the power demand may not be desirable under normal circumstances, as this will automatically decrease the total revenue of the operation. This can also give rise to a lower net revenue if the decrease in the total revenue is greater than the decrease in the fuel cost. This is even more undesirable. What is needed is an OPF solution that does not decrease the total power demand in order to achieve a minimum fuel cost. The standard OPF solution satisfies this criterion. However, given a fair number of loads that are fed by fixed tap transformers, the standard OPF solution can be significantly different from the practically observed version of this solution.

Before attempting to find an OPF solution incorporating load models that satisfies the required criterion, we deal with the reason for the problem. In a standard OPF formulation, the total revenue is constant and independent of the solution. Hence, we can define net revenue R_N, which is linearly related to the total fuel cost F_C by the formula:

$$R_N = -F_C + \textbf{constant} \tag{23.22}$$

The constant term is the total revenue dependent on the total power demand and the unit price of electricity charged to the customers. From this relationship we see that a solution with minimum fuel cost will automatically give maximum net revenue. Now, when load models are incorporated at some buses, the total power demand is not a constant, and hence the total revenue will also not be constant. As a result,

$$R_N = -F_C + R_T \tag{23.23}$$

where "R_T" is the total demand revenue and is no longer a constant.

If instead of minimizing the fuel cost, we now maximize the net revenue, we will definitely avoid the difficulties encountered earlier. This is equivalent to minimizing the difference between the fuel cost and the total revenue. Hence we see that, in the standard OPF, the required maximum net revenue is implied, and the equivalent minimum fuel cost is the only function that enters the computations.

23.3.4 Security Constrained OPF Including Load Models

A conventional OPF result can have optimal but insecure states during certain contingencies. This can be avoided by using a SCOPF. Unlike in the former, for a SCOPF, we can incorporate load models in a variety of ways. For example, we can consider the loads as independent of voltage for the intact system, but dependent on the voltage during contingencies. This can be justified by saying that the voltage deviations encountered during a standard OPF and modeled OPF are small compared to those that can be encountered during contingencies. Since the total power demand for the intact system is not changed, fuel cost comparisons between this case and a standard SCOPF seem more reasonable. We can also incorporate load models for the intact system as well as during contingencies, while minimizing the fuel cost. However, we then encounter the problem discussed in the previous section regarding net earnings. Another approach is to incorporate load models for the intact case as well as during contingencies, while minimizing the total fuel cost minus the total revenue.

23.3.5 Inaccuracies of Standard OPF Solutions

It was stated earlier that the standard OPF (or standard SCOPF) solution can give results not compatible with practical observations (i.e., using the control variable values from these solutions) when a fair number of loads are fed by fixed tap transformers. The discrepancies between the simulated and observed results will be due to discrepancies between the voltage at a bus feeding a load through a fixed tap transformer, and the voltage at which the specified power demand for that load occurs. The observed results can be simulated approximately by performing a power flow incorporating load models. The effects of load modeling in power flow studies have been treated in a few cases (Dias and El-Hawary, 1990; El-Hawary and Dias, 1987a–c). In all these studies, the specified power demand of the modeled loads was assumed to occur at a bus voltage of 1.0 per unit. The simulated modeled power flow solution will be same as the practically observed version only when exact model parameters are utilized.

23.4 SCOPF Including Load Modeling

SCOPF takes into account outages of certain transmission lines or equipment (Alsac and Stott, 1974; Schnyder and Glavitsch, 1987). Due to the computational complexity of the problem, more work has been devoted to obtaining faster solutions requiring less storage, and practically no attention has been paid to incorporating load models in the formulations. A SCOPF solution is secure for all credible contingencies or can be made secure by corrective means. In a secure system (level 1), all load is supplied, operating limits are enforced, and no limit violations occur in a contingency. Security level 2 is one where all load is supplied, operating limits are satisfied, and any violations caused by a contingency can be corrected by control action without loss of load. Level 1 security is considered in Dias and El-Hawary (1991a).

Studies of the effects of load voltage dependence in PF and OPF (Dias and El-Hawary, 1989) concluded that for PF incorporating load models, the standard solution gives more conservative results with respect to voltages in most cases. However, exceptions have been observed in one test system. Fuel costs much lower than those associated with the standard OPF are obtained by incorporating load models with all voltages free to vary within bounds. This is due to the decrease in the power demand by the reduction of the voltages at buses whose loads are modeled. When quite a few loads are modeled, the minimum fuel costs may be much lower than the corresponding standard OPF fuel cost with a significant decrease in power demand.

A similar effect can be expected when load models are incorporated in SCOPF studies. The decrease in the power demand when load models are incorporated in OPF studies may not be desirable under normal operating conditions. This problem can be avoided in a SCOPF by incorporating load models during contingencies only. This not only gives results that are more comparable with standard OPF results, but may also give lower fuel costs without lowering the power demand of the intact system. The modeled loads are assumed to be fed by fixed tap transformers and are modeled using an exponential type of load model.

In Dias and El-Harawy (1990), some selected buses were modeled using an exponential type of load model in three cases. In the first, the specified load at modeled buses is obtained with unity voltage. In the second case, the transformer taps have been adjusted to give all industrial-type consumers 1.0 per unit at the low-voltage panel when the high-side voltage corresponds to the standard OPF solution. In the third case, the specified power demand is assumed to take place when the high-side voltages correspond to the intact case of the standard SCOPF solution. It is concluded that a decrease in fuel cost can be obtained in some instances when load models are incorporated in SCOPF studies during contingencies only. In situations where a decrease in fuel cost is obtained in this manner, the magnitude of decrease depends on the total percentage of load fed by fixed tap transformers and the sensitivity of these loads to modeling. The tap settings of these fixed tap transformers influence the results as well. An increase in fuel cost can also occur in some isolated cases. However, in either case, given accurate load models, OPF solutions that are more accurate than the conventional OPF solutions can be obtained. An alternate approach for normal OPF as well as SCOPF is also suggested.

23.4.1 Influence of Fixed Tap Transformer Fed Loads

A standard OPF assumes that all loads are independent of other system variables. This implies that all loads are fed by ULTC transformers that hold the load-side voltage to within a very narrow bandwidth sufficient to justify the assumption of constant loads. However, when some loads are fed by fixed tap transformers, this assumption can result in discrepancies between the standard OPF solution and its observed version. In systems where the average voltage of the system is reasonably above 1.0 per unit (specifically where the loads fed by fixed tap transformers have voltages greater than the voltage at which the specified power demand occurs), the practically observed version of the standard OPF solution will have a higher total power demand, and hence a higher fuel cost, and total revenue, and net revenue. Conversely, where such voltages are lower than the voltage at which the specified power demand occurs, the total power demand, fuel cost, total and net revenues will be lower than expected. For the former case, the system voltages will usually be slightly less than expected, while for the latter case they will usually be slightly higher than expected.

The changes in the power demand at some buses (in the observed version) will alter the power flows on the transmission lines, and this can cause some lines to deliver more power than expected. When this occurs on transmission lines that have power flows near their upper limit, the observed power flows may be above the respective upper limit, causing a security violation. Where the specified power demand occurs at the bus voltages obtained by a standard OPF solution, the observed version of the standard OPF solution will be itself, and there will ideally be no security violations in the observed version.

Most of the above conclusions apply to SCOPF as well (Dias and El-Hawary, 1991b). However, since a SCOPF solution will in general have higher voltages than its normal counterpart (in order to avoid low voltage limit violations during contingencies), the increase in power demand, and total and net revenues will be more significant while the decrease in the above quantities will be less significant. Also, the security violations due to line flows will now be experienced mainly during contingencies, as most line flows will now usually be below their upper limits for the intact case. For SCOPF solutions that incorporate load models only during contingencies, the simulated and observed results will mainly differ in the intact case. Also, with loads modeled during contingencies, the average voltage is lower than for the standard SCOPF solution and hence there will be more cases with a decrease in the power demand, fuel cost, and total and net revenues in the observed version of the results than for its standard counterpart.

23.5 Operational Requirements for Online Implementation

The most demanding requirements on OPF technology are imposed by online implementation. It was argued that OPF, as expressed in terms of smooth nonlinear programming formulations, produces results that are far too approximate descriptions of real-life conditions to lead to successful online implementations. Many OPF formulations do not have the capability to incorporate all operational considerations into the solutions. Moreover, some operating practices are occasionally incompatible with such OPF formulations. Consequently, many proposed "theoretical optimal solutions" are of little value to the operators who are almost constantly presented with simultaneous events that are outside the scope of OPF definition. These limitations, if properly addressed, do not have to prevent OPF programs from being used in practice, especially when the operational optimal solution may also not be known. Papalexopoulos (1996) offers some of the requirements that need to be met so that OPF applications are useful to, and usable by, the dispatchers in online applications.

23.5.1 Speed Requirements

Fast OPF programs designed for online application are needed because under normal conditions, the state of the power system changes continuously and can change abruptly during emergency conditions. The changes involve the evolution of bus active and reactive power generation and loads with time, control variables moving to and off their limits as time changes, and topology changes due to switching operations and other planned or forced outages. The need for fast OPF solutions is especially true when an excessive amount of calculations due to modeling of contingency constraints or repeated OPF runs is involved.

In general, an online OPF calculation should have been completed before the state of the power system has changed to another state that is appreciably different from the earlier state. Determining the optimal execution frequency to maximize the benefits of the computations depends on the specific situation and is limited by finite computing resources. It may be preferable to develop incrementally correct and flexible algorithms to offer fast and more frequent scheduling. This leads us to conclude that conventional formulations and algorithms characterized with quadratic convergence that give very accurate and "mathematically optimal" solutions, but neglect operational realities are not appropriate for online implementation. Fast and frequent scheduling requires "hot start" OPF capabilities developed to take advantage of the optimal status of previously optimized operating points. The hot start option is significant when the rate of change of system state is small and previously optimized points are still "relevant" to the current operating conditions.

23.5.2 Robustness of OPF Solutions with Respect to Initial Guess Point

An OPF program needs to produce consistent solutions and thus must not be sensitive to the selected initial guess used. In addition, changes in the OPF solutions between operating states need to be consistent with the changes in the power system operating constraints. The OPF solutions will never be exactly the same when starting from different initial guess points because the solution process is iterative. Any differences should be within the tolerances specified by the convergence criterion, and of a magnitude that would be considered insignificant to the operator. First-order OPF solution methods were not well received because noticeably different solutions could be obtained when an OPF algorithm was initialized from different initial guess points, with only one (or even none) of the solutions actually constituting a local optimum. Theoretically, if the objective function and the feasible region can be shown to be convex, then the optimal solution will be unique (Gill et al., 1981). Unfortunately, the complexity of the nonlinear equations and inequality constraints involved in OPF problems make it untenable to rigorously prove convexity. If multiple local minima actually exist, then additional computational or heuristic methods must be used to resolve the issue.

A normally feasible OPF solution space may become nonconvex (thus leading to multiple OPF solutions) due to two considerations. The first is due to use of discontinuous techniques to model specific operating practices and preferences, and the second is due to modeling of local controls. The conventional power flow problem with local control capability, whose implicit objective is feasible with respect to a limited set of inequalities, does not have a unique solution. Nevertheless, solutions of the same problem from different starting conditions usually match quite closely. Occasionally, different initial guess solutions can lead to different solutions. This takes place when two or more feasible voltage levels can satisfy nonlinear loads. OPF applications, however, should be able to overcome this type of ambiguity.

23.5.3 Discrete Modeling

Discrete control is widely used in the electric network. For example, transformers are used for voltage control, shunt capacitors and reactors are switched on or off to correct voltage profiles and to reduce active power transmission losses, and phase shifters are used to regulate the MW flows of transmission lines. An efficient and effective OPF discretization procedure is needed to assist the operators in utilizing discrete controls in a realistic and optimal or near-optimal manner. Discrete elements to be included in the OPF formulation are branch switching; prohibited zones of generator cost curves; and priority sequence levels for unfeasibility handling. OPF algorithms designed for online applications should be able to appropriately handle the discrete aspects of the problem.

Using both discrete and continuous controls converts the OPF into a mixed discrete-continuous optimization problem. A possible accurate solution using a method such as mixed-integer nonlinear programming would be orders of magnitude slower than ordinary nonlinear programming methods (Gill et al., 1981). Linear programming-based OPF algorithms allow substantial recognition of discrete controls by setting the cost curve segment break points at discrete control steps. However, most methods that solve for a nonseparable objective function by nonlinear programming methods do not properly model discrete controls.

Current OPF algorithms treat all controls as continuous variables during the initial solution process. Once the continuous solution is obtained, each discrete variable is moved to the nearest discrete setting. This produces acceptable solutions, assuming that the step sizes for the discrete controls are sufficiently small, which is usually the case for transformer taps and phase shifter angles (Papalexopoulos et al., 1989). Approximate solutions that can produce near-optimal results appear to be a reasonable alternative to rigorous solution methods. One such scheme (Liu et al., 1991) uses penalty functions for discrete controls. The object is to penalize the continuous approximations of discrete control variables for movements away from their discrete steps. This scheme is well suited for Newton-based OPF algorithms. The scheme consists of a set of rules to determine the timing of introduction and criteria of updating the penalties in the optimization process. This heuristic algorithm is of limited scope. Substantially more work is needed to effectively resolve all problems associated with the discrete nature of controls and other discrete elements of the OPF problem.

23.5.4 Detecting and Handling Infeasibility

As the requirements for satisfactory system operation increase, the region of feasible solutions that satisfy all constraints simultaneously may become too small. In this case, there is a need to establish criteria to prioritize the constraints. For OPF applications, this means that when a feasible solution cannot be found, it is still very important for the algorithm to suggest the "best optimal" engineering solution in some sense, even though it is infeasible. This is even more critical for OPF applications that incorporate contingency constraints.

There are several approaches to deal with this problem. In one approach, all power flow equations are satisfied and only the soft constraints that truly cause the bottlenecks are allowed to be violated using a least squares approximation process. An LP approach introduces a weighted slack variable for each binding constraint. If a constraint can be enforced, the slack variable will be reduced to zero and the constraint will be satisfied. The constraints causing infeasibility will have nonzero slack variables whose magnitudes are proportional to the amounts they need to be relaxed to achieve feasibility. Usually, all binding constraints of a particular type are modeled as if they have identical infeasibility characteristics. That is, all slack variables corresponding to these binding constraints share the same cost curve, and their sensitivities are scaled by a weighting factor associated with the type of the corresponding constraint. Using Newton's method, if the OPF does not converge in the first specified set of iterations, the constraint weighting factors, corresponding to the penalty functions associated with the load bus voltage limits and the branch flow limits, will be reduced successively until a solution is reached. This normally results in all constraints being met except for those load bus voltage and branch flow limits that contribute to infeasibility. Special care should be taken in selecting the proper weighting factors to avoid numerical problems and produce acceptable solutions.

Another approach develops hierarchical rules that operate on the controls and constraints of the OPF problem. The rules introduce discontinuous changes in the original OPF formulation. These changes include using a different set of control/constraint limits, expansion of the control set by class or individually, branch switching, load shedding, etc. They are usually implemented in a predefined priority sequence to be consistent with utility practices. The decision as to when to proceed to the next priority level of modifications to achieve feasibility is critical, especially when it involves radial overloads, normally overloaded constraints and constraints known to have "soft" limits. The selection of a final optimal solution among all the others in the set is achieved with the implementation of a "preference index." An application of the preference index approach that minimizes postcontingency line overloads due to generator outages is given in Yokoyama et al. (1988).

23.5.5 Consistency of OPF Solutions with Other Online Functions

Online OPF is implemented in either study or closed-loop mode. In study mode, the OPF solutions are presented as recommendations to the operator. In closed-loop mode, control actions are implemented in the system via the SCADA system of the EMS (*IEEE Trans.*, 1983). In closed-loop mode, OPF is triggered by a number of events, including an operator request, the execution of the real-time sequence and security analysis, structural change, large load change, etc. A major concern for an OPF in closed-loop mode is the design of its interface with the other online functions, which are executed at different frequencies. Some of these functions are unit commitment, ED, real-time sequence, security analysis, automatic generation control (AGC), etc. To reduce the discrepancy between ideal and realistic OPF solutions, emphasis should be placed on establishing consistency between these functions and static optimal solutions produced by OPF. This requires proper interfacing and integration of OPF with these functions. The integration design should be flexible enough to allow OPF formulation modifications consistent with the ever dynamic and sometimes ill-defined security problem definition.

23.5.6 Ineffective "Optimal" Rescheduling

Production-grade OPF algorithms use all available control actions to obtain an optimal solution, but for many applications it is not practical to execute more than a limited number of control actions. The OPF problem then becomes one of selecting the best set of actions of a limited size out of a much larger set of possible actions. The problem was identified but no concrete remedies were offered. It is not possible to select the best and most effective set of a given size from existing OPF solutions that use all controls to solve each problem. The control actions cannot be ranked and the effectiveness of an action is not related to its magnitude. Each control facility participates in both minimization of the objective function and enforcement of the constraints. Separation of the two effects for evaluation purposes is not feasible. The problem is difficult to define analytically and existing conventional technologies are not adequate. It is important to note that emerging computational intelligence tools such as fuzzy reasoning and neural networks may offer some resolution. The problem of ineffective rescheduling is related to but is not identical to the "minimum number of controls" objective. It is also closely linked to the problem of discrete control variables, since methods that recognize the discrete nature of some control facilities tend to decrease the number of control actions by keeping inefficient discrete controls at their initial settings.

23.5.7 OPF-Based Transmission Service Pricing

OPF programs are capable of computing marginal costs. Information about the optimal states with respect to changes, such as load variations, operating limit changes, or constraint parameter changes, can be used in many practical applications. Specifically, the sensitivities of the production cost of generation with respect to changes in the bus active power injections are called Bus Incremental Costs (BICs). BICs can be used as nodal prices for pricing transmission services, as they reflect the transmission loss and the congestion components for transferring power from one point to another. In a lossless network with no binding constraints, all BICs should be equal. However, when an operating limit is reached, the congestion component takes effect and all BICs in the network can be different. This means that nodal price differences across uncongested lines can be much larger than marginal losses. Extensive experience has shown that it is possible for power to flow from a bus with higher nodal price to a bus with lower nodal price, resulting in negative transmission charges. Failure to properly account for this effect can lead to unacceptable incentives for transmission users. The same applies in the case of transmission reinforcements to mitigate congestion. If as a result of the upgrades, the incremental transmission rights (positive or negative) are not accounted for properly, similar distortions are possible.

23.6 Conclusions

A review of recent developments in optimal economic operation of electric power systems with emphasis on the OPF formulation was given. We dealt with conventional formulations of ED, conventional OPF, and accounting for the dependence of the power demand on voltages in the system. Challenges to OPF formulations and solution methodologies for online application were also outlined.

References

Abadie, J. and Carpentier, J., Generalization of the Wolfe reduced gradient method to the case of nonlinear constraints, *Optimization*, R. Fletcher, Ed., Academic Press, New York, 1969, pp. 37–47.

Alsac, O. and Stott, B., Optimal load flow with steady-state security, *IEEE Trans. Power App. Syst.*, PAS-93, 745–751, May/June 1974.

Benthall, T.P., Automatic load scheduling in a multiarea power system, *Proc. IEE*, 115, 592–596, April 1968.

Burchett, R.C., Happ, H.H., and Vierath, D.R., Quadratically convergent optimal power flow, *IEEE Trans. Power App. Syst.*, PAS-103, 3267–3275, November 1984.

Burchett, R.C., Happ, H.H., Vierath, D.R., and Wirgau, K.A., Developments in optimal power flow, *IEEE Trans. Power App. Syst.*, PAS-101, 406–414, February 1982.

Carpentier, J., Contribution a l'etude du dispatching economique, *Bull. Soc. Francaise Electriciens*, 8, 431–447, 1962.

Carpentier, J., Differential injections method: A general method for secure and optimal load flows, in *IEEE PICA Conference Proceedings*, Minneapolis, MN, June 1973, pp. 255–262.

Concordia, C. and Ihara, S., Load representation in power system stability studies, *IEEE Trans. Power App. Syst.*, PAS-101, 969–977, 1982.

Contaxis, G.C., Delkis, C., and Korres, G., Decoupled optimal power flow using linear or quadratic programming, *IEEE Trans. Power Syst.*, PWRS-1, 1–7, May 1986.

Dias, L.G. and El-Hawary, M.E., Effects of active and reactive modelling in optimal load flow studies, *IEE Proc.*, 136, Part C, 259–263, September 1989.

Dias, L.G. and El-Hawary, M.E., A comparison of load models and their effects on the convergence of Newton power flows, *Int. J. Electr. Power Energy Syst.*, 12, 3–8, 1990.

Dias, L.G. and El-Hawary, M.E., Effects of load modeling in security constrained OPF studies, *IEEE Trans. Power Syst.*, 6(1), 87–93, February 1991a.

Dias, L.G. and El-Hawary, M.E., Security constrained OPF: Influence of fixed tap transformer loads, *IEEE Trans. Power Syst.*, 6, 1366–1372, November 1991b.

Dommel, H.W. and Tinney, W.F., Optimal power flow solutions, *IEEE Trans. Power App. Syst.*, PAS-87, 1866–1876, October 1968.

El-Din, H.M.Z., Burns, S.D., and Graham, C.E., Voltage/var control with limited control actions, in *Proceedings of the Canadian Electrical Association Spring Meeting*, Toronto, Ontario, Canada, March 1989.

El-Hawary, M.E., Power system load modeling and incorporation in load flow solutions, in *Proceedings of the Third Large Systems Symposium*, University of Calgary, Calgary, Alberta, Canada, June 1982.

El-Hawary, M.E. and Dias, L.G., Bus sensitivity to model parameters in load-flow studies, *IEE Proc.*, 134, Part C, 302–305, July 1987a.

El-Hawary, M.E. and Dias, L.G., Incorporation of load models in load flow studies: Form of model effects, *IEE Proc.*, 134, Part C, 27–30 January 1987b.

El-Hawary, M.E. and Dias, L.G., Selection of buses for detailed modeling in load flow studies, *Electr. Mach. Power Syst.*, 12, 83–92, 1987c.

El-Kady, M.A., Bell, B.D., Carvalho, V.F., Burchett, R.C., Happ, H.H., and Vierath, D.R., Assessment of real-time optimal voltage control, *IEEE Trans. Power Syst.*, PWRS-1, 98–107, May 1986.

Fiacco, A.V. and McCormick, G.P., Computational algorithm for the sequential unconstrained minimization technique for nonlinear programming, *Manage. Sci.*, 10, 601–617, 1964.

Fletcher, R. and Powell, M.J.D., A rapidly convergent descent method for minimization, *Comput. J.*, 6, 163–168, 1963.

Gill, P.E., Murray, W., and Wright, M.H., *Practical Optimization*, Academic Press, New York, 1981.

Glavitsch, H. and Spoerry, M., Quadratic loss formula for reactive dispatch, *IEEE Trans. Power App. Syst.*, PAS-102, 3850–3858, December 1983.

Happ, H.H. and Vierath, D.R., The OPF for operations planning and for use on line, in *Proceedings of the Second International Conference on Power Systems Monitoring and Control*, University of Durham, U.K., July 1986, pp. 290–295.

IEEE Committee Report, System load dynamic simulation, effects and determination of load constants, *IEEE Trans. Power App. Syst.*, PAS-92, 600–609, 1973.

IEEE Current Operating Problems Working Group Report, On-line load flows from a system operator's viewpoint, *IEEE Trans. Power App. Syst.*, PWRS-102, 1818–1822, June 1983.

IEEE Working Group Report, The effect of frequency and voltage on power system loads, *IEEE Winter Meeting*, New York, Paper 31 CP 66-64, 1966.

Iliceto, F., Ceyhan, A., and Ruckstuhl, G., Behavior of loads during voltage dips encountered in stability studies, *IEEE Trans. Power App. Syst.*, PAS-91, 2470–2479, 1972.

Karmarkar, N., New polynomial-time algorithm for linear programming, *Combinatorica*, 4, 373–395, 1984.

Kuhn, H.W. and Tucker, A.W., Nonlinear programming, in *Proceedings of Second Berkeley Symposium on Mathematical Statistics and Probability*, University of California, Berkeley, CA, 1951, pp. 481–492.

Lindqvist, A., Bubenko, J.A., and Sjelvgren, D., A generalized reduced gradient methodology for optimal reactive power flows, in *Proceedings of the 8th PSCC*, Helsinki, Finland, 1984.

Liu, E., Papalexopoulos, A.D., and Tinney, W.F., Discrete shunt controls in a Newton optimal power flow, *IEEE Winter Power Meeting 1991*, New York, Paper 91 WM 041-4 PWRS, 1991.

Lootsma, F.A., Logarithmic programming: A method of solving nonlinear programming problems, *Philips Res. Rep.*, 22, 329–344, 1967.

Maria, G.A. and Findlay, J.A., A Newton optimal power flow program for Ontario hydro EMS, *IEEE Trans. Power Syst.*, PWRS-2, 576–584, August 1987.

Merlin, A., On optimal generation planning in large transmission systems (The Maya Problem), in *Proceedings of 4th PSCC*, Grenoble, France, Paper 2.1/6, September 1972.

Momoh, J.A., Application of quadratic interior point algorithm to optimal power flow, EPRI Final report RP 2473-36 II, Palo Alto, CA, EPRI, March 1992.

Papalexopoulos, A.D., Challenges to an on-line OPF implementation, in *IEEE Tutorial Course Optimal Power Flow: Solution Techniques, Requirements, and Challenges*, Publication 96 TP 111-0, IEEE Power Engineering Society, Piscataway, NJ, 1996.

Papalexopoulos, A.D., Imparato, C.F., and Wu, F.F., Large-scale optimal power flow: Effects of initialization, decoupling and discretization, *IEEE Trans. Power App. Syst.*, PWRS-4, 748–759, May 1989.

Peschon, J., Bree, D.W., and Hajdu, L.P., Optimal solutions involving system security, in *Proceedings of the 7th PICA Conference*, Boston, MA, 1971, pp. 210–218.

Polak, E. and Ribiere, G., Note sur la convergence de methods de directions conjugees, *Rev. Fr. Inform. Rech. Operation*, 16-R1, 35–43, 1969.

Sasson, A.M., Combined use of the Powell and Fletcher-Powell nonlinear programming methods for optimal load flows, *IEEE Trans. Power App. Syst.*, PAS-88, 1530–1537, October 1969.

Sasson, A.M., Viloria, F., and Aboytes, F., Optimal load flow solution using the Hessian matrix, *IEEE Trans. Power App. Syst.*, PAS-92, 31–41, January/February 1973.

Schnyder, G. and Glavitsch, H., Integrated security control using an optimal power flow and switching concepts, in *IEEE PICA Conference Proceedings*, Montreal, Quebec, Canada, May 18–22, 1987, pp. 429–436.

Shen, C.M. and Laughton, M.A., Power system load scheduling with security constraints using dual linear programming, *IEE Proc.*, 117, 2117–2127, November 1970.

Stott, B., Alsac, O., and Monticelli, A.J., Security analysis and optimization, *Proc. IEEE*, 75, 1623–1644, December 1987.

Stott, B. and Hobson, E., Power system security control calculations using linear programming. Parts 1 and 2, *IEEE Trans. Power App. Syst.*, PAS-97, 1713–1731, September/October 1978.

Sun, D.I., Ashley, B., Brewer, B., Hughes, A., and Tinney, W.F., Optimal power flow by Newton approach, *IEEE Trans. Power App. Syst.*, PAS-103, 2864–2880, October 1984.

Vaahedi, E. and El-Din, H.M.Z., Considerations in applying optimal power flow to power systems operation, *IEEE Trans. Power Syst.*, PWRS-4, 694–703, May 1989.

Vaahedi, E., El-Kady, M., Libaque-Esaine, J.A., and Carvalho, V.F., Load models for large scale stability studies from end user consumption, *IEEE Trans. Power Syst.*, PWRS-2, 864–871, 1987.

Vargas, L.S., Quintana, V.H., and Vannelli, A., A tutorial description of an interior point method and its applications to security-constrained economic dispatch, *IEEE Trans. Power Syst.*, 8, 1315–1323, 1993.

Wells, D.W., Method for economic secure loading of a power system, *IEE Proc.*, 115, 1190–1194, August 1968.

Wolfe, P., Methods of nonlinear programming, *Nonlinear Programming*, J. Abadie, Ed., North Holland, Amsterdam, the Netherlands, 1967, pp. 97–131.

Yan, X. and Quintana, V.H., Improving an interior-point-based OPF by dynamic adjustments of step sizes and tolerances, *IEEE Trans. Power Syst.*, 14, 709–717, 1999.

Yokoyama, R., Bae, S.H., Morita, T., and Sasaki, H., Multiobjective optimal generation dispatch based on probability security criteria, *IEEE Trans. Power App. Syst.*, PWRS-3, 317–324, 1988.

Yu, D.C., Fagan, J.E., Foote, B., and Aly, A.A., An optimal load flow study by the generalized reduced gradient approach, *Electr. Power Syst. Res.*, 10, 47–53, 1986.

Zangwill, W.I., Non-linear programming via penalty functions, *Manage. Sci.*, 13, 344–358, 1967.

24
Security Analysis

24.1 Definition 24-2
24.2 Time Frames for Security-Related Decision 24-2
24.3 Models 24-3
24.4 Determinist vs. Probabilistic 24-5
   Security under Deregulation
24.A Appendix A 24-6
   Standards
24.B Appendix B 24-6
   Shift Factor Derivation
References 24-7

Nouredine Hadjsaid
Institut National Polytechnique de Grenoble

The power system as a single entity is considered the most complex system ever built. It consists of various equipment with different levels of sophistication, complex and nonlinear loads, various generations with a wide variety of dynamic responses, a large-scale protection system, a wide-area communication network, and numerous control devices and control centers. This equipment is connected with a large network (transformers, transmission lines) where a significant amount of energy transfer often occurs. This system, in addition to the assurance of good operation of its various equipment, is characterized by an important and simple rule: electricity should be delivered to where it is required in due time and with appropriate features such as frequency and voltage quality. Environmental constraints, the high cost of transmission investments and low/long capital recovery, and the willing of utilities to optimize their network for more cost effectiveness makes it very difficult to expand or oversize power systems. These constraints have pushed power systems to be operated close to their technical limits, thus reducing security margins.

On the other hand, power systems are continuously subjected to random and various disturbances that may, under certain circumstances, lead to inappropriate or unacceptable operation and system conditions. These effects may include cascading outages, system separation, widespread outages, violation of emergency limits of line current, bus voltages, system frequency, and loss of synchronism (Debs and Benson, 1975). Furthermore, despite advanced supervisory control and data acquisition systems that help the operator to control system equipment (circuit breakers, on-line tap changers, compensation and control devices, etc.), changes can occur so fast that the operator may not have enough time to ensure system security. Hence, it is important for the operator not only to maintain the state of the system within acceptable and secure operating conditions but also to integrate preventive functions. These functions should allow him enough time to optimize his system (reduction of the probability of occurrence of abnormal or critical situations) and to ensure recovery of a safe and secure situation.

Even though for small-scale systems the operator may eventually, on the basis of his experience, prevent the consequences of most common outages and determine the appropriate means to restore a secure state, this is almost impossible for large systems. It is therefore essential for operators to have at their disposal, efficient tools capable of handling a systematic security analysis. This can be achieved through the diagnosis of all contingencies that may have serious consequences. This is the concern of *security analysis*.

The term contingency is related to the possibility of losing any component of the system, whether it is a transmission line, a transformer, or a generator. Another important event that may be included in this definition concerns busbar faults (bus split). This kind of event is, however, considered rare but with (serious) dangerous consequences. Most power systems are characterized by the well-known N − 1 security rules where N is the total number of system components. This rule is the basic requirement for the planning stage where the system should be designed in order to withstand (or to remain in a normal state) any single contingency. Some systems also consider the possibility of N − 2/k (k is the number of contingencies), but mostly for selected and specific cases.

24.1 Definition

The term security as defined by NERC (1997) is the ability of the electric systems to withstand sudden disturbance such as electric short-circuits or unanticipated loss of system elements. (See Appendix A.)

Security analysis is usually handled for two time frames: static and dynamic. For the static analysis, only a "fixed picture" or a snapshot of the network is considered. The system is supposed to have passed the transient period successfully or be dynamically stable. Therefore, the monitored variables are line flows and bus voltages. Hence, all voltages should be within a predefined secure range, usually around ±5% of nominal voltage (for some systems, such as distribution networks, the range may be wider). In fact, if bus voltages drop below a certain level, there will be a risk of voltage collapse in addition to high losses. On the other hand, if bus voltages are too high compared to nominal values, there will be equipment degradation or damage. Furthermore, overload of transmission lines may be followed by unpredictable line tripping that accelerates the degradation of the voltage profile.

Line flows are related to circuit overload (lines and transformers) and should keep below a maximum limit, usually settled according to line thermal limits. The dynamic security is related to loss of synchronism (transient stability) and oscillatory swings or dynamic instability. In that case the evolution of essential variables are monitored based upon a required time frame (transient period).

Normally, system security is analyzed differently whether it is considered for planning studies or for monitoring and operational purposes. The difference comes from the type of action that should be initiated in case of expected harmful contingencies. However, for both stages, all variables should remain within the bounded domain defining or determining system normal state (Fink and Carlsen, 1978).

24.2 Time Frames for Security-Related Decision

There are generally three different time frames for security-related decisions. In operations, the decision-maker is the operator, who must continuously monitor and operate his system economically in such a way that the normal state is appropriately preserved (maintained). For this purpose, he has specific tools for diagnosing his system and operating rules that allow the required decisions to be made in due time. In operational planning, the operating rules are developed recognizing that the bases for the decision are reliability/security criteria specifying minimum operating requirements, which define acceptable performance for the credible contingencies. In facility planning, the planner must determine the best way to reinforce the transmission system, based on reliability/security criteria for system design, which generally adhere to the same disturbance-performance criteria specified by minimum operating requirements.

One may think that since these systems are designed to operate "normally" or in "a secure state" for a given security rule (N − k), there is nothing to worry about during operations. The problem is that, during the planning stage and for a set of given economical constraints, a number of assumptions are made for operating conditions that concern topology, generation, and consumption. Since there may be several years between the planning stage and the operations, the uncertainties in the system's security may be very significant. Therefore, security analysis is supplemented by operational planning and operations studies.

The decision following any security analysis can be placed in one of two categories: preventive or corrective actions. For corrective actions, once a contingency or an event is determined as potentially dangerous, the operator should be confident that in case of that event, he will be able to correct the system by means of appropriate actions on system conditions (generation, load, topology) in order to keep the system in a normal state and even away from the insecure region. The operator should also prepare a set of preventive actions that may correct the effect of the expected dangerous event.

In operations, the main constraint is the time required for the analysis of the system's state and for the required decision to be made following the security analysis results. The security analysis program should be able to handle all possible contingencies, usually on the N – 1 basis or on specific N – 2. For most utilities, the total time window considered for this task is between 10 and 30 min. Actually for this time window, the system's state is considered as constant or quasi-constant allowing the analysis to be valid within this time frame. This means that changes in generation or in consumption are considered as negligible.

For large systems, this time frame is too short even with very powerful computers. Since it is known that only a small number of contingencies may really cause system violations, it has been realized that it is not necessary to perform a detailed analysis on all possible contingencies, which may be on the order of thousands. For this purpose, the operator may use his engineering judgment to select those contingencies that are most likely to cause system violation. This procedure has been used (and is still in use) for many years in many control centers around the world. However, as system conditions are characterized by numerous uncertainties, this approach may not be very efficient especially for large systems.

The concept of contingency selection has arisen in order to reduce the list of all possible contingencies to only the potentially harmful. The selection process should be very fast and accurate enough to identify dangerous cases (Hadjsaid et al., 1992). This process has existed for many years, and still is a major issue in all security studies for operations whether for static or dynamic and transient purposes.

24.3 Models

The static security analysis is mainly based on load flow equations. Usually, active/angle and reactive/voltage problems are viewed as decoupled. The active/angle subproblem is expressed as (Stott and Alsac, 1974):

$$\Delta\theta = \left[\frac{dP}{d\theta}\right]^{-1} \Delta P \tag{24.1}$$

where
  $\Delta\theta$ is a vector of angular changes with a dimension of Nb – 1 (Nb = number of buses)
  ΔP a vector of active injection changes (Nb – 1)
  $[dP/d\theta]$ is a part of the Jacobian matrix

In the DC approach, this Jacobian is approximated by the B′ (susceptance) matrix representing the imaginary part of the Ybus matrix. This expression is used to calculate the updated angles following a loss of any system component. With appropriate numerical techniques, it is straightforward to update only necessary elements of the equation. Once the angles are calculated, the power flows of all lines can be deducted. Hence, it is possible to check for line limit violation.

Another approach that has been, and still is used in many utilities for assessing the impact of any contingency on line flows is known as shift factors. The principle used recognizes that the outage of any line will result in a redistribution of the power previously flowing through this line on all the remaining lines. This distribution is mainly affected by the topology of the network. Hence, the power flow of any line ij following an outage of line km can be expressed as (Galiana, 1984) (see Appendix B for more details):

$$P_{ij/km} = P_{ij} + \alpha_{ij/km} * P_{km} \quad (24.2)$$

where
$P_{ij/km}$ is the active power flow on line ij after the outage of line km
P_{ij}, P_{km} is the active power previously flowing respectively on line ij and km (before the outage)
$\alpha_{ij/km}$ is the shift factor for line ij following the outage of line km

Equation 24.2 shows that the power flow of line ij ($P_{ij/km}$) when line km is tripped, is determined as the initial power flow on line ij (P_{ij}) before the outage of line km plus a proportion of the power flow previously flowing on line km. This proportion is defined by the terms $\alpha_{ij/km} * P_{km}$.

The shift factors are determined in a matrix form. The important features of these factors are the simplicity of computing and their dependency on network topology. Therefore, if the topology does not change, the factors remain constant for any operating point. The main drawback of these factors is that they are determined on the basis of DC approximation and the shift factor matrix should be updated for any change in the topology. In addition, for some complex disturbances such as bus split, updating these factors becomes a complicated task.

A similar method based on reactive power shift factors has been developed. Interested readers may refer to Ilic-Spong and Phadke (1986) and Taylor and Maahs (1991) for more details.

The reactive/voltage subproblem can be viewed as (Stott and Alsac, 1974):

$$\Delta V = \left[\frac{dQ}{dV}\right]^{-1} \Delta Q \quad (24.3)$$

where
ΔV is the vector of voltages change (Nb – Ng, Ng is the number of generators)
ΔQ is the vector of reactive power injections change (Nb – Ng, Ng is the number of generators)
[dQ/dV] is the Jacobean submatrix

In the well-known Fast Decoupled Load Flow (FDLF) model (Stott and Alsac, 1974), the Jacobian submatrix is replaced by the B″ (susceptance) matrix representing the imaginary part of the Ybus matrix with a dimension of Nb – Ng, where Ng is the number of voltage regulated (generator) buses. In addition, the vector ΔQ is replaced by $\Delta Q/V$.

Once bus voltages are updated to account for the outage, the limit violations are checked and the contingency effects on bus voltages can be assessed.

The most common framework for the contingency analysis is to use approximate models for the selection process, such as the DC model, and use the AC power flow model for the evaluation of the actual impact of the given contingency on line flows and bus voltages (Figure 24.1).

Concerning the dynamic security analysis, the framework is similar to the one in static analysis in terms of selection and evaluation. The selection process uses simplified models, such as Transient Energy Functions (TEF), and the evaluation one uses detailed assessing tools such as time domain simulations. The fact that the dynamic aspect is more related to transient/dynamic stability technique makes the process much more complicated than for the static problem. In fact, in addition to the number of contingencies to be analyzed, each analysis will require detailed stability calculations with an appropriate network and system component model such as the generator model (park, saturation, etc.),

Security Analysis

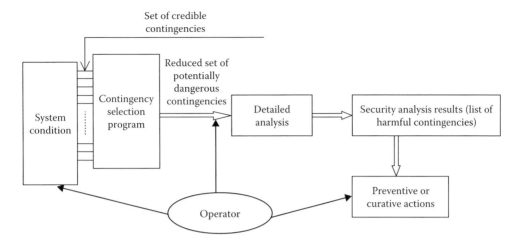

FIGURE 24.1 Contingency analysis procedure.

exciter (AVR: Automatic Voltage Regulator; PSS: Power System Stabilizer), governor (nuclear, thermal, hydroelectric, etc.), or loads (nonlinear, constant power characteristics, etc.). In addition, integration and numerical solutions are an important aspect for these analyses.

24.4 Determinist vs. Probabilistic

The basic requirement for security analysis is to assess the impact of any possible contingency on system performance. For the purpose of setting planning and operating rules that will enable the system to be operated in a secure manner, it is necessary to consider all credible contingencies, different network configurations, and different operating points for given performance criteria. Hence, in the deterministic approach, these assessments may involve a large number of computer simulations even if there is a selection process at each stage of the analysis. The decision in that case is founded on the requirement that each outage event in a specified list, the contingency set, results in system performance that satisfies the criteria of the chosen performance evaluation (Fink and Carlsen, 1978). To handle these assessments for all possible situations by an exhaustive study is generally not reasonable. Since the resulting security rules may lead to the settlement and schedule of investment needs as well as operating rules, it is important to optimize the economical impact of security measures that have to be taken in order to be sure that there is no unnecessary or unjustified investment or operating costs. This has been the case for many years, since the emphasis was on the most severe, credible event leading to overly conservative solutions.

One way to deal with this problem is the concept of the probability of occurrence (contingencies) in the early stage of security analysis. This can be jointly used with a statistical approach (Schlumberger et al., 1999) that allows the generation of appropriate scenarios in order to fit more with the reality of the power system from the technical point of view as well as from the economical point view.

24.4.1 Security under Deregulation

With deregulation, the power industry has pointed out the necessity to optimize the operations of their systems leading to less investment in new facilities and pushing the system to be exploited closer to its limits. Furthermore, the open access has resulted in increased power exchanges over the interconnections. In some utilities, the number of transactions previously processed in 1 year is now managed in 1 day. These increased transactions and power exchanges have resulted in increased parallel flows leading to unpredictable loading conditions or voltage problems. A significant number of these transactions are non-firm and volatile. Hence, the security can no longer be handled on a zonal basis but rather on large interconnected systems.

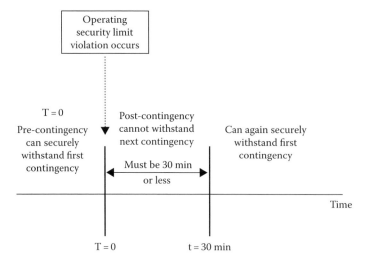

FIGURE 24.2 Current NERC basic reliability requirement. (From Pope, J.W., Transmission reliability under restructuring, in *Proceedings of IEEE SM'99*, Edmonton, Alberta, Canada, pp. 162–166, July 18–22, 1999. With permission.)

24.A Appendix A

The current NERC basic reliability requirement from NERC Policy 2-transmission (Pope, 1999) is (Figure 24.2):

24.A.1 Standards

1. Basic reliability requirement regarding single contingencies: All control areas shall operate so that instability, uncontrolled separation, or cascading outages will not occur as a result of the most severe single contingency.
    1.1 Multiple contingencies: Multiple outages of credible nature, as specified by regional policy, shall also be examined and, when practical, the control areas shall operate to protect against instability, uncontrolled separation, or cascading outages resulting from these multiple outages.
    1.2 Operating security limits: Define the acceptable operating boundaries.
2. Return from operating security limit violation: Following a contingency or other event that results in an operating security limit violation, the control area shall return its transmission system to within operating security limits soon as possible, but no longer than 30 min.

24.B Appendix B

24.B.1 Shift Factor Derivation

Consider a DC load flow for a base case (Galiana, 1984):

$$[B']\underline{\theta} = \underline{P}$$

where
  $\underline{\theta}$ is the vector of phase angles for the base case
  $[B']$ is the susceptance matrix for the base case
  $\underline{P}$ is the vector of active injections for the base case

Suppose that the admittance of line jk is reduced by ΔY_{jk} and the vector ΔP is unchanged, then:

$$\left[[B'] - \Delta Y_{jk} \underline{e}_{jk} \underline{e}_{jk}^T\right] \underline{\theta} = \underline{P}$$

where $\underline{e}_{jk}$ is the vector (Nb – 1) containing 1 in the position j, –1 in the position k and 0 elsewhere T is the Transpose

Now we can compute the power flow on an arbitrary line lm when line jk is outaged:

$$\underline{P}_{lm/jk} = Y_{lm}(\underline{\theta}_l - \underline{\theta}_m) = Y_{lm} \underline{e}_{lm}^T \underline{\theta}$$

$$= Y_{lm} \underline{e}_{lm}^T \left[[B'] - \Delta Y_{jk} \underline{e}_{jk} \underline{e}_{jk}^T\right]^{-1} P$$

By using the matrix inversion lemma, we can compute:

$$\underline{P}_{lm/jk} = Y_{lm} \underline{e}_{lm}^T \left[[B'] + \left([B']^{-1} \underline{e}_{jk} \underline{e}_{jk}^T [B']^{-1}\right) \Big/ \left((\Delta Y_{jk}) - 1 - \underline{e}_{jk}^T [B']^{-1} \underline{e}_{jk}\right)\right] P$$

Finally:

$$\underline{P}_{lm/jk} = \underline{P}_{lm} + \alpha_{jk/jk} * P_{jk}$$

where

$$\alpha_{jk/jk} = Y_{lm} * (\Delta Y_{jk}/Y_{jk}) * \left(\underline{e}_{lm}^T [B']^{-1} \underline{e}_{jk}\right) \Big/ \left(1 - \Delta Y_{jk} \underline{e}_{jk}^T [B']^{-1} \underline{e}_{jk}\right)$$

References

Debs, A.S. and Benson, A.R., Security assessment of power systems, in *System Engineering for Power: Status and Prospects*, Henniker, NH, pp. 144–178, August 17–22, 1975.

Fink, L. and Carlsen, K., Operating under stress and strain, *IEEE Spectr.*, 15, 48–53, March 1978.

Galiana, F.D., Bound estimates of the severity of line outages in power system contingency analysis and ranking, *IEEE Trans. Power App. Syst.*, PAS-103(9), 2612–2624, September 1984.

Hadjsaid, N., Benahmed, B., Fandino, J., Sabonnadiere, J.-C., and Nerin, G., Fast contingency screening for voltage-reactive considerations in security analysis, in *IEEE Winter Meeting*, New York, WM 185-9 PWRS, 1992.

Ilic-Spong, M. and Phadke, A., Redistribution of reactive power flow in contingency studies, *IEEE Trans. Power Syst.*, PWRS-1(3), 266–275, August 1986.

McCaulley, J.D., Vittal, V., and Abi-Samra, N., An overview of risk based security assessment, in *Proceedings of IEEE SM'99*, Edmonton, Alberta, Canada, pp. 173–178, July 18–22, 1999.

Pope, J.W., Transmission reliability under restructuring, in *Proceedings of IEEE SM'99*, Edmonton, Alberta, Canada, pp. 162–166, July 18–22, 1999.

Schlumberger, Y., Lebrevelec, C., and De Pasquale, M., Power system security analysis: New approaches used at EDF, in *Proceedings of IEEE SM'99*, Edmonton, Alberta, Canada, pp. 147–151, July 18–22, 1999.

Stott, B. and Alsac, O., Fast decoupled load flow, *IEEE Trans. Power App. Syst.*, PAS-93, 859–869, May/June 1974.

Taylor, D.G. and Maahs, L.J., A reactive contingency analysis algorithm using MW and MVAR distribution factors, *IEEE Trans. Power Syst.*, 6, 349–355, February 1991.

The North American Electric Reliability Council, NERC Planning Standards, approved by NERC Board of Trustees, September 1997.

Index

A

Adjustable speed (doubly fed) synchronous machines, 13-14
Advanced encryption algorithm (AES), 15-5
Anti-motoring protection, 2-9
Armature current relay, 17-12
Artificial neural networks (ANNs), 5-18–5-19
Asynchronous transfer mode (ATM), 4-9
Automatic generation control (AGC)
    ED, *see* economic dispatch (ED)
    interchange transaction scheduling, 20-5
    LFC, *see* Load frequency control (LFC)
    reserve monitoring, 20-5
Automatic trend relay (ATR), 13-9
Automatic voltage regulator (AVR), 14-6

B

Battery energy storage system (BESS), 19-25
Bits per second (BPS), 4-8
Bonneville power administration (BPA), 13-10
Boundary controlling UEP (BCU), 12-10–12-11

C

CERTS, *see* Consortium for Electric Reliability Technology Solutions (CERTS)
Computer relays
    analog, 3-12
    COMTRADE and SYNCHROPHASOR, 3-13
    high-performance microprocessors, 3-12–3-13
    subsystem, 3-12
    transmission line and transformer/bus protection, 3-13
COMTRADE, 3-13
Consortium for Electric Reliability Technology Solutions (CERTS), 15-29, 15-35
Continuation power flow (CPF), 16-8–16-9
Convertible static compensator (CSC), 13-13
CPF, *see* Continuation power flow (CPF)
CSC, *see* Convertible static compensator (CSC)
CTs, *see* Current transformers (CTs)
Current actuated relays
    directional overcurrent
        defined, 3-8
        polarizing quantities, 3-8
    fuses
        characteristics, 3-5
        defined, 3-5
        load-current values, 3-5
        TCT, 3-5
    instantaneous overcurrent
        description, 3-7
        effects, 3-7
        maximum and minimum values, 3-7
    inverse TDOC, *see* Inverse time-delay overcurrent (TDOC) relays
Current transformers (CTs)
    description, 1-6
    mismatch, current, 1-6
    rolled, 6-11
    saturation, 1-6–1-7, 6-6–6-7

D

Damping
    SVCs
        auxiliary signals choice, 19-10
        single machine infinite bus system, 19-9
        two-area study system, 19-10–19-11
    TCSC
        enhancement, 19-16, 19-17
        supplementary signals, 19-15
DampMon, 15-39–15-40
Data and network management task team (DNMTT), 15-12, 15-13
DDSPSO, *see* Device-dependent supersynchronous oscillations (DDSPSO)
Decision trees (DTs), 15-27–15-28
Device-dependent supersynchronous oscillations (DDSPSO)
    countermeasure, 17-21
    events, 17-20

Index-2

HVDC converter controls, 17-17
physical principles, 17-20–17-21
power system stabilizers, 17-17–17-18
renewable energy projects, 17-18
variable speed motor controllers, 17-17
DFT, *see* Discrete Fourier transform (DFT)
Digital relaying
    antialiasing filters
        discrete sequences, decimated, 5-2
        Fourier transform, 5-2
        SWC, 5-2
    DFT, 5-9–5-10
    differential equations
        line protection algorithms, 5-10–5-12
        transformer protection algorithms, 5-12–5-13
    Kalman filter, 5-13–5-15
    least squares fitting, 5-9
    microprocessor protection, 5-1
    neural networks, 5-18–5-19
    parameter estimation
        current and voltage waveforms, 5-7
        fundamental frequency phasor, 5-8
        power system transients, 5-8–5-9
    phasors, samples
        fundamental frequency component, DFT, 5-5
        recursive phasor calculation, 5-6
        reference axis, 5-5
        representation, 5-4
        sinusoidal quantity, 5-4
    sampling, 5-2
    Sigma-Delta A/D converters, 5-3–5-4
    symmetrical components
        defined, 5-6
        negative and zero sequence components, 5-6
        positive sequence calculation, 5-7
    Wavelet transform, 5-15–5-18
Direct stability methods
    BCU, *see* Boundary controlling UEP (BCU)
    controlling UEP determination
        procedure, 12-10
        steps, 12-9
    power system model, *see* Power system model
    TEF, *see* Transient energy functions (TEF)
    transient stability assessment
        degree of stability, 12-9
        energy margin, 12-9
Discrete Fourier transform (DFT), 4-7, 5-5
Distance relays
    admittance
        electromechanical design, 3-9
        pilot schemes and backup, 3-8
    characteristics, 3-8
    description, 3-8
    impedance, 3-8
    reactance
        defined, 3-9
        ground, 3-10
    HV and EHV system, 3-10
    three-phase power system, 3-9
    three-zone steps, 3-9
Distributed network protocol (DNP), 4-9
DNMTT, *see* Data and network management task team (DNMTT)
DSA, *see* Dynamic security assessment (DSA)
DTs, *see* Decision trees (DTs)
Dynamic braking, 9-10, 13-11
Dynamic security assessment (DSA)
    analysis methods, 16-8–16-11
    defined, 16-2
    elements, 16-3
    load models, 16-7
    modeling work steps, 16-5
    off-line
        base case and transfer analysis, 16-11
        two-dimensional transfer limit monogram, 16-11, 16-12
    online
        active and reactive power reserves, 16-15
        calibration and validation, models, 16-15–16-16
        distributed and variable generation, 16-14–16-15
        equipment maintenance, 16-15
        functional modules, 16-12–16-13
        monitor system security, 16-14
        postmortem analysis, incidents, 16-16
        preventative and corrective control actions, 16-14
        protection systems, 16-15
        stability limits, 16-14
        system restoration, 16-16
        system studies, 16-16
        transactions, power market, 16-15

E

Economic dispatch (ED)
    algorithm, 20-5
    defined
        economic dispatch calculation (EDC), 21-1
        independent system operator (ISO), 21-2
    EDC, *see* Economic dispatch calculation (EDC)
    generation level, unit, 20-4–20-5
    optimum generation pattern, 20-5
Economic dispatch calculation (EDC)
    and auctions, 21-6–21-7
    cost-minimizing
        generator data and solution, 21-5
        marginal cost curve, 21-5
        marginal costs, unit, 21-5, 21-6
    factors
        generation cost, 21-2
        price, 21-2
        quantity supplied, 21-2
    mathematical formulation
        demand constraint, 21-3

fuel costs, 21-3
fuel input and power output, 21-4
GENCO, 21-4–21-5
marginal cost, 21-4
objective, 21-3
solution techniques, 21-5
system limitations, 21-3
UC solution, 21-11
ED, *see* Economic dispatch (ED)
EHV, *see* Extra high voltage (EHV)
Electrical transformer protection
differential
defined, restrained differential relay, 1-2
faults, 1-3
inrush current waveforms, 1-2, 1-3
misoperation, 1-2
substation configurations, 1-3–1-4
use, 1-4
voltage and current phase shift, 1-3
fuse, 1-2
overcurrent, 1-2
overexcitation, 1-4
Electromagnet transient program (EMTP)
generator model, 17-10
peak transient shaft torque, 17-10
power system model, 17-10
turbine-generator mechanical model, 17-10
Electromechanical relays
construction, induction disk, 3-12
defined, 3-11
Plunger-type, 3-11
EMTP, *see* Electromagnet transient program (EMTP)
Energy management
AGC
ED, 20-4–20-5
interchange transaction scheduling, 20-5
LFC, 20-4
reserve monitoring, 20-5
defined, 20-1
functions, 20-7
load management
distribution automation equipment, 20-6
distribution control center functionality, 20-6
Manitoba Hydro Control Center, 20-2
OTS, *see* Operator training simulator (OTS)
power system data acquisition and control
SCADA, 20-3
SCADA systems, 20-1–20-2
security control, 20-7–20-8
trends
grid integrity, 20-9
predictive decision making, 20-10
self-healing grid, 20-10
water, hydro plants reservoir, 20-6
Energy management systems (EMSs), 16-12, 20-2
applications layer, modern, 20-2
state estimator, 22-7

Excitation system modeling
AVR, 14-6
model structure, 14-5
PSS, 14-6
stability programs, 14-5
Extra high voltage (EHV), 3-3, 3-10

F

FACTS, *see* Flexible AC transmission systems (FACTS)
Fast Decoupled Load Flow (FDLF) model, 24-4
Fatigue life expenditure (FLE), 17-6–17-7
Field ground protection
AC injection method, 2-6
DC injection method, 2-6
description, 2-6
voltage divider method, 2-6
FLE, *see* Fatigue life expenditure (FLE)
Flexible AC transmission systems (FACTS)
controllers classification, 19-2
controllers, energy storage
BESS, 19-25
SMES, 19-25
coordination, long-term voltage-var management
AVR and AQR, 19-26
capacitor bank selection, 19-30–19-31
fast voltage control, 19-28–19-29
installation, 19-27–19-28
load-tap changers (LTCs), 19-27
primary control, 19-26
reserve capacity control, 19-30
power system dynamic performance, 19-31
power transfer
capacity, 19-3
transmission line, 19-2–19-3
reactive power compensation, transmission lines
midline voltage, 19-4
voltage profile, 19-3, 19-4
SSSC
applications, 19-21–19-22
operating principle, 19-21
STATCOM
applications, 19-20
operating principle, 19-18–19-20
SVC, *see* Static var compensator (SVC)
TCSC, *see* Thyristor-controlled series compensation (TCSC)
transmission systems, 19-2
UPFC, *see* Unified power flow controller (UPFC)
Frequency-scanning analysis, SSR
induction generator effect, 17-8
Navajo Project generator, 17-7
torque amplification, 17-9
torsional interaction, 17-8–17-9

Frequency stability
    classification, power system, 8-3, 8-7
    defined, 8-6
    excursions, 8-6–8-7

G

Generalized reduced gradient (GRG) method, 23-5
Generation control
    economic dispatch (ED), *see* Economic dispatch (ED)
    unit commitment (UC) problem, *see* Unit commitment (UC) problem
Generator-load model, voltage stability
    limit-induced bifurcation, 11-3
    maximum system loadability, 11-3
    power flow model, 11-2
    PV curves, 11-3
    QV curves, 11-4, 11-5
    saddle-node bifurcation, 11-2
    system loadability margin, 11-4
Generator tripping, 9-10, 13-9
Generic object oriented substation event (GOOSE), 15-14
Genetic algorithm (GA)
    description, 21-12
    price-based UC
        modifications, 21-12
        population, schedule, 21-13
        reproduction process, 21-13
        schedule format, 21-12, 21-13
        two-point crossover, 21-14
    price-based UC-GA results
        control parameters, 21-15, 21-16
        costs and average costs, 21-15
        exhaustive search methods, 21-16
        forecasted load and prices, 21-14
        maximum, minimum and average fitnesses, 21-17
        schedules, population, 21-16
        unit data, two-generator case, 21-15
Global navigation satellite system (GLONASS), 4-7
Global positioning satellite (GPS) system, 4-7, 5-7
GLONASS, *see* Global navigation satellite system (GLONASS)
GOOSE, *see* Generic object oriented substation event (GOOSE)

H

HHT, *see* Hilbert-Huang transform (HHT)
High voltage (HV), 3-3, 3-10
High voltage DC (HVDC), 4-6, 14-11
Hilbert-Huang transform (HHT), 10-19
Hot-spot temperature, 1-4
HV, *see* High voltage (HV)
HVDC, *see* High voltage DC (HVDC)

I

IEEE Power Engineering Society Power System Dynamic Performance Committee, 10-10
Inter control center protocol (ICCP), 4-9
Interline power flow controller (IPFC), 13-13
International telecommunications union (ITU), 4-9
Inverse time-delay overcurrent (TDOC) relays
    coordination, 3-7
    description, 3-5
    family, characteristics, 3-6
    pickup and time delay, 3-6
IPFC, *see* Interline power flow controller (IPFC)
ITU, *see* International telecommunications union (ITU)

K

Kalman filter, 5-13–5-15

L

Large blackouts
    cascading failure, 7-1
    economic damage, 7-1
    power transmission grids, 7-2
    protection systems, 7-3
    reduction, 7-3
    risk, defined, 7-2
    size probability distribution, North American power transmission, 7-2
LFC, *see* Load frequency control (LFC)
Linear programming, state estimation, 22-6
Line protection algorithms, 5-10–5-12
Load dynamics, voltage stability
    large-disturbance
        air-conditioning motors, 11-8–11-9
        load center, 11-7
        monotonic voltage collapse, 11-8
        simulation, voltage collapse, 11-8
    small-signal, 11-9
Load frequency control (LFC)
    ACE and UCE, 20-4
    objectives, 20-4
Load models
    dynamic, 16-7
    OPF
        inaccuracies, standard OPF solution, 23-9
        minimum fuel cost, 23-8, 23-9
        modeling, 23-7
        Ontario Hydro energy management system, 23-8
        security constrained OPF, 23-8, 23-9
    static, 16-7
Load shedding, 9-10–9-11
Load tripping, 13-11
LoopAssist, 15-39

Index-5

Loss-of-excitation protection
  offset-mho relay, 2-6, 2-7
  positive sequence impedance
    trajectory, 2-7
  situations, 2-6

M

Mechanical transformer protection
  accumulated gases, 1-4
  pressure relay, 1-4
Microprocessor differential relay installation, 6-10, 6-11
Monitoring, WAMS
  angle monitoring and alarming
    angles, pair of buses, 15-16
    areas, 15-16
    double circuit line, 15-15
    internal and external area stress, 15-16
  improved state estimation
    bus voltages and line currents, 15-32
    calibration problem, 15-33–15-34
    digital line relays, 15-35
    extra-high-voltage (EHV) state estimation problem, 15-34
    Q–R algorithm, 15-33
    three-phase estimator, 15-33
  small-signal stability monitoring
    modal frequencies and damping, 15-17
    mode estimation, 15-21–15-22
    mode shape estimation, 15-22–15-23
    optimization, 15-17
    response types, 15-20–15-21
    signal-processing methods, modes estimation, 15-21
    Western North American Power System, 15-17–15-20
  transient stability monitoring, 15-29–15-32
  voltage stability, 15-24–15-29

N

NASPI, see North American SynchroPhasor Initiative (NASPI)
NASPInet
  components, 15-12
  concept, 15-35
North American Electric Reliability Corporation (NERC), 16-3, 16-5
North American SynchroPhasor Initiative (NASPI)
  CERTS, 15-35
  description, 15-35
  NASPInet, 15-12, 15-35
  Smart Grid Investment Grant (SGIG) projects, 15-36
  system baselining, 15-36

O

Offshore WPPs
  AC collector system, 18-9
  AC transmission system, 18-9
  DC collector system and transmission
    aggregated WTG units, 18-10
    centralized DC–DC conversion, 18-11
    distributed DC–DC converters, 18-10
  power transmission system, 18-8
  total power, 18-8
Operator training simulator (OTS)
  energy control system (ECS), 20-8
  instructional system, 20-9
  power system dynamic simulation, 20-9
  subsystems, 20-8, 20-9
OPF, see Optimal power flow (OPF)
Optimal power flow (OPF)
  defined, 23-1
  economic scheduling, 23-2–23-3
  formulation
    conjugate gradient method, 23-5
    control variables, 23-3, 23-4
    equality and inequality constraints, 23-3
    GRG optimization method, 23-5
    Hessian-based methods, 23-5–23-6
    interior point methods, 23-7
    linear programming-based methods, 23-6
    minimization, 23-3
    network equality constraints, 23-4
    Newton OPF, 23-6
    polar form, 23-4
    quadratic programming methods, 23-6–23-7
    rectangular form, 23-4
    reduced gradient method, 23-5
  load models
    inaccuracies, standard OPF solution, 23-9
    minimum fuel cost, 23-8, 23-9
    modeling, 23-7
    Ontario Hydro energy management system, 23-8
    security constrained OPF, 23-8, 23-9
    static, 23-7–23-8
  objectives, 23-2
  operational requirements, online implementation
    consistency, 23-13
    discrete modeling, 23-12
    ineffective "optimal" rescheduling, 23-13–23-14
    infeasibility, detection and handling, 23-12–23-13
    OPF-based transmission service pricing, 23-14
    robustness, OPF solutions, 23-11–23-12
    speed, 23-11
  SCOPF, see Security constrained optimal power flow (SCOPF)

Oscillograph records, system performance
  analysis, 6-2
  bus/line potentials, 6-5
  carrier performance, critical transmission
      lines, 6-2
  corrected connection, 6-10, 6-11
  current reversals, 6-7
  current transformer (CT) saturation, 6-6–6-7
  dead time, 6-5
  delayed carrier response
      adjacent line, 6-3, 6-4
      internal fault, 6-3, 6-4
  good breaker operations, 6-6
  line fault, sequence steps, 6-8–6-9
  maintenance tool, 6-3
  microprocessor differential relay installation, 6-10
  oscillograms, 6-9
  questionable carrier response, internal fault,
      6-3, 6-4
  relays, 6-1
  restrike
      breakers, 6-9
      interrupter, 6-9
      lightning strikes, 6-10
  waveforms, 6-3
Overexcitation protection
  ANSI/IEEE, 2-10
  combined definite and inverse-time
      characteristics, 2-10
  defined, volts per hertz, 2-9–2-10
  dual definite-time characteristics, 2-10
  Lenz Law, 2-9
Overvoltage protection, 2-10–2-11

P

Pacific HVDC intertie (PDCI), 15-19
PDCI, see Pacific HVDC intertie (PDCI)
PDCs, see Phasor data concentrators (PDCs)
Phasor data concentrators (PDCs)
  anti-aliasing filters, 15-10
  functions
      control center, 15-11
      local/substation, 15-10–15-11
      super-PDC, 15-11
  levels, 15-10
Phasor gateway and NASPInet
  Data Bus, 15-14
  data exchange, 15-12
  data multicast, 15-14
  DNMTT, 15-12
  information assurance functions, 15-14
  NASPI, 15-12
  NASPInet DB services, 15-12
  performance monitoring, 15-13
  signal registration process, 15-13
  WAMPACS, 15-12

Phasor measurement units (PMUs)
  generic PMU, 15-9
  IEEE standards, 15-10
  phasors and synchrophasors, 15-8–15-9
  positive sequence measurements, 15-9
  transients and off-nominal frequency signals,
      15-9–15-10
Phasors, 15-8–15-9
Pilot protection
  current differential, 3-11
  directional comparison, 3-10
  microwave/fiber-optic channels, 3-10
  phase comparison, 3-11
  power line carrier, 3-10
  three-zone step distance relay, 3-9, 3-10
  transfer tripping, 3-10
  wire, 3-11
Pilot wire, 3-11
PMUs, see Phasor measurement units (PMUs)
Power system analysis, modal identification,
    10-18–10-19
Power system dynamic interaction, turbine
    generators
  DDSPSO, see Device-dependent supersynchronous
      oscillations (DDSPSO)
  device-dependent subsynchronous oscillations
      HVDC converter controls, 17-17
      power system stabilizers, 17-17–17-18
      renewable energy projects, 17-18
      variable speed motor controllers, 17-17
  fatigue damage and monitoring
      plastic deformation, 17-14
      shaft torque monitoring techniques, 17-14
      turbine-generator shafts, 17-13
  SPSR
      countermeasures, 17-20
      description, 17-18–17-19
      physical principles, 17-19–17-20
      turbine-generator, 17-19
  SSR, see Subsynchronous resonance (SSR)
  transient shaft torque oscillations, 17-21–17-22
Power system dynamic modeling
  equivalents, 14-12
  excitation system modeling, see Excitation system
      modeling
  generator
      electrical model, 14-3–14-4
      rotor mechanical model, 14-2–14-3
      saturation modeling, 14-4–14-5
  prime mover modeling, see Prime mover
      modeling
  requirements
      computational models, 14-1
      timescale, 14-1, 14-2
  transmission device models
      static VAr systems, 14-11–14-12
      transformers, 14-11

Power system model
  generator
    angles and speed, 12-5
    equations of motion, 12-4
  stability theory
    domain of attraction, 12-6
    equilibrium point, 12-5–12-6
    "local" approximation, boundary, 12-7
    postdisturbance/postfault period, 12-6
Power system network
  description, 16-5
  modeling considerations, 16-6
Power system oscillations
  critical damping, 10-2
  damping, 10-7
  generator voltage regulators, 10-3
  hydrogenerators, 10-1
  inadequate damping, 10-4
  interaction characteristics
    electromechanical oscillations, 10-5
    interarea modes, 10-4
    power-angle relationship, AC systems, 10-5
    power-swing mode, 10-6
    subsynchronous frequency oscillations, 10-4
    synchronizing component, torque, 10-5
    torsional mode instability, 10-4–10-5
  interconnected power systems, 10-3
  mitigation
    adverse side effects, 10-14–10-15
    closed-loop control design, 10-11–10-12
    control algorithm, 10-13
    control objectives, 10-11
    control output limits, 10-14
    gain selection, 10-13
    input-signal filtering, 10-12–10-13
    input-signal selection, 10-12
    performance evaluation, 10-14
    power system stabilizer tuning example, 10-15–10-18
    siting, 10-10
  observations, 10-6
  reappearance, 10-2
  study procedure
    analysis and verification, 10-9–10-10
    modeling requirements, 10-8
    objectives, 10-7–10-8
    performance requirements, 10-8
    system condition setup, 10-9
Power system stability
  advantages, 8-9
  analysis method
    decision tree methods, 16-11
    direct methods, 16-10
    eigenvalue analysis, 16-10
    expert system method, 16-11
    neural network methods, 16-11
    pattern recognition, 16-11
    power flow analysis, 16-8
    probabilistic methods, 16-11
    P–V analysis, 16-8–16-9
    time-domain simulations, 16-9–16-10
    V–Q analysis for voltage stability assessment, 16-11
  classification
    considerations, 8-2
    descriptions, 8-2–8-3
  complexity, 8-7
  control and enhancements, 16-11
  description, 8-1
  design and operation, 8-9
  DSA, see Dynamic security assessment (DSA)
  dynamic characteristics, 8-8–8-9
  electric system, 8-7
  frequency, 8-6–8-7
  improvements, 8-8
  load/generation mismatch, 8-2
  modeling
    advanced transmission technologies, 16-7
    generators, 16-6
    loads, 16-6–16-7
    model validation, 16-7–16-8
    power system network, 16-5–16-6
    protective devices, 16-7
  operation states, 16-2
  phenomena, interest, 16-3
  rotor angle, 8-3–8-4
  security criteria
    N-2 contingency, 16-4
    NERC transmission planning standards, 16-3
    physical response types, 16-4
    steady-state and transient voltage, 16-4
    thermal loading, 16-4
    transient frequency, 16-5
    voltage and transient stability, 16-5
  SSA and DSA, 16-2
  stability problem, 8-2
  transient disturbance, 8-1
  transient stability simulation programs, 8-8
  voltage, 8-5–8-6
Power system stability controls
  actuators, 13-7
  dynamic security assessment, 13-14
  effectiveness and robustness, 13-6–13-7
  feedback controls
    bang–bang discontinuous control, 13-6
    discontinuous controls, 13-5
  feedforward controls, 13-6
  industry restructuring, 13-16
  "intelligent" controls, 13-14
  power failures, 13-16
  reliability, 13-7
  synchronizing and damping torques, 13-6

synchronous
    damping, 13-4, 13-5
    generator electrical power, 13-3
    generator electromechanical dynamics, 13-2–13-3
    loss of synchronism, 13-4
    power–angle curve, 13-4
    remote power plant, 13-3
    "swing equation", 13-2
  types
    adjustable speed (doubly fed) synchronous machine, 13-14
    controlled separation and underfrequency load shedding, 13-14
    current injection, voltage sourced inverters, 13-12
    dynamic braking, 13-11
    excitation control, 13-8
    fast fault clearing, 13-9–13-10
    fast voltage phase angle control, 13-13
    generator tripping, 13-9
    high-speed reclosing, and single-pole switching, 13-9–13-10
    HVDC links, 13-13
    load tripping and modulation, 13-11
    prime mover control, 13-8–13-9
    reactive power compensation switching or modulation, 13-11–13-12
  wide-area stability controls, 13-15
Power system stabilizer (PSS)
  input and output, 14-6
  tuning
    AC exciter and parameters, 10-15
    eigenvalue analysis, 10-17
    local mode, tuned PSS, 10-18
    phase characteristics, 10-16–10-17
    PSS and parameters, 10-15
    unstable mode, simulation, 10-16
Primary–secondary phase shift, 1-7
Prime mover modeling
  fossil fuel steam plant, 14-7
  gas (combustion) turbines, 14-7
  hydro plants, 14-7
  load modeling
    composite load model, 14-9–14-10
    composition-based modeling, 14-10
    induction motor dynamic model, 14-9
    staged testing, load feeders, 14-10
    static load model, 14-9
    system disturbance monitoring, 14-10
    transient and oscillatory stability analysis, 14-8
  transient stability analysis, 14-6
  types, 14-6
  wind-turbine-generator systems, *see* Wind-turbine-generator systems

Prony analysis
  description, 15-21
  power-system ring-down analysis, 15-21
Protection, synchronous generators
  vs. accidental energization, 2-16–2-17
  anti-motoring, 2-9
  breaker failure, 2-17–2-18
  current imbalance
    defined, 2-7
    digital relays, 2-9
    reasons, 2-7
    static/digital time-inverse 46 curve, 2-8, 2-9
    thermal law, 2-8
  digital multifunction relays-
    protective functions, 2-19–2-20
    signal processing, 2-18–2-19
  field ground, 2-6
  frequency operation, turbine, 2-15–2-16
  functions
    generator-transformer protection scheme, 2-2, 2-3
    relays, 2-2
  loss-of-excitation, 2-6–2-7
  out-of-step, 2-14–2-15
  overexcitation, 2-9–2-11
  overvoltage, 2-10–2-11
  stator faults
    CT saturation, 2-3
    description, 2-2
    differential characteristics, 2-3, 2-4
    generator-transformer protection scheme, 2-3, 2-4
    winding current configuration, 2-4
  vs stator winding ground fault
    description, 2-4
    stator-to-ground neutral overvoltage scheme, 2-5
    techniques, 2-5
    third harmonic, neutral and terminals, 2-5
  system backup, 2-12–2-13
  tripping principles, 2-18
  voltage imbalance, 2-11–2-12
PSS, *see* Power system stabilizer (PSS)

R

Relay designs
  computer, 3-12–3-13
  electromechanical, 3-11–3-12
  solid-state, 3-12
Relay operations, 1-11
Relays
  current actuated, 3-5–3-8
  designs, 3-11–3-13
  distance, 3-8–3-10
  nature, 3-2–3-4

Response, power system
  modal frequency and damping estimation algorithms, 15-21
  power flowing, 15-20
Restoration
  date, manufacture, 1-10
  history, 1-10
  magnetizing inrush
    differential relays, 1-10
    relay harmonic restraint circuit, 1-11
  power transformers, 1-10
  relay operations, 1-11
  system monitoring equipments, 1-10
  system operators, 1-10
Rotor angle stability
  classification, power system, 8-3, 8-4
  components, electrical torque, 8-3–8-4
  defined, first swing, 8-4
  description, 8-3
  electromechanical oscillations, 8-3
  small and large disturbance, 8-4

S

Sample value (SV) datasets, 15-14–15-15
SCADA, *see* Supervisory control and data acquisition (SCADA)
SCOPF, *see* Security constrained optimal power flow (SCOPF)
SDH, *see* Synchronous digital hierarchy (SDH)
Secure hash algorithm (SHA), 15-15
Security analysis
  definition, 24-2
  determinist *vs.* probabilistic, 24-5–24-6
  models
    active/angle subproblem, 24-3
    contingency analysis procedure, 24-5
    DC approach, 24-3
    FDLF, 24-4
    power flow, 24-4
    reactive/voltage subproblem, 24-4
    shift factors, 24-4
  NERC basic reliability requirement, 24-6
  shift factor derivation, 24-6–24-7
  time frame, security-related decision
    corrective actions, 24-3
    large systems, 24-3
    operational planning, 24-2
Security constrained optimal power flow (SCOPF)
  fixed tap transformer fed load, 23-10
  load models, 23-10
  secure system, 23-9
Sensitivity-based voltage stability indices, 15-27
SHA, *see* Secure hash algorithm (SHA)
Shunt reactors, 1-8

Sigma-Delta A/D converters
  low pass filter (LPF), 5-3
  modulator and error, 5-3
  quantization, 5-3
  resolution, 5-4
Single-machine infinite-bus (SMIB) system, 16-10
Single pole switching and reclosing, 9-10
Singular value decomposition (SVD), 15-27
Situational awareness
  action, 15-6
  comprehension, 15-5
  decision making, 15-6
  defined, 15-5
  objective, 15-5
  perception, 15-5
  power grid operations, 15-6
  projection, 15-6
Small-signal stability analysis
  computational burden, 10-18
  higher-order modes, 10-18
SMES, *see* Superconducting magnetic energy storage (SMES)
SMIB system, *see* Single-machine infinite-bus (SMIB) system
Solid-state relays
  circuit configuration, 3-12
  description, 3-12
SONET, *see* Synchronous optical network (SONET)
Special protection schemes (SPS)
  advantages, 4-7
  defined, 4-6
  design procedures, 4-6
  FACTS and HVDC, 4-6
  properties, 4-6
  types, 4-6–4-7
SPS, *see* Special protection schemes (SPS)
SPSR, *see* Supersynchronous resonance (SPSR)
SSSC, *see* Static synchronous series compensator (SSSC)
STATCOM
  applications, 19-20
  operating principle
    capacitive and inductive operation, 19-19
    converter, 19-19
    DC source, 19-19
    power circuit, 19-18
    reactive power exchange, 19-18
    V–I characteristic, 19-20
State estimation
  assumptions
    normal probability distribution curve, 22-2
    standard deviation, 22-3
    telemetered measurements, 22-2
  EMS, 22-7
  error identification
    "bad" data, 22-8
    parameter data, 22-8–22-9

    telemetered data, 22-8
    topology data, 22-9
  measurement representations
    bus-admittance matrix, 22-4
    bus injection equation, 22-3
    error component, 22-3
    transmission line representation, 22-3
  network topology assessment, 22-7
  observability criterion, 22-2
  solution methods
    linear programming, 22-6
    WLS, 22-4–22-6
  state of system, defined, 22-2
  terms, definition, 22-12
  three-bus system, system description, 22-9–22-10
  unobservability, 22-9
  WLS, 22-10–22-11
Static compensator (STATCOM), 13-13
Static security assessment (SSA), 16-2
Static synchronous series compensator (SSSC)
  applications
    alleviation, voltage instability, 19-22
    damping, power oscillations, 19-21–19-22
    mitigation, SSR, 19-22
    power flow control, 19-21
  operating principle, 19-21
  power circuit and operating modes, 19-21
  valve-side voltage rating, 19-20
Static var compensator (SVC)
  applications
    alleviation, voltage instability, 19-11
    grid integration, wind power generation systems, 19-12
    HVDC converters, 19-11–19-12
    mitigation, subsynchronous resonance, 19-11
    power transmission capacity, increase, 19-8
    system damping enhancement, 19-9–19-11
    system transient stability limit improvement, 19-8–19-9
  capacitors, 19-5
  control inputs, 19-6
  elements, 19-4
  operating principle
    dynamic reactive power support, disturbances, 19-6
    V–I characteristics, 19-6, 19-7
  single-phase TCR, 19-4
  TSC, 19-5
  TSC-TCR, 19-5
  voltage control
    features, 19-8
    Thevenin's equivalent, power system, 19-7
Static VAr systems (SVSs)
  models, 14-11
  TCR-based, 14-11, 14-12

StressMon, 15-39
Subsynchronous resonance (SSR)
  countermeasures
    nonunit-tripping, 17-13
    TCSC, see Thyristor-controlled series capacitor (TCSC)
    unit-tripping, 17-12
  data, analysis
    life expenditure curves, 17-12
    system data, 17-11
    turbine-generator data, 17-11–17-12
  eigenvalue analysis, 17-9–17-10
  frequency-scanning, see Frequency-scanning analysis, SSR
  mitigation
    accurate studies, 17-6
    California–Oregon transmission system, 17-5
    countermeasure requirements, 17-6–17-7
    interim protection, 17-6
    screening studies, 17-6
    steam-driven turbine generator, 17-5
    tests, 17-6
  physical principles
    induction generator effect, 17-4–17-5
    series-compensated transmission line, 17-3, 17-4
    torque amplification, 17-5
    torsional interaction, 17-5
  terms and definitions, 17-3
  testing
    countermeasure, 17-15–17-16
    modal damping tests, 17-14–17-15
    torsional mode frequency tests, 17-14
  torsional interaction, 17-3
  transient analysis
    EMTP, 17-10
    fatigue life expenditure, 17-10–17-11
    torque amplification, 17-10
Superconducting magnetic energy storage (SMES), 13-12, 19-25
Supersynchronous resonance (SPSR)
  description, 17-18–17-19
  events, 17-19
  physical principles
    negative sequence current flow, 17-19
    ramp test, 17-20
    turbine-generator, 17-19–17-20
Supervisory control and data acquisition (SCADA) systems
  asynchronous measurements, 15-4
  dual computer system, 20-3
  functions, 20-3
  remote terminal units (RTUs), 20-3
  traditional systems, 15-7
SVC, see Static var compensator (SVC)
SVD, see Singular value decomposition (SVD)
SVSs, see Static VAr systems (SVSs)

Swing equation, transient stability
    accelerating torque, 9-2
    angular velocity, rotor, 9-3
    inertia, 9-2
    synchronous machine, 9-2
    transiently stable system, 9-3
SynchAssist, 15-39
Synchronous digital hierarchy (SDH), 4-8, 4-9
Synchronous optical network (SONET), 4-8, 4-9
SYNCHROPHASOR, 3-13
Synchrophasor, 15-9
System protection
    communication technology
        defined, 4-8
        power line carrier, 4-8
        requirements and factors, 4-8
        ring topology, 4-9
        self-healing capability, 4-9
        SONET/SDH and ATM, 4-8–4-9
        UCA/MMS protocol, 4-9
    description, 4-1
    disturbances
        equal-area criterion and variation, 4-2
        frequencies, 4-2
        out-of-step, 4-1
        voltage instability, 4-2
    hierarchical structure, 4-10
    methods and algorithm determination, 4-10
    out-of-step relays, 4-9
    overload and underfrequency load shedding
        description, 4-3
        multistep action, 4-3
        oscillations, 4-3–4-4
        problems, frequency change, 4-4
    phasor measurement technology
        computer-based relay, 4-7
        description, 4-7
        DFT, 4-7
        GPS and GLONASS, 4-7
        PMU-PDC, 4-7–4-8
        WAMS concept, 4-8
    protection/control, system-wide disturbance, 4-9
    SPS, *see* Special protection schemes (SPS)
    transient stability and out-of-step
        distance relays, 4-3
        frequency, 4-2
        stable oscillations, 4-2
        trajectories, stable and unstable swings, 4-2, 4-3
    voltage stability and undervoltage load shedding
        defined, 4-4
        fast and slow disturbances, 4-4
        parameters, 4-5
        problems, 4-4
        relationship, load and voltage collapse, 4-4–4-5
        remedial measures, 4-5–4-6
        sensitivity matrix, 4-5
    wide area monitoring and control, 4-1

T

TCSC, *see* Thyristor-controlled series capacitor (TCSC); Thyristor-controlled series compensation (TCSC)
TCT, *see* Total clearing time (TCT)
TEB, *see* Transient excitation boosting (TEB)
TEF, *see* Transient energy functions (TEF)
Tennessee Valley Authority (TVA), 15-11
Thermal transformer protection
    current harmonic content
        eddy-current loss, 1-5
        harmonics, 1-5
        nonsinusoidal, 1-5
    heating, overexcitation, 1-5
    hot-spot temperature, 1-4
    load tap-changer overheating, 1-6
    solar-induced currents, 1-6
Thyristor-controlled series capacitor (TCSC), 17-13
Thyristor-controlled series compensation (TCSC)
    damping
        enhancement, 19-16, 19-17
        supplementary signals, 19-15
    description, 19-12
    mitigation, subsynchronous resonance, 19-16
    module, 19-12
    operating principle
        inductive impedance, 19-13
        modes, 19-13
        reactance characteristic, 19-14
        waveforms, capacitive mode, 19-15–19-16
    reactance, increase, 13-11
    remote power plant integration, 13-12
    system power transfer capacity improvement, 19-14–19-15
    voltage instability prevention, 19-18
Thyristor switched capacitor (TSC), 19-5
Time-domain simulations
    applications, 16-10
    description, 16-9
    input data, 16-10
Timescale decomposition, voltage stability
    system response, break, 11-12
    time windows, 11-11
Torsional motion relay, 17-12
Total clearing time (TCT), 3-5
Transformer protection
    algorithms, 5-12–5-13
    backup, 1-8
    CTs, *see* Current transformers (CTs)
    electrical
        differential, 1-2–1-4
        fuse, 1-2
        overcurrent, 1-2
        overexcitation, 1-4

magnetizing inrush
  initial, 1-7
  recovery, 1-7
  sympathetic, 1-7
mechanical
  accumulated gases, 1-4
  pressure relay, 1-4
phase angle and voltage regulators, 1-9
primary–secondary phase shift, 1-7
restoration, 1-10–1-11
shunt reactors, 1-8
single-phase, 1-9
sustained voltage unbalance, 1-9
thermal, 1-4–1-6
through faults, 1-8
turn-to-turn faults, 1-7–1-8
types, faults
  core and bushings, problems, 1-1
  CTs, 1-1
unit systems
  backup protection, 1-9
  description, 1-9
  volts-per-hertz relay, 1-9
zigzag, 1-8–1-9
Transient energy functions (TEF)
  applications, 12-11
  kinetic energy, 12-8
  power system model, 12-7
Transient excitation boosting (TEB), 13-6
Transient shaft torque oscillations
  cracks, 17-21–17-22
  shaft damage, 17-21
  turbine-generator design, 17-21
Transient stability
  analytical methods, 9-8
  factors, 9-9
  modeling
    generators, 9-7
    HVDC lines, 9-8
    loads, 9-8
    synchronous machines, 9-7
  monitoring
    defined, 15-29
    energy functions, 15-30–15-31
    U.S. Western North American Power System, 15-31–15-32
  simulation studies
    input data, 9-9
    output data, 9-9
  system design
    actions, stability improvement, 9-10–9-11
    time-domain simulations, 9-10
  system operation, 9-11–9-12
  theory
    equal area criterion, 9-5–9-7
    generator rotor angle *vs.* time, 9-1–9-2

    power-angle relationship, 9-3–9-5
    "swing curves", 9-1
    swing equation, *see* Swing equation, transient stability
Transmission line protection
  current actuated relays
    directional overcurrent, 3-8
    fuses, 3-5
    instantaneous overcurrent, 3-7
    inverse TDOC, 3-5–3-7
  distance relays
    admittance, 3-8–3-9
    impedance, 3-8
    reactance, 3-9–3-10
  nature, relaying
    primary and backup, 3-3
    reclosing, 3-3–3-4
    reliability, 3-2
    speed, 3-3
    system configuration, 3-4
    zones, 3-2
  pilot, 3-10–3-11
  relay designs, *see* Relay designs
TSC, *see* Thyristor switched capacitor (TSC)
Turbine generators, *see* Power system dynamic interaction, turbine generators
Turn-to-turn faults, 1-7–1-8
TVA, *see* Tennessee Valley Authority (TVA)

U

UC problem, *see* Unit commitment (UC) problem
Unified power flow controller (UPFC)
  applications
    power transfer and improvement, 19-23, 19-24
    study system, 19-24
  implementation, 19-22
  operating principle
    power flow control functions, 19-23, 19-24
    series voltage injection, 19-23
    shunt-connected converter, 19-23
Unit commitment (UC) problem
  and auctions, 21-17–21-18
  defined, 21-7
  EDC, 21-11
  electric energy source
    additional generator, 21-8
    demand curve, 21-9
  electricity supplier, 21-8
  genetic-based UC algorithm, *see* Genetic algorithm (GA)
  mathematical formulation
    cost-minimizing paradigm, 21-11
    monopolist cost-minimization, 21-9
    solution process, 21-9–21-10

Index

objective, 21-7
quantity to supply
    competitive electric markets, 21-8
    vertically integrated monopolies, 21-7
schedule, 21-7
solution methods, 21-11–21-12
Unit-tripping logic scheme, 17-12
UPFC, *see* Unified power flow controller (UPFC)

V

Voltage imbalance protection
    causes, 2-11
    conditions, time delay, 2-12
    examples, voltage balance relay, 2-11
    fuse failure detection, 2-12
    VTs signals, 2-11
Voltage security assessment (VSA) tools, 15-28
Voltage sourced inverters, 13-12
Voltage stability
    continuation methods
        "perpendicular intersection" method, 11-11
        power flow, 11-10
        steps, 11-10
    defined, 8-5
    definition, 11-1
    EHV transmission lines, 8-5
    generator-load example, *see* Generator-load model, voltage stability
    load dynamics
        large-disturbance, 11-7–11-9
        small-signal, 11-9
    load instability, 11-1
    load modeling
        aggregate load response, step-voltage change, 11-5
        network response, 11-4
        steady-state power demand, 11-6
        transient output, 11-6
    mitigation
        generator reactive overload capability, 11-13
        lower power factor generators, 11-13
        must-run generation, 11-12
        operating, higher voltage, 11-12
        secondary voltage regulation, 11-13
        series and shunt capacitors, 11-12
        static compensators, 11-12
        undervoltage load shedding, 11-13
    optimization/direct methods
    power flow analysis
        QV relation, 11-9–11-10
        reactive power limit, generator, 11-9
    progressive drop, 8-5
    short-term and long-term, 8-6
    small and large disturbance, 8-5–8-6
    static analysis techniques, 8-6
    timescale decomposition, 11-11–11-12

Voltage stability monitoring
    active and reactive powers, 15-25
    advantages, 15-24
    causes, 15-24
    collapse incidents, 15-24
    critical point, 15-26
    instability detection
        control center implementations, 15-28, 15-29
        DTs, 15-27–15-28
        indices, 15-26–15-27
        values, indices threshold, 15-29
    power–voltage relationship, 15-25
    two-bus system and power-voltage characteristics, 15-24, 15-25
Voltage transformer (VT) failure, 2-11

W

WACS, *see* Wide-area control systems (WACS)
WAMPACS, *see* Wide-area monitoring, protection, and control system (WAMPACS)
WAMS, *see* Wide area measurement system (WAMS); Wide-area monitoring system (WAMS)
Wavelet transform, 5-15–5-18
WECC, *see* Western Electricity Coordinating Council (WECC)
Weighted least squares (WLS)
    branch values, 22-11
    convergence, 22-6
    iterative solution, 22-5
    measurement vector, defined, 22-10
    performance index, 22-5
    solution, 22-11
    system values, 22-11
    weighting factor, 22-4
Western Electricity Coordinating Council (WECC), 14-9
Western North American power system
    boundary power plant oscillation, 15-18–15-19
    forced inter-area power oscillation, 15-18
    PDCI, 15-19
    power oscillations, 15-17
    U.S. Western interconnection
        "close call", 15-18
        unstable oscillation, 15-17–15-18
Wide-area control systems (WACS), 15-5, 15-12
Wide area measurement system (WAMS), 4-8
Wide-area monitoring, protection, and control system (WAMPACS), 15-12
Wide-area monitoring system (WAMS)
    Brazil
        hydro generation park, 15-37
        LoopAssist, 15-39
        SynchAssist, 15-39
        synchrophasor application, 15-40
        system oscillations monitoring, 15-39–15-40
        system stress monitoring, 15-39

deployment roadmap
　　benefit categories, 15-40
　　near-term and medium-term group, 15-41
　　PMU technology, 15-40
　　SynchroPhasor applications, 15-40, 15-41
description, 15-3
drivers, 15-2–15-5
electric utility industry, 15-3
Europe
　　research and development projects, 15-36
　　Swissgrid WAMS links, 15-37, 15-38
　　TSO applications, 15-36–15-37
infrastructure
　　phasor data concentrator, 15-10–15-11
　　phasor gateway and NASPInet, 15-11–15-14
　　PMUs, see Phasor measurement units (PMUs)
　　protocols and standards, 15-14–15-15
monitoring applications
　　angle monitoring and alarming, 15-15–15-16
　　improved state estimation, 15-32–15-35
　　small-signal stability monitoring, 15-17–15-23
　　transient stability monitoring, 15-29–15-32
　　voltage stability, 15-24–15-29
North America, NASPI, 15-35–15-36
situation awareness, see Situational awareness
Wide-area stability and voltage control system (WACS), 13-15
Wind power integration
　　WPP
　　　　control, 18-14–18-16
　　　　description, 18-7
　　　　models, 18-11–18-14
　　　　offshore, 18-8–18-11
　　　　onshore, 18-8
　　WTG
　　　　fixed-speed, 18-2–18-3
　　　　type-3 WTG system, 18-5–18-6
　　　　type-4 WTG system, 18-6–18-7
　　　　variable-speed, 18-3–18-5
Wind power plants (WPPs)
　　control
　　　　frequency, 18-16
　　　　level reactive-power control, 18-15–18-16
　　　　level real-power control, 18-14–18-15
　　description, 18-7
　　models
　　　　developmental challenges, 18-12
　　　　dynamic analysis subgroups, 18-11
　　　　generic models, 18-13–18-14
　　　　mathematical models, 18-12
　　　　power system study categories, 18-12
　　　　type-3/type-4 WTG, 18-13
　　offshore, see Offshore WPPs
　　onshore, 18-8
Wind turbine generator (WTG)
　　fixed-speed
　　　　features, 18-3
　　　　rotor speed, 18-2
　　type-3 WTG system
　　　　angle error, 18-6
　　　　DFAG rotor circuit, 18-5
　　　　flux magnitude and angle control, 18-5
　　type-4 WTG system
　　　　control functions, 18-6
　　　　dq-frame-based control, 18-6, 18-7
　　　　grid-side VSC, 18-6
　　　　high-level control, 18-6
　　　　VSC control, 18-6
　　variable-speed
　　　　fully rated converter system, 18-4–18-5
　　　　limited variable-speed, 18-3–18-4
　　　　partially rated converter system, 18-4
Wind-turbine-generator systems
　　generation technologies, 14-8
　　wind farm, 14-8
WLS, see Weighted least squares (WLS)
WPPs, see Wind power plants (WPPs)
WTG, see Wind turbine generator (WTG)

Y

Yule–Walker algorithm, 15-21

Z

Zigzag transformers
　　description, 1-8
　　neutral CT, 1-9
　　overcurrent relay, 1-9